ITCH
Mechanisms and Treatment

FRONTIERS IN NEUROSCIENCE

Series Editor
Sidney A. Simon, Ph.D.

Published Titles

Apoptosis in Neurobiology

Yusuf A. Hannun, M.D., Professor of Biomedical Research and Chairman, Department of Biochemistry and Molecular Biology, Medical University of South Carolina, Charleston, South Carolina

Rose-Mary Boustany, M.D., tenured Associate Professor of Pediatrics and Neurobiology, Duke University Medical Center, Durham, North Carolina

Neural Prostheses for Restoration of Sensory and Motor Function

John K. Chapin, Ph.D., Professor of Physiology and Pharmacology, State University of New York Health Science Center, Brooklyn, New York

Karen A. Moxon, Ph.D., Assistant Professor, School of Biomedical Engineering, Science, and Health Systems, Drexel University, Philadelphia, Pennsylvania

Computational Neuroscience: Realistic Modeling for Experimentalists

Eric DeSchutter, M.D., Ph.D., Professor, Department of Medicine, University of Antwerp, Antwerp, Belgium

Methods in Pain Research

Lawrence Kruger, Ph.D., Professor of Neurobiology (Emeritus), UCLA School of Medicine and Brain Research Institute, Los Angeles, California

Motor Neurobiology of the Spinal Cord

Timothy C. Cope, Ph.D., Professor of Physiology, Wright State University, Dayton, Ohio

Nicotinic Receptors in the Nervous System

Edward D. Levin, Ph.D., Associate Professor, Department of Psychiatry and Pharmacology and Molecular Cancer Biology and Department of Psychiatry and Behavioral Sciences, Duke University School of Medicine, Durham, North Carolina

Methods in Genomic Neuroscience

Helmin R. Chin, Ph.D., Genetics Research Branch, NIMH, NIH, Bethesda, Maryland

Steven O. Moldin, Ph.D., University of Southern California, Washington, D.C.

Methods in Chemosensory Research

Sidney A. Simon, Ph.D., Professor of Neurobiology, Biomedical Engineering, and Anesthesiology, Duke University, Durham, North Carolina

Miguel A.L. Nicolelis, M.D., Ph.D., Professor of Neurobiology and Biomedical Engineering, Duke University, Durham, North Carolina

The Somatosensory System: Deciphering the Brain's Own Body Image

Randall J. Nelson, Ph.D., Professor of Anatomy and Neurobiology, University of Tennessee Health Sciences Center, Memphis, Tennessee

The Superior Colliculus: New Approaches for Studying Sensorimotor Integration

William C. Hall, Ph.D., Department of Neuroscience, Duke University, Durham, North Carolina

Adonis Moschovakis, Ph.D., Department of Basic Sciences, University of Crete, Heraklion, Greece

Animal Models of Cognitive Impairment
Edward D. Levin, Duke University Medical Center, Durham, North Carolina
Jerry J. Buccafusco, Medical College of Georgia, Augusta, Georgia

The Role of the Nucleus of the Solitary Tract in Gustatory Processing
Robert M. Bradley, University of Michigan, Ann Arbor, Michigan

Brain Aging: Models, Methods, and Mechanisms
David R. Riddle, Wake Forest University, Winston-Salem, North Carolina

Neural Plasticity and Memory: From Genes to Brain Imaging
Frederico Bermudez-Rattoni, National University of Mexico, Mexico City, Mexico

Serotonin Receptors in Neurobiology
Amitabha Chattopadhyay, Center for Cellular and Molecular Biology, Hyderabad, India

TRP Ion Channel Function in Sensory Transduction and Cellular Signaling Cascades
Wolfgang B. Liedtke, M.D., Ph.D., Duke University Medical Center, Durham, North Carolina
Stefan Heller, Ph.D., Stanford University School of Medicine, Stanford, California

Methods for Neural Ensemble Recordings, Second Edition
Miguel A.L. Nicolelis, M.D., Ph.D., Professor of Neurobiology and Biomedical Engineering,
   Duke University Medical Center, Durham, North Carolina

Biology of the NMDA Receptor
Antonius M. VanDongen, Duke University Medical Center, Durham, North Carolina

Methods of Behavioral Analysis in Neuroscience
Jerry J. Buccafusco, Ph.D., Alzheimer's Research Center, Professor of Pharmacology and Toxicology,
   Professor of Psychiatry and Health Behavior, Medical College of Georgia, Augusta, Georgia

In Vivo Optical Imaging of Brain Function, Second Edition
Ron Frostig, Ph.D., Professor, Department of Neurobiology, University of California,
Irvine, California

Fat Detection: Taste, Texture, and Post Ingestive Effects
Jean-Pierre Montmayeur, Ph.D., Centre National de la Recherche Scientifique, Dijon, France
Johannes le Coutre, Ph.D., Nestlé Research Center, Lausanne, Switzerland

The Neurobiology of Olfaction
Anna Menini, Ph.D., Neurobiology Sector International School for Advanced Studies, (S.I.S.S.A.),
   Trieste, Italy

Neuroproteomics
Oscar Alzate, Ph.D., Department of Cell and Developmental Biology, University of North
   Carolina, Chapel Hill, North Carolina

Translational Pain Research: From Mouse to Man
Lawrence Kruger, Ph.D., Department of Neurobiology, UCLA School of Medicine, Los Angeles,
   California
Alan R. Light, Ph.D., Department of Anesthesiology, University of Utah, Salt Lake City, Utah

Advances in the Neuroscience of Addiction
Cynthia M. Kuhn, Duke University Medical Center, Durham, North Carolina
George F. Koob, The Scripps Research Institute, La Jolla, California

Neurobiology of Huntington's Disease: Applications to Drug Discovery
Donald C. Lo, Duke University Medical Center, Durham, North Carolina
Robert E. Hughes, Buck Institute for Age Research, Novato, California

ITCH
Mechanisms and Treatment

Edited by

E. Carstens
University of California, Davis, USA

Tasuku Akiyama
University of California, Davis, USA

CRC Press
Taylor & Francis Group
Boca Raton London New York

CRC Press is an imprint of the
Taylor & Francis Group, an **informa** business

Cover photos: Upper and lower left-hand photos from Tominaga & Takamori, Ch. 17. Upper right photo from Mishra & Hoon, *Science* 340:968-971, 2013, Fig. 4F, with permission from the American Association for the Advancement of Science. Lower right photo: unpublished images courtesy of Drs. Alex Papoiu and Gil Yosipovitch.

First published in paperback 2024

First published 2014 by CRC Press
2385 NW Executive Center Drive, Suite 320, Boca Raton FL 33431

and by CRC Press
4 Park Square, Milton Park, Abingdon, Oxon, OX14 4RN

CRC Press is an imprint of Taylor & Francis Group, LLC

© 2014, 2024 Taylor & Francis Group, LLC

Library of Congress Cataloging-in-Publication Data

Itch (Carstens)
    Itch : mechanisms and treatment / editors, Earl E. Carstens, Tasuku Akiyama.
    p. ; cm. -- (Frontiers in neuroscience)
    Includes bibliographical references and index.
    ISBN 978-1-4665-0543-8 (hardcover : alk. paper)
    I. Carstens, Earl, editor of compilation. II. Akiyama, Tasuku, editor of compilation. III. Title. IV. Series: Frontiers in neuroscience (Boca Raton, Fla.)
    [DNLM: 1. Pruritus--physiopathology. 2. Pruritus--therapy. 3. Antipruritics--therapeutic use. WR 282]

RL721
616.5--dc23   2013048169

ISBN: 978-1-4665-0543-8 (hbk)
ISBN: 978-1-03-291931-7 (pbk)
ISBN: 978-0-429-09834-5 (ebk)

DOI: 10.1201/b16573

Visit the Taylor & Francis Web site at
http://www.taylorandfrancis.com

and the CRC Press Web site at
http://www.crcpress.com

Contents

Series Preface

The Frontiers in Neuroscience Series presents the insights of experts on emerging experimental technologies and theoretical concepts that are, or will be, at the vanguard of neuroscience.

The books cover new and exciting multidisciplinary areas of brain research and describe breakthroughs in fields like visual, gustatory, auditory, olfactory neuroscience as well as aging biomedical imaging. Recent books cover the rapidly evolving fields of multisensory processing, depression, and different aspects of reward.

Each book is edited by experts and consists of chapters written by leaders in a particular field. The books have been richly illustrated and contain comprehensive bibliographies. The chapters provide substantial background material relevant to the particular subject.

The goal is for these books to be the references every neuroscientist uses in order to acquaint themselves with new information and methodologies in brain research. My task as series editor is to produce outstanding products that contribute to the broad field of neuroscience. Now that the chapters are available online, the effort put in by us—the publisher, the book editors, and individual authors—will contribute to the further development of brain research. To the extent that you learn from these books, we will have succeeded.

Sidney A. Simon, PhD
Series Editor

Preface

Itch is gaining increasing recognition as a widespread and costly medical and socio-economic problem. It has been estimated that medical costs for itch associated with atopic dermatitis in the United States exceeded $3 billion in 2009. While the biological basis of pain has been intensively investigated over the past several decades, much less research has been devoted toward itch until quite recently. This gap is now quickly being filled, however, as numerous basic research and clinical laboratories have begun to investigate itch mechanisms and treatment. The progressively growing interest in itch has resulted in numerous spectacular discoveries concerning the molecular genetics of itch, identification of itch transduction and signaling pathways, skin biology, and central neural processing of itch. These discoveries have identified new targets for the development of drugs and other strategies to treat the myriad forms of chronic itch. Our book represents an attempt to capture the excitement, discovery, and challenges in this rapidly advancing field. The contributing authors to this volume are leaders in basic research and clinical investigation and treatment of itch, and we are grateful for their participation in this stimulating endeavor. We wish to thank Series Editor Prof. Sidney A. Simon for inviting us to edit this book as part of the Frontiers in Neuroscience series. Finally, we wish to acknowledge the outstanding assistance of the editorial and production staff at Taylor and Francis for seeing this project through to completion.

E. Carstens and Tasuku Akiyama

Editors

E. Carstens is currently a distinguished professor of neurobiology, physiology, and behavior at the University of California, Davis. Dr. Carstens earned his BS degree in biological sciences from Cornell University (1972) and his PhD degree from the University of North Carolina, Chapel Hill, in neurobiology (1977). He then conducted postdoctoral research on descending modulation of pain at the University of Heidelberg (Germany) School of Medicine from 1977 to 1980 with fellowships from the National Institutes of Health and the Alexander von Humboldt Foundation. He joined the faculty at the University of California, Davis in 1980, where his research interests have focused on somatosensory mechanisms and particularly itch, pain, and chemesthesis. He was awarded Fulbright Senior Professor Research Awards in 1987 and again in 1995 and served as the vice chair of his home department (2009–2012). He is currently the vice president of the International Forum for the Study of Itch (2011–present), and has served as section editor for *Acta Dermato-Venerologica* in the areas of neurodermatology and itch since 2006. Dr. Carstens has coauthored numerous research articles, book chapters, and reviews on itch, pain, chemesthesis, and mechanisms of anesthetic action.

Tasuku Akiyama is an assistant project scientist at the University of California, Davis. He earned his PhD in 2008 in pharmacology at Toyama University. From 2008 to 2012, he was a postdoctoral fellow at the University of California, Davis. Dr. Akiyama's research focuses on the investigation of neuronal mechanisms of itch and pain. His published work includes studies on the spinal and trigeminal processing of itch and pain, the neuronal mechanisms of itch in mosquito bite and dry skin, and the mechanisms of itch sensitization under chronic itch conditions.

Contributors

Tasuku Akiyama, PhD
Department of Neurobiology,
    Physiology and Behavior
University of California
Davis, California

Tsugunobu Andoh, PhD
Department of Applied Pharmacology
Graduate School of Medicine and
    Pharmaceutical Sciences
University of Toyama
Toyama, Japan

Diana M. Bautista, PhD
Department of Molecular and Cell
    Biology
and
Helen Wills Neuroscience Institute
University of California
Berkeley, California

Nora V. Bergasa, MD, FACP
Department of Medicine
Metropolitan Hospital Center
and
New York Medical College
New York, New York

Paul L. Bigliardi, MD
University Medicine Cluster
and
National University Hospital
and
Institute of Medical Biology
A* STAR
Singapore

Mei Bigliardi-Qi, PhD
Senior Research Scientist
and
COO Clinical Research Unit for Skin,
    Allergy and Regeneration
Institute of Medical Biology
A* STAR
Singapore

Joerg Buddenkotte, PhD
Department of Dermatology
University Hospital of Muenster
Muenster, Germany

Timo Buhl, MD
Department of Dermatology
University of California
San Francisco, California

E. Carstens, BS, PhD
Department of Neurobiology,
    Physiology and Behavior
University of California
Davis, California

Ferda Cevikbas, PhD
Department of Dermatology
University of California
San Francisco, California

Sandra R. Chaplan, MD
Janssen Research and Development,
    L.L.C.
San Diego, California

Ulf Darsow, MD
Department of Dermatology and
    Allergy Biederstein
Technische Universität München
and
ZAUM—Center of Allergy and
    Environment
Munich, Germany

Steve Davidson, PhD
Pain Center and Department of
    Anesthesiology
Washington University School of
    Medicine
St. Louis, Missouri

Xinzhong Dong, PhD
Solomon H. Snyder Department of
    Neuroscience
Johns Hopkins University School of
    Medicine
Baltimore, Maryland

Clemens Forster, PhD
Department of Physiology and
    Pathophysiology
University of Erlangen-Nurnberg
Erlangen, Germany

Glenn Giesler, PhD
Department of Neuroscience
University of Minnesota
Minneapolis, Minnesota

Andrew J. Greenspan, MD
Janssen Research and Development,
    L.L.C.
San Diego, California

Junichi Hachisuka, MD, PhD
Department of Neurobiology and
    The Center for Pain Research
University of Pittsburgh
Pittsburgh, Pennsylvania

Hermann O. Handwerker, MD, PhD
Department of Physiology and
    Pathophysiology
University of Erlangen-Nurnberg
Erlangen, Germany

Ru-Rong Ji, PhD
Sensory Plasticity Laboratory
Department of Anesthesiology
Duke University Medical Center
Durham, North Carolina

Kayvan Kazerouni
Janssen Research and Development, L.L.C.
San Diego, California

Cordula Kempkes, PhD
Department of Dermatology
University of California
San Francisco, California

Mei-Chuan Ko, PhD
Department of Physiology and
    Pharmacology
School of Medicine
Wake Forest University
Winston-Salem, North Carolina

Yasushi Kuraishi, PhD
Department of Applied Pharmacology
Graduate School of Medicine and
    Pharmaceutical Sciences
University of Toyama
Toyama, Japan

Tong Liu, PhD
Sensory Plasticity Laboratory
Department of Anesthesiology
Duke University Medical Center
Durham, North Carolina

Qiufu Ma, PhD
Dana-Farber Cancer Institute and
    Department of Neurobiology
Harvard Medical School
Boston, Massachusetts

Uwe Matterne, PhD
Department Clinical Social Medicine
Occupational and Environmental
  Dermatology
University Hospital Heidelberg
Heidelberg, Germany

Benjamin McNeil, PhD
Solomon H. Snyder Department of
  Neuroscience
Johns Hopkins University School of
  Medicine
Baltimore, Maryland

Christian Mess
Department of Dermatology
University Hospital of Muenster
Muenster, Germany

Thomas Mettang, MD, FASN
Department of Nephrology
German Clinic for Diagnostics
Wiesbaden, Germany

Richard Meyer, MS
Department of Neurosurgery
School of Medicine
Johns Hopkins University
Baltimore, Maryland

Laurent Misery, MD, PhD
Laboratory of Neurosciences
University of Brest
and
Department of Dermatology
University Hospital of Brest
Brest, France

Hideki Mochizuki, PhD
Department of Dermatology
and
Temple Itch Center
Temple University Medical School
Philadelphia, Pennsylvania

Hannah Moser, PhD
Department of Neuroscience
University of Minnesota
Minneapolis, Minnesota

**Anne Louise Oaklander, MD, PhD,
FAAN, FANA**
Departments of Neurology and
  Pathology (Neuropathology)
Massachusetts General Hospital
Harvard Medical School
Boston, Massachusetts

Alexandru D.P. Papoiu, MD, PhD
Department of Dermatology
Wake Forest School of Medicine
Winston-Salem, North Carolina

Ulrike Raap, MD
Department of Dermatology and
  Allergology
Hannover Medical University
Hannover, Germany

Adam Reich, MD, PhD
Department of Dermatology,
  Venereology and Allergology
Wroclaw Medical University
Wroclaw, Poland

Matthias Ringkamp, MD, PhD
Department of Neurosurgery
School of Medicine
Johns Hopkins University
Baltimore, Maryland

Sarah E. Ross, PhD
Department of Neurobiology and
  The Center for Pain Research
University of Pittsburgh
Pittsburgh, Pennsylvania

Martin Schmelz, MD
Department of Anesthesiology and
    Intensive Care Medicine
Medical Faculty Mannheim
Heidelberg University
Mannheim, Germany

Sonja Ständer, MD
Competence Center Chronic Pruritus
    and Department of Dermatology
University Hospital Münster
Münster, Germany

Martin Steinhoff, MD, PhD
Department of Dermatology
University of California
San Francisco, California

Jacek C. Szepietowski, MD, PhD
Department of Dermatology,
    Venereology and Allergology
Wroclaw Medical University
Wroclaw, Poland

Kenji Takamori, MD, PhD
Institute for Environmental and
    Gender Specific Medicine
Juntendo University Graduate School of
    Medicine
and
Department of Dermatology
Juntendo University Urayasu Hospital
Chiba, Japan

Robin L. Thurmond, PhD
Janssen Research and Development,
    L.L.C.
San Diego, California

Andrew J. Todd, MBBS, PhD
Spinal Cord Group
Institute of Neuroscience and
    Psychology
University of Glasgow
Glasgow, Scotland, United Kingdom

Mitsutoshi Tominaga, PhD
Institute for Environmental and
    Gender Specific Medicine
Juntendo University Graduate School of
    Medicine
Chiba, Japan

Elke Weisshaar, MD
Department of Clinical Social
    Medicine
Occupational and Environmental
    Dermatology
University Hospital Heidelberg
Heidelberg, Germany

Sarah R. Wilson, BA
Department of Molecular and
    Cell Biology
University of California
Berkeley, California

Gil Yosipovitch, MD
Department of Dermatology
and
Temple Itch Center
Temple University Medical School
Philadelphia, Pennsylvania

1 Itch Hypotheses

From Pattern to Specificity and to Population Coding

Hermann O. Handwerker

In the late nineteenth century, physiologists began to regard pain as a sensation. At first this seemed to be a strange idea, because pain is so closely associated with suffering and hence with the realm of emotions. One of the first scientists relating pain to certain mechanisms of transmission of peripheral stimuli into the central nervous system was René Descartes in the seventeenth century. He outlined his idea of nerve conduction on the example of heat pain, thus placing it alongside the classical sense of vision (Descartes 1969). In the early nineteenth century, at the dawn of a naturalistic physiology, Johannes Mueller formulated his "Gesetz der spezifischen Sinnesenergien," i.e., the postulate that each sense is subserved by its specific neuronal sensory and nervous apparatus (Mueller 1837). In the last decade of the nineteenth century, when physiology had reached a first peak as a natural science, several researchers made the observation that the skin, our largest sensory organ, can be regarded as a mosaic of sensory spots that can be stimulated separately with fine probes to evoke pure sensations of touch, warmth, or cold. Pioneers of this research were the Swedish physiologist Blix, the German Goldscheider, and the Austrian/German von Frey. Max von Frey was the most systematic and important of this group. He developed an easy and reliable method to stimulate the "touch points" with "Tasthaar" stimulators, nowadays called "von Frey filaments." By employing thin and sharp plant spicules he found locations between the touch points where such a fine spicule could induce pain without accompanying touch sensation. From this finding he deduced his famous "specificity theory" of pain. One can still have a feeling of his enthusiasm for this discovery when he wrote (von Frey 1896; see also Handwerker and Brune 1987, p. 95)

> "It is possible to stimulate the skin in such a way as to produce a painful sensation with no preceeding or accompanying pressure sensation. That this can be done ... leads to the conclusion that pain is the result of exciting special organs ... the pain points are a sign of the irregular distribution of specifically pain sensitive organs over the skin."

With this hypothesis, pain was accepted in the realm of the senses. The specificity hypothesis was a basis for searching a dedicated peripheral and central nervous apparatus. For the peripheral nervous system this hypothesis converged with the line of thinking of C. S. Sherrington, who had found in animal experiments that withdrawal reflexes, the correlate of pain sensations in humans, are elicited by the excitation of a special kind of nerve fibers that he called "nociceptors" (Sherrington 1906).

The "specificity theory" was controversial from the beginning. von Frey's colleague Goldscheider (who was a famous clinician besides being a physiologist) used the same kind of stimulators but came to contradictory results. By inserting the spicules not perpendicularly but slantwise, he found always a touch before a pain sensation. He formulated an "intensity hypothesis" of pain: weak stimuli applied to the skin induce always touch-like sensations at first, and only when becoming more intense this sensation turns to pain (Goldscheider 1898). At this point I have to leave the saga of the development of pain concepts, because the topic of this article is itch, not pain.

Itch came later in the focus of physiologists searching for nervous mechanisms of sensations. A common notion was that itch is kind of a minor form of pain. Max von Frey adopted this hypothesis when he published a paper on itch two decades after his groundbreaking studies on pain (von Frey 1922). He had observed that stimulation of "pain points" with spicules often leads to itchy aftersensations. The notions that there are no individual "itch points" that can be excited by mechanical or thermal stimuli led him to an "intensity hypothesis" for itch. According to this hypothesis, itch is not an independent "sensory modality" in Johannes Mueller's sense but a quality of pain. Support of this hypothesis came from clinical observations that hemisection of the spinal cord or surgical disconnection of the anterolateral ascending tract (cordotomy; in former times a neurosurgical treatment of intractable pain) leads to simultaneous loss of itch and pain on the contralateral side, but to ipsilateral impairment of touch perception. Itch and pain are apparently conducted in ascending pathways along the same trajectories, whereas mechano-sensation follows a different path. Later it was found that both, pain and itch are mediated by slowly conducting nerve fibers.

The intensity hypothesis is, however, at variance with many other observations: there are many forms of pain that do not turn to itch when the stimuli get weaker. Intense itch can be more troublesome than many forms of pain without swelling to pain. In contrast to pain, itch is restricted to the skin and adjacent mucous membranes. Opioids suppress pain but have a tendency to enhance itch. Finally, the related reflexes are completely different: pain leads to withdrawal reflexes and other types of avoidance behavior, whereas itch induces the urge to scratch. All this indicates an independent neuronal apparatus for itch processing.

An important step in the development of our understanding of itch was the discovery of the endogenous compound histamine by Henry Dale, around 1910. In the following decades it became clear that histamine released from basophiles and mast cells is a potent vasodilator and a key substance in allergic and other inflammatory processes. Sir Thomas Lewis discovered the "triple response" of the skin to local histamine release: a local reaction, followed by an edema (weal) and an extended erythema (flare) around the affected skin site (Lewis and Harmer 1927). Thomas Lewis was not so much interested in the itch sensation but in the local cutaneous reaction that is dependent on afferent nerve fibers that are excited by histamine and release another vasoactive substance from their axon collaterals (axon reflex), which was later identified as CGRP. To apply histamine reliably to the superficial layers of the skin, the "blister base technique" was developed. To this purpose a cantharidin plaster was affixed to the skin, which leads to a debonding of the epidermis and forming of a blister. Low concentrations of mediators can be applied to the exposed nerve endings at the blister base. Keele and Armstrong found with this technique that low concentrations of histamine induced pure itch, but higher concentrations lead to itch mixed with pain (Keele and Armstrong 1964). Together with the observation that subcutaneous injections of histamine induced no itch, but weak pain, this lead to the conclusion that histamine is "the" itch mediator, albeit only in the superficial layers of the skin. Also, deeper in the body, nerve fibers with an affinity to histamine exist that do not mediate itch sensations but preferentially pain. Later a specific membrane receptor for histamine (H1) was discovered, a prerequisite for drug development, and nowadays antihistaminic drugs are still among the most common medications for the treatment of itch.

With histamine a means was available to study the responsiveness of nerve endings with a rather specific itching stimulus. Another form of pure pruritic stimulation came in the focus of itch researchers in form of the spicules of the cowhage bean (Shelley and Arthur 1955). The active agent mucunain has been synthetized recently (Reddy et al. 2008). One might have assumed that with these stimulus paradigms it should have been easy to characterize the kind of peripheral nervous input leading to itchy sensations. However, in an extensive study on myelinated (Tuckett and Wei 1987a) and unmyelinated (Tuckett and Wei 1987b) nerve fibers in the cat hairy skin, these authors found little evidence for nerve fibers responding specifically to cowhage stimulation, but also the evidence for a characteristic pattern in a population of unspecific fibers remained inconclusive. Our group used microneurography to study single human skin nerve fibers and iontophoresis for histamine stimulation. We encountered weak excitation of polymodal C-nociceptors with the exception of an accidental finding of a strong histamine response of a C-fiber that we took for a polymodal nociceptor at that time (Handwerker et al. 1991). Afterward we suspected that we had studied a biased sample of nerve fibers by employing mechanical search stimuli for identifying receptive nerve endings in the skin. Together with H. E. Torebjoerk we developed a bias-free microneurography technique in which the electrical responsiveness of nerve endings sufficed for electrophysiological characterization of an afferent nerve fiber ("marking technique"). With this technique we were able to characterize a type of mechanically insensitive afferent C-fiber, which is not uncommon in human skin, namely, the mechano-insensitive, or "sleeping" nociceptors (sleeping because they become mechanosensitive in inflamed skin). It turned out that these C-units are distinct from the more common "polymodal nociceptors" by several features, e.g., slower conduction velocities, higher transcutaneous electrical thresholds, higher average heat thresholds, and, in particular, more pronounced slowing of the axonal conduction upon repetitive use (Schmidt et al. 1995). In this class of newly identified C-afferents (CMI for C-fibers, insensitive for mechanical stimuli) we found a subtype that was particularly sensitive to histamine (CMI, his+ units). Furthermore, the discharge patterns of these units were running strikingly parallel to the long lasting itch sensations induced by short periods of histamine iontophoresis (Schmelz et al. 1997). Another group found secondary neurons with a similar response spectrum in the superficial dorsal horn of the cat, which projected to the shell region of the somatosensory thalamus (Andrew and Craig 2001). In microneurography experiments on patients with prurigo nodularis suffering from severe itching, spontaneously active CMI (his+) units were found. These findings indicate that this fiber class plays also a role in some forms of pathological pruritus (Schmelz et al. 2003a).

With this pathway we came close to a specificity hypothesis of histaminergic itch. It soon became clear, however, that the histamine-sensitive C-fibers were also responsive to other endogenous inflammatory mediators, e.g., to prostaglandin E and to acetylcholine. They were insensitive, however, to bradykinin, which provokes a burning sensation when brought into the skin. Like the pain-mediating nociceptors, itch fibers are apparently endowed with a mosaic of membrane receptors for endogenous and exogenous agents in their terminals, though this mosaic is different to some degree from that of the more common types of nociceptors. Most remarkably, however, the "itch fibers" are also responsive to capsaicin, the ligand of the TRPV1 receptor (Schmelz et al. 2003b). Because capsaicin induces strong burning pain, even in low concentrations, this finding seems to exclude again a "pure" specificity hypothesis for itch.

For an interpretation of these apparently contradictory results, one has to consider the ambiguity of the concept of a "specific primary afferent unit." (1) Such a unit could be specific in a molecular sense by being endowed with a specific terminal receptor apparatus, ideally, with only one type of transduction molecule. Most probably, these units do not exist among the slowly conducting primary cutaneous afferents. (2) It could also be specific, however, by projecting into one but not into the other central neuronal pool constituting a projection pathway. In case of the itch fibers this would imply that a majority of nociceptive primary afferents endowed with the TRPV1 receptor project into the pain pathway, but a minority project into the itch pathway.

When all TRPV1 positive units are excited by capsaicin, pain will arise, because the pain pathway is dominant and, in addition, pain inhibits itch. Inhibitory interneurons play a crucial role in this concept that has been called "*selectivity hypothesis*" or "*population coding*" (Akiyama et al. 2009; Patel and Dong 2010). For a review, see Ma (2010). Recent findings of molecular mechanisms of itch transduction and processing have boosted this line of thinking. For example, it has been shown that histamine H1 receptors and TRPV1 receptors are often expressed together in nerve terminals. The H1 receptor is G-protein coupled and controls second messenger pathways, which activate TRPV1 receptors (Ross 2011).

Before turning to this last development in itch research, I have to deal with the problem of "nonhistaminergic itch." It has long been known that many clinical pruritic diseases, most notably atopic dermatitis, are not dependent on the histaminergic pathway, and hence antihistaminic drugs are ineffective in their treatment. It has been mentioned above that the spicules of the cowhage plant (mucuna pruriens) induce strong itch. However, it has been known since the studies of Arthur and Shelley (1955, 1959) that this itch is not accompanied by a flare response, and hence it is not related to histamine release. When cowhage spicules were employed in single fiber experiments in monkeys (Johanek et al. 2008) and in human microneurography (Namer et al. 2008), it turned out that this type of itch is not due to the excitation of the CMI (his+) fibers described by Schmelz et al. (1997) but related to excitation of the common form of polymodal nociceptors. In humans, there is no overlap between the two peripheral itch pathways (Namer et al. 2008). The finding that most polymodal nociceptors in the superficial layers of the skin are excited by cowhage apart from being responsive to mechanical, heat, and chemical pain stimuli poses a new problem that is hard to resolve, even with a selectivity hypothesis (see below). In subhuman mammalian species the differentiation is less complete, and, in particular, in mice, there is considerable overlap between histamine and cowhage sensitive primary afferent units (Ma et al. 2012). However, in primates, a separate pathway exists for histamine and cowhage induced itch, as proven by the finding of separate groups of projection neurons in the monkey thalamus (Davidson et al. 2007). It is quite possible that more than two central pathways exist for different kinds of itch, though up until now, no conclusive evidence has been found. The differentiation into different neuronal subsystems for histamine-dependent and histamine-independent itch raises new difficulties for a unifying theory. It has to be stated that the two forms of itch are subserved by different peripheral neuronal populations, though both belong to the realm of unmyelinated and possibly thinly myelinated nerve fibers (Ringkamp et al. 2011). Both kinds of itch induce the urge to scratch and are temporarily suppressed by scratching (Kosteletzky et al. 2009). Both are probably enhanced by opioids.

Figure 1.1 shows a simplified schedule of itch and pain processing in the periphery and in the spinal cord. Two itch pathways are delineated with one pain pathway receiving the overwhelming majority of nociceptive afferents.

The itch pathways are inhibited by pain input (e.g., from scratching) through interneurons. According to this hypothesis these inhibitory interneurons are again inhibited by opioidergic interneurons and descending pathways that also inhibit the pain processing. "Inhibition of the inhibition" by the "itch interneurons" enhances itching. This is basically the *selectivity hypothesis* of itch (Ma 2010). In a way this hypothesis is a synthesis of the older "pattern" and "specificity" hypotheses. This picture has been greatly enhanced by recent molecular biology as shown in Figure 1.1. The details in this diagram are mainly composed with the help of recent reviews (Ross 2011; Ma 2010). "Knock-out" experiments have shown that elimination of certain receptors or neuronal populations can lead to a complete loss of scratching behavior in mice. Sun and coworkers showed that out-knocking of the neurons expressing the GRPR–membrane receptor show no longer scratching behavior to a variety of itching compounds (Sun et al. 2009; see also Handwerker and Schmelz 2009). On the other hand, Liu and coworkers have shown that knocking out the VGLUT2-dependent glutamate release from those primary afferents that contain NaV 1.8 receptors in mice leads to a switch from protective wiping (pain defense) to scratching (itch behavior) after injection of capsaicin in the skin (Liu et al. 2010). Recently, Han et al. 2012 have shown that any excitation of the MrgprA3 neurons evokes scratching behavior in mice, irrespective of how the neurons are activated. To this purpose, they expressed the capsaicin receptor TRPV1 exclusively in the MrGprA3 positive neurons leaving it deleted in all other types of neurons. These are spectacular examples of selective populations responsible for itch processing. Slowly, a new biology of itch is appearing. Much of this will be covered in this volume. Therefore, in this historical review, no further details will be reported. However, one has to keep in mind that the molecular biology of itch at present is virtually restricted to mice. In the light of the known species differences, much work has to be done to translate it into human neurobiology.

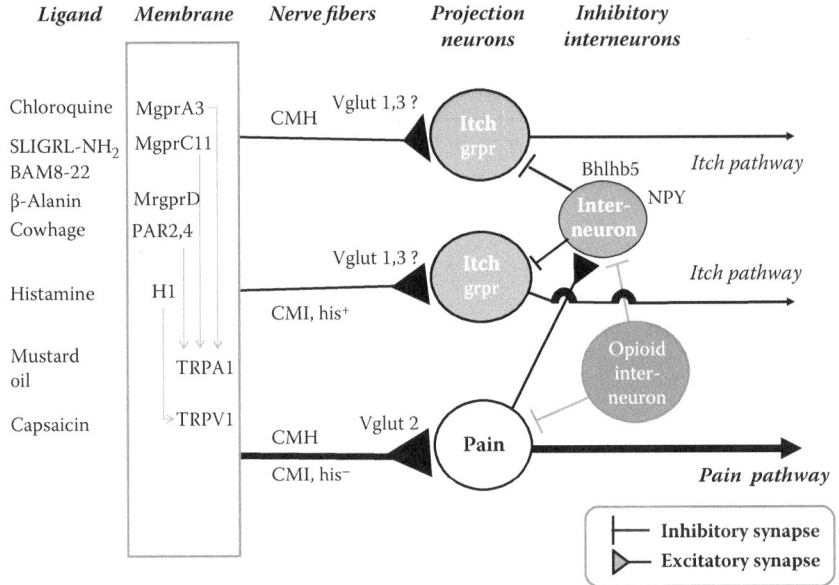

FIGURE 1.1 **(See color insert.)** Diagram of hypothetical itch processing in the peripheral nervous system and in the spinal cord. Two itch pathways and a pain pathway (containing a larger neuronal population) are assumed. Grey and white circles symbolize the transmission neurons in the spinal cord, and blue and red circles show inhibitory interneurons. The schedule shows from left to right the mediators binding to the afferent terminals, the membrane receptors, the types of nerve fibers, and the central neurons with the molecular characteristics recently described (see Ross 2011). For further details, see text.

Is the hypothesis of "population coding" the last step in the development of itch hypotheses? From this short review it should be clear that not all features of the various itch phenomena can be explained at present. One additional line of reasoning might solve some of the remaining problems: Taking into account that itch can be evoked only from the superficial skin layers and that most itching stimuli are punctual and excite a limited number of skin receptors, one might speculate that these few units are characterized by large central receptive fields because of the lack of lateral inhibition. To my knowledge, this hypothesis was first formulated by Greaves and Wall (1996). Recently, it was renewed in the light of the paradox that cowhage excites most polymodal nociceptors in human skin. Itch might arise from the sharp contrast between individual polymodal nociceptors firing while the surrounding units remain silent and hence do not exert the usual lateral inhibition (Namer and Reeh 2013).

REFERENCES

Akiyama, T., Merrill, A. W., Carstens, M. I., and Carstens, E. (2009). Activation of superficial dorsal horn neurons in the mouse by a PAR-2 agonist and 5-HT: Potential role in itch. *J. Neurosci.* 20(29): 6691–6699.

Andrew, D., and Craig, A. D. (2001). Spinothalamic lamina I neurons selectively sensitive to histamine: A central neural pathway for itch. *Nat. Neurosci.* 4: 72–77.

Arthur, R. P., and Shelley, W. B. (1955). Experimental evidence for an enzymatic basis for itching in man. *Nature* 175: 901–902.

Arthur, R. P., and Shelley, W. B. (1959). The innervation of human epidermis. *J. Invest. Dermatol.* 32: 397–411.

Davidson, S., Zhang, X., Yoon, C. H., Khasabov, S. G., Simone, D. A., and Giesler, G. J., Jr. (2007). The itch-producing agents histamine and cowhage activate separate populations of primate spinothalamic tract neurons. *J. Neurosci.* 27: 10007–10014.

Descartes, R. (1969). *De homine—Ueber den Menschen*. Translation by K. E. Rothshuh from "*L'Homme.*" Heidelberg: Lambert Schneider.

Goldscheider, A. (1898). *Physiologie der Hautsinnesnerven*. Leipzig: Johann Ambrosius Barth.

Greaves, M. W., and Wall, P. D. (1996). Pathophysiology of itching. *Lancet* 348: 938–940.

Han, L., Ma, C., Liu, Q., Weng, H. J., Cui, Y., Tang, Z., Kim, Y. et al. (2012). A subpopulation of nociceptors specifically linked to itch. *Nat. Neurosci.* 16: 174–182.

Handwerker, H. O., and Brune, K. (1987). *Classical German Contributions to Pain Research*. Heidelberg: German Pain Society.

Handwerker, H. O., Forster, C., and Kirchhoff, C. (1991). Discharge patterns of human C-fibers induced by itching and burning stimuli. *J. Neurophysiol.* 66: 307–315.

Handwerker, H. O., and Schmelz, M. (2009). Pain: Itch without pain—a labeled line for itch sensation? *Nat. Rev. Neurol.* 5: 640–641.

Johanek, L. M., Meyer, R. A., Friedman, R. M., Greenquist, K. W., Shim, B., Borzan, J., Hartke, T., LaMotte, R. H., and Ringkamp, M. (2008). A role for polymodal C-fiber afferents in nonhistaminergic itch. *J. Neurosci.* 28: 7659–7669.

Keele, C. A., and Armstrong, D. (1964). *Substances Producing Pain and Itch*. London: Edward Arnold.

Kosteletzky, F., Namer, B., Forster, C., and Handwerker, H. O. (2009). Impact of scratching on itch and sympathetic reflexes induced by cowhage (Mucuna pruriens) and histamine. *Acta Derm. Venereol.* 89: 271–277.

Lewis, T., and Harmer, I. M. (1927). Vascular reactions of the skin to injury. Tyrt IX. Further evidence of the release of a histamine-like substance from the injured skin. *Heart* 14: 19–26.

Liu, Y., Abdel, S. O., Zhang, L., Duan, B., Tong, Q., Lopes, C., Ji, R. R., Lowell, B. B., and Ma, Q. (2010). VGLUT2-dependent glutamate release from nociceptors is required to sense pain and suppress itch. *Neuron* 68: 543–556.

Ma, C., Nie, H., Gu, Q., Sikand, P., and LaMotte, R. H. (2012). In vivo responses of cutaneous C-mechanosensitive neurons in mouse to punctate chemical stimuli that elicit itch and nociceptive sensations in humans. *J. Neurophysiol.* 107: 357–363.

Ma, Q. (2010). Labeled lines meet and talk: Population coding of somatic sensations. *J. Clin. Invest.* 120: 3773–3778.

Mueller, J. (1837). *Handbuch der Physiologie des Menschen*. Coblenz: J. Hoelscher.

Namer, B., Carr, R., Johanek, L. M., Schmelz, M., Handwerker, H. O., and Ringkamp, M. (2008). Separate peripheral pathways for pruritus in man. *J. Neurophysiol.* 100: 2062–2069.

Namer, B., and Reeh, P. W. (2013). Scratching an itch. *Nat. Neurosci.* 16: 117–118.

Patel, K. N., and Dong, X. (2010). An itch to be scratched. *Neuron* 68: 334–339.

Reddy, V. B., Iuga, A. O., Shimada, S. G., LaMotte, R. H., and Lerner, E. A. (2008). Cowhage-evoked itch is mediated by a novel cysteine protease: A ligand of protease-activated receptors. *J. Neurosci.* 28: 4331–4335.

Ringkamp, M., Schepers, R. J., Shimada, S. G., Johanek, L. M., Hartke, T. V., Borzan, J., Shim, B., LaMotte, R. H., and Meyer, R. A. (2011). A role for nociceptive, myelinated nerve fibers in itch sensation. *J. Neurosci.* 31(42): 14841–14849.

Ross, S. E. (2011). Pain and itch: Insights into the neural circuits of aversive somatosensation in health and disease. *Curr. Opin. Neurobiol.* 21: 880–887.

Schmelz, M., Hilliges, M., Schmidt, R., Orstavik, K., Vahlquist, C., Weidner, C., Handwerker, H. O., and Torebjork, H. E. (2003a). Active "itch fibers" in chronic pruritus. *Neurology* 61: 564–566.

Schmelz, M., Schmidt, R., Bickel, A., Handwerker, H. O., and Torebjörk, H. E. (1997). Specific C-receptors for itch in human skin. *J. Neurosci.* 17: 8003–8008.

Schmelz, M., Schmidt, R., Weidner, C., Hilliges, M., Torebjork, H. E., and Handwerker, H. O. (2003b). Chemical response pattern of different classes of C-nociceptors to pruritogens and algogens. *J. Neurophysiol.* 89: 2441–2448.

Schmidt, R., Schmelz, M., Forster, C., Ringkamp, M., Torebjork, E., and Handwerker, H. (1995). Novel classes of responsive and unresponsive C nociceptors in human skin. *J. Neurosci.* 15: 333–341.

Shelley, W. B., and Arthur, R. P. (1955). Studies on cowhage (mucuna pruriens) and its pruritogenic proteinase, mucunain. *Arch. Dermatol.* 72: 399–406.

Sherrington, C. S. (1906). *The Integrative Action of the Nervous System.* New Haven, CT: Yale University Press.

Sun, Y. G., Zhao, Z. Q., Meng, X. L., Yin, J., Liu X. Y., and Chen, Z. F. (2009). Cellular basis of itch sensation. *Science* 325: 1531–1534.

Tuckett, R. P., and Wei, J. Y. (1987a). Response to an itch-producing substance in cat. I. Cutaneous receptor populations with myelinated axons. *Brain Res.* 413: 87–94.

Tuckett, R. P., and Wei, J. (1987b). Response to an itch-producing substance in cat. II. Cutaneous receptor population with unmyelinated axons. *Brain Res.* 413: 95–103.

von Frey, M. (1896). *Untersuchungen ueber die Sinnesfunctionen der menschlichen Haut,* vol. 1. *Druckempfindung und Schmerz.* Leipzig: S. Hirzel.

von Frey, M. (1922). Zur Physiologie der Juckempfindung. *Arch. Neerl. Physiol.* 7: 142–145.

2 Epidemiology of Itch

Elke Weisshaar and Uwe Matterne

CONTENTS

2.1 PREFACE/STARTING POINT

Dermatoepidemiology is an important emerging field in dermatology. For a long time, epidemiological issues seemed to have played only a minor role within dermatology. One of the reasons may be that dermatology is a relatively young specialty in medicine having evolved as a distinct specialty around 1850 (Apfelbacher et al. 2011). The term dermatoepidemiology was shaped only a few years ago and is now closely linked to the principles of evidence-based medicine. Its importance has increased over time in recognition of the fact that many skin diseases are not only highly prevalent (e.g., eczema) but also on the rise (e.g., skin cancer) (Apfelbacher et al. 2011) to the point that it is said to have flourished over the past decade (Barzilai et al. 2005).

Evidence suggests that at least one quarter of the European population suffers from a skin disease at least once in their lifetime (Williams et al. 2006). Epidemiological data about the frequency and attributes of pruritus exist to a minor extent only. These data, however, are more and more appreciated. For instance in order to improve the quality of care or make the right decisions about resource allocation, it is not only necessary to consider the findings of efficacy studies, but also studies reporting patient reported outcomes, cost-utility analyses, and, of course, have knowledge about how widespread a particular symptom or condition is, what determines the latter and how it can be controlled.

2.2 IMPORTANT EPIDEMIOLOGICAL PARAMETERS AND HOW TO OBTAIN THEM

From the viewpoint of an epidemiologist and depending on the study design different epidemiological parameters can be obtained. A cross-sectional survey, asking a representative sample of participants about whether they have the particular characteristic in question, yields prevalence estimates. Depending on the reference period, this can be a point, 12-months, or lifetime (ever) prevalence. These measures will tell you the proportion of individuals within a population having the symptom at present (point), during the last 12 months (12 month), or having ever had it (lifetime).

To obtain information about how many individuals become afflicted with the symptom within a specified period of time, or, in other words, to obtain an estimate of the incidence, a prospective research design is necessary. A cohort of individuals in whom the symptom is not present, is followed-up over a specific period of time. Occurrence of the symptom over the specified time period is then recorded and counted.

2.3 INTRODUCTION

While it is often stated that pruritus is the most frequent symptom in dermatology and that it may also occur in internal, neurological, and psychiatric/somatoform diseases as well as that it may be drug-induced, there are surprisingly few studies about the prevalence or incidence of pruritus in particular diseases or patient populations. The situation is made worse by the following:

- Many patients with acute pruritus but also with chronic pruritus do not present to a physician (Weisshaar et al. 2003; Matterne et al. 2009). This may be the result, for instance, of how the local health system is organized or how access to specialists is regulated (the question of whether a patient can see a specialist directly or only by referral). This explains the need for population-based studies on prevalence and incidence of chronic pruritus.
- The symptom "pruritus" often receives only minor attention in non-dermatological specialties and is often not considered a variable worth assessing (Weisshaar et al. 2003).
- The findings of studies on pruritus are often difficult to compare or interpret due to the fact that parameters are not operationalized in a consistent manner (e.g., lack of unanimously accepted definition and classification of pruritus). Clinical studies often use outcome measures that are not clearly defined. For instance, it is important to distinguish between the point (Are you currently suffering from…), period (During the last 12 months have you suffered from…), and lifetime (Have you ever in your life suffered from…) prevalence and clearly communicate the particular kind of prevalence estimate that was measured in a publication.
- The symptom whether acute or chronic may fluctuate; e.g., it may not occur daily. The most recent classification provided by International Forum for the Study of Itch (IFSI) appears the most comprehensive so far but is far from being perfect.

2.4 PRURITUS IN SPECIFIC AGE GROUPS

There appear to be no studies about the frequency of occurrence of pruritus in children in general. Nevertheless, pruritus is a dominant and cardinal symptom of atopic eczema (atopic dermatitis) (Rajka 1989), whose prevalence in the Western world is estimated to be between 6.1% and 22.3% (Shaw et al. 2011; Odhiambo et al. 2009). As atopic eczema is the most frequent skin disease in childhood, its prevalence can be used as a point of reference. In a population-based Norwegian study, 8.8% of the teenagers/adolescents suffered from pruritus (not precisely defined). The prevalence of pruritus was significantly associated with atopic eczema and mental distress (Halvorsen et al. 2009). In small samples of teenagers with acne, 13.8% to 70%, depending on ethnic origin, suffered from episodes of acute pruritus (Reich et al. 2008; Lim et al. 2008).

Chronic pruritus is often observed during pregnancy. While it is thought that approximately 18% of all pregnant women encounter pruritus (Black et al. 2008; Weisshaar and Dalgard 2009), there are hardly any epidemiological studies having investigated the prevalence of the symptom during pregnancy. In a French prospective study in 3192 pregnant women, 1.6% reported pruritus (Roger et al. 1994). All dermatoses of pregnancy are accompanied by pruritus. They comprise atopic eruption of pregnancy, polymorphic eruption of pregnancy, pemphigoid gestationis, and intrahepatic cholestasis of pregnancy (Ambros-Rudolph et al. 2006). It is possible that some pruritic dermatoses coincide by chance with pregnancy while preexisting pruritic dermatoses may exacerbate during pregnancy. The lack of consistent epidemiological and clinical data highlights again the need for more research in the area.

Elderly people are believed to be particularly prone to chronic pruritus. Previous studies looked at pruritic skin diseases. Depending on region and sample size, their prevalence was estimated to be between 11.5% and 41% (Weisshaar and Dalgard 2009). There appears to be no general agreement in the literature as to what constitutes the elderly (i.e., no precise cut-off score for age) or what precise clinical picture is to be expected in patient populations (Weisshaar and Dalgard 2009; Beauregard and Gilchrest 1987). One population-based study found a significant association between age and lifetime prevalence of chronic pruritus. The trend appeared to be nonlinear, but in a bimodal fashion with peaks for age groups between 31 and 40 and 51 and 60 years, respectively. No significant association of age was observed with the point or 12-month prevalence, respectively, nor was there a higher likelihood of chronic pruritus occurrence in individuals aged >65 years (Matterne et al. 2011). Another working population survey found that current chronic pruritus increased with age from 12.3% (16–30 years) to 20.3% (61–70 years) (Ständer et al. 2010). Several other studies were conducted in very small or highly select samples of elderly patients (Yalcin et al. 2006; Beauregard and Gilchrest 1987; Norman 2003) and as such do not provide general population estimates. All this highlights the need for more well-designed epidemiological research in order to establish an evidence base for the claim that pruritus is more frequent in the elderly.

2.5 PRURITUS IN SPECIFIC DISEASES

Pruritus is a defining feature of atopic eczema (Rajka 1989, p. 20). Daily pruritus is described in 87% to 91% of patients with atopic eczema (Yosipovitch et al. 2002; Dawn et al. 2009). There are no symptom-related epidemiological studies on the prevalence and the course of the predominantly chronic pruritus in atopic eczema.

According to a patients survey, atopic eczema patients were more likely to experience current pruritus and reported a higher pruritus intensity compared to psoriasis patients (O'Neill et al. 2011).

According to a questionnaire-based study in 17,488 **psoriasis** patients, pruritus was the second most frequently reported symptom and was reported by 79% of the interviewed patients (Krueger et al. 2001). Generalized pruritus was a feature of psoriasis in 84% of psoriasis patients 77% of whom reported to experience it on a daily basis (Yosipovitch et al. 2000). Regarding several studies on chronic pruritus, the symptom appears to occur in psoriasis more frequently (Weisshaar and Dalgard 2009; Krueger et al. 2001; Yosipovitch et al. 2000; O'Neill et al. 2011) than previously believed. The underestimation of pruritus in daily clinical practice may be due to lower intensity compared to what is observed in other pruritic skin diseases and patient underreporting (Weisshaar and Dalgard 2009).

Chronic idiopathic **urticaria** is defined as the occurrence of daily, or almost daily, wheals and pruritus for at least 6 weeks, with no obvious cause. No detailed epidemiological studies regarding its frequency appear to exist (Greaves 2003). However, 68% of chronic idiopathic urticaria patients were found to suffer from chronic pruritus that occurred daily (Yosipovitch et al. 2002).

Infectious diseases of the skin such as, e.g., scabies, pediculosis, bacterial and viral infections, and mycoses, often go along with pruritus that is usually not of chronic character.

Recent evidence suggests that pruritus is still common in **HIV**. Thirty-one percent of patients on combination antiretroviral therapy reported pruritus (Blanes et al. 2012). Pruritic dermatoses as, for instance, PPE (pruritic papular eruption) are a frequently seen cutaneous manifestation in 11% to 46% of HIV patients (Eisman 2006).

Pruritus is said to occur frequently in **systemic diseases**, but there is no epidemiological data on the frequency of pruritus in systemic disease in general and data for specific systemic disease refers mainly to uraemic pruritus (see below). Instead, there is a fairly large body of research regarding systemic causes of chronic pruritus. However, no clear picture can be drawn owing to diverging variables, methods, and study populations used in these studies. Generalized pruritus it is attributed to a systemic disease in 16% to 50% of individuals; however, by doing so any potential multifactorial causation may be disregarded (Weisshaar and Dalgard 2009). One study took multifactorial causation into account and reported that in 13.3% of the sample, pruritus was attributable to a single systemic cause, while in 24.7% a multifactorial causation was established (Sommer et al. 2007).

In a German sample of patients attending a chronic pruritus clinic, 36% suffered from a systemic cause, while in an Ugandan sample of chronic pruritus patients, no patient was diagnosed to have an underlying systemic disease (Weisshaar et al. 2006). It is important to note that both groups varied largely in age (German sample: M_{age} = 54.5 years; Ugandan sample: M_{age} = 28.0 years) and that life expectancy in Uganda is lower compared to Germany, and most systemic conditions in which pruritus may co-occur are related to higher age.

While a systemic cause is often implicated in the aetiology of pruritus no studies exist quantifying the extent to which pruritus exists over all systemic diseases. However, there is some data describing the frequency of pruritus in discrete systemic diseases.

The prevalence of pruritus in **chronic renal failure and in patients undergoing dialysis** varies worldwide between 10% and 70% (Pisoni et al. 2006). Eighty to 100% of patients with **cholestasis** contract pruritus. Twenty-five to 70% of the patients with primary biliary cirrhosis, and 15% of the patients with hepatitis C suffer from chronic pruritus (Weisshaar and Dalgard 2009). It is estimated that about 30% of patients with **Morbus Hodgkin**, 10% of patients with non-Hodgkin lymphoma, 40% of patients with polycythemia vera, and about 5% of leukemia patients suffer from chronic pruritus. Regarding other malignant diseases, the prevalence of paraneoplastic pruritus is reported to range between 2% and 26%. It should be borne in mind that these studies are methodologically diverse and that there is no clear definition of paraneoplastic pruritus; hence direct comparisons are difficult to make (Weisshaar and Dalgard 2009).

Diabetes mellitus is the most common **endocrinological** disease. Although studies suggest that the condition is accompanied by dermatological diseases and cutaneous manifestations in up to 71% of patients (Yosipovitch et al. 1998; Mahajan et al. 2003), and other studies report the symptom pruritus to be common in diabetes (Mahajan et al. 2003; Al-Mutairi et al. 2006), the empirical basis is insufficient at best. While generalized pruritus was once considered to be a frequent symptom in diabetes mellitus, no confirmative data are actually available. First, there are no well-designed studies providing valid estimates of the prevalence of the symptom among diabetes patients (Sreedevi et al. 2002). Second, the findings of another study suggest that generalized pruritus does not occur more often among diabetes patients compared to other patients matched for age and gender (Neilly et al. 1986). It is hence not possible at present to quantify the extent to which pruritus occurs in diabetes mellitus.

Pruritus is also said to be commonly associated with drug intake (Bigby et al. 1986; deShazo and Kemp 1997). **Drug-induced pruritus** without a rash may occur as acute pruritus (<6 weeks) caused, for instance by chloroquine (George 2004) or opioids (Swegle and Logemann 2006). Hydroxyethyl starch (HES) induced pruritus is described as occurring in 12% to 42% of patients treated with HES (Bork 2005). Opioid-induced pruritus is a common problem after epidural and intrathecal administration of opioids. The likelihood of developing pruritus with opioid use ranges from 2% to 10% and is substantially increased when opioids are given by epidural or intraspinal injections (Swegle and Logemann 2006). Although there are a number of drugs (ranging from antihypertensive to antiepileptic) that can be considered to induce pruritus there are no studies on how frequently chronic pruritus is caused by a particular drug in patient populations.

It is also assumed by some clinicians that there is an association between pruritus and **psychiatric** morbidity. However, only a few studies report on the prevalence of pruritus in psychiatric disorders and all of these are stricken with methodological flaws. Small opportunity samples instead of a random selection of participants are studied, and a variety of information biases may also exist. Between 17% and 32% of the investigated inpatient psychiatric patients have been reported to suffer from pruritus according to these studies (Mazeh et al. 2008; Pacan et al. 2009). Whether pruritus constitutes a common problem within psychiatric disease cannot be answered based on the available empirical basis.

2.6   PRURITUS IN THE GENERAL POPULATION

Studying the symptoms of pruritus in the general population can provide valuable information on hitherto unknown associations with demographic factors, psychosocial factors, or other disease. Until a few years ago, there was very little information about how frequently the symptom is encountered in the general population. It has to be noted, though, that the above mentioned patient populations are part of the general population and as such the sampling strategy should take account of this. In other words, obtaining representative samples is of paramount importance. Further, in order to arrive at estimates that can be compared across studies, a standardized assessment of the symptom appears necessary. However, as is clear from a look at the few available studies reporting epidemiological parameters, the operationalization of the symptom varied tremendously. Only of late have there been systematic attempts at providing more clear-cut criteria for the assessment of the symptom (Ständer et al. 2007; Matterne et al. 2009).

One of the first studies to look into epidemiological parameters of pruritus in a wider sense was the so-called "Lambeth Study" (Rea et al. 1976). This population-based study sought to establish prevalence estimates of skin diseases and was able to demonstrate a (presumably point) prevalence of pruritic and similar diseases ("prurigo and allied conditions") of 8.2% in an urban adult population. Many years have passed before another attempt at quantifying pruritus at the population level has been made. A population-based French study (Wolkenstein et al. 2003), again aiming at providing data on the frequency of occurrence of skin disease in the population, also reported one prevalence estimate of pruritus. While it was claimed that chronic pruritus was measured, no descriptions were given of how chronic pruritus was defined. In other words, we do not know whether this group used the >6 week minimum duration criteria as suggested by the IFSI (Ständer et al. 2007). The authors reported a 2-year prevalence of 12.4%.

Several years later a population-based study conducted in Oslo, Norway, in 2000 and 2001 assessed among many other variables self-reported skin complaints. It was found that 8.4% of the surveyed individuals (n = 18,747) aged 30 to 76 had suffered from acute pruritus during the past two weeks (Dalgard et al. 2004). In a study among employees voluntarily attending an early skin cancer detection programme, the point prevalence of chronic pruritus (>6 weeks) was 16.7% (Ständer et al. 2010).

The Heidelberg Pruritus Prevalence Study also aimed at assessing the prevalence of chronic pruritus (<6 weeks) in two urban and one rural population. A previously validated questionnaire developed for the assessment of the prevalence and characteristics of chronic pruritus in the general population (Matterne et al. 2009) was sent out by mail. A maximum of three contact attempts was made. The response rate was 57.8%. The point prevalence was estimated to be 13.5%, the 12-month prevalence 16.4%, and the lifetime prevalence 22% (Matterne et al. 2011). Women were affected more often than men, but a significant gender difference was only found for the lifetime prevalence. Occurrence of current chronic pruritus and during the past 12 months appeared unrelated to increasing age while the aforementioned bimodal pattern was found for the lifetime prevalence. Ethnic origin was significantly associated with the occurrence of chronic pruritus. Individuals with a non-German background reported significantly more chronic pruritus than Germans (Matterne et al. 2011).

2.7 NEW DEVELOPMENTS AND CURRENT RESEARCH

While the epidemiological database regarding the prevalence of pruritus is growing, little is known about the rate of new cases of people suffering from the symptom. According to a recent population-based study, the 12-month cumulative incidence of chronic pruritus in the general population may be as high as 7% (Matterne et al. 2013). This study used a cohort of 1190 participants who had participated in the cross-sectional study (Heidelberg Pruritus Prevalence Study) on the prevalence of chronic pruritus (baseline study in 2008–2009, $n = 2540$ [Matterne et al. 2011]) and consented to be contacted again. These individuals were again approached for follow-up one year later (response rate then 83.1%). Incident chronic pruritus was assessed in 651 individuals who had been free of the symptom at baseline. The whole cohort of 1190 participants was also analyzed with regard to the prevalence of chronic pruritus. The point prevalence was 15.4%, the 12-month prevalence 18.2%, and the lifetime prevalence 25.5%. The prevalence estimates are very similar to those obtained in our baseline survey (Matterne et al. 2011) and in the survey conducted with employees seeking early detection cancer screenings (Ständer et al. 2010) and again highlight a substantial prevalence of the symptom in the population.

The follow-up data (Matterne et al. 2013) was also longitudinally linked to baseline sociodemographic data. Increasing age and being retired significantly increased the risk for incident chronic pruritus as shown in univariate analyses. Female gender was associated with an increased but nonsignificant risk for incident chronic pruritus during the past 12 months.

The follow-up survey also systematically assessed the ever presence of medical conditions, as well as numerous lifestyle (among others, e.g., body mass index, sun protection behavior, smoking, and exercise) and psychosocial factors (e.g., social support, signs of depression and anxiety, and health anxiety). The question of whether these medical conditions, lifestyle, and psychosocial factors influence the incidence of chronic pruritus cannot be answered at the moment as these variables were assessed at the first follow-up. However, another follow-up 3 years after the first follow-up may shed more light on the question of whether certain medical conditions, lifestyle factors, and psychosocial variables are longitudinally associated with the incident of chronic pruritus. Meanwhile, these three groups of variables were analyzed to assess their cross-sectional associations with prevalent lifetime pruritus. Significant correlates of prevalent lifetime chronic pruritus in multivariate analyses were liver disease, asthma, eczema, and dry-skin within the medical domain, an elevated body mass index within the lifestyle domain and higher anxiety scores within the psychosocial domain.

2.8 CONCLUDING REMARKS

Reliable evidence as to how common a symptom of pruritus is, still rare. On the one hand, there are some exceptions. For instance, with regard to some dermatological conditions it can be said with a high degree of certainty that pruritus is very common in eczema, psoriasis, or chronic idiopathic urticaria. On the other hand, although claims have been made that pruritus is common in diabetes or psychiatric conditions, no reliable data as to how prevalent pruritus is in these conditions appears to exist.

More high-quality research is needed in order to be able to assess the burden of pruritus in specific diseases and patient populations. The empirical basis for an assessment in the general population is growing but replications of the available findings in independent samples would be desirable. Nevertheless, these recent population-based studies have revealed various new insights regarding the epidemiological underpinnings of for instance chronic pruritus (e.g., prevalence, incidence, and determinants).

In daily clinical practice, consideration of coexisting disease and overlapping symptoms across diseases may help in the attribution of the cause of pruritus. The role of mixed origin has rarely been studied. This is, nevertheless, of growing importance in elderly individuals who are more likely to suffer from several diseases (multimorbidity) and regularly require medication. Taking this multimorbidity into account will, however, make the design of epidemiological research attempting to quantify the frequency of pruritus in patient populations more complex.

REFERENCES

Al-Mutairi, N., Zaki, A., Sharma, A. K., and Al-Sheltawi, M. (2006). Cutaneous manifestations of diabetes mellitus. Study from Farwaniya hospital, Kuwait. *Med. Princ. Pract.* 15: 427–430.

Ambros-Rudolph, C. M., Mullegger, R. R., Vaughan-Jones, S. A., Kerl, H., and Black, M. M. (2006). The specific dermatoses of pregnancy revisited and reclassified: Results of a retrospective two-center study on 505 pregnant patients. *J. Am. Acad. Dermatol.* 54: 395–404.

Apfelbacher, C. J., Diepgen, T. L., and Weisshaar, E. (2011). [Dermato-epidemiology]. *Hautarzt* 62: 859–868; quiz 869–870.

Barzilai, D. A., Freiman, A., Dellavalle, R. P., Weinstock, M. A., and Mostow, E. N. (2005). Dermatoepidemiology. *J. Am. Acad. Dermatol.* 52: 559–573; quiz 574–578.

Beauregard, S., and Gilchrest, B. A. (1987). A survey of skin problems and skin care regimens in the elderly. *Arch. Dermatol.* 123: 1638–1643.

Bigby, M., Jick, S., Jick, H., and Arndt, K. (1986). Drug-induced cutaneous reactions. A report from the Boston Collaborative Drug Surveillance Program on 15,438 consecutive inpatients, 1975 to 1982. *JAMA* 256: 3358–3363.

Black, M. M., Lynch, P. J., Edwards, L., and Ambros-Rudolph, C. (2008). *Obstetric and Gynecologic Dermatology*. London: Mosby/Elsevier.

Blanes, M., Belinchon, I., Portilla, J., Betlloch, I., Reus, S., and Sanchez-Paya, J. (2012). Pruritus in HIV-infected patients in the era of combination antiretroviral therapy: A study of its prevalence and causes. *Int. J. STD AIDS* 23: 255–257.

Bork, K. (2005). Pruritus precipitated by hydroxyethyl starch: A review. *Br. J. Dermatol.* 152: 3–12.

Dalgard, F., Svensson, A., Holm, J. O., and Sundby, J. (2004). Self-reported skin morbidity in Oslo. Associations with sociodemographic factors among adults in a cross-sectional study. *Br. J. Dermatol.* 151: 452–457.

Dawn, A., Papoiu, A. D., Chan, Y. H., Rapp, S. R., Rassette, N., and Yosipovitch, G. (2009). Itch characteristics in atopic dermatitis: Results of a web-based questionnaire. *Br. J. Dermatol.* 160: 642–644.

deShazo, R. D., and Kemp, S. F. (1997). Allergic reactions to drugs and biologic agents. *JAMA* 278: 1895–1906.

Eisman, S. (2006). Pruritic papular eruption in HIV. *Dermatol. Clin.* 24: 449–457.

George, A. O. (2004). Chloroquine induced pruritus—Questionnaire based epidemiological study. *Afr. J. Health Sci.* 11: 87–92.

Greaves, M. W. (2003). Chronic idiopathic urticaria. *Curr. Opin. Allergy Clin. Immunol.* 3: 363–368.

Halvorsen, J. A., Dalgard, F., Thoresen, M., Bjertness, E., and Lien, L. (2009). Itch and mental distress: A cross-sectional study among late adolescents. *Acta Derm. Venereol.* 89: 39–44.

Krueger, G., Koo, J., Lebwohl, M., Menter, A., Stern, R. S., and Rolstad, T. (2001). The impact of psoriasis on quality of life: Results of a 1998 National Psoriasis Foundation patient-membership survey. *Arch. Dermatol.* 137: 280–284.

Lim, Y. L., Chan, Y. H., Yosipovitch, G., and Greaves, M. W. (2008). Pruritus is a common and significant symptom of acne. *J. Eur. Acad. Dermatol. Venereol.* 22: 1332–1336.

Mahajan, S., Koranne, R. V., and Sharma, S. K. (2003). Cutaneous manifestation of diabetes mellitus. *Indian J. Dermatol. Venereol. Leprol.* 69: 105–108.

Matterne, U., Apfelbacher, C. J., Loerbroks, A., Schwarzer, T., Bűttner, M., Ofenloch, R. Diepgen, T. L., and Weisshaar, E. (2011). Prevalence, correlates and characteristics of chronic pruritus: A population-based cross-sectional study. *Acta Derm. Venereol.* 91: 674–679.

Matterne, U., Apfelbacher, C. J., Vogelgsang, L., Loerbroks, A., and Weisshaar, E. (2013). Incidence and determinants of chronic pruritus: A population-based cohort study. *Acta Derm. Venereol.* 93: 532–537.

Matterne, U., Strassner, T., Apfelbacher, C. J., Diepgen, T. L., and Weisshaar, E. (2009). Measuring the prevalence of chronic itch in the general population: Development and validation of a questionnaire for use in large-scale studies. *Acta Derm. Venereol.* 89: 250–256.

Mazeh, D., Melamed, Y., Cholostoy, A., Aharonovitzch, V., Weizman, A., and Yosipovitch, G. (2008). Itching in the psychiatric ward. *Acta Derm. Venereol.* 88: 128–131.

Neilly, J. B., Martin, A., Simpson, N., and MacCuish, A. C. (1986). Pruritus in diabetes mellitus: Investigation of prevalence and correlation with diabetes control. *Diabetes Care* 9: 273–275.

Norman, R. A. (2003). Xerosis and pruritus in the elderly: Recognition and management. *Dermatol. Ther.* 16: 254–259.

Odhiambo, J. A., Williams, H. C., Clayton, T. O., Robertson, C. F., and Asher, M. I. (2009). Global variations in prevalence of eczema symptoms in children from ISAAC Phase Three. *J. Allergy Clin. Immunol.* 124: 1251–1258.e23.

O'Neill, J. L., Chan, Y. H., Rapp, S. R., and Yosipovitch, G. (2011). Differences in itch characteristics between psoriasis and atopic dermatitis patients: Results of a web-based questionnaire. *Acta Derm. Venereol.* 91: 537–540.

Pacan, P., Grzesiak, M., Reich, A., and Szepietowski, J. C. (2009). Is pruritus in depression a rare phenomenon? *Acta Derm. Venereol.* 89: 109–110.

Pisoni, R. L., Wikstrom, B., Elder, S. J., Akizawa, T., Asano, Y., Keen, M. L., Saran, R., Mendelssohn, D. C., Young, E. W., and Port, F. K. (2006). Pruritus in haemodialysis patients: International results from the Dialysis Outcomes and Practice Patterns Study (DOPPS). *Nephrol. Dial. Transplant* 21: 3495–3505.

Rajka, G. (1989). *Essential Aspects of Atopic Dermatitis*. Berlin, New York: Springer.

Rea, J. N., Newhouse, M. L., and Halil, T. (1976). Skin disease in Lambeth. A community study of prevalence and use of medical care. *Br. J. Prev. Soc. Med.* 30: 107–114.

Reich, A., Trybucka, K., Tracinska, A., Samotij, D., Jasiuk, B., Srama, M., and Szepietowski, J. C. (2008). Acne itch: Do acne patients suffer from itching? *Acta Derm. Venereol.* 88: 38–42.

Roger, D., Vaillant, L., Fignon, A., Pierre, F., Bacg, Y., Brechot, J. F., Grangeponte, M. C., and Lorette, G. (1994). Specific pruritic diseases of pregnancy. A prospective study of 3192 pregnant women. *Arch. Dermatol.* 130: 734–739.

Shaw, T. E., Currie, G. P., Koudelka, C. W., and Simpson, E. L. (2011). Eczema prevalence in the United States: Data from the 2003 National Survey of Children's Health. *J. Invest. Dermatol.* 131: 67–73.

Sommer, F., Hensen, P., Böckenholt, B., Metze, D., Luger, T. A., and Ständer, S. (2007). Underlying diseases and, co-factors in patients with severe chronic pruritus: A 3-year retrospective study. *Acta Derm. Venereol.* 87: 510–516.

Sreedevi, C., Car, N., and Pavlic-Renar, I. (2002). Dermatologic lesions in diabetes mellitus. *Diabetol. Croatica* 31: 147–159.

Ständer, S., Schafer, I., Phan, N. Q., Blome, C., Herberger, K., Heigel, H., and Augustin, M. (2010). Prevalence of chronic pruritus in Germany: Results of a cross-sectional study in a sample working population of 11,730. *Dermatology* 221: 229–235.

Ständer, S., Weisshaar, E., Mettang, T., Szepietowski, J. C., Carstens, E., Ikoma, A., Bergasa, N. V. et al. (2007). Clinical classification of itch: A position paper of the International Forum for the Study of Itch. *Acta Derm. Venereol.* 87: 291–294.

Swegle, J. M., and Logemann, C. (2006). Management of common opioid-induced adverse effects. *Am. Fam. Physician* 74: 1347–1354.

Weisshaar, E., Apfelbacher, C., Jager, G., Zimmermann, E., Bruckner, T., Diepgen, T. L., and Gollnick, H. (2006). Pruritus as a leading symptom: Clinical characteristics and quality of life in German and Ugandan patients. *Br. J. Dermatol.* 155: 957–964.

Weisshaar, E., and Dalgard, F. (2009). Epidemiology of itch: Adding to the burden of skin morbidity. *Acta Derm. Venereol.* 89: 339–350.

Weisshaar, E., Kucenic, M. J., and Fleischer, A. B., Jr. (2003). Pruritus: A review. *Acta Derm. Venereol. Suppl. (Stockh)* 213: 5–32.

Williams, H., Svensson, A., Diepgen, T., Naldi, L., Coenraads, P. J., Elsner, P., Brob, J. J., and Bouwes Bavinck, J. N. (2006). Epidemiology of skin diseases in Europe. *Eur. J. Dermatol.* 16: 212–218.

Wolkenstein, P., Grob, J. J., Bastuji-Garin, S., Ruszczynski, S., Roujeau, J. C., and Revuz, J. (2003). French people and skin diseases: Results of a survey using a representative sample. *Arch. Dermatol.* 139: 1614–1619.

Yalcin, B., Tamer, E., Toy, G. G., Oztas, P., Hayran, M., and Alli, N. (2006). The prevalence of skin diseases in the elderly: Analysis of 4099 geriatric patients. *Int. J. Dermatol.* 45: 672–676.

Yosipovitch, G., Ansari, N., Goon, A., Chan, Y. H., and Goh, C. L. (2002). Clinical characteristics of pruritus in chronic idiopathic urticaria. *Br. J. Dermatol.* 147: 32–36.

Yosipovitch, G., Goon, A., Wee, J., Chan, Y. H., and Goh, C. L. (2000). The prevalence and clinical characteristics of pruritus among patients with extensive psoriasis. *Br. J. Dermatol.* 143: 969–973.

Yosipovitch, G., Goon, A. T., Wee, J., Chan, Y. H., Zucker, I., and Goh, C. L. (2002). Itch characteristics in Chinese patients with atopic dermatitis using a new questionnaire for the assessment of pruritus. *Int. J. Dermatol.* 41: 212–216.

Yosipovitch, G., Hodak, E., Vardi, P., Shraga, I., Karp, M., Sprecher, E., and David, M. (1998). The prevalence of cutaneous manifestations in IDDM patients and their association with diabetes risk factors and microvascular complications. *Diabetes Care* 21: 506–509.

3 Atopic Dermatitis

Ulf Darsow, Ulrike Raap, and Sonja Ständer

CONTENTS

3.1   ATOPIC ECZEMA

Atopic dermatitis (AD; atopic eczema, eczema) is an inflammatory, chronically relapsing, and intensely pruritic skin disease occurring often in families with atopic diseases (atopic dermatitis, bronchial asthma, and/or allergic rhino-conjunctivitis). AD is a noncontagious inflammation of the epidermis and dermis with characteristic clinical (itch, erythema, papule, seropapule, vesicle, squames, crusts, lichenification, in synchronous, or metachronous polymorphy) and dermatopathological (spongiosis, acanthosis, hyper- and parakeratosis, lymphocytic infiltrates, and exocytosis, eosinophils) signs. With a prevalence of 2% to 5% (in children and young adults around 15%), AD is one of the most common skin diseases. The varying etiologic concepts of this disease are mirrored by the different names that are or have been used: "neurodermatitis," "neurodermitis," and "endogenous eczema" are just a few examples of current terms. Atopy is a strikingly common finding in these patients (Hanifin and Rajka 1980). It can be defined as familial hypersensitivity of the skin and the mucosa to environmental substances, associated with increased production of immunoglobulin E (IgE) and/or altered pharmacologic reactivity (Ring et al. 2006; Ring 2004). More recently, a new definition for atopy, restricted to IgE production, has been proposed: "a personal or familial tendency to produce IgE antibodies in response to low doses of allergens, usually proteins, and to develop typical symptoms such as asthma, rhinoconjunctivitis, or eczema/dermatitis" (Johansson et al. 2004, pp. 832–836).

The atopic diseases are genetically linked, and the concordance in monozygotic twins is 80% versus 30% in dizygotic twins (Schultz-Larsen 1985). A multifactorial trait involving numerous gene loci on different chromosomes has been proposed (Cookson and Moffatt 2002). Described genetic polymorphisms in AD involve mediators of atopic inflammation on different chromosomes, and some of these may also play a role in respiratory atopy. To date, the highest associations were shown with mutations in the filaggrin gene also associated with ichthyosis vulgaris, highlighting the predisposing barrier defect in AD patients (review in Weidinger et al. 2008).

In the first months of life, a yellowish desquamation on the scalp, known as "cradle cap," may be a presentation of AD. The disease may then spread to the face and extensor surfaces of the arms and legs of toddlers, sometimes showing extensive oozing and crusting. Later on, the typical preferential pattern develops with eczematous involvement of flexures, neck, and hands, accompanied by dry skin and skin barrier dysfunction reflected by an increased transepidermal water loss. Lichenification is a result of scratching and rubbing, and most frequently in adults, this may result in the prurigo type of AD with predominant excoriated nodular lesions. Exacerbations often start as increased itch without visible skin lesions. This is followed by erythema, papules, and infiltration.

The histopathology of acute AD lesions is characterized by epidermal hyperplasia, spongiosis occasionally leading to vesicle formation, by a marked inflammatory infiltrate composed of lymphocytes and histiocytes, variable number of eosinophils and mast cells in the upper dermis, and exocytosis of lymphocytes into the epidermis. Chronic, lichenified lesions show hyper- and parakeratosis, irregular epidermal hyperplasia, a moderate superficial dermal infiltrate of lymphocytes, histiocytes and some eosinophils, and increased numbers of mast cells. Moreover, thickening of the papillary dermis and venular changes including endothelial hyperplasia and basement membrane thickening are observed (Mihm et al. 1976).

As these features are not specific for AD, routine histology is not a useful tool for diagnosing AD. In the absence of a specific diagnostic laboratory marker, mostly cutaneous stigmata of atopy have been used as diagnostic signs (Hanifin and Rajka 1980; Ring et al. 2006), the diagnosis of AD is made clinically. Hanifin and Rajka stated three of four main criteria to be necessary: pruritus, typical morphology and distribution, chronic or chronically relapsing course, and atopic personal or family history, in addition to three minor criteria among a list of 21 (Hanifin and Rajka 1980). According to the UK working party (Williams et al. 1996) who developed criteria especially suitable for epidemiological purposes, but not in small children, itchy skin changes have to be diagnosed in the last 12 months, in addition to at least three of the following criteria: onset of the disease under the age of 2 years, history of involvement of skin folds, generalized dry skin, other atopic diseases, and visible flexural eczema.

Management of exacerbated AD is a therapeutic challenge, as it requires efficient short-term control of acute symptoms, without compromising the overall management plan that is aimed at long-term stabilization, flare prevention, and avoidance of side effects (Darsow et al. 2010; Ring et al. 2012).

3.2   ATOPIC ITCH

Itch is one of the most important symptoms in inflammatory skin diseases and allergic disorders. It is defined as "unpleasant sensation, eliciting the urge to scratch" (Hafenreffer 1660, pp. 98–102). This very old definition still holds true after the last 50 years of neurophysiological research (which was, however, mainly focused on pain perception). AD is one of the most pruritic skin diseases. Often, pruritus is the first symptom of eczema relapse. In severe cases, patients scratch the involved skin areas until bleeding excoriations result. Nocturnal prolonged scratching with sleep loss is a common problem in these patients. In fact, itch is an essential diagnostic feature of AD (in association with other clinical criteria: age-related eczematous appearance and localization, history and clinical signs of atopy, and IgE-mediated sensitization [Ring 2004]).

3.3   FEATURES OF ATOPIC ITCH SHOWN WITH SCALES AND QUESTIONNAIRES

In experimental itch in healthy volunteers, interindividual differences of itch sensation in response to histamine were high (Bromm et al. 1995; Darsow et al. 1996). The clinical features of itch in different pruritic skin diseases reveal a range of diversity in the perception of this symptom. In many clinical trials, a quantification of subjective itch intensity by visual analog scale (VAS) (Hägermark and Wahlgren 1992) is the only measure or itch is even omitted in symptom scores. Murray and Rees (2011) recently demonstrated that VAS ratings and nocturnal actigraphy as covariate of itch in patients with AE were not correlated. The only use of VAS may lead to an incomplete registration of the sensation, because the influence of qualitative factors on quantitative scales is already known in pain research. This has led to the development of several questionnaire instruments for pain psychophysiology and for the measurement of quality of life outcomes.

In cooperation of Dermatology and Neurophysiology Departments, a multi-dimensional questionnaire was developed, the Eppendorf Itch Questionnaire (EIQ) (Darsow et al. 1997). It is available in English (Darsow et al. 2001) and can be used for clinical and study purposes. The EIQ is designed in analogy to the established McGill Pain Questionnaire (Melzack 1975) in pain research. The EIQ was used in therapy assessment and evaluated in an atopy patch test model for AD (Weissenbacher et al. 2005). The German version of this questionnaire was used in 108 patients with acute AD in comparison to the SCORAD (scoring atopic dermatitis, severity instrument) (Darsow et al. 2001). Atopic itch was described as increased warmth, localizable, tingling, hot, and burning with many adverse affective descriptors. A principal component analysis with varimax rotation identified main factors of clinical itch. It was shown that atopic itch is a multidimensional sensation with 12 clusters of descriptors, but on a more general level, descriptors could be integrated in three main components (explaining 58% of total variance) that describe the atopic itch (Table 3.1). Component A, "suffering," described the decrease in quality of life, which is caused by pruritus. Main component B contained the quality of the sensation itself (wave formed and prickling, some further descriptors were chosen here).

TABLE 3.1
Eppendorf Itch Questionnaire and SCORAD

	Correlation Coefficients		
Main Component	SCORAD	VAS Itch	Area
A. "Suffering"	0.59*	0.52*	0.47*
B. "Phasic intensity factor"	0.38*	n.s.	n.s.
C. "Compulsive/active reaction"	n.s.	n.s.	n.s.

Source: Data from Darsow, U. et al., *Hautarzt*, 48, 730–733, 2001.

Note: Correlation between main components A–C of subjective description of itch (Eppendorf Itch Questionnaire) and parts of the SCORAD severity index in atopic eczema. Components A–C together explain 58% of the total variance. VAS, visual analog scale; area, extent of eczema. $N = 108$ patients with atopic eczema.

* = $p < 0.05$.

The third component was a compulsive component describing loss of control and warm feelings. Surprisingly, it comprised also positive emotional descriptors which were chosen by the patients, and it was the only main component that was not significantly related to the eczema severity (SCORAD). We suggested that this component may represent an important factor of the so-called "itch-scratch-cycle" in AD.

We compared the itch sensation of acute inpatients with chronic urticaria and AD with VAS and the EIQ. Table 3.2 shows that the difference between these pruritic diseases is not a function of pure itch intensity itself (VAS) but of the differentiated perception of the symptom: the mean total EIQ score in patients with AD was markedly higher. This was partially due to higher loads in affective items chosen by patients with atopic eczema. It may be speculated that this phenomenon is a feature of chronification of the itch sensation. Features of AE itch in different questionnaire studies are summarized in Table 3.3.

Recently, the relationship between itch and psychological status of patients with AE was investigated in Poland (Chrostowska-Plak et al. 2009, 2012). Intensity of

TABLE 3.2
Itch in Atopic Eczema and Chronic Urticaria

		N	Mean	Standard Deviation	Standard Error of the Mean
VAS%	Atopic eczema	62	74.47	20.667	2.625
	Urticaria	58	75.12	18.013	2.365
Score EIQ	Atopic eczema	62	231.68	91.157	11.577
	Urticaria	58	175.24	72.257	9.488

Note: Two groups of inpatients with a chronic pruritic disease were investigated with regard to itch intensity (VAS) and qualitative total scores of the Eppendorf Itch Questionnaire. Mean VAS ratings were comparable, whereas the questionnaire score was higher in atopic eczema.

TABLE 3.3
Features of Atopic Itch

Feature	Reference
Sharp, stinging, burning, tickling, pricking	Darsow et al. (2001), Dawn et al. (2009)
High emotional burden	Darsow et al. (2001), Dawn et al. (2009), Chrostowska-Plak et al. (2009)
28% daily, only in 11% no itch > 1 month	Chrostowska-Plak et al. (2009)
91% daily (web questionnaire)	Dawn et al. (2009)
VAS mean 62–79 ± 23%, evening and night	Darsow et al. (2001), Chrostowska-Plak et al. (2009)
Acute itch with AE lesions versus chronic itch	Chrostowska-Plak et al. (2009)
Often no response to antihistamines	Chrostowska-Plak et al. (2009)

Note: Questionnaire studies on atopic eczema itch point to distinct clinical descriptions of the symptom.

pruritus was significantly related to the stress experienced by the patients prior to disease exacerbation. A significant correlation between pruritus and a validated quality of life index was also found. Patients with symptoms suggesting depression had more intense pruritus compared with the rest of patients.

All recent questionnaire studies consistently showed a high emotional burden and chronicity as features of atopic itch (e.g., Table 3.3). Such studies and clinical observations also point to a marked central nervous processing and modulation of the itch sensation in AE patients, which is now also investigated in neuroimaging studies (Pfab et al. 2010, 2012).

3.4   MECHANISM OF ATOPIC ITCH

Itch is a common problem that can severely impair the quality of life of affected patients. In the skin and nervous system, complex cellular interaction mechanisms have a pivotal impact on the pathophysiology of itch. In the following we will describe the possible underlying mechanisms of itch in AD skin with special emphasis on peripheral nerves, eosinophils, neuropeptides, neurotrophins, and IL-31.

3.5   NERVES AND ITCH

In AD, irritants including wools can immediately lead to intense itch suggesting a hyperreactivity of the surrounding nerves (Wahlgren et al. 1991). Indeed, the distribution density of cutaneous nerve fibers was found to be much higher in AD patients than in normal control skin. Further, the diameter of these fibers was much larger, because of the large number of axons in each nerve fiber (Urashima and Mihara 1998). The number of peripheral nerves in AD skin has been described to be increased in acute lesions compared to unaffected skin of patients with AD (Sugiura et al. 1997). Also, the activity of these peripheral nerves was found to be increased in AD patients as assessed with electron microscopy revealing bulging of axons with many mitochondria and a loss of their surrounding sheath of Schwann cells (Sugiura et al. 1997). This finding suggests that the free nerve endings in skin lesions of AD are in an active state of excitation, assuming one mechanism why the skin seems to be hyperreactive with regard to unspecific activation via wools in AD patients (Sugiura et al. 1997).

In the skin, itch is mediated via free nerve endings of nonmyelinated C-type nerve fibers that are located at the dermoepidermal junction and within the epidermis. There is a subpopulation of itch-specific nonmyelinated C nerve fibers that respond only to histamine. In addition, histamine-independent itch specific nerve fibers have been described to play a role in itch regulation (Nakano et al. 2008). In this regard, specific G protein-coupled receptors that mediate chloroquine-induced itch have been identified in peripheral sensory neurons (Liu et al. 2009). These neurons also respond to other itch-inducing signals including capsaicin. Recently, it has been shown that spinal neurons express the gastrin-releasing peptide receptor (GRPR). In mice, this receptor was involved in the transmission of itch, indicating that receptor expressing neurons may be itch specific (Sun and Chen 2007). It is assumed that the GRPR-positive, itch-sensitive neurons are involved in atopic itch (Tominaga et al. 2009).

3.6   SUBSTANCE P AND ITCH

In mice, substance P (SP)-positive nerve fibers have been shown to be increased in AD models (Tominaga et al. 2009). In rats, dorsal horn neurons were demonstrated to express the SP-specific neurokinin-1 receptor, suggesting that the neurokinin/neuropeptide SP may also play a role as a spinal transmitter for itch (Carstens et al. 2010). In human skin, SP-positive nerve fibers were shown to be increased in density in dermal nerve fibers in chronic itch of patients with prurigo nodularis, underlining its role in itch regulation in humans (Haas et al. 2010). Because SP is capable of inducing the neurotrophin nerve growth factor (NGF) (Toyoda et al. 2002), SP is a very interesting neuropeptide in the neuroimmune interaction circus with regard to the mechanistic pathway of itch. Indeed, targeting the NK-1 receptor in AD pruritus has recently been shown to be effective (Stände r et al. 2010).

3.7   EOSINOPHILS AND NERVES IN ATOPIC DERMATITIS

Tissue eosinophilia is a regular finding in AD skin (Kiehl et al. 2001). An increased quantity of eosinophilic granule protein, indicating complete eosinophil activation and degranulation of eosinophils is found in acute spontaneous lesions of AD when compared to normal skin (Kiehl et al. 2001). Interestingly, eosinophils have been found in close vicinity to nerves, assuming a role in their activation (Costello et al. 1997). Recently, it was shown that eosinophils dramatically increased branching of sensory neurons isolated from the dorsal root ganglia of mice (Foster et al. 2011). Thus, it is very likely that nerves are also activated at least in part by the interplay with proinflammatory cells including eosinophils in the skin, leading to the exacerbation of pruritus in AD (Raap and Kapp 2010).

3.8 EOSINOPHILS AND NEUROTROPHINS/ NEUROPEPTIDES IN THE REGULATION OF ITCH

Eosinophils constitutively express mRNA for NGF and NT-3 (Solomon et al. 1998; Kobayashi et al. 2002). This is interesting because neurotrophins are capable of inducing the cutaneous nerve sprouting and myelinization of nerves. Direct neuro-immune interactions between eosinophils and nerves have been shown in an *in vitro* model in which stimulated eosinophils released NGF and induced neurite outgrowth that was abolished by anti-NGF neutralizing antibodies (Kobayashi et al. 2002).

NGF and BDNF contents are higher in eosinophils of patients with AD compared with healthy controls and can be released accordingly (Toyoda et al. 2003; Raap et al. 2005). This is of interest because increased BDNF levels correlate with disease severity assessed by SCORAD score in adults (Raap et al. 2006) and in children with AD (Hon et al. 2007). Further, BDNF levels correlate with scratching activity as assessed in children with AD that had been monitored during the night with a Digi-Trac model to assess the scratching activities (Hon et al. 2007).

In mice, the neurotrophin receptor antagonist for tyrosinkinase A (trkA) led to an inhibition of scratching and skin inflammation, indicating a role of neurotrophins in the mechanism of itch (Takano et al. 2005).

BDNF, which is increased in peripheral blood and plasma of AD patients (Raap et al. 2005), was found to induce chemotaxis, but only of eosinophils derived from patients with AD and not from healthy controls, revealing a vicious circle with enhancement of the inflammatory cell infiltrate in the skin of patients with AD (Raap et al. 2005, 2008). Further, neurotrophin receptor expression is highly increased in eosinophils of AD patients, assuming a higher responsiveness of eosinophils to the per se increased neurotrophin levels in AD (Raap et al. 2008). In addition to the release of neurotrophins, eosinophils have been assessed to release the neuropeptide vascular endothelial growth factor, which has a role in pruritus induction in urticaria (Tedeschi et al. 2009).

Further, eosinophils express histamine receptors including the H4 receptor (O'Reilly et al. 2002). Thus, the capability of eosinophils to respond to various trigger factors via the production of neurotrophins, neuropeptides, and other cytokines displays a novel pathophysiological aspect of pruritus in AD skin.

3.9 IL-31 AND ITCH IN ATOPIC DERMATITIS

The T-cell cytokine named interleukin (IL)-31 was shown to have a pivotal role in severe itch and AD, as assessed in mice overexpressing IL-31 (Dillon et al. 2004). In Nc/Nga, mice IL-31 levels correlated with scratching behavior (Takaoka et al. 2005), which could be ameliorated by the use of anti-IL-31 Ab (Grimstad et al. 2009). In AD patients, IL-31 was found to be increased compared to the skin of healthy controls (Raap et al. 2008). Also, in children with AD, IL-31 serum levels correlated with disease severity and were increased compared to age matched skin of healthy children (Raap et al. 2012; Ezzat et al. 2011). In addition, to the correlation with SCORAD score, IL-31 levels correlated with TH2 cytokines including IL-4 and IL-13 (Raap et al.

2012), assuming that besides its role on the regulation of itch, IL-31 also seems to play an important role in the regulation of the inflammatory infiltrate.

In human skin, IL-31 mRNA was found to be increased in AD when compared to skin healthy controls (Sonkoly et al. 2006). In AD, staphylococcal superantigens represent a trigger factor for worsening of eczema. In this regard, superantigens were shown to induce IL-31 mRNA expression in the skin and in PBMC of atopic individuals (Sonkoly et al. 2006), indicating a role in itch sensation. The cellular source of IL-31 is represented by skin infiltrating CLA+ T-cells, CD4+ T cells, and peripheral blood CD45R0 CLA+ T-cells and mast cells (Bilsborough et al. 2006; Szegedi et al. 2012; Niyonsaba et al. 2010). Interestingly, IL-31 receptor expression has been assessed on keratinocytes and on dorsal root ganglia, indicating that regulation of itch indeed could be transmitted via IL-31 receptors (Sonkoly et al. 2006; Heise et al. 2009).

In mice, anti-IL-31 treatment lead to a significant inhibition of itch (Grimstad et al. 2009), although it displayed no impact on the remaining inflammatory infiltrate in affected skin. Thus, it still needs to be clarified as to which IL-31 contributes to itch and inflammation in humans. Trials with patients have currently been undertaken.

3.10 THERAPY OF AD ITCH

According to the complex pathophysiology of atopic itch, the treatment of AD patients is challenging. Several steps have to be considered such as the application of emollients against dry skin, disinfectants against bacterial infections, and topical steroids or calcineurininhibitors against eczema/inflammation. Despite this, itch may persist and necessitates symptomatic antipruritic therapies. Application of topicals containing urea, camphor, or menthol preparations, wet wrap dressings, wrappings with black tea, or lukewarm showers may induce short-term relief of AD (Ring et al. 2012). Also, a combination of polidocanol and 5% urea lead to the relief of AD itch and may be applied during the night or in itch attacks (Ring et al. 2012).

For long-term relief, antihistamines are usually applied, but their efficacy is limited.

Furthermore, because scratching also represents a trigger factor of AD itch contributing to the itch-scratch cycle, the control of itch as well as scratching is important in AD patients. Scratch lesions also need a specific therapy. In summary, multimodal therapy regimens are necessary in AD itch to date.

3.11 ANTI-INFLAMMATORY THERAPIES

The use of anti-inflammatory therapies often results in cessation of pruritus in those patients with acute flaring up of AD and dense skin inflammatory cell infiltrate (Ständer and Luger 2010). Systemic and topical immunomodulators such as glucocorticoids, cyclosporine A, tacrolimus, pimecrolimus, and ultraviolet light therapy continue to be consistently the most effective antipruritic agents (Ständer and Luger 2010; Wahlgren et al. 1990; Hanifin et al. 2001; Jekler and Larkö 1990; Luger et al. 2001). The topical immunomodulators *tacrolimus* and *pimecrolimus* were frequently demonstrated to reduce erythema as well as pruritus and excoriations (Hanifin et al. 2001; Luger et al. 2001). Cyclosporin A (CyA) has been reported to have an itch-relieving effect in various diseases including AD. In a randomized study, CyA was demonstrated to significantly reduce itch intensity (Ständer and Luger 2010; Wahlgren et al. 1990). A case series reported relief of itch and scratch lesions in prurigo forms of AD (Siepmann et al. 2008). Upon clinical experience, CyA is one of the most potent drugs to relief atopic itch in a short amount of time.

3.12 ANTIHISTAMINES

Although various symptomatic treatments are employed to relieve pruritus and scratching in patients with AD, no target-specific therapies for AD itch are available as of yet. Because several studies have demonstrated that different mechanisms are involved in AD, it is not surprising that conventional therapeutic modalities like antihistamines often fail to ameliorate pruritus in AD (Klein and Clark 1999). Placebo-controlled studies concerning the antipruritic effect of oral antihistamines have shown conflicting results in AD. In some studies, antihistamines demonstrated no superior effect compared to placebo, while in others, they showed a significant antipruritic effect (Ring et al. 2012; Ständer and Luger 2010; Hannuksela et al. 1993). For example, cetirizine showed some benefit (Hannuksela et al. 1993), while an evidence-based review of the efficacy of antihistamines in relieving pruritus in AD concluded that little objective evidence exists in general for the antipruritic efficacy of H1-antihistamines in AD (Klein and Clark 1999).

3.13 TARGET-SPECIFIC THERAPY

There are some therapies that may target aspects of the complex pathophysiology of AD pruritus. For example, PUVA was described to reduce the neuronal hyperplasia and increased levels of NGF in AD patients (Tominaga et al. 2009). Skin neuropeptides are targeted by topical capsaicin, tacrolimus, or systemic application of SP-antagonists (Ständer and Luger 2010; Inagaki et al. 2010). Besides this, some new targets have been identified in AD itch as for example IL-31, the histamine 4 receptor, the neurokinin 1 receptor (NK1R), mu-opioid-receptors, proteinase activated receptor 2, nerve growth factor, and prostaglandin D2. It is currently unknown which mediator or receptor is the most important one. There is already some clinical evidence that targeting the NK1R leads to clinical relevant reduction of itch (Ständer et al. 2010). However, the development of new therapies against AD itch is urgently needed (Ständer and Weisshaar 2012). Fortunately, some targets are currently evaluated in clinical trials that selected as indication AD itch. This refers, for example, to new studies addressing the antipruritic potency of topical mu-opioid-receptor antagonist, systemical interleukin 31 antagonist, systemic proteinase activated receptor 2 antagonists, or prostaglandin D2 inductor (Ständer and Weisshaar 2012).

REFERENCES

Bilsborough, J., Leung, D. Y., Maurer, M., Howell, M., Boguniewicz, M., Yao, L., Storey, H. et al. (2006). IL-31 is associated with cutaneous lymphocyte antigen-positive skin homing T cells in patients with atopic dermatitis. *J. Allergy Clin. Immunol.* 117(2): 418–425.

Bromm, B., Scharein, E., Darsow, U., and Ring, J. (1995). Effects of menthol and cold on histamine-induced itch and skin reactions in man. *Neurosci. Lett.* 187: 157–160.

Carstens, E. E., Carstens, M. I., Simons, C. T., and Jinks, S. L. (2010). Dorsal horn neurons expressing NK-1 receptors mediate scratching in rats. *Neuroreport* 21(4): 303–308.

Chrostowska-Plak, D., Reich, A., and Szepietowski, J. C. (2012). Relationship between itch and psychological status of patients with atopic dermatitis. *J. Eur. Acad. Dermatol. Venereol.* 27: 339–342.

Chrostowska-Plak, D., Salomon, J., Reich, A., and Szepietowski, J. C. (2009). Clinical aspects of itch in adult atopic dermatitis patients. *Acta Derm. Venereol.* 89: 379–383.

Cookson, W. O., and Moffatt, M. F. (2002). The genetics of atopic dermatitis. *Curr. Opin. Allergy Immunol.* 2: 383–387.

Costello, R. W., Schofield, B. H., Kephart, G. M., Gleich, G. J., Jacoby, D. B., and Fryer, A. D. (1997). Localization of eosinophils to airway nerves and effects on neuronal M2 muscarinic receptor function. *Am. J. Physiol.* 273(1 Pt 1): L93–L103.

Darsow, U., Mautner, V. F., Bromm, B., Scharein, E., and Ring, J. (1997). The Eppendorf Pruritus Questionnaire. *Hautarzt* 48: 730–733.

Darsow, U., Ring, J., Scharein, E., and Bromm, B. (1996). Correlations between histamine-induced wheal, flare and itch. *Arch. Dermatol. Res.* 288: 436–441.

Darsow, U., Scharein, E., Simon, D., Walter, G., Bromm, B., and Ring, J. (2001). New aspects of itch pathophysiology: Component analysis of atopic itch using the "Eppendorf Itch Questionnaire." *Int. Arch. Allergy Immunol.* 124: 326–331.

Darsow, U., Wollenberg, A., Simon, D., Taïeb, A., Werfel, T., Oranje, A., Gelmetti, C. et al. for the European Task Force on Atopic Dermatitis/EADV Eczema Task Force. (2010). ETFAD/EADV Eczema Task Force 2009 position paper on diagnosis and treatment of atopic dermatitis. *J. Eur. Acad. Dermatol. Venereol.* 24: 317–328.

Dawn, A., Papoiu, A. D., Chan, Y. H., Rapp, S. R., Rassette, N., and Yosipovitch, G. (2009). Itch characteristics in atopic dermatitis: Results of a web-based questionnaire. *Br. J. Dermatol.* 160: 642–644.

Dillon, S. R., Sprecher, C., Hammond, A., Bilsborough, J., Rosenfeld-Franklin, M., Presnell, S. R., Haugen, H. S. et al. (2004). Interleukin 31, a cytokine produced by activated T cells, induces dermatitis in mice. *Nat. Immunol.* 5(7): 752–760.

Ezzat, M. H. M., Hasan, Z. E., and Shaheen, K. Y. A. (2011). Serum measurement of interleukin-31 (IL-31) in paediatric atopic dermatitis: Elevated levels correlate with severity scoring. *J. Eur. Acad. Dermatol. Venereol.* 25(3): 334–349.

Foster, E. L., Simpson, E. L., Fredrikson, L. J., Lee, J. J., Lee, N. A., Fryer, A. D., and Jacoby, D. B. (2011). Eosinophils increase neuron branching in human and murine skin and in vitro. *Plos One* 6(7): e22029.

Grimstad, O., Sawanobori, Y., Vestergaard, C., Bilsborough, J., Olsen, U. B., Grønhøj-Larsen, C., and Matsushima, K. (2009). Anti-interleukin-31-antibodies ameliorate scratching behaviour in NC/Nga mice: A model of atopic dermatitis. *Exp. Dermatol.* 18(1): 35–43.

Haas, S., Capellino, S., Phan, N. Q., Bohm, M., Luger, T. A., Straub, R. H., and Ständer, S. (2010). Low density of sympathetic nerve fibers relative to substance P-positive nerve fibers in lesional skin of chronic pruritus and prurigo nodularis. *J. Dermatol. Sci.* 58(3): 193–197.

Hafenreffer, S. (1660). *Nosodochium, in quo cutis, eique adaerentium partium, affectus omnes, singulari methodo, et cognoscendi et curandi fidelissime traduntur*. Kühnen, B. (Ed.). Ulm: Kůln, pp. 98–102.

Hägermark, O., and Wahlgren, C. F. (1992). Some methods for evaluating clinical itch and their application for studying pathophysiological mechanisms. *J. Dermatol. Sci.* 4: 55–62.

Hanifin, J. M., Ling, M. R., Langley, R., Breneman, D., and Rafal, E. (2001). Tacrolimus ointment for the treatment of atopic dermatitis in adult patients. Part I, Efficacy. *J. Am. Acad. Dermatol.* 44: S28–S38.

Hanifin, J. M., and Rajka, G. (1980). Diagnostic features of atopic dermatitis. *Acta Derm. Venereol. Suppl.* 92: 44–47.

Hannuksela, M., Kalimo, K., Lammintausta, K., Turjanmaa, K., Vajionen, E., and Coulie, P. J. (1993). Dose ranging study: Cetirizine in the treatment of atopic dermatitis in adults. *Ann. Allergy* 70: 127–133.

Heise, R., Neis, M. M., Marquardt, Y., Joussen, S., Heinrich, P. C., Merk, H. F., Hermanns, H. M., and Baron, J. M. (2009). IL-31 receptor alpha expression in epidermal keratinocytes is modulated by cell differentiation and interferon gamma. *J. Invest. Dermatol.* 129(1): 240–243.

Hon, K. L., Lam, M. C., Wong, K. Y., Leung, T. F., and Ng, P. C. (2007). Pathophysiology of nocturnal scratching in childhood atopic dermatitis: The role of brain-derived neurotrophic factor and substance P. *Br. J. Dermatol.* 157(5): 922–925.

Inagaki, N., Shiraishi, N., Igeta, K., Nagao, M., Kim, J. F., Chikumoto, T., Itoh, T., Katoh, H., Tanaka, H., and Nagai, H. (2010). Depletion of substance P, A mechanism for inhibition of mouse scratching behavior by tacrolimus. *Eur. J. Pharmacol.* 626: 283–289.

Jekler, J., and Larkö, O. (1990). Combined UVA-UVB versus UVB phototherapy for atopic dermatitis: A paired-comparison study. *J. Am. Acad. Dermatol.* 22: 49–53.

Johansson, S. G. O., Bieber, T., Dahl, R., Friedmann, P. S., Lanier, B. Q., Lockey, R. F., Motala, C. et al. (2004). Revised nomenclature for allergy for global use: Report of the Nomenclature Review Committee of the World Allergy Organization, October 2003. *J. Allergy Clin. Immunol.* 113: 832–836.

Kiehl, P., Falkenberg, K., Vogelbruch, M., and Kapp, A. (2001). Tissue eosinophilia in acute and chronic atopic dermatitis: A morphometric approach using quantitative image analysis of immunostaining. *Br. J. Dermatol.* 145(5): 720–729.

Klein, P. A., and Clark, R. A. F. (1999). An evidence-based review of the efficacy of antihistamines in relieving pruritus in atopic dermatitis. *Arch Dermatol.* 135: 1522–1525.

Kobayashi, H., Gleich, G. J., Butterfield, J. H., and Kita, H. (2002). Human eosinophils produce neurotrophins and secrete nerve growth factor on immunologic stimuli. *Blood* 99(6): 2214–2220.

Liu, Q., Tang, Z. X., Surdenikova, L., Kim, S., Patel, K. N., Kim, A., Ru, F. et al. (2009). Sensory neuron-specific GPCR Mrgprs are itch receptors mediating chloroquine-induced pruritus. *Cell* 139(7): 1353–1365.

Luger, T., van Leent, E. J. M., Graeber, M., Hedgecock, S., Thurston, M., Kandra, A., Berth-Jones, J. et al. (2001). SDZ ASM 981: An emerging safe and effective treatment for atopic dermatitis. *Br. J. Dermatol.* 144: 788–794.

Melzack, R. (1975). The McGill Pain Questionnaire: Major properties and scoring methods. *Pain* 1: 277–299.

Mihm, M. C., Soter, N. A., Dvorak, H. F., and Austen, K. F. (1976). The structure of normal skin and the morphology of atopic eczema. *J. Invest. Dermatol.* 67: 305–312.

Murray, R., and Rees, J. L. (2011). Are subjective accounts of itch to be relied on? The lack of relation between visual analogue itch scores and actigraphic measures of scratch. *Acta Derm. Venereol.* 91: 18–23.

Nakano, T., Andoh, T., Lee, J. B., and Kuraishi, Y. (2008). Different dorsal horn neurons responding to histamine and allergic itch stimuli. *Neuroreport* 19(7): 723–726.

Niyonsaba, F., Ushio, H., Hara, M., Yokoi, H., Tominaga, M., Takamori, K., Kajiwara, N. et al. (2010). Antimicrobial peptides human beta-defensins and cathelicidin LL-37 induce the secretion of a pruritogenic cytokine IL-31 by human mast cells. *J. Immunol.* 184(7): 3526–3534.

O'Reilly, M., Alpert, R., Jenkinson, S., Gladue, R. P., Foo, S., Trim, S., Peter, B., Trevethick, M., and Fidock, M. (2002). Identification of a histamine H-4 receptor on human eosinophils—Role in eosinophil chemotaxis. *J. Recept. Signal Transduct. Res.* 22(1–4): 431–448.

Pfab, F., Valet, M., Napadow, V., Tölle, T., Behrendt, H., Ring, J., and Darsow, U. (2012). Itch and the brain. In *Allergy and the Nervous System*. Bienenstock, J. (ed.). *Chem. Immunol. Allergy* 98: 254–266.

Pfab, F., Valet, M., Sprenger, T., Huss-Marp, J., Athanasiadis, G. I., Baurecht, H. J., Konstantinow, A. et al. (2010). Temperature modulated histamine-itch in lesional and non-lesional skin in atopic eczema—A combined psychophysical and neuroimaging study. *Allergy* 65: 84–94.

Raap, U., Deneka, N., Bruder, M., Kapp, A., and Wedi, B. (2008). Differential up-regulation of neurotrophin receptors and functional activity of neurotrophins on peripheral blood eosinophils of patients with allergic rhinitis, atopic dermatitis and nonatopic subjects. *Clin. Exp. Allergy* 38(9): 1493–1498.

Raap, U., Goltz, C., Deneka, N., Bruder, M., Renz, H., Kapp, A., and Wedi, B. (2005). Brain-derived neurotrophic factor is increased in atopic dermatitis and modulates eosinophil functions compared with that seen in nonatopic subjects. *J. Allergy Clin. Immunol.* 115(6): 1268–1275.

Raap, U., and Kapp, A. (2010). Neurotrophins in healthy and diseased skin. *G. Ital. Dermatol. Venereol.* 145(2): 205–211.

Raap, U., Weissmantel, S., Gehring, M., Eisenberg, A. M., Kapp, A., and Folster-Holst, R. (2012). IL-31 significantly correlates with disease activity and Th2 cytokine levels in children with atopic dermatitis. *Pediatr. Allergy Immunol.* 23(3): 285–288.

Raap, U., Werfel, T., Goltz, C., Deneka, N., Langer, K., Bruder, M., Kapp, A., Schmid-Ott, G., and Wedi, B. (2006). Circulating levels of brain-derived neurotrophic factor correlate with disease severity in the intrinsic type of atopic dermatitis. *Allergy* 61(12): 1416–1418.

Raap, U., Wichmann, K., Bruder, M., Stander, S., Wedi, B., Kapp, A., and Werfel, T. (2008). Correlation of IL-31 serum levels with severity of atopic dermatitis. *J. Allergy Clin. Immunol.* 122(2): 421–423.

Ring, J. (2004). *Allergy in Practice*. Heidelberg: Springer.

Ring, J., Alomar, A., Bieber, T., Deleuran, M., Fink-Wagner, A., Gelmetti, C., Gieler, G. et al. (2012). Guidelines for treatment of atopic eczema (atopic dermatitis) part I. *J. Eur. Acad. Dermatol. Venereol.* 26: 1045–1060.

Ring, J., Przybilla, B., and Ruzicka, T. (eds.) (2006). *Handbook of Atopic Eczema*, 2nd ed. Heidelberg: Springer.

Schultz-Larsen, F. (1985). *Atopic Dermatitis. Etiological Studies Based on a Twin Population*. Kopenhagen: Laegeforeningens.

Siepmann, D., Luger, T. A., and Ständer, S. (2008). Antipruritic effect of cyclosporine micro-emulsion in prurigo nodularis: Results of a case series. *J. Dtsch. Detmatol. Ges.* 6: 941–946.

Solomon, A., Aloe, L., Pe'er, J., Frucht-Pery, J., Bonini, S., Bonini, S., and Levi-Schaffer, F. (1998). Nerve growth factor is preformed in and activates human peripheral blood eosinophils. *J. Allergy Clin. Immunol.* 102(3): 454–460.

Sonkoly, E., Muller, A., Lauerma, A. I., Pivarcsi, A., Soto, H., Kemeny, L., Alenius, H. et al. (2006). IL-31: A new link between T cells and pruritus in atopic skin inflammation. *J. Allergy Clin. Immunol.* 117(2): 411–417.

Ständer, S., and Luger, T. (2010). Itch in atopic dermatitis—Pathophysiology and treatment. *Acta Dermatovenerol. Croat.* 18: 289–296.

Ständer, S., Siepmann, D., Herrgott, I., Sunderkotter, C., and Luger, T. A. (2010). Targeting the neurokinin receptor 1 with aprepitant: A novel antipruritic strategy. *Plos One* 5(6): e10968.

Ständer, S., and Weisshaar, E. (2012). Medical treatment of pruritus. *Expert Opin. Emerg. Drugs* 17: 335–345.

Sugiura, H., Omoto, M., Hirota, Y., Danno, K., and Uehara, M. (1997). Density and fine structure of peripheral nerves in various skin lesions of atopic dermatitis. *Arch Dermatol. Res.* 289(3): 125–131.

Sun, Y. G., and Chen, Z. F. (2007). A gastrin-releasing peptide receptor mediates the itch sensation in the spinal cord. *Nature* 448(7154): 700–703.

Szegedi, K., Kremer, A. E., Kezic, S., Teunissen, M. B., Bos, J. D., Luiten, R. M., Res, P. C., and Middelkamp-Hup, M. A. (2012). Increased frequencies of IL-31-producing T cells are found in chronic atopic dermatitis skin. *Exp. Dermatol.* 21(6): 431–416.

Takano, N., Sakurai, T., and Kurachi, M. (2005). Effects of anti-nerve growth factor antibody on symptoms in the NC/Nga mouse, an atopic dermatitis model. *J. Pharmacol. Sci.* 99(3): 277–286.

Takaoka, A., Arai, I., Sugimoto, M., Yamaguchi, A., Tanaka, M., and Nakaike, S. (2005). Expression of IL-31 gene transcripts in NC/Nga mice with atopic dermatitis. *Eur. J. Pharmacol.* 516(2): 180–181.

Tedeschi, A., Asero, R., Marzano, A. V., Lorini, M., Fanoni, D., Berti, E., and Cugno, M. (2009). Plasma levels and skin-eosinophil-expression of vascular endothelial growth factor in patients with chronic urticaria. *Allergy* 64(11): 1616–1622.

Tominaga, M., Ogawa, H., and Takamori, K. (2009). Histological characterization of cutaneous nerve fibers containing gastrin-releasing peptide in NC/Nga mice: An atopic dermatitis model. *J. Invest. Dermatol.* 129: 2901–2905.

Tominaga, M., Tengara, S., Kamo, A., Ogawa, H., and Takamori, K. (2009). Psoralen-ultraviolet A therapy alters epidermal Sema 3A and NGF levels and modulates epidermal innervation in atopic dermatitis. *J. Dermatol. Sci.* 55: 40–46.

Toyoda, M., Nakamura, M., Makino, T., Hino, T., Kagoura, M., and Morohashi, M. (2002). Nerve growth factor and substance P are useful plasma markers of disease activity in atopic dermatitis. *Br. J. Dermatol.* 147(1): 71–79.

Toyoda, M., Nakamura, M., Makino, T., and Morohashi, M. (2003). Localization and content of nerve growth factor in peripheral blood eosinophils of atopic dermatitis patients. *Clin. Exp. Allergy* 33(7): 950–955.

Urashima, R., and Mihara M. (1998). Cutaneous nerves in atopic dermatitis. A histological, immunohistochemical and electron microscopic study. *Virchows Arch.* 432(4): 363–370.

Wahlgren, C. F., Hagermark, O., and Bergstrom, R. (1991). Patients' perception of itch induced by histamine, compound 48/80 and wool fibres in atopic dermatitis. *Acta Derm. Venereol.* 71(6): 488–494.

Wahlgren, C. F., Scheynius, A., and Hägermark, Ö. (1990). Antipruritic effect of oral cyclosporin A in atopic dermatitis. *Acta Derm. Venereol.* 70: 323–329.

Weidinger, S., O'Sullivan, M., Illig, T., Baurecht, H., Depner, M., Rodriguez, E., Ruether, A. et al. (2008). Mutations, atopic eczema, hay fever, and asthma in children. *J. Allergy Clin. Immunol.* 121: 1203–1209.

Weissenbacher, S., Bacon, T., Targett, D., Behrendt, H., Ring, J., and Darsow, U. (2005). Atopy patch test—Reproducibility and elicitation of itch in different application sites. *Acta Derm. Venereol.* 85: 147–151.

Williams, H. C., Burney, P. J. G., Pembroke, A. C., and Hay, R. J. (1996). Validation of the UK diagnostic criteria for atopic dermatitis in a population setting. *Br. J. Dermatol.* 135: 12–17.

4 Clinical Aspects of Itch
Psoriasis

Adam Reich and Jacek C. Szepietowski

CONTENTS

4.1 INTRODUCTION

Psoriasis is a chronic, inflammatory skin disease affecting about 1% to 3% of the Caucasian population and slightly less frequently occurring also in other races. The most common variant of psoriasis, namely, plaque-type psoriasis, is clinically characterized by the presence of well-demarcated papules and plaques covered by silvery scales, which classically demonstrate symmetric distribution involving most commonly scalp, sacral area, and extensor surfaces of elbows and knees (Figure 4.1). Less often skin lesions may occur within the flexures and on the face. Other clinical subtypes include guttate, erythrodermic, and generalized or localized pustular psoriasis. Many patients (up to 80%) have nail abnormalities, and some of them (about 5%–30%) develop psoriatic arthritis. The disease may occur at any age, but two peaks of morbidity can be observed: the first one between 20 and 30 years of age, and the second one between 50 and 60 years of age (van de Kerkhof 2003). The pathogenesis of psoriasis is still not completely understood. The genetic background seems to be the most important factor, and many genes have been identified to predispose to this skin disease so far (van de Kerkhof 2003; Reich and Szepietowski 2007). However, environmental factors like infections, stress, some drugs, smoking, or alcohol also play a role. Altogether, genetic and extrinsic factors lead to abnormal keratinocyte proliferation, cutaneous inflammation, and skin vessel disturbances finally resulting in clinical features of psoriasis (Reich and Szepietowski 2007).

To date, a number of treatment options of psoriasis have been developed, but none is a curative one. Patients with psoriasis frequently experience relapses of skin lesions, causing a need of a lifelong therapy. This may also lead to discouragement and abandonment of treatment. Because of its chronicity and visibility, psoriasis is responsible for significant distress, suffering, decrease of quality of life level, and

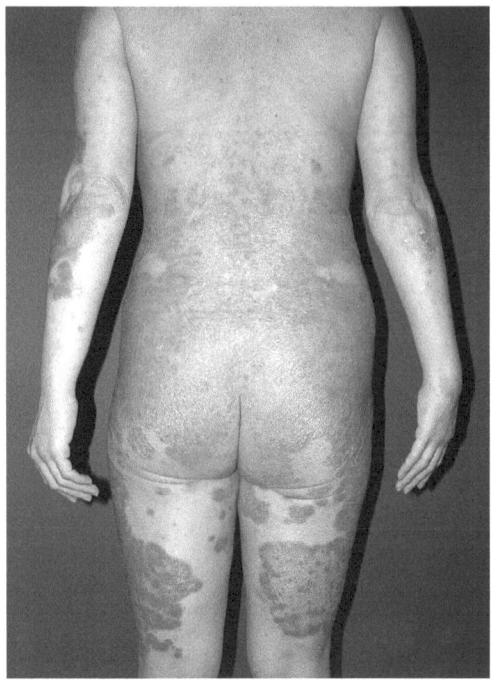

FIGURE 4.1 **(See color insert.)** Clinical presentation of psoriasis.

stigmatization (Böhm et al. 2013; Hrehorów et al. 2012; Raho et al. 2012). Several studies documented that patients with psoriasis often suffer from chronic pruritus, which further contributes to lowering of psoriatic patients' well-being (Yosipovitch et al. 2000; Reich et al. 2010a).

4.2 PREVALENCE, SEVERITY, AND CLINICAL PRESENTATION OF ITCH IN PSORIASIS

For years, psoriasis was handled as a skin disease, which only occasionally was accompanied by pruritus. However, studies being performed over the last 30 years have clearly documented that pruritus in psoriasis must be considered as a frequent phenomenon (Weisshaar 2012). In 1977, Newbold noted pruritus in 92% of 200 consecutive hospitalized patients with psoriasis. Other authors confirmed high frequency of itching among psoriatic subjects, which has ranged from 64% to 97% of studied individuals (Gupta et al. 1988; Yosipovitch et al. 2000; Szepietowski et al. 2002; Reich et al. 2003, 2007a; Sampogna et al. 2004; Chang et al. 2007; Amatya et al. 2008; Bilac et al. 2009; Prignano et al. 2009) (Figure 4.2). Importantly, itch was the most common subjective symptom reported by psoriatic patients (Sampogna et al. 2004; Bilac et al. 2009). Usually, pruritus was limited to lesional skin, but in some patients (in about 20%–30%) itch was also present within the uninvolved skin (Yosipovitch et al. 2000; Szepietowski et al. 2002; Prignano et al. 2009). In addition, women with psoriasis often reported vulvar pruritus that did not necessarily correlate with the

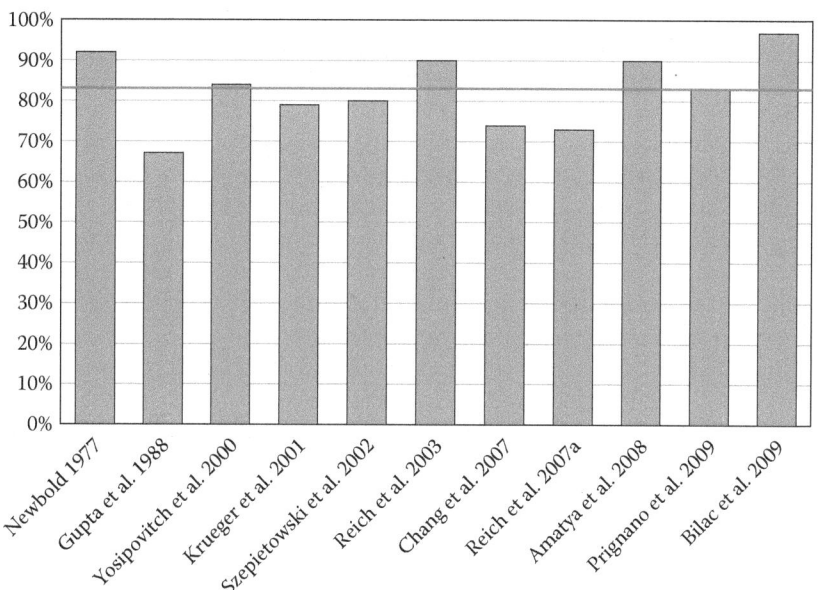

FIGURE 4.2 Prevalence of pruritus in psoriasis (the horizontal line indicates the mean prevalence from all studies).

presence of vulvar psoriatic plaques (Zamirska et al. 2008). Itch may involve all areas of the body, most commonly occurring on the extremities, trunk, and scalp, while the face and neck being only occasionally pruritic. However, a subgroup of patients was shown to suffer from generalized pruritus (Yosipovitch et al. 2000; Szepietowski et al. 2002; Amatya et al. 2008). In one study the mean pruritus-affected body surface area was estimated to be 18.1% ± 16.3% (Amatya et al. 2008). The itch location was unrelated to handedness (Yosipovitch et al. 2000).

The mean severity of pruritus assessed according to visual analog scale (VAS) ranged from 4.2 to 6.4 points (Yosipovitch et al. 2000; Reich et al. 2003, 2010a; Amatya et al. 2008; Wiśnicka et al. 2004; Remröd et al. 2007). Despite VAS not being the only way of itch assessment, it correlated well with other methods of itching measurement (Szepietowski et al. 2002; Amatya et al. 2008). On the basis of the recent grouping of VAS scores (Reich et al. 2012) it can be assumed that, in general, pruritus in psoriasis is of moderate severity; however, individual variance may of course occur. Although not all, some studies demonstrated a significant correlation between pruritus intensity and psoriasis severity (Szepietowski et al. 2002; Sampogna et al. 2004; Chang et al. 2007; Kou et al. 2012). However, the correlation was rather weak, indicating that other factors also must play an important role in evoking this symptom in psoriasis. In most studies the frequency and intensity of itching was independent of age, marital status, family history of psoriasis or atopy, alcohol or smoking habits, education level, and duration of psoriasis (Yosipovitch et al. 2000; Szepietowski et al. 2002; Sampogna et al. 2004; Amatya et al. 2008; Bilac et al. 2009; Prignano et al. 2009). Some authors observed that pruritus was more frequent and more intense in women compared to men (Sampogna et al. 2004; Amatya et al. 2008). Interestingly, Prignano et al. (2009) found that presence of pruritus correlated with BMI scoring: 40% of patients with pruritus were overweight (BMI between 25 and 30) and 10% obese (BMI > 30). Itch severity was rated as greater at night and during the winter time (O'Neill et al. 2011). Itch in psoriasis was most commonly described by the patients as related to stinging, tickling, and crawling sensations (Amatya et al. 2008). Some patients reported pain and heat sensation during itch episodes (Amatya et al. 2008). Many patients described it also as bothersome, annoying, or unbearable (Amatya et al. 2008). In about three quarters of patients with psoriasis, pruritus appeared on the daily basis and had prolonged duration (Yosipovitch et al. 2000; Amatya et al. 2008). According to patient's evaluation, the most intensive itching was observed during the skin lesions appearance or the extension of psoriatic lesions, while significant relief of itching was usually associated with complete resolving of psoriatic lesions (Szepietowski et al. 2002). Most important factors exacerbating itch in psoriasis were ambient heat, skin dryness, sweating, and emotional stress, while sleep and cold showers were often mentioned as itch alleviating measures (Yosipovitch et al. 2000).

4.3   MECHANISM OF ITCH IN PSORIASIS

The pathomechanism of pruritus in psoriasis is still not fully elucidated. Nevertheless, it seems to be rather a complex phenomenon, at least partially related to neurogenic inflammation ongoing in the skin. A relationship between pruritus and emotional stress might support this hypothesis (Reich et al. 2003; Verhoeven et al. 2009a,b; Reich et al. 2010b). Neuropeptides, i.e., peptides and small proteins released from dermal nerve endings, which possess various immunomodulatory properties, might be one of possible mediators of pruritus in psoriasis. They may degranulate mast cells, activate dendritic cells, lymphocytes, macrophages and neutrophils, exert hyperproliferative effect on keratinocytes, induce angiogenesis, expression of vascular adhesion molecules, and dilatation of vessels as well as stimulate synthesis of nitric oxide (Saraceno et al. 2006; Reich et al. 2007b, 2010b). Several studies demonstrated abnormal expression and/or distribution within the psoriatic skin of various neuropeptides and their receptors including substance P, calcitonin gene-related peptide (CGRP), somatostatin, beta-endorphin, vasoactive intestinal peptide (VIP), or pituitary adenylate cyclase activating polypeptide (PACAP) (Eedy et al. 1991; Naukkarinen et al. 1993; Glinski et al. 1994; Al'Abadie et al. 1995; Chan et al. 1997; Jiang et al. 1998; Raychaudhuri et al. 1998; Staniek et al. 1999; Steinhoff et al. 1999; He et al. 2000; Saraceno et al. 2006; Reich et al. 2007b).

Regarding pruritus in psoriasis, Nakamura et al. (2003) found a significantly increased number of substance P-positive nerves in the perivascular area as well as decreased expression of neutral endopeptidase (i.e., enzyme responsible for neuropeptide degradation) in the epidermal basal layer and in endothelial cells of cutaneous blood vessels in psoriatics with pruritus. These findings were accompanied in pruritic skin by significantly increased number of nerve growth factor (NGF) immunoreactive keratinocytes, elevated NGF content in lesional skin, higher expression of high-affinity receptor for NGF (Trk-A) in the epidermis and dermal nerve fibers, and more numerous protein gene product 9.5 (PGP-9.5) immunoreactive nerve fibers in the epidermis and in the upper dermal areas, indicating that abnormal cutaneous innervation may be, at least in part, connected with pruritus in psoriasis. The pruritus intensity in psoriatic patients significantly correlated with the number of PGP-9.5-immunoreactive intraepidermal nerve fibers, number of NGF-immunoreactive keratinocytes, and expression level of Trk-A in the epidermis (Nakamura et al. 2003). Recently, it was also shown that expression of semaphorin-3A, an axon-guidance molecule that inhibits neurite outgrowth of sensory C-fibers, was decreased in psoriasis with pruritus, and the level of semaphorin-3A expression negatively correlated with itch intensity assed by VAS (Taneda et al. 2011; Kou et al. 2012). As suggested by authors, downregulation of semaphorin-3A and upregulation of NGF in psoriatic skin may trigger hyperinnervation of C-fibers in the epidermis, leading to increased itch (Kou et al. 2012). Furthermore, Nakamura et al. (2003) found an increased number of activated mast cells in the papillary dermis with the presence of free mast cell granules in the close connection to unmyelinated nerve fibers, a phenomenon never observed in the skin from patients without pruritus. However, no differences between pruritic and nonpruritic psoriatics were found regarding the skin expression of brain derived neurotrophic factor, neurotrophin 3, VIP, neuropeptide Y, somatostatin, low-affinity receptor for NGF, and angiotensin converting enzyme (Nakamura et al. 2003). Similarly, Chang et al. (2007) analyzing 154 patients with psoriasis, observed that lesional keratinocytes of subjects with pruritus not only demonstrated increased expression of Trk-A but also showed higher expression of receptors for substance P and CGRP, while the immunoreactivity for substance P, CGRP, VIP, and PACAP did not significantly differ between pruritic and nonpruritic individuals. Amatya et al. (2011) also found a significant correlation between the pruritus intensity and the number of substance P-positive nerve fibers and number of neurokinin-2 receptor immunoreactive cells in the lesional skin. Despite in one other study the authors did not find any relevant correlation between pruritus severity and the number of substance P-positive nerve fibers nor cells, these results cannot be taken as of great relevance because of the limited number of analyzed patients ($n = 13$) (Remröd et al. 2007). Our group observed decreased neuropeptide Y plasma level in patients with pruritus compared to those without pruritus, while CGRP plasma level was elevated, and, in addition, CGRP plasma levels correlated with itching intensity in selected subgroups of patients with psoriasis (Wiśnicka et al. 2004; Reich et al. 2007a). The important role of altered innervation and neuropeptide imbalance in pruritus accompanying psoriasis may also be supported by the observation that topically applied capsaicin, a potent substance P depletory, effectively relieved pruritus in psoriatics (Bernstein et al. 1986; Ellis et al. 1993) (Table 4.1).

Regarding other possible mechanisms of pruritus in psoriasis it seems that histamine, a well-known and potent pruritogen, does not play a significant role, as no differences in histamine plasma level between patients suffering from psoriasis with and without pruritus were noted, as well as the plasma level of histamine was not correlated with pruritus intensity (Wiśnicka et al. 2004). Furthermore, antihistamines are rather of limited value. Although some patients may sometimes benefit from these drugs, it is rather due to their sedative properties and not due to histamine blockade (Szepietowski et al. 2002; Dawn and Yosipovitch 2006).

TABLE 4.1

Possible Mechanisms Involved in the Pathomechanism of Pruritus in Psoriasis

Suggested Mechanism	Study	Observations in Psoriasis
Abnormal cutaneous innervation, neurogenic inflammation, and neuropeptide misbalance	Nakamura et al. (2003)	Increased number of nerve growth factor (NGF) immunoreactive keratinocytes, elevated NGF content in lesional skin, higher expression of high-affinity receptor for NGF (Trk-A) in the epidermis and dermal nerve fibers, and more numerous protein gene product 9.5 (PGP-9.5) immunoreactive nerve fibers in the epidermis and in the upper dermal areas of patients with pruritus
		Increased number of substance P-positive nerves in the perivascular area in the skin of patients with pruritus
		Decreased expression of neutral endopeptidase in the epidermal basal layer and in endothelial cells of cutaneous blood vessels in psoriatics with pruritus
		Correlation between pruritus intensity and the number of PGP-9.5-immunoreactive intraepidermal nerve fibers, number of NGF-immunoreactive keratinocytes, and expression level of Trk-A in the epidermis
	Taneda et al. (2011)	Expression of semaphorin-3A, an axon-guidance molecule that inhibits neurite outgrowth of sensory C-fibers, decreased in psoriasis with pruritus
	Kou et al. (2012)	Expression of semaphorin-3A decreased in psoriasis with pruritus
		Semaphorin-3A mRNA expression negatively correlated with itch intensity
	Chang et al. (2007)	Increased expression of Trk-A and higher expression of receptors for substance P and calcitonin gene-related peptide (CGRP) in lesional keratinocytes of subjects with pruritus
	Amatya et al. (2011)	Significant correlation between pruritus intensity and the number of substance P positive nerve fibers and number of neurokinin-2 receptor immunoreactive cells in the lesional skin

	Reich et al. (2007a)	Decreased neuropeptide Y plasma level in patients with pruritus compared to those without pruritus
	Wiśnicka et al. (2004)	Elevated CGRP plasma level in patients with pruritus
		Correlation between CGRP plasma level and itching intensity in selected subgroups of patients
	Bernstein et al. (1986) and Ellis et al. (1993)	Topically applied capsaicin, a potent substance P depletory, effectively relieved pruritus in psoriatics
Abnormal functioning of peripheral opioid system	Taneda et al. (2011)	The levels of κ-opioid receptor and dynorphin A significantly decreased in the epidermis of psoriatic patients with itch compared with healthy controls
	Kupczyk et al. (2012)	A reduced expression of κ-opioid receptors in the skin of psoriatic subjects without pruritus
		Inverse correlation between pruritus intensity and epidermal expression of κ-opioid receptors (Kupczyk et al. 2012)
Pruritus mediated by cytokines released by immune cells	Nigam et al. (2010)	The number of GABA-positive and $GABA_A$ receptor positive inflammatory cells significantly correlated with pruritus intensity
	Nakamura et al. (2003)	Increased number of IL-2 immunoreactive cells in pruritic compared to nonpruritic lesions of psoriasis
	Narbutt et al. (2012)	Interleukin 31, a pruritogenic cytokine, unregulated in psoriasis
Pruritus mediated by vascular abnormalities	Nakamura et al. (2003)	Significant increase of the density of endothelial leukocyte adhesion molecule 1 (ELAM-1) positive venules in patients with pruritus
		Significant correlation of itching intensity with the density of E-selectin immunoreactive vessels
	Madej et al. (2007)	Elevated serum level of soluble vascular adhesion protein 1 (VAP-1) in patients with pruritus

Recently, two other interesting mechanisms of pruritus origin have been postulated. Taneda et al. (2011) observed that lesional epidermis of psoriatic patients suffering from pruritus showed decreased expression of κ-opioid receptors, while the expression of μ-opioid receptors remained unaltered. Reduced expression of κ-opioid receptor was accompanied by the decreased expression of its agonist, dynorphin A (Taneda et al. 2011). These observations have recently been confirmed by our study showing a reduced expression of κ-opioid receptors in the skin of subjects with pruritus, while patients without pruritus had expression levels similar to healthy controls (Kupczyk et al. 2012). Furthermore, pruritus intensity inversely correlated with the expression of κ-opioid receptors (Kupczyk et al. 2012). As it is believed that activation of μ-opioid receptors may evoke itching, while activation of κ-opioid receptors exerts antipruritic effect, it seems highly probable that misbalance of the cutaneous opioid system in psoriasis may partake in pruritus origin (Bigliardi et al. 2009). However, the exact mechanism why the κ-opioid pathway is depressed in psoriasis remains unclear and requires further investigations.

Another interesting observation was reported by Nigam et al. (2010), who demonstrated that lesional psoriatic skin showed increased number of inflammatory cells with immunoreactivity for gamma-aminobutyric acid (GABA) and its receptor $GABA_A$. Interestingly, the number of GABA-positive and $GABA_A$ receptor positive inflammatory cells significantly correlated with pruritus intensity (Nigam et al. 2010). GABA is an important neurotransmitter that regulates neuronal activation in the nervous system and primarily GABAergic neurons inhibit neuronal transmission. GABA is responsible, e.g., for controlling hypothalamic–pituitary–adrenocortical axis in rats (Cullinan et al. 2008; Mikkelsen et al. 2008). However, GABA may also modulate activity of various immune cells (Tian et al. 1999; Rane et al. 2005). Thus, the pruritogenic activity of GABA may be explained by controlling the neuroendocrine functions or by stimulating synthesis of some mediators by immune cells. Future studies should demonstrate whether any of these hypotheses is valid. A possible pruritogenic molecule could be e.g., interleukin 2 as Nakamura et al. (2003) found an increased number of IL-2 immunoreactive cells in pruritic in comparison to nonpruritic lesions of psoriasis. Another interesting option would be interleukin 31, which was shown to be unregulated in psoriasis (Narbutt et al. 2012). Interleukin 31 is believed to be the most important cytokine engaged in pruritus pathogenesis (Sonkoly et al. 2006). Other cytokines, like interferon γ, tumor necrosis factor α, and interleukins 1α, 1β, 4, 5, 6, 8, 10, or 12 seem to be unrelated to pruritus in psoriasis (Nakamura et al. 2003) (Table 4.1).

Some data indicate that pruritus is also partially driven by vascular abnormalities. Nakamura et al. (2003) observed significant increase of the density of endothelial leukocyte adhesion molecule 1 (ELAM-1) positive venules in patients with pruritus and the itching intensity significantly correlated with the density of E-selectin immunoreactive vessels. Our group found that psoriatic patients with pruritus showed an elevated serum level of soluble vascular adhesion protein 1 (VAP-1) (Madej et al. 2007). The cutaneous expression of other adhesion molecules (intercellular cell adhesion molecule 1, ICAM-1; vascular cell adhesion molecule 1, VCAM-1; or platelet endothelial cell adhesion molecule 1, PECAM-1) does not seem to be important for itch perception in psoriasis (Nakamura et al. 2003) (Table 4.1).

4.4 BURDEN OF ITCH IN PSORIASIS

Many patients consider pruritus as the most bothersome symptom of psoriasis, even though the severity of itch in psoriasis seems to be lower than in other pruritic skin conditions (Gupta et al. 1988; Welz-Kubiak et al. 2012). Patients with pruritus showed more reduced health-related quality of life (on average a large decrease is noted), compared to those without pruritus, who usually had moderately decreased quality of life and the pruritus intensity correlated with the degree of quality of life impairment (Reich et al. 2011). Psoriatic subjects with pruritus also demonstrated more depressive symptoms. The pruritus intensity significantly influenced the severity of depressive symptoms, as well as the level of stigmatization (Reich et al. 2011). In addition, many patients with pruritus demonstrated problems in falling asleep and more frequent awakenings (Yosipovitch et al. 2000; Gowda et al. 2010). Because of pruritus, 35% of patients became more agitated, 24% depressed, 30% showed concentration difficulties, 23% changed their eating habits, and 35% reported their sexual function to be decreased or nonexistent (Yosipovitch et al. 2000). Amatya et al. (2008) also documented that a majority of patients with psoriasis consider pruritus as a symptom negatively affecting their quality of life, with mood, concentration, sleep, sexual desire, and appetite being the most impaired daily life aspects. Psoriasis patients also reported more embarrassment associated with itch than did patients with atopic dermatitis (O'Neill et al. 2011). Pruritus significantly altered the work ability in psoriasis patients: it was mentioned by 48% of subjects as the most important symptom of psoriasis interfering with the work activity, and only scaling was more frequently reported. Furthermore, significant inverse correlation was found between itch intensity and work ability ($R = -0.31$, $p < 0.01$) (Zimoląg et al. 2009).

4.5 TREATMENT OF ITCH IN PSORIASIS

Treatment of pruritus in psoriasis is challenging. To date, no antipruritic therapy has been properly tested in a randomized way in psoriasis. Furthermore, causative treatment is difficult to be rationally developed, as the pathogenesis of this symptom is still not completely understood.

Usually, some relief is provided by proper skin moisturizing with emollients and moisturizers; however, only a minority of psoriatic patients (less than 20%) considered them as highly effective (Szepietowski et al. 2002; Dawn and Yosipovitch 2006). Usually, patients claimed that the disappearance of pruritus is related to the complete resolution of skin lesions (Szepietowski et al. 2002; Reich et al. 2011). It is in contrast to, e.g., lichen planus, in which pruritus reliefs within a couple of days after treatment initiation (Reich et al. 2011). Thus, the most effective antipsoriatic therapies should lead to the improvement of pruritus, if this symptom is really of psoriasis origin. However, a number of patients cannot achieve complete long-lasting remission of skin plaques, and therefore many of them may suffer from a chronic itch. In the review by Dawn and Yosipovitch (2006), several types of antipsoriatic treatment modalities were described as possibly helpful in treating patients with psoriasis who simultaneously suffered from itch, such as tar products, topical corticosteroids, topical salicylates, vitamin D analogs, topical immunomodulators, phototherapy, methotrexate, and biologics. One study documented efficacy of narrowband ultraviolet B (UVB) therapy in treating psoriatic itch (Gupta et al. 1999). Recently, Narbutt et al. (2012) have shown that treatment with narrowband UVB decreased the level of interleukin 31 in the sera of psoriatic patients, a finding that might explain the antipruritic effect of phototherapy. However, it must be underlined that in some patients at the beginning of the treatment, narrowband UVB may even aggravate itch owing to increasing of skin dryness, and therefore concomitant use of moisturizers or emollients throughout phototherapy is highly recommended (Dawn and Yosipovitch 2006).

As mentioned in the pathogenesis section, histamine seems not to be involved in transmission of itch stimuli in psoriasis; thus, antihistamines seem to be ineffective unless they cause sedation. On the basis of our own practice, hydroxyzine (10–25 mg two to three times a day) or clemastine (1–2 mg twice daily) are very useful for pruritus diminishment, but this suggestion has not confirmed in any controlled study so far. In more severe cases, oral antidepressants like mirtazapine (15 mg at night), doxepin (10–20 mg three times daily) or paroxetine (20 mg/d) can be tried. Mirtazapine was shown to relieve itch even in severe pruritus associated with erythrodermic psoriasis (Dawn and Yosipovitch 2006). Mirtazapine exert a sedative effect due to its H_1-antihistamine properties, but it also acts as an antagonist on noradrenergenic $\alpha 2$-receptors and $5\text{-}HT_2$ and $5\text{-}HT_3$ serotonin receptors (Dawn and Yosipovitch 2006). Furthermore, support given by the family and/or health professionals may increase the ability of patients to cope with itching (van Os-Medendorp et al. 2008).

It seems that in patients unresponsive to any kind of the above mentioned modalities, one may try therapeutic strategies used in recalcitrant chronic itch patients of different origin. On the basis of recent studies on pathogenesis of itch in psoriasis, drugs acting on peripheral opioid system (e.g., nalfurafine, a kappa receptor agonist) (Kumagai et al. 2012; Reich and Szepietowski 2012) or activating GABA receptors (e.g., pregabalin and gabapentin) seems to be the most promising ones (Solak et al. 2012; Ahuja and Gupta 2013); however, no studies confirming their efficacy in psoriasis exist. In addition, it would be of interest to prove whether endocannabinoids could be effective as well (Kupczyk et al. 2009; Tóth et al. 2011).

Summarizing, the therapy of psoriatic itch remains a real and urgent challenge. No single therapy exists that is effective in all psoriasis patients with itch. In seems that some patients may benefit from a combination of different methods of pruritus treatment. Complete clearance of psoriatic lesions usually is accompanied by itch resolution, but the majority of patients require antipruritic therapy prior to clearance of visible skin lesions (Dawn and Yosipovitch 2006).

REFERENCES

Ahuja, R. B., and Gupta, G. K. (2013). A four arm, double blind, randomized and placebo controlled study of pregabalin in the management of post-burn pruritus. *Burns* 39: 24–29.

Al'Abadie, M. S., Senior, H. J., Bleehen, S. S., and Gawkrodger, D. J. (1995). Neuropeptides and general neuronal marker in psoriasis—An immunohistochemical study. *Clin. Exp. Dermatol.* 20: 384–389.

Amatya, B., Wennersten, G., and Nordlind, K. (2008). Patients' perspective of pruritus in chronic plaque psoriasis: A questionnaire-based study. *J. Eur. Acad. Dermatol. Venereol.* 22: 822–826.

Amatya, B., El-Nour, H., Holst, M., Theodorsson, E., and Nordlind, K. (2011). Expression of tachykinins and their receptors in plaque psoriasis with pruritus. *Br. J. Dermatol.* 164: 1023–1029.

Bernstein, J. E., Parish, L. C., Rapaport, M., Rosenbaum, M. M., and Roenigk, H. H., Jr. (1986). Effects of topically applied capsaicin on moderate and severe psoriasis vulgaris. *J. Am. Acad. Dermatol.* 15: 504–507.

Bigliardi, P. L., Tobin, D. J., Gaveriaux-Ruff, C., and Bigliardi-Qi, M. (2009). Opioids and the skin—Where do we stand? *Exp. Dermatol.* 18: 424–430.

Bilac, C., Ermertcan, A. T., Bilac, D. B., Deveci, A., and Horasan, G. D. (2009). The relationship between symptoms and patient characteristics among psoriasis patients. *Indian J. Dermatol. Venereol. Leprol.* 75: 551.

Böhm, D., Stock Gissendanner, S., Bangemann, K., Snitjer, I., Werfel, T., Weyergraf, A., Schulz, W., Jäger, B., and Schmid-Ott, G. (2013). Perceived relationships between severity of psoriasis symptoms, gender, stigmatization and quality of life. *J. Eur. Acad. Dermatol. Venereol.* 27(2): 220–6.

Chan, J., Smoller, B. R., Raychauduri, S. P., Jiang, W. Y., and Farber, E. M. (1997). Intraepidermal nerve fiber expression of calcitonin gene-related peptide, vasoactive intestinal peptide and substance P in psoriasis. *Arch. Dermatol. Res.* 289: 611–616.

Chang, S.-E., Han, S.-S., Jung, H.-J., and Choi, J.-H. (2007). Neuropeptides and their receptors in psoriatic skin in relation to pruritus. *Br. J. Dermatol.* 156: 1272–1277.

Cullinan, W. E., Ziegler, D. R., and Herman, J. P. (2008). Functional role of local GABAergic influences on the HPA axis. *Brain Struct. Funct.* 213: 63–72.

Dawn, A., and Yosipovitch, G. (2006). Treating itch in psoriasis. *Dermatol. Nurs.* 18: 227–233.

Eedy, D. J., Johnston, C. F., Shaw, C., and Buchanan, K. D. (1991). Neuropeptides in psoriasis: An immunocytochemical and radioimmunoasay study. *J. Invest. Dermatol.* 96: 434–438.

Ellis, C. N., Berberian, B., Sulica, V. I., Dodd, W. A., Jarratt, M. T., Katz, H. I., Prawer, S., Krueger, G., Rex, I. H., Jr., and Wolf, J. E. (1993). A double-blind evaluation of topical capsaicin in pruritic psoriasis. *J. Am. Acad. Dermatol.* 29: 438–442.

Glinski, W., Brodecka, H., Glinska-Ferenz, M., and Kowalski, D. (1994). Neuropeptides in psoriasis: Possible role of beta-endorphin in the pathomechanism of the disease. *Int. J. Dermatol.* 33: 356–360.

Gowda, S., Goldblum, O. M., McCall, W. V., and Feldman, S. R. (2010). Factors affecting sleep quality in patients with psoriasis. *J. Am. Acad. Dermatol.* 63: 114–123.

Gupta, G., Long, J., and Tillman, D. M. (1999). The efficacy of narrowband ultraviolet B phototherapy in psoriasis using objective and subjective outcome measures. *Br. J. Dermatol.* 140: 887–890.

Gupta, M. A., Gupta, A. K., Kirkby, S., Weiner, H. K., Mace, T. M., Schork, N. J., Johnson, E. H., Ellis, C. N., and Voorhees, J. J. (1988). Pruritus in psoriasis. A prospective study of some psychiatric and dermatologic correlates. *Arch. Dermatol.* 124: 1052–1057.

He, Y., Ding, G., Wang, X., Zhu, T., and Fan, S. (2000). Calcitonin gene-related peptide in Langerhans cells in psoriatic plaque lesions. *Chin. Med. J.* 113: 747–751.

Hrehorów, E., Salomon, J., Matusiak, L., Reich, A., and Szepietowski, J. C. (2012). Patients with psoriasis feel stigmatized. *Acta Derm. Venereol.* 92: 67–72.

Jiang, W. -Y., Raydchaudhuri, S. P., and Farber, E. M. (1998). Double-labelled immunofluorescence study of cutaneous nerves in psoriasis. *Int. J. Dermatol.* 37: 572–574.

Kou, K., Nakamura, F., Aihara, M., Chen, H., Seto, K., Komori-Yamaguchi, J., Kambara, T., Nagashima, Y., Goshima, Y., and Ikezawa, Z. (2012). Decreased expression of semaphorin-3A, a neurite-collapsing factor, is associated with itch in psoriatic skin. *Acta Derm. Venereol.* 92: 521–528.

Krueger, G., Koo, J., Lebwohl, M., Menter, A., Stern, R. S., and Rolstad, T. (2001). The impact of psoriasis on quality of life: Results of a 1998 National Psoriasis Foundation patient-membership survey. *Arch. Dermatol.* 137: 280–284.

Kumagai, H., Ebata, T., Takamori, K., Miyasato, K., Muramatsu, T., Nakamoto, H., Kurihara, M., Yanagita, T., and Suzuki, H. (2012). Efficacy and safety of a novel κ-agonist for managing intractable pruritus in dialysis patients. *Am. J. Nephrol.* 36: 175–183.

Kupczyk, P., Reich, A., and Szepietowski, J. C. (2009). Cannabinoid system in the skin—A possible target for future therapies in dermatology. *Exp. Dermatol.* 18: 669–679.

Kupczyk, P., Reich, A., Hołysz, M., and Szepietowski, J. (2012). Expression of MOR and KOR opioid receptors and OPRK1 gene in psoriasis patients. *Przegl. Dermatol.* 99: 467–468.

Madej, A., Reich, A., Orda, A., and Szepietowski, J. C. (2007). Vascular adhesion protein-1 (VAP-1) is overexpressed in psoriatic patients. *J. Eur. Acad. Dermatol. Venereol.* 21: 72–78.

Mikkelsen, J. D., Bundzikova, J., Larsen, M. H., Hansen, H. H., and Kiss, A. (2008). GABA regulates the rat hypothalamic-pituitary-adrenocortical axis via different GABA-A receptor alpha-subtypes. *Ann. N. Y. Acad. Sci.* 1148: 384–392.

Nakamura, M., Toyoda, M., and Morohashi, M. (2003). Pruritogenic mediators in psoriasis vulgaris: Comparative evaluation of itch-associated cutaneous factors. *Br. J. Dermatol.* 149: 718–730.

Narbutt, J., Olejniczak, I., Sobolewska-Sztychny, D., Sysa-Jedrzejowska, A., Slowik-Kwiatkowska, I., Hawro, T., and Lesiak, A. (2012). Narrow band ultraviolet B irradiations cause alteration in interleukin-31 serum level in psoriatic patients. *Arch. Dermatol. Res.* 305(5): 191–195.

Naukkarinen, A., Harvima, I., Paukkonen, K., Aalto, M. L., and Horsmanheimo, M. (1993). Immunohistochemical analysis of sensory nerves and neuropeptides, and their contacts with mast cells in developing and mature psoriatic lesions. *Arch. Dermatol. Res.* 285: 341–346.

Newbold, P. C. H. (1977). Pruritus in psoriasis. In *Psoriasis. Proceedings of the Second International Symposium*, Farber, E. M., and Cox, A. J. (Eds.). New York: Yorke Medical Books, pp. 334–336.

Nigam, R., El-Nour, H., Amatya, B., and Nordlind, K. (2010). GABA and GABA(A) receptor expression on immune cells in psoriasis: A pathophysiological role. *Arch. Dermatol. Res.* 302: 507–515.

O'Neill, J. L., Chan, Y. H., Rapp, S. R., and Yosipovitch, G. (2011). Differences in itch characteristics between psoriasis and atopic dermatitis patients: Results of a web-based questionnaire. *Acta Derm. Venereol.* 91: 537–540.

Prignano, F., Ricceri, F., Pescitelli, L., and Lotti, T. (2009). Itch in psoriasis: Epidemiology, clinical aspects and treatment options. *Clin. Cosmet. Investig. Dermatol.* 2: 9–13.

Raho, G., Koleva, D. M., Garattini, L., and Naldi, L. (2012). The burden of moderate to severe psoriasis: An overview. *Pharmacoeconomics* 30: 1005–1013.

Rane, M. J., Gozal, D., Butt, W., Gozal, E., Pierce, W. M., Jr., Guo, S. Z., Wu, R. et al. (2005). Gamma-amino butyric acid type B receptors stimulate neutrophil chemotaxis during ischemia-reperfusion. *J. Immunol.* 174: 7242–7249.

Raychaudhuri, S. P., Jiang, W.-Y., and Farber, E. M. (1998). Psoriatic keratinocytes express high levels of nerve growth factor. *Acta Derm. Venereol.* 78: 84–86.

Reich, A., and Szepietowski, J. (2007). Genetic and immunological aspects of the pathogenesis of psoriasis. *Wiad. Lek.* 60: 270–276.

Reich, A., and Szepietowski, J. C. (2012). Non-analgesic effects of opioids: Peripheral opioid receptors as promising targets for future anti-pruritic therapies. *Curr. Pharm. Des.* 18: 6021–6024.

Reich, A., Szepietowski, J. C., Wiśnicka, B., and Pacan, P. (2003). Does stress influence itching in psoriatic patients? *Dermatol. Psychosom.* 4: 151–155.

Reich, A., Orda, A., Wiśnicka, B., and Szepietowski, J. C. (2007a). Plasma neuropeptides and perception of pruritus in psoriasis. *Acta Derm. Venereol.* 87: 299–304.

Reich, A., Orda, A., Wiśnicka, B., and Szepietowski, J. C. (2007b). Plasma concentration of selected neuropeptides in patients suffering from psoriasis. *Exp. Dermatol.* 16: 421–428.

Reich, A., Hrehorów, E., and Szepietowski, J. C. (2010a). Pruritus is an important factor negatively influencing the well-being of psoriatic patients. *Acta Derm. Venereol.* 90: 257–263.

Reich, A., Wójcik-Maciejewicz, A., and Slominski, A. T. (2010b). Stress and the skin. *G. Ital. Dermatol. Venereol.* 145: 213–219.

Reich, A., Welz-Kubiak, K., and Szepietowski, J. C. (2011). Pruritus differences between psoriasis and lichen planus. *Acta Derm. Venereol.* 91: 605–606.

Reich, A., Heisig, M., Phan, N. Q., Taneda, K., Takamori, K., Takeuchi, S., Furue, M. et al. (2012). Visual analogue scale: Evaluation of the instrument for the assessment of pruritus. *Acta Derm. Venereol.* 92: 497–501.

Remröd, C., Lonne-Rahm, S., and Nordlind, K. (2007). Study of substance P and its receptor neurokinin-1 in psoriasis and their relation to chronic stress and pruritus. *Arch. Dermatol. Res.* 299: 85–91.

Sampogna, F., Gisondi, P., Melchi, C. F., Amerio, P., Girolomoni, G., Abeni, D., and IDI Multipurpose Psoriasis Research on Vital Experiences Investigators (2004). Prevalence of symptoms by patients with different clinical types of psoriasis. *Br. J. Dermatol.* 151: 594–599.

Saraceno, R., Kleyn, C. E., Terenghi, G., and Griffiths, C. E. (2006). The role of neuropeptides in psoriasis. *Br. J. Dermatol.* 155: 876–882.

Solak, Y., Biyik, Z., Atalay, H., Gaipov, A., Guney, F., Turk, S., Covic, A., Goldsmith, D., and Kanbay, M. (2012). Pregabalin versus gabapentin in the treatment of neuropathic pruritus in maintenance haemodialysis patients: A prospective, crossover study. *Nephrology* 17: 710–717.

Sonkoly, E., Muller, A., Lauerma, A. I., Pivarcso, A., Soto, H., Kemeny, L., Alenius, H. et al. (2006). IL-31: A new link between T cells and pruritus in atopic skin inflammation. *J. Allergy Clin. Immunol.* 117: 411–417.

Staniek, V., Doutremepuich, J., Schmitt, D., Claudy, A., and Misery, L. (1999). Expression of substance P receptors in normal and psoriatic skin. *Pathobiology* 67: 51–54.

Steinhoff, M., McGregor, G. P., Radleff-Schlimme, A., Steinhoff, A., Jarry, H., and Schmidt, W. E. (1999). Identification of pituitary adenylate cyclase activating polypeptide (PACAP) and PACAP type 1 receptor in human skin: Expression of PACAP-38 is increased in patients with psoriasis. *Regul. Pept.* 80: 49–55.

Szepietowski, J. C., Reich, A., and Wiśnicka, B. (2002). Itching in patients suffering from psoriasis. *Acta Derm. Venereol. Croat.* 10: 216–221.

Taneda, K., Tominaga, M., Negi, O., Tengara, S., Kamo, A., Ogawa, H., and Takamori, K. (2011). Evaluation of epidermal nerve density and opioid receptor levels in psoriatic itch. *Br. J. Dermatol.* 165: 277–284.

Tian, J., Chau, C., Hales, T. G., and Kaufman, D. L. (1999). GABA(A) receptors mediate inhibition of T cell responses. *J. Neuroimmunol.* 96: 21–28.

Tóth, B. I., Dobrosi, N., Dajnoki, A., Czifra, G., Oláh, A., Szöllosi, A. G., Juhász, I., Sugawara, K., Paus, R., and Biró, T. (2011). Endocannabinoids modulate human epidermal keratinocyte proliferation and survival via the sequential engagement of cannabinoid receptor-1 and transient receptor potential vanilloid-1. *J. Invest. Dermatol.* 131: 1095–1104.

van de Kerkhof, P. C. M. (2003). Psoriasis. In *Dermatology*, Bologna, J. L., Jorizzo, J. L., and Rapini, R. P. (Eds.). London: Mosby, pp. 125–149.

van Os-Medendorp, H., Guikers, C. L., Eland-de Kok, P. C., Ros, W. J., Bruijnzeel-Koomen, C. A., and Buskens, E. (2008). Costs and cost-effectiveness of the nursing programme "Coping with itch" for patients with chronic pruritic skin disease. *Br. J. Dermatol.* 158: 1013–1021.

Verhoeven, E. W., Kraaimaat, F. W., de Jong, E. M., Schalkwijk, J., van de Kerkhof, P. C., and Evers, A. W. (2009a). Individual differences in the effect of daily stressors on psoriasis: A prospective study. *Br. J. Dermatol.* 161: 295–299.

Verhoeven, E. W., Kraaimaat, F. W., Jong, E. M., Schalkwijk, J., van de Kerkhof, P. C., and Evers, A. W. (2009b). Effect of daily stressors on psoriasis: A prospective study. *J. Invest. Dermatol.* 129: 2075–2077.

Weisshaar, E. (2012). Pruritus and psoriasis: An important but frequently underestimated relation (in German). *Hautarzt* 63: 547–552.

Welz-Kubiak, K., Rams, Ł., and Reich, A. (2012). Patient insights into psoriasis. *Przegl. Dermatol.* 99: 553–554.

Wiśnicka, B., Szepietowski, J. C., Reich, A., and Orda, A. (2004). Histamine, substance P and calcitonin gene-related peptide plasma concentration and pruritus in patients suffering from psoriasis. *Dermatol. Psychosom.* 5: 73–78.

Yosipovitch, G., Goon, A., Wee, J., Chan, Y. H., and Goh, C. L. (2000). The prevalence and clinical characteristics of pruritus among patients with extensive psoriasis. *Br. J. Dermatol.* 143: 969–973.

Zamirska, A., Reich, A., Berny-Moreno, J., Salomon, J., and Szepietowski, J. C. (2008). Vulvar pruritus and burning sensation in women with psoriasis. *Acta Derm. Venereol.* 88: 132–135.

Zimolag, I., Reich, A., and Szepietowski, J. C. (2009). Influence of psoriasis on the ability to work. *Acta Derm. Venereol.* 89: 575–576.

5 Pruritus in Renal Disease

Thomas Mettang

CONTENTS

5.1 INTRODUCTION

Uremic pruritus, now better named "chronic kidney disease-associated pruritus" (CKD-aP), remains a frequent and compromising symptom in patients with advanced or end-stage renal disease (Mettang et al. 1996). Most therapeutic trials have shown only limited success. Several times in the past a new treatment option has been reported to be effective, but very soon thereafter conflicting results appear (De Marchi et al. 1992; Balaskas and Uldall 1992; Peer et al. 1996; Pauli-Magnus et al. 2000b). The main obstacle in the effort to create effective treatment modalities is the incomplete knowledge of the underlying pathophysiological mechanisms. Furthermore, given the great clinical heterogeneity of patients with kidney failure, systematically performed studies are hard to undertake and therefore sparse.

5.2 CLINICAL FEATURES OF CKD-aP

The intensity and spatial distribution of pruritus vary significantly over time, and some patients are affected to a varying degree throughout the duration of their renal disease. The intensity of CKD-aP ranges from sporadic discomfort to complete restlessness during day- and nighttime. The skin of hemodialysis patients with chronic itch looks quite similar to that of patients without itch. However, there is evidence of secondary skin changes, most likely due to scratching. Excoriation by scratching with or without impetigo can occur as a secondary phenomenon and rarely even prurigo nodularis is observed (Figure 5.1a through c). There are interindividual differences in the spatial distribution of CKD-aP: 25%–50% of patients with CKD-aP complain about generalized pruritus (Morvay and Marghescu 1988; Ponticelli and Bencini 1992). In the remaining patients, CKD-aP seems to affect predominantly the back, the face, and the forearm, respectively (Gilchrest et al. 1982). In about 25% of patients with CKD-aP, pruritus was most severe during or immediately after dialysis (Dawn and Yosipovitch 2006). Once patients develop CKD-aP, it can last for months or years (Mathur et al. 2010).

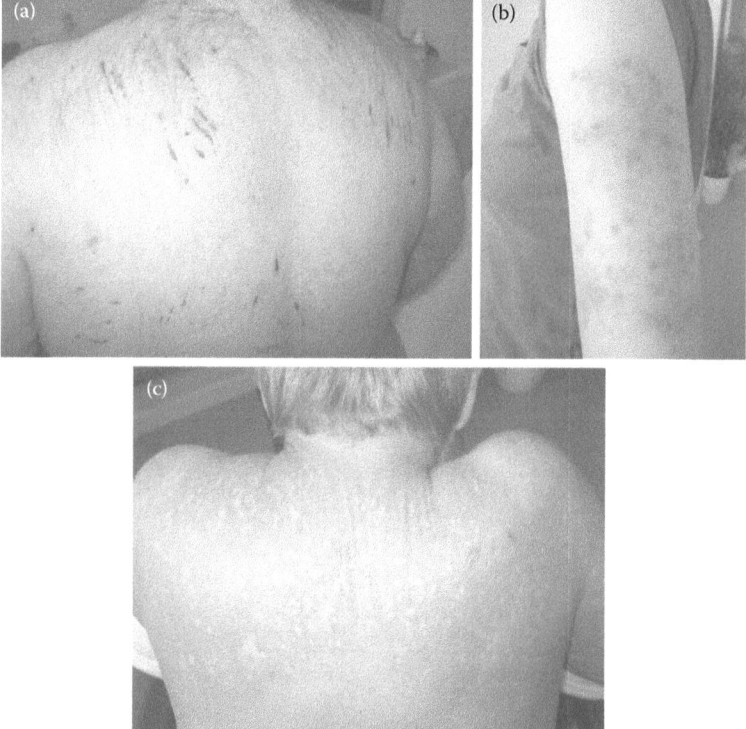

FIGURE 5.1 (**See color insert.**) Typical skin changes due to uremic pruritus. (a) Scratch marks on the back with excoriations. (b) Typical hyperkeratotic partly excoriated nodules (prurigo nodularis) on the arm. (c) Deep scars on the shoulders and back of a female patient on hemodialysis.

5.2.1 EPIDEMIOLOGY

Whereas during the early days of dialysis treatment, CKD-aP was a very common problem, it appears that its incidence has declined over the past 20 years. In the early 1970s, Young et al. (1973) reported that about 85% of patients were affected by CKD-aP. This number decreased to 50% to 60% in the late 1980s (Mettang et al. 1990). An investigation in Germany showed that only 22% of all dialysis patients complained about moderate to severe pruritus at the time they were studied (Pauli-Magnus et al. 2000b). On the basis of a large-scale investigation published a few years ago, more than 40% of hemodialysis patients suffer from chronic pruritus (Pisoni et al. 2006) (Figure 5.2).

Interestingly, severe pruritus is very rare in pediatric patients on dialysis. This could be shown by a systematic review of all German pediatric dialysis centers involving 199 children, where only 9.1% of the children on dialysis complained of pruritus (Figure 5.3). Moreover, the intensity was not very severe in the affected patients (Schwab et al. 1999).

Data on the prevalence of CKD-aP in peritoneal dialysis patient are rather scarce. The few reports available, however, permit the conclusion that patients undergoing peritoneal dialysis are affected as much by pruritus as hemodialysis patients are (Tessari et al. 2009).

The prevalence of CKD-aP is obviously underestimated by German nephrologists. We were able to corroborate this by a recent poll among German nephrologists. There are many reasons for this. First, the great variation in the intensity of symptoms may be responsible, and second, it is classified as a minor problem (not acutely life-threatening) (Weisshaar et al. 2009).

5.2.2 ETIOPATHOLOGY

So far, there have been no clear ideas regarding the pathogenesis of CKD-aP (Mettang et al. 2002). Among others, *parathyroid hormone* and *histamine* were investigated more closely as presumed pruritogenic factors. Parathyroid hormone is believed to

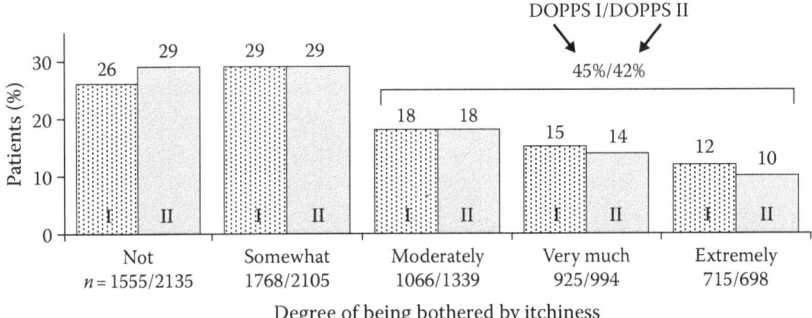

FIGURE 5.2 Prevalence and intensity of uremic pruritus according to DOPPS-data from 1996–1999 (I) and 2002–2003 (II); DOPPS, Dialysis Outcome and Practice Patterns Study. (After Pisoni, R. L. et al., *Nephrol. Dial. Transplant*, 21, 3495–3505, 2006.)

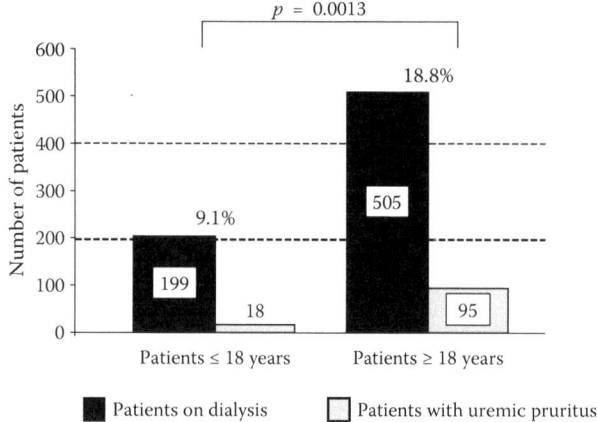

FIGURE 5.3 Prevalence of uremic pruritus in children on dialysis (18 years or younger) and in adult dialysis patients (older than 18 years). Prevalence of uremic pruritus in children is significantly lower than in adult patients (chi-squared test). (From Schwab, M. et al., *Monatszeitschrift Kinderheilkunde,* 147, 232, 1999.)

be a possible pathogenetic factor based on multiple observations that persistent pruritus in patients with secondary hyperparathyroidism was substantially improved after parathyroidectomy (Massry et al. 1968; Hampers et al. 1968). However, substantial data argue against parathyroid hormone being the likely trigger of CKD-aP (Stahle-Bäckdahl et al. 1989).

It has also been hypothesized that histamine, which accumulates in renal failure and classically mediates pruritus, might also account for the itching in patients with end-stage renal failure (Stockenhuber et al. 1990). Nevertheless, the literature contains contradictory data as to the impact of histamine (Mettang et al. 1990). The concept of histamine playing a significant part in the generation of pruritus is challenged by the absence of typical skin changes such as wheals and by the usually observed therapeutic failure of antihistaminics (Weisshaar et al. 2004).

The impact of xenobiotic agents and uremic toxins has not yet been established. There is also some controversy regarding the influence of other factors, i.e., increased tissue concentrations of vitamin A and metastatic microcalcifications caused by calcium and magnesium salts (Blachley et al. 1985). In a study including 10 hemodialysis patients with or without pruritus, we wanted to find out whether there might be a difference regarding the calcium-binding protein fetuin. This admittedly small study population revealed no discrepancy in the serum levels of fetuin, calcium, or 25-hydroxy-vitamin D (Mettang et al. 2010).

Xerosis of the skin, described in a large number of patients with renal failure, was at least partially ruled out as pathogenetically relevant (Szepietowski et al. 2004).

Special attention has focused not only on neuropathic disorders but also on receptor proliferation of pruritus-mediating cells and on central nervous alterations. Enhanced stimulation of central μ-opioid receptors by accumulated endorphins or endogenous morphine-like compounds is possibly responsible for increased pruritus. This hypothesis is supported by the observation that uremic patients will experience a substantial amelioration of itching after the oral application of naltrexone (a μ-opioid receptor antagonist) (Peer et al. 1996). The positive action of this substance, however, could not be demonstrated in a large-scale study by our group on dialysis patients presenting with pruritus (Pauli-Magnus et al. 2000b; see Treatment Options section). According to another hypothesis, pruritus in patients with reduced kidney function is brought about by an imbalance between the activities of the mostly antagonistic acting μ- and κ-opioid receptors in favor of μ-receptor activation. For this reason, the application of a κ-agonist was recommended for treatment (see Treatment Options section) (Umeuchi et al. 2003).

More recent research points to another etiopathology of CKD-aP, namely, a microinflammation in skin and probably at the systemic level. Proof of elevated CRP values was established in hemodialysis patients with chronic pruritus (Virga et al. 2002; Kimmel et al. 2006); a relative increase in proinflammatory-relevant TH 1-cells and raised IL-6 levels were detected as well by our study group (Kimmel et al. 2006).

5.2.3 TREATMENT OPTIONS

Therapeutic options in CKD-aP are limited. Most studies on this subject are not truly valid as a result of inadequate documentation of the basics, of concomitant diseases and therapeutic measures taken, and of very low case numbers. The most consequential approaches to treatment are presented herewith:

- Topical treatment
- Systemic treatment with μ-opioid receptor antagonists and κ-agonists
- Gabapentin
- Drugs with an anti-inflammatory action
- Ultraviolet phototherapy
- Acupuncture

5.3 TOPICAL TREATMENT

5.3.1 TACROLIMUS OINTMENTS

Treatment with tacrolimus-containing unguents or creams for skin lesions in atopic dermatitis have in a number of cases resulted in complete healing and remission of pruritus (Gianni and Sulli 2001). Back in the late 1990s, we had three patients with particularly bothersome CKD-aP, and within the framework of an individual treatment trial, we gave them a 0.03% tacrolimus-containing unguent that was to be applied twice daily. The patients were also instructed to keep an "itch diary" with daily documentation of the intensity of their pruritus (visual analog scale 1–10) to define the degree of itching 1 week before, during and after treatment. Two of these patients felt they were completely cured, and the third patient reported substantial relief (reduction of VAS score from 7 to 3).

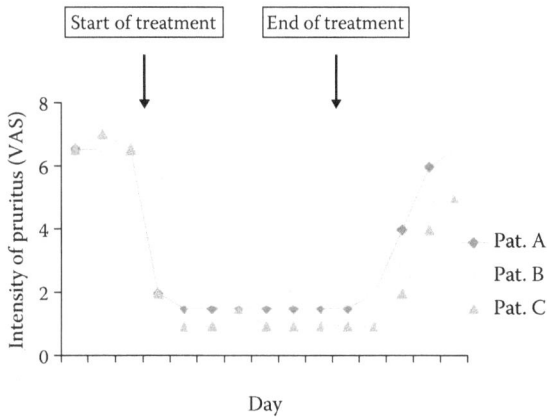

FIGURE 5.4 **(See color insert.)** Course of symptoms on treatment with tacrolimus oint-ment 0.03% in three patients with intractable uremic itch. (After Pauli-Magnus, C. et al., *Perit. Dial. Int.*, 6, 802–803, 2000a.)

Unfortunately, pruritus relapsed in all three of them right after the termination of treatment. Major side effects under this short-term treatment were not observed (Pauli-Magnus et al. 2000a) (Figure 5.4).

A few years later, Kuypers and coworkers initiated a "proof of concept" study enrolling 25 patients afflicted with CKD-aP and were able to prove the efficacy of tacrolimus ointments (Kuypers et al. 2004). A visual analog scale (VAS) was used again for the assessment of pruritus and the prospective outcome of treatment. At the end of the 6 weeks course of treatment, they found a mean reduction of itching by 42.9%.

The results of a double-blind, placebo-controlled study conducted in the US on 22 hemodialysis patients with pruritus (Duque et al. 2005) are difficult to interpret. Treatment with both the active agent as well as vehicle turned out to be highly effec-tive (itch was reduced by 80% versus baseline). A difference between tacrolimus and vehicle could not be demonstrated. However, the authors were at loss to explain the surprising result that the vehicle itself was highly effective. It should be noted that serious adverse reactions were not observed in this study.

5.3.2 UNGUENTS CONTAINING GAMMA LINOLENIC ACID

In a study published a few years ago, Chen and coworkers tested the efficacy of topi-cal treatment with gamma linolenic acid at high concentrations (essential fatty acid that also contains various plant seeds, evening primrose oil, etc.).

Seventeen patients with severe refractory CKD-aP were studied for 6 weeks in a crossover study (washout period 2 weeks). A VAS (score of 0–100) was used. The intensity of pruritus could be brought down from 75 to 30 when the active agent was applied, whereas the pruritus scores merely dropped from 72.5 to 67.5 in the patients receiving placebo (Chen et al. 2006). The mechanism of action of this therapeu-tic approach remains obscure. It can be speculated that gamma linolenic acid may serve as a "prodrug" in the formation of anti-inflammatory-acting prostaglandins. Considering that no side effects were encountered, this substance might be a valuable add-on in the treatment of patients with CKD-aP.

5.4 μ-OPIOID RECEPTOR ANTAGONISTS

5.4.1 NALTREXONE

In a double-blind crossover study, Peer et al. (1996) were able to demonstrate that one week of treatment with the orally available μ-receptor antagonist naltrexone had brought on nearly complete remission of itching in 15 patients with severe hemodialysis-related pruritus. Subsequently, our group tried to reproduce these results in a similarly designed study (placebo controlled, double blind, crossover trial). Twenty-three patients with CKD-aP in whom other approaches had failed thus far received 50 mg naltrexone or placebo for a total of 4 weeks. Sixteen of these 23 patients completed the study. During the naltrexone treatment period, pruritus was eased by 29% (according to ratings using a VAS of 1–10) and by 17.6% according to a pruritus score (based on a questionnaire focusing on the frequency, intensity, and nocturnal occurrence of itching). Compared to this, pruritus declined by 16.9% (VAS) and by 22.3% (pruritus score) during the period of placebo treatment (Pauli-Magnus et al. 2000b). The differences between the naltrexone and the placebo periods were statistically insignificant for both VAS ratings and pruritus scores.

The results of these two studies are contradictory and confusing. The disparate results are neither explained by dissimilarities in compliance, the doses of naltrexone, or the study design, which was crossover, double blind, and placebo controlled in both cases. The studies merely differed as to the intensity of pruritus in the evaluated groups. Whereas the trial by Peer and coworkers exclusively concentrated on patients most seriously afflicted with pruritus (mean VAS 10), the mean intensity of pruritus in our patients was found to be VAS 6. This and other differences, for instance, regarding hemodialysis treatment, the material used, divergent lifestyle, eating habits, and environmental circumstances in various parts of the world may well have been involved in producing these contradictory results.

5.5 κ-OPIOID RECEPTOR AGONISTS

Because κ-opioid receptors primarily mediate μ-opioid receptor-antagonistic effects and because κ-agonistic drugs are known to suppress morphine-induced itch, the assumption that these substances might be capable of alleviating pruritus did not seem far-fetched (Umeuchi et al. 2003).

5.5.1 NALFURAFINE

A meta-analysis of two randomized double-blind and placebo-controlled studies on a total of 144 hemodialysis patients with CKD-aP corroborated the antipruritic effect of nalfurafine, a potent κ-receptor agonist. The substance was administered as a short infusion, 3 times weekly for a period of 4 weeks following hemodialysis. The rate of side effects was low. Compared to placebo, a significant effectiveness could be confirmed (Wikström et al. 2005).

In a randomized prospective placebo-controlled phase III study with a total of 337 hemodialysis patients with pruritus, a Japanese work group applied nalfurafine hydrochloride orally, the dose being 2.5 and/or 5 μg daily over 2 weeks (Kumagai et al. 2010). The effects on itching were registered by a horizontal VAS scale from 0 to 100 mm. On treatment with the test substance, the intensity of pruritus decreased by 22 (5 μg) and 23 mm (2.5 μg), respectively, after 7 days of application, while the intensity of itching dropped by only 13 mm in the placebo group. The differences were statistically significant. However, the incidence of undesired drug actions (insomnia in particular) was substantially higher in both treatment groups (35.1% with 5 μg and 25% with 2.5 μg) than in the placebo group (16.2%). Moreover, the effect of medication wore off quickly once treatment was completed.

5.6   OTHER MEDICAL TREATMENT OPTIONS

5.6.1   PENTOXIPHYLLINE

Pentoxiphylline is a weak TNF-alpha inhibitor; its immunomodulating action has been verified in various studies.

Under the presumption that CKD-aP had a (micro-)inflammatory origin, we infused seven hemodialysis patients exhibiting pronounced pruritus with 600 mg pentoxiphylline for 4 weeks, subsequent to each hemodialysis treatment that they underwent (Mettang et al. 2007). Only three out of these seven patients were able to tolerate treatment for the entire duration of the study. Because of side effects or concomitant development of other illnesses, four patients withdrew from the trial. The patients who completed the entire course of therapy reported considerable amelioration of itching.

This effect was sustained until several weeks after the end of treatment. Considering the rather modest tolerance of the agent (at least with the dose chosen), this approach can only be recommended in exceptional cases.

5.6.2   THALIDOMIDE

Thalidomide is a substance used for immune modulation in the treatment of graft versus host reactions after bone marrow transplantation or in the treatment of multiple myeloma, and which—among other things—is capable of reducing the lymphocytic TNF-alpha production. It could be shown that thalidomide also helps in the management of CKD-aP (Silva et al. 1994).

In a placebo-controlled randomized double-blind crossover study on the management of CKD-aP resistant to treatment, thalidomide proved to lower the pruritus scores by approximately 50%. The precise mechanism of action of this substance has not yet been elucidated. Aside from the previously described suppression of TNF-alpha production, a central abating effect might be involved as well (Daly and Shuster 2000).

5.6.3 GABAPENTIN

The substance gabapentin was originally developed as an anticonvulsant. Researchers later discovered that this agent exerts a pain-modulating effect, especially in patients with diabetic neuropathy. In a randomized, placebo-controlled study with hemodialysis patients, Gunal et al. (2004) showed that the 3 times weekly oral application of 300 mg gabapentin led to a dramatic diminution of pruritus in these patients (reduction of the intensity of itch, determined by a VAS of 0–10, from 8.4 prior to treatment to 1.2 at the end of the 4 weeks' study period).

Razeghi et al. (2009) performed another double-blind crossover study with 34 hemodialysis patients who received 100 mg gabapentine 3 times weekly with similar results.

The rate of side effects in these two studies was trivial. Against this background, gabapentin appears to be a highly interesting, promising substance for the management of CKD-aP and might come to be one of the most important tools in the combat against pruritus in patients on dialysis.

5.6.4 ULTRAVIOLET PHOTOTHERAPY

Back in the 1970s, Gilchrest et al. (1979) had already reported on the efficacy of UV-phototherapy in patients with CKD-aP. Whereas the treatment by long-wave ultraviolet radiation remained largely ineffective, UVB radiation resulted in significantly improved itching in nine out of 10 patients.

A series of studies followed suit to determine the efficacy of UV radiation in patients with CKD-aP. According to a meta-analysis by Tan et al. (1991), UVB therapy was the most encouraging mode of treatment, whereas UVA therapy turned out to be a failure. More recent research suggests that narrowband UVB radiation is just as effective as treatment with broadband UV rays, with less frequent side effects (Baldo et al. 2002). In a recently published study, however, this effect could not be verified (Ko et al. 2011).

The fear that UVB radiation enhances the development of malignant skin tumors could not be totally dispelled. Some caution is thus required in the treatment of patients who are scheduled on long-term immunosuppressive therapy, e.g., after kidney transplantation.

5.6.5 ACUPUNCTURE

A very interesting approach to treat CKD-aP is acupuncture. In a study by Duo (1987), electro-acupuncture or sham-electro-stimulation was applied to six patients on hemodialysis in a blinded manner. Patients receiving acupuncture showed a significantly higher reduction in pruritus score than the sham-treated patients. In another study a few years ago, 40 patients with CKD-aP were treated with acupuncture either at the Quchi (LI11) acupoint or at a nonacupoint 2 cm lateral, three times weekly for 1 month. Patients treated at the Quchi acupoint revealed a substantial reduction in pruritus score regarding severity, distribution, and sleep disturbance ending up with maximum 45 points (38.3 ± 4.3, 17.3 ± 5.5, and 16.5 ± 4.9 start versus 4 weeks versus

12 weeks later), whereas pruritus in patients receiving sham-acupuncture did not change substantially (38.3 ± 4.3, 37.5 ± 3.2, and 37.1 ± 5 start versus 4 weeks versus 12 weeks later) (Che-yi et al. 2005).

Given these results, acupuncture in experienced hands might be a useful tool in the treatment of CKD-aP.

Figure 5.5 shows a potential therapeutic algorithm of CKD-aP. Considering that pruritus is generally not immediately life-threatening and that its treatment is often associated with considerable adverse reactions, the decision should always be carefully balanced as to benefits and risks. Preference should be given to therapeutic procedures with a favorable or benign profile of side effects. In desperate cases, patients principally eligible for a kidney transplant may be declared "high urgency," which will decrease their waiting time. In most cases, a successful kidney transplantation will stop CKD-aP in patients treated (Altmeyer et al. 1986).

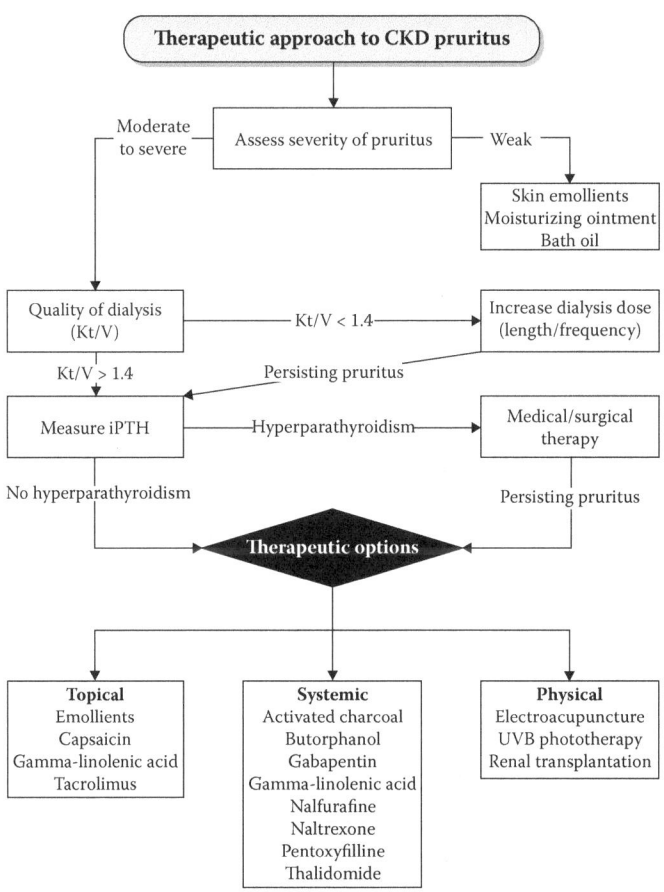

FIGURE 5.5 Therapeutic algorithm in uremic pruritus: Kt/V, urea clearance in relation to urea distribution volume; iPTH, intact parathormone. (After Mettang, T. and Weisshaar, E., *Skin Therapy Lett.*, 15, 1–5, 2010.)

REFERENCES

Altmeyer, P., Kachel, H. G., Schäfer, G., and Faßbinder, W. (1986). Normalisierung der urämischen Hautveränderungen nach Nierentransplantation. *Hautarzt* 37: 217–221.

Balaskas, E. V., and Uldall, R. P. (1992). Erythropoietin does not improve uremic pruritus. *Perit. Dial. Int.* 12: 330–331.

Baldo, A., Sammarco, E., Plaitano, R., Martinelli, V., and Monfrecola, G. (2002). Narrowband (TL-01) ultraviolet B phototherapy for pruritus in polycythaemia vera. *Br. J. Dermatol.* 147: 979–981.

Blachley, J. D., Blankenship, D. M., Menter, A., Parker, T. F., III, and Knochel, J. P. (1985). Uremic pruritus: Skin divalent ion content and response to ultraviolet phototherapy. *Am. J. Kidney Dis.* 5: 237–241.

Chen, Y. C., Chiu, W. T., and Wu, M. S. (2006). Therapeutic effect of topical gamma-linolenic acid on refractory uremic pruritus. *Am. J. Kidney Dis.* 48: 69–76.

Che-yi, C., Wen, C. W., Min-Tsung, K., and Chiu-Ching, H. (2005). Acupuncture in haemodialysis patients at the Quchi (LI11) acupoint for refractory uraemic pruritus. *Nephrol. Dial. Transplant* 20: 912–915.

Daly, B. M., and Shuster, S. (2000). Antipruritic action of thalidomide. *Acta Derm. Venereol.* 80(1): 24–25.

Dawn, A. G., and Yosipovitch, G. (2006). Butorphanol for treatment of intractable pruritus. *Am. Acad. Dermatol.* 54: 527–531.

De Marchi, S., Cecchin, E., Villalta, D., Sepiacci, G., Santini, G., and Bartoli, E. (1992). Relief of pruritus and decreases in plasma histamine concentrations during erythropoietin therapy in patients with uremia. *New Engl. J. Med.* 326: 969–974.

Duo, L. J. (1987). Electrical needle therapy of uremic pruritus. *Nephron* 47(1987): 179–183.

Duque, M. I., Yosipovitch, G., Fleischer, A. B., Willard, J., and Freedman, B. I. (2005). Lack of efficacy of tacrolimus ointment 0.1% for treatment of hemodialysis-related pruritus: A randomized, double-blind vehicle-controlled study. *J. Am. Acad. Dermatol.* 52: 519–521.

Gianni, L. M., and Sulli, M. M. (2001). Topical tacrolimus in the treatment of atopic dermatitis. *Ann. Pharmacother.* 35: 943–946.

Gilchrest, B. A., Rowe, J. W., Brown, R. S., Steinman, T. I., and Arndt, K. A. (1979). Ultraviolet phototherapy of uremic pruritus. Long-term results and possible mechanisms of action. *Ann. Intern. Med.* 91: 17–21.

Gilchrest, G. A., Stern, R. S., Steinman, T. I., Brown, R. S., Arndt, K. A., and Anderson, W. W. (1982). Clinical features of pruritus among patients undergoing maintenance hemodialysis. *Arch. Dermatol.* 118: 154–156.

Gunal, A. I., Ozalp, G., Yoldas, T. K., Gunal, S. Y., Kirciman, E., and Celiker, H. (2004). Gabapentin therapy for pruritus in haemodialysis patients: A randomized, placebo-controlled, double-blind trial. *Nephrol. Dial. Transplant* 19: 3137–3139.

Hampers, C. L., Katz, A. I., Wilson, R. E., and Merrill, J. P. (1968). Disappearance of uremic itching after subtotal parathyroidectomy. *New Engl. J. Med.* 279: 695–697.

Kimmel, M., Alscher, D. M., Dunst, R., Braun, N., Machleidt, C., Kiefer, T., Stülten, C. et al. (2006). The role of micro-inflammation in the pathogenesis of uraemic pruritus in haemodialysis patients. *Nephrol. Dial. Transplant* 21: 749–755.

Ko, M. J., Yang, J. Y., Wu, H. Y., Hu, F. C., Chen, S. I., Tsai, P. J., Jee, S. H., and Chiu, H. C. (2011). Narrowband ultraviolet B phototherapy for patients with refractory uremic pruritus: A randomized controlled trial. *Br. J. Dermatol.* 165: 633–639.

Kumagai, H., Ebata, T., Takamori, K., Muramatsu, T., Nakamoto, H., and Suzuki, H. (2010). Effect of a novel kappa-receptor agonist, nalfurafine hydrochloride, on severe itch in 337 haemodialysis patients: A phase III, randomized, double-blind, placebo-controlled study. *Nephrol. Dial. Transplant* 25: 1251–1257.

Kuypers, D. R., Claes, K., Evenepoel, P., Maes, B., and Vanrenterghem, Y. (2004). A prospective proof of concept study of the efficacy of tacrolimus ointment on uraemic pruritus (UP) in patients on chronic dialysis therapy. *Nephrol. Dial. Transplant* 19: 1895–1901.

Massry, S., Popovzer, M. M., Coburn, J. M., Mokoff, D. L., Maxwell, M. H., and Kleeman, C. R. (1968). Interactable pruritus as a manifestation of secondary hyperparathyroidism in uremia. *New Engl. J. Med.* 279: 697–700.

Mathur, V. S., Lindberg, J., Germain, M., Block, G., Tumlin, J., Smith, M., Grewal, M., and McGuire, D. (2010). ITCH National Registry Investigators. A longitudinal study of uremic pruritus in hemodialysis patients. *Clin. J. Am. Soc. Nephrol.* 5: 1410–1419.

Mettang, T., Fischer, F. P., and Kuhlmann, U. (1996). Urämischer Pruritus—Pathophysiologische und therapeutische Konzepte. *Dtsch. Med. Wschr.* 121: 1025–1031.

Mettang, T., Fritz, P., Weber, J., Machleidt, C., Hübel, E., and Kuhlmann, U. (1990). Uremic pruritus in patients on hemodialysis or continuous ambulatory peritoneal dialysis (CAPD). The role of plasma histamine and skin mast cells. *Clin. Neprol.* 34: 136–141.

Mettang, T., Krumme, B., Bohler, J., and Roeckel, A. (2007). Pentoxifylline as treatment for uraemic pruritus—An addition to the weak armentarium for a common clinical symptom? *Nephrol. Dial. Transplant* 22: 2727–2728.

Mettang, T., Matterne, U., Roth, H. J., and Weisshaar, E. (2010). Lacking evidence for calcium-binding protein fetuin-A to be linked with chronic kidney disease-related pruritus (CKD-rP). *NDTPlus* 3: 104–105.

Mettang, T., Pauli-Magnus, C., and Alscher, D. M. (2002). Uraemic pruritus—New perspectives and insights from recent trials. *Nephrol. Dial. Transplant* 17: 1558–1563.

Mettang, T., and Weisshaar, E. (2010). Pruritus: Control of itch in patients on dialysis. *Skin Therapy Lett.* 15: 1–5.

Morvay, M., and Marghescu, S. (1988). Hautveränderungen bei Haemodialysepatienten. *Med. Klin.* 83: 507–510.

Pauli-Magnus, C., Klumpp, S., Alscher, D., Kuhlmann, U., and Mettang, T. (2000a). Short-term efficacy of tacrolimus ointment in severe uremic pruritus. *Perit. Dial. Int.* 6: 802–803.

Pauli-Magnus, C., Mikus, G., Alscher, D. M., Kirschner, T., Nagel, W., Gugeler, N., Risler, T., Berger, E. D., Kuhlmann, U., and Mettang, T. (2000b). Naltrexone does not relieve uremic pruritus: Results of a randomized, placebo-controlled crossover-study. *J. Am. Soc. Nephrol.* 11: 514–515.

Peer, G., Kivity, S., Agami, O., Fireman, E., Silverberg, D., Blum, M., and Iaina, A. (1996). Randomised crossover trial of naltrexone in uraemic pruritus. *Lancet* 348: 1552–1554.

Pisoni, R. L., Wikström, B., Elder, S. J., Akizawa, T., Asano, Y., Keen, M. L., Saran, R., Mendelssohn, D. C., Young, E. W., and Port, F. K. (2006). Pruritus in haemodialysis patients: International results from the Dialysis Outcomes and Practice Patterns Study (DOPPS). *Nephrol. Dial. Transplant* 21: 3495–3505.

Ponticelli, C., and Bencini, P. L. (1992). Uremic pruritus: A review. *Nephron* 60: 1–5.

Razeghi, E., Eskandari, D., Ganji, M. R., Meysamie, A. P., Togha, M., and Khashayar, P. (2009). Gabapentin and uremic pruritus in hemodialysis patients. *Ren. Fail.* 31: 85–90.

Schwab, M., Mikus, G., Mettang, T., Pauli-Magnus, C., Kuhlmann, U., and Arbeitsgemeinschaft für Pädiatrische Nephrologie (1999). Urämischer Pruritus im Kindes- und Jugendalter. *Monatszeitschrift Kinderheilkunde* 147: 232.

Silva, S. R., Viana, P. C., Lugon, N. V., Hoette, M., Ruzany, F., and Lugon, J. R. (1994). Thalidomide for the treatment of uremic pruritus: A crossover randomized double-blind trial. *Nephron* 67(3): 270–273.

Stahle-Bäckdahl, M., Hägermark, O., Lins, L. E., Törring, O., Hilliges, M., and Johansson, O. (1989). Experimental and immunohistochemical studies on the possible role of parathyroid hormone in uremic pruritus. *J. Intern. Med.* 225(1989): 411–415.

Stockenhuber, F., Kurz, R. W., Sertl, K., Grimm, G., and Balcke, P. (1990). Increased plasma histamine in uremic pruritus. *Clin. Sci.* 79(5): 477–482.

Szepietowski, J. C., Reich, A., and Schwartz, R. A. (2004). Uraemic Xerosis. *Nephrol. Dial. Transplant* 19: 2709–2712.

Tan, J. K. L., Haberman, H. F., and Coldman, A. J. (1991). Identifying effective treatments for uremic pruritus. *J. Am. Acad. Dermatol.* 25: 811–824.

Tessari, G., Dalle Vedove, C., Loschiavo, C., Tessitore, N., Ruguiu, C., Lupo, A., and Girolomoni, G. (2009). The impact of pruritus on the quality of life of patients undergoing dialysis: A single centre cohort study. *J. Nephrol.* 22(2): 241–248.

Umeuchi, H., Togashi, Y., Honda, T., Nakao, K., Okano, K., Tanaka, T., and Nagase, H. (2003). Involvement of central mu-opioid system in the scratching behavior in mice, and the suppression of it by the activation of κ-opioid system. *Eur. J. Pharmacol.* 477: 29–35.

Virga, G., Visentin, I., La Milia, V., and Bonadonna, A. (2002). Inflammation and pruritus in haemodialysis patients. *Nephrol. Dial. Transplant* 17: 2164–2169.

Weisshaar, E., Dunker, N., Röhl, F. W., and Gollnick, H. (2004). Antipruritic effects of two different 5-HT3 receptor antagonists and an antihistamine in haemodialysis patients. *Exp. Dermatol.* 13: 298–304.

Weisshaar, E., Matterne, U., and Mettang, T. (2009). How do nephrologists in haemodialysis units consider the symptom of itch? Results of a survey in Germany. *Nephrol. Dial. Transplant* 24: 1328–1330.

Wikström, B., Gellert, R., Ladefoged, S. D., Danda, Y., Akai, M., Ide, K., Ogasawara, M. et al. (2005). Kappa-opioid system in uremic pruritus: Multicenter, randomized, double-blind, placebo-controlled clinical studies. *J. Am. Soc. Nephrol.* 16: 3742–3747.

Young, A. W., Sweeney, E. W., David, D. S., Cheigh, J., Hochgelerent, E. L., Sakai, S., Stenzel, K. H., and Rubin, A. L. (1973). Dermatologic evaluation of pruritus in patients on hemodialysis. *N. Y. State J. Med.* 73: 2670–2674.

6 Pruritus of Cholestasis

Nora V. Bergasa

CONTENTS

Cholestasis is defined as impaired secretion of bile.[1] Pruritus is a complication of cholestasis including that associated with mutations in genes that code for transporters in the hepatocyte[2,3] and from inflammatory liver diseases.[4,5] In this regard, pruritus is more common in conditions characterized by bile duct inflammatory destruction and ductopenia including primary bile cirrhosis primary biliary cirrhosis (PBC)[5] than in those characterized by hepatocellular injury such as chronic viral hepatitis.[4] The pruritus of cholestasis tends to be generalized. It leads to scratching, sometimes violent, resulting in excoriations and prurigo nodularis. This type of pruritus can lead to sleep deprivation, and in some patients, to suicidal ideations. Intractable pruritus from liver disease is an indication for liver transplantation even in the absence of liver failure.[6,7] Accordingly, pruritus is a complication of liver disease that requires specific management and intense research in an effort to design effective antipruritic medications.

Women with cholestasis of pregnancy can experience severe pruritus especially in the third trimester.[8] The pruritus from cholestasis of pregnancy resolves itself after delivery; if it does not, investigations to rule out liver disease are indicated.

In patients with PBC the pruritus has not been consistently reported to correlate with biological markers of disease.[9] Also, in PBC the presence of intrahepatic florid bile duct lesions and granulomata were significantly related to the severity of pruritus (and fatigue), which was assessed subjectively, on the day of the biopsy.[10] This observation is interesting because, although it is difficult to infer the degree of cholestasis from a liver biopsy, as there is sampling error, the fact that patients with stage I PBC can experience severe pruritus may suggest an important release of pruritogens in association with active inflammatory destruction of biliary epithelial cells, which characterizes this stage. The pruritus can persist and remit throughout the course of the disease, and as the disease progresses to liver failure, the pruritus tends to cease,[11] as if a certain degree of liver function were necessary for pruritogen(s) or its cofactors to be produced or for the sensation to be perceived. In relation with PBC, it was also reported that presentation of the disease at less than 50 years of age correlated with an increased likelihood of reporting symptoms (and not responding to ursodeoxycholic acid, the drug approved for the treatment of this disease).[12]

An Internet survey via the Web site supported by the PBCers organization (http://pbcers.org/) was conducted to understand how patients perceived pruritus.[13] Two hundred and thirty-nine (239) subjects with a diagnosis of PBC responded to the survey; of these, 232 were women, and 69% of all subjects reported itch. Seventeen percent of the respondents reported that the itch was "relentless" or so severe that it led to wanting to "tear (the) skin off," and 3.6% of the subjects stated that they scratched until they bled. Seventy-four percent of the 162 respondents who addressed the question reported that the itch affected sleep, 65% that the itch was worse at night, and 11% reported that nothing relieved the itch.[13] The minority of respondents to the survey reported that the pruritus was addressed by their physician;[13] this underscores the need to educate physicians and patients on the association between pruritus and liver disease, and the importance of investigating the cause of pruritus in patients who do not have an apparent skin condition that can explain the itch.

The prevalence of pruritus in patients with PBC determined from the survey reported above[13] was 69%, close to what has been reported by others.[14] From a retrospective study from a tertiary referral research institute in the United States, 5% of patients with chronic hepatitis C presented with pruritus;[4] considering the high prevalence of this type of liver disease, the relevance of the pruritus of cholestasis in medicine is substantial. This inference is supported by a recent publication, also retrospective in nature, from one center in Europe reporting that 40% of the patients with chronic pruritus from liver disease had chronic viral hepatitis B or C.[15]

6.1 PATHOGENESIS OF THE PRURITUS OF CHOLESTASIS

It has been inferred that the pruritogen(s) that mediates the pruritus is made in the liver and excreted in bile and that, as a result of cholestasis, accumulates in body tissues and by some mechanism triggers the sensation of itch. The following observations support a liver origin of the pruritogen(s): (1) in patients with cholestasis, the cessation of itch portends liver failure,[11] (2) patients with pruritus report the disappearance of their symptom after liver transplantation,[16] and (3) relief of cholestasis from mechanical obstruction (e.g., common duct stones) is associated with decrease or disappearance of pruritus.[17] The nature of the pruritogen, however, is unknown.

Bile acids accumulate in the tissues of patients with cholestasis.[11,18] Under experimental conditions, bile acids were reported to trigger local "itch" when injected intracutaneously in normal volunteers;[19] this, however, is not a model of the pruritus of cholestasis. The oral administration of cholylsarcosine, a synthetic bile acid, to four patients with PBC was reported to be associated with pruritus in one patient and worsening pruritus in another.[20] As pruritus is intermittent in cholestasis, this observation cannot be interpreted as evidence in support of a role of bile acids in the pruritus of cholestasis because pruritus tends to be intermittent; in addition, the intake of certain foodstuffs, usually patient specific, is sometimes reported to increase or to worsen the pruritus. Three observations do not support a role of bile acids in the mediation of the pruritus of cholestasis: (1) in liver failure, when bile acids are maximally elevated, pruritus tends to disappear,[11] (2) not all patients with cholestasis report pruritus, in spite of marked elevations of serum bile acids, and (3) pruritus can fluctuate independently from the serum concentration of bile acids. It is possible, however, that a certain profile of serum bile acids is necessary for these substances to mediate the pruritus. A recent clinical study published in abstract form reported that the administration of obeticholic acid, in contrast to the placebo drug, to patients with PBC was associated with pruritus, was severe in some patients, for which the drug had to be stopped, and in others sufficiently inconvenient to require dose reduction. Obeticholic acid is a synthetic derivative of chenodeoxycholic acid that is an agonist at the farsenoid nuclear receptor (FXR) and that has choleretic properties.[21] FXR is a bile acid sensor associated with a decrease in bile acid production.[22] The relevance of this observation in the pathogenesis of the pruritus of cholestasis is unknown.

Histamine was reported to be increased in patients with liver disease and pruritus;[23] indeed, 16% of patients with PBC who participated in the Internet survey conducted through the PBCers[13] reported that hydroxyzine, an antihistamine, was frequently prescribed to treat their pruritus, and provided some relief in some patients.[13] The skin of patients with cholestasis and pruritus is devoid of the classic histamine mediated reaction consisting of erythema and edema, which does not support a major role of this pruritogen in the mediation of this type of pruritus; however, the sedative effect of antihistamines, and not an antipruritic effect, may be responsible for some of the relief in association with this type of drugs.

Substance P is an excitatory neurotransmitter that acts through the NK-1 receptor synthesized by primary afferent nociceptors and released into the spinal cord after noxious stimuli.[24] The central administration of substance P is associated with scratching behavior.[25] In addition, in animal studies, increased opioidergic tone secondary to the administration of morphine activates mechanisms that promote pain (i.e., nociception), instead of analgesia, mediated, in part, by the NK-1 receptor;[26] furthermore, the expression of substance P in the dorsal root ganglia, which are involved in the transmission of nociceptive stimuli,[24] is increased in association with prolonged administration of opiates.[24] Analogous to the activation of mechanisms that promote pain mediated by substance P associated with chronic opiate administration, the increased opioidergic tone, which is associated with cholestasis,[27] may also contribute to a state of enhanced nociception that may be perceived as pruritus mediated, in part, by substance P. In this context, the mean serum concentration of substance P was 12-fold higher in patients with liver disease and pruritus than in patients with liver disease without pruritus, and that of a control group of subjects.[28] These data suggest that substance P may mediate some of the manifestations of liver disease, including pruritus; however, in primates, which seem to respond to the central administration of pruritogens similarly to human beings, the administration of substance P was not associated with scratching.[29]

In animal studies, it has been reported that lipophosphatidic acid (LPA) induces nociception through substance P release from peripheral nerve endings,[30] and it has been implicated as a mediator in animal and in *in vitro* neural models that explore signals associated with pain.[31–33] A role of LPA and the enzyme that generates its production, autotaxin, has been recently proposed in the pruritus of cholestasis.[34,35] The concentration of several LPA species was reported to be significantly higher in the sera from patients with cholestasis of pregnancy than in the sera from pregnant women matched for gestation term.[34] In C57BL/6J female mice, LPA injections were reported to cause scratching behavior, which was interpreted as evidence for a pruritogenic role of LPA in cholestasis;[34] however, this finding is not relevant in the study of the pruritus of cholestasis as scratching behavior in association with the intradermal injection, as given in this study, of substances, including LPA, is not a model of scratching in cholestasis.[34] The activity of autotaxin was reported to be higher in the serum from women with cholestasis of pregnancy, and that of patients with cholestasis and pruritus from liver diseases and not in the serum of patients with pruritus from uremia, Hodgkin's "disease," and atopic dermatitis.[35] The serum activity of autotaxin was also reported to have decreased in the serum of patients with liver disease and pruritus who had responded to the antibiotic rifampicin.[35] Autotaxin activity was not decreased in the serum from patients with cholestasis and pruritus who had participated in a controlled trial of colesevalan, a nonabsorbable resin that binds bile acids in the gut, versus placebo, in which the resin was reported not to relieve pruritus. The activity of autotaxin was also reported to correlate with a decrease in the severe pruritus of patients who had undergone dialysis with Molecular Adsorbent Recirculating System (MARS™), or nasobiliary drainage.[35] Nasobiliary drainage facilitates bile flow, and the MARS™ procedure is associated with removal of vasoactive substances and with important changes in blood flow, which may also increase bile flow;[36] the effects of these two interventions thus, may decrease the degree of cholestasis and hence the pruritus associated with it. Autotaxin was not found in the bile of patients treated with nasobiliary drainage and who had responded to the treatment with a decrease in pruritus, although it was reported to have returned to pretreatment serum levels when the pruritus recurred.[35] The changes in the activity of autotaxin in relation with improvement or worsening of the pruritus may reflect changes in the degree of cholestasis and do not necessarily imply a pruritogenic effect of this enzymatic pathway. The antibiotic rifampicin induces drug-metabolizing enzymes and transporters, through activation of the pregnane X receptor (PXR).[37,38] The idea that the reported antipruritic effect of this drug results from either the enhanced metabolism of the "pruritogens" or its transport has been suggested.[39] The inhibition of the expression of autotaxin by rifampicin *in vitro*, in HepG2 hepatoma cells, and in hepatoma cells overexpressing PXR, and not on cells in which PXR had been knocked out, was interpreted to suggest that a decrease in autotaxin by rifampicin may be the mechanism by which this antibiotic decreases pruritus in patients with cholestasis;[40,41] however, the lack of effect of rifampicin in some patients was not explained in this context.[42] The effect of rifampicin *in vitro* as measured in this study[35] cannot be interpreted in the context of the pruritus of cholestasis as the model does not reflect the cholestatic milieu in human beings. It was concluded from these studies that autotaxin may be a therapeutic target for the treatment of the pruritus of liver disease.[35] Whether autotaxin or LPA mediate the pruritus of cholestasis and the potential neurophysiologic and/or neuropathophysiologic mechanisms by which these substance may mediate pruritus are unknown.

The landmark publication by Thornton and Loswoski in 1988 focused on the role of the endogenous opioid system in the clinical manifestations of patients with liver disease by assessing their response to the opiate antagonist nalmefene.[43] It was reported that the administration of nalmefene to patients with cholestasis was associated with an opiate withdrawal-like reaction; in addition, the patients experienced a disappearance of their pruritus.[43] The reaction included high blood pressure, insomnia, and abdominal pain, in sharp contrast to the absence of any reaction in normal volunteers who, in other studies, had taken 60 times the dose of nalmefene than that used in this study.[43] The reaction that the patients experienced after the intake of nalmefene was interpreted to result from cerebral dependence to opioids due to their high serum concentration, which allowed them to cross the blood brain barrier. The relief of the pruritus was interpreted to result from the opiate antagonist inhibiting opioid peptides from causing pruritic substances to be liberated.[43] Indeed, the opiate withdrawal-like reaction experienced by the patients with cholestasis suggests that, in the absence to exposure to opiate drugs, they had endogenously increased opioidergic tone.[27] Furthermore, the reported relief of pruritus in association with nalmefene by the patients in this study suggested a relationship between central opioid receptors and pruritus.[27] This alternate interpretation is supported by the naloxone reversible pruritus in human beings[44,45] and the scratching behavior in laboratory animals[46–48] in association with the pharmacological increase in opioidergic tone by morphine and other drugs with agonist activity at the opioid receptors. These observations provided support for a hypothesis: *the pruritus of cholestasis is mediated by increased central opioidergic tone.*[27] Three lines of evidence suggest that in cholestasis there is increased opioidergic tone: (1) the opiate withdrawal like reaction triggered by an opiate antagonist in patients with cholestasis,[43,49–52] (2) the state of stereospecific, opiate antagonist reversible antinociception in an animal model of cholestasis,[53] and (3) the downregulation of mu (and delta) opioid receptor in brain membranes of rats with cholestasis.[54] The reason for increased opioidergic tone is cholestasis is unknown; however, several findings provide potential explanations: (1) the *de novo* production of endogenous opioids by the liver in a rat model of cholestasis is suggested by the expression of the mRNA of the gene that codes for Met and Leu-enkephlins and for enkephalin containing endogenous opioids, preproenkephalin mRNA,[55] (2) the concentration of two pentapeptide endogenous opioids, Met-enkephalin and Leu-enkephalin, and that of Met-enkephalin containing opioids is significantly higher in liver extracts from rats with cholestasis secondary to bile duct resection than from those of sham resected controls,[56] (3) the serum concentration of Met-enkephalin is increased in animal models of cholestasis[57] and in some patients with liver disease,[43,58] and (4) endogenous opioids accumulate in the liver of patients with liver disease including PBC and chronic hepatitis C, as evidenced by the expression of hepatic

Met-enkephalin immunoreactivity[59,60] and by its detection in bile.[61] One question that arises is whether the enhanced expression of Met-enkephalin immunoreactivity in liver disease is related in any way to the increased central opioidergic tone. In this context, some transport proteins that can transport opiates *in vitro* are common to the basolateral domain of the hepatocyte, to the choroid plexus, and to the blood–brain barrier;[62] accordingly, it is plausible that they can also transport periphery-derived opioids into the central nervous system (CNS). Furthermore, it has been reported that increased availability of opioid peptides in the periphery may facilitate their entrance into the CNS.[63] The hypothesis that increased opioidergic tone contributes to the pruritus of cholestasis,[27] however, does not depend on serum levels of endogenous opioids including Met-enkephalin, or their hepatic expression; it is, in contrast, supported by (1) the opiate withdrawal-like reaction that can be precipitated by opiate antagonists,[43,49–52] and (2) the ameliorating effect of opiate antagonists on the pruritus of cholestasis.[43,49–52,64–68] Indeed, in short-term studies, opiate antagonists were associated with a decrease in hourly scratching activity in an average of 89% of the patients (range, 82.8%–100%).[49,50,64,65] Opiate antagonists were also associated with a decrease in the visual analogue scale for pruritus in an average of 91.5% of patients from a series of clinical studies (range, 65.5%–100%).[43,49–51,64–67] In a placebo-controlled study, the opiate antagonist naltrexone was associated with a 50% reduction in the visual analogue scale for pruritus in 13 of 34 patients (38%);[52] although changes smaller than 50% were not discernible from the publication.[52] The results of a meta-analysis that included five trials of opiate antagonists; three that tested the effect of opiate antagonists administered orally (i.e., naltrexone and nalmefene) and two that tested the effects of intravenous naloxone with a reported total of 84 participants also support an ameliorating effect of this type of drug on the pruritus of cholestasis.[69] Opiate antagonists were associated with a decrease in pruritus more significantly than the control intervention, with a standard mean difference (SMD) of -0.68, 95% -1.19 to -0.17, and to decrease scratching activity with a SMD of -0.64, 95% -1.28 to 0.01.[69] Thus, increased opioidergic tone contributes to the pathogenesis of the pruritus of cholestasis and opiate antagonists are available to treat this type of pruritus.[70]

6.2 INDIVIDUAL PREDISPOSITION TO EXPERIENCE PRURITUS BY PATIENTS WITH CHOLESTASIS

Not all patients with cholestasis report pruritus;[4,13] this observation may suggest a genetic predisposition to experience this sensation in the context of cholestasis. In this regard, if the liver is the source of the pruritogen(s) and the pruritogen(s) is excreted in bile, the transporters that participate in the transport of compounds into the biliary canalicula may be relevant on how these substances are excreted. MRP2 (ABCC2) is a member of the family of ATP binding cassette (ABC) transporters expressed in various organs including the liver and the blood–brain barrier.[71,72] MRP2 mediates the transport of several organic anions[73] in the hepatocyte; in addition, and pertinent to the pruritus of cholestasis, opioid ligands are MRP2 substrates.[74] A single nucleotide polymorphism (SNP) characterized by the substitution of valine for glutamate (V1188E) in exon 25 of the MRP2 gene was found in heterozygosity in a group of patients with PBC.[75] V1188E was significantly associated with the presence of pruritus; thus, V1188E in MRP2 may alter the pruritogen(s) or its cofactors from being transported from the hepatocyte into the bile canalicula or into the central nervous system and may favor the experience of pruritus.

The manner in which the endogenous opioid system responds to opioid ligands may also be relevant in the way the pruritus of cholestasis is perceived. The stimulation of the mu opioid receptor is the mechanism by which opiate (and morphine) induced pruritus is generated. Accordingly, the presence of genetic polymorphisms in the gene that codes for the mu opioid receptor may be relevant in the sensation of itch. In this context, the SNP A118G in exon1 of opioid receptor mu 1 (OPRM1) gene, which predicts an Asn-to-Asp change in amino acid residue 40 in the extracellular domain of the receptor, at a putative N-glycosylation site,[76] has been associated with changes in the behavioral expression from the stimulation of the mu opioid receptor in human beings,[77,78] and changes in the binding profile of at least one endogenous opioid, beta-endorphin, to the receptor.[79] A study from patients with PBC found A118G in heterozygosity 1.5 times more frequently in DNA samples from a group of patients without pruritus from the United States than in the rest of the samples, which included those from patients from Italy.[80] These preliminary findings suggested that patients with the A118G SNP in the OPRM1 may be protected from the perception of itch from cholestasis.[80] Indeed, this interpretation may be supported by the results of a study of the incidence of pruritus in patients after 24 hours of having received epidural morphine for postcesarean section analgesia that revealed that the G allele in the A118G polymorphism was associated with a decrease in the incidence of pruritus with genotype AA being associated with significant pruritus in 53% of patients, AG in 41.9%, and GG in 4.8%.[81] In addition, ethnicity may inform the experience of side effects from the administration of intravenous morphine; in this regard, in a study of children pruritus (and vomiting) was significantly more common in subjects identified as Latino than in subjects identified as non-Latino Caucasian.[82]

The intact skin of patients with atopic dermatitis, a condition characterized by chronic pruritus, allows for the enhanced penetration of chemical solutes.[83] In this regard, mutations in the gene that codes for filaggrin, a protein necessary for the formation of the stratum corneum, and hence instrumental in the maintenance of the skin as a barrier, may be relevant in the pathogenesis of this disease.[83,84] Skin permeability may also be relevant in the vulnerability of patients with liver disease to experience pruritus.

It seems reasonable, accordingly, to conduct genetic studies that may identify mutations in genes including those that code for transport proteins in the hepatobiliary system and CNS, those involved in the activation of sensory neurons such as the phospholipase C B3 pathway,[85] and those that code for proteins that determine skin integrity[83] as certain polymorphisms may inform how patients experience itch in cholestasis.

6.3 RECEPTORS SPECIFICITY IN THE MEDIATION OF OPIOID-INDUCED PRURITUS

There is strong evidence to support that the stimulation of opioid agonist ligands is associated with pruritus and scratching that can be ameliorated and prevented by opiate antagonists in human beings[44,45] and in laboratory animals.[46–48] Studies exploring the role of specific opioid receptors[47,48,86,87] have identified an antiscratching effect of kappa agonists; in this regard, nalfurafine, an agonist at the kappa opioid receptor has been reported to decrease the sensation of pruritus in patients with uremia.[88–90] In this context, scratching behavior in a rat model of estrogen-induced cholestasis was prevented by the administration of nalfurafine, suggesting that stimulation of the kappa opioid receptor may be a therapeutic option in the treatment of the pruritus of cholestasis.[87] In this context, butorphanol, is an antagonist at the mu opioid receptor and an agonist at the kappa receptor.[91] In a patient with intractable pruritus from liver disease secondary to chronic hepatitis C, butorphanol in spray form (1 mg per application) was associated with prolonged relief of pruritus.[92] It was subsequently reported that in a series of patients with PBC, butorphanol had been associated with relief of the pruritus.[93] The potential addiction to butorphanol, initially considered to be low,[94] is important;[95] thus, the use of this drug for the treatment of the pruritus of cholestasis has to be considered carefully, i.e., short courses, in light of addiction complications. Anecdotal reports of the use of codeine, an agonist at the mu opioid receptor,[96] and buprenorphine, an agonist at the mu and an antagonist at the kappa receptor,[97] in association with amelioration of the pruritus of cholestasis have been published. These reports have to be reconciled with the effects of butorphanol in regards to its receptor preference.

Gastrin-releasing peptide receptor (GRPR) was proposed to mediate scratching behavior from pruritogenic stimuli in mice.[98] Morphine-induced scratching behavior does not occur in genetically engineered mice lacking GRPR or the mu opioid receptor (MOR), to which morphine binds. By the use of exon-specific knockdown, an MOR 1 D isoform was identified.[99] MOR 1 D was reported to be essential for scratching behavior, interpreted as a response to itch, whereas the MOR 1 was involved in the mediation of antinociception (i.e., analgesia); no cross-reactivity between the two behaviors, scratching and anticociception, was appreciated.[99] Double immunostaining revealed coexpression of GRPR and of MOR 1 D, defined as constitutive heterodimers; there was no coexpression of GRPR and MOR. The addition of morphine to the experimental preparation was associated with internalization of GRPR and of MOR 1 D; the addition of gastrin releasing peptide was associated with internalization of GRPR but not with the internalization of MOR 1 D or of MOR. The results of these experiments were interpreted as evidence in support of a unidirectional cross-activation of GRPR signaling with MOR 1 D by formation of heterodimers, and also that the scratching due to morphine is an independent effect from the antinociception effect (i.e., analgesia).[99] It is unknown whether these interesting findings relate to the pruritus of cholestasis; however, the revelation of MOR 1 D as the responsible receptor in the mediation of opioid-mediated scratching in the animals studied and its connection with the GRPR may suggest that GRPR may also be relevant in pruritus in human beings and may further support the role of the endogenous opioid system in the pruritus of cholestasis. The demonstration that MOR 1 D can combine with GRPR seems to support the idea that opioid receptors can form complexes that, when activated by specific ligands, mediate synergistic effects, including potentiation of analgesia[100] and, as now suggested,[99] the specific transmission of itch signals. In addition, the combination of opioid receptors with other receptors may be a way by which the binding of nonpruritic substances to the complex may induce pruritus; this possibility may be relevant in cholestasis, as a result of which numerous substances accumulate in various tissue compartments.

6.4   PRURITUS OF CENTRAL ORIGIN

There is strong evidence in support of pruritus of central origin, including the pruritus from increased opioidergic tone.[44,45] It is also considered that the pruritus complicating neurological disease including cerebrovascular accidents[101,102] and multiple sclerosis[103] is of central origin. In the context of centrally mediated pruritus, it has been suggested that conditions associated with chronic pruritus may also be associated with central sensitization for itch. Studies in patients with pruritus secondary to atopic dermatitis experienced noxious stimuli like heat and pressure, and electrical and chemical stimulations as itch, in contrast to the pain experienced by the control subjects.[104] It was proposed that the central sensitization for itch results from constant pruritogenic (i.e., pruriceptive) input, enabling nociceptive stimuli to facilitate itch instead of inhibiting it. Microneurographic recordings in a patient with chronic pruritus and prurigo nodularis secondary to chronic scratching revealed spontaneously active itch fibers;[105] thus, in addition to a genesis of central opioid-mediated pruritus in cholestasis, it is conceivable that the chronicity of the pruritus in liver disease and uremia, for example, results in central sensitization for pruritus, leading to nonpruritogenic stimuli resulting from the constant stimuli of C-pruriceptors (e.g., by toxic substances retained as a result of cholestasis) to be perceived as pruritus.

In this regard, the results of brain scans by single photon emission computed tomography and functional magnetic resonance imaging methodology in patients with pruritus of cholestasis during periods of itch and no itch have been published in abstract form.[106] It was reported that itch was not associated with sensory cortex activation; there was a correlation between increasing itch severity with activity in the prefrontal cortex, orbital frontal cortex, putamen, globus pallidus, insular cortex, and orbital anterior and posterior cingulate cortices. It was concluded, based on the pattern of activation, that the limbic system is the primary central nervous system pathway involved in the perception of itch and that the findings support a central origin for this type of pruritus or itch.[106] These preliminary studies are consistent with the hypothesis stating that the pruritus of cholestasis is centrally mediated.[27]

The pruritus of cholestasis may also have a peripheral component mediated by the stimulation of peripheral receptors that respond to the pruritogens that may accumulate in body tissues as a result of cholestasis.

6.5   SCRATCHING ACTIVITY IN PATIENTS WITH THE PRURITUS OF CHOLESTASIS: LESSONS FROM BEHAVIORAL STUDIES

The problem with studies of the pruritus of cholestasis is, in part, the methodology that had been used traditionally, i.e., the measurements of substances that accumulate in plasma as a result of cholestasis. Evidence to explain how any of the proposed substances mediate the neurophysiological changes required for the sensation of the pruritus of cholestasis to be transmitted has not been provided[18,23,28,35] except for what has been documented for substances with agonist activity at the mu opioid receptor.[44–48]

Oral reports from patients on the effect of drugs on their pruritus are extremely relevant. In the context of clinical trials, for example, patients can say yes or no in response to the question of relief, with responses to be interpreted as a measure of effectiveness versus lack of effect; however, that is not how response to study drugs is assessed in studies of pruritus, as numbers, generated by questionnaires and visual analogue scales for pruritus have a certain appeal. The problem with the use of this type of methodology is that pruritus is a perception, and as a perception, it cannot be directly quantitated; accordingly, the meaning of the data collected by the above stated methods is uncertain. In contrast, scratching, the behavior that results from the sensation of pruritus, can be quantitated by appropriate methodology, providing the unique opportunity to study this well-conserved protective reflex that results from a symptom of disease. A scratching activity monitoring system was specifically developed to record scratching behavior independent from gross body movement in patients with cholestasis,[107] a need that has been identified by other investigators in the field of pruritus.[108,109] The use of this scratching activity monitoring system allowed for continued recording of scratching from patients with cholestasis and pruritus over consecutive 24-hour periods in clinical trials that evaluated treatment of patients with pruritus from liver disease. In addition to providing evidence for a decrease in scratching by opiate antagonists,[49,50,64,65] unique information was revealed by these recordings: (1) 96 hours of recording revealed a 24-hour rhythm in scratching behavior in some patients with maximum activity between 1200 and 1800 hours, and the nadir during hours of sleep (Figure 6.1).[65] This finding underscores the uncertainty of randomly collected data, as time of day or night can affect the data themselves; in addition, the presence of a rhythm suggested the possibility of circadian regulation of itch and scratching, (2) the effect of bright-light phototherapy (10,000 lux) indirectly projected toward the eyes was associated with a decrease in the outbursts of scratching,[110] suggesting some value of this type of therapy on the pruritus of cholestasis, perhaps in combination with other interventions; these results also offered support for a potential role of light, which regulates circadian

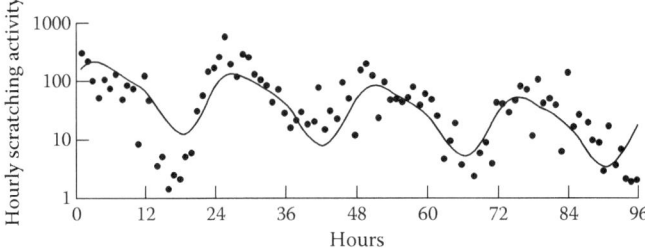

FIGURE 6.1 Mean hourly scratching activity during the 96-hour study period of a patient with benign recurrent intrahepatic cholestasis who participated in a study of naloxone infusions for the pruritus of cholestasis. The continuous line indicates the 24-hour rhythm that best fits the observation. The line has a significant downward linear trend (slope = −0.0081 ± 0.0021) ($P < 0.001$), which is consistent with the sequence of infusion (placebo, placebo, naloxone, naloxone), indicating that the patient scratched less on the study drug than on placebo.[65]

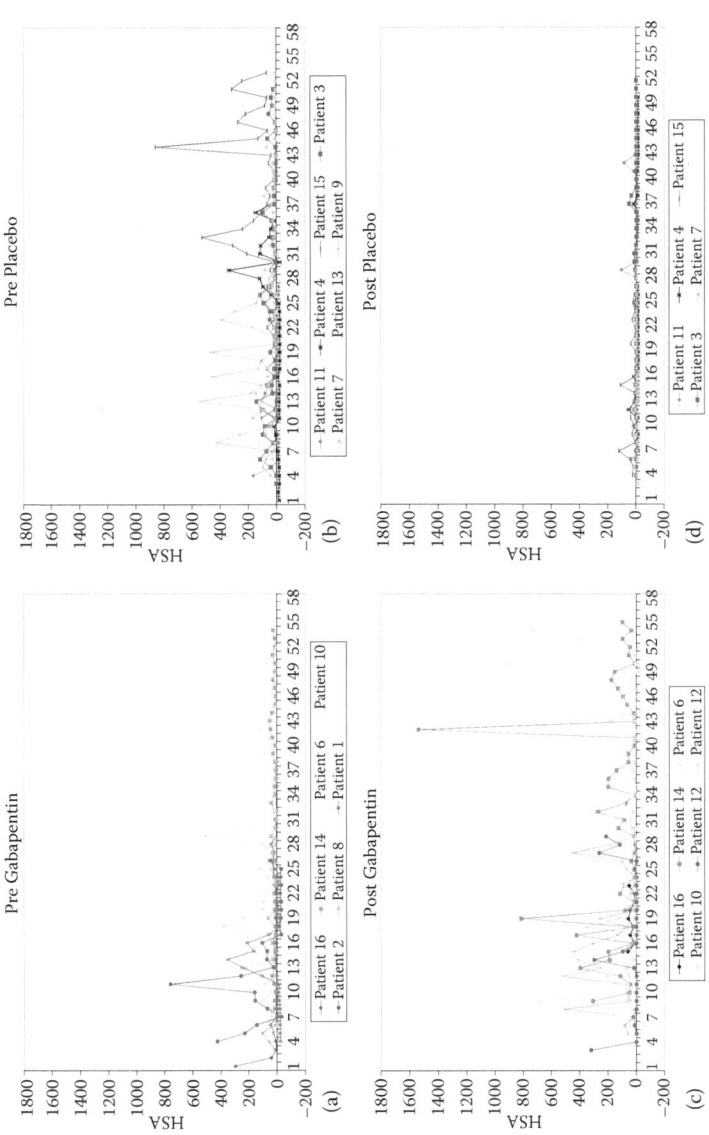

FIGURES 6.2 (See color insert.) Hourly scratching activity (HSA) from patients with pruritus secondary to liver disease who participated in a randomized, double-blind, placebo-controlled study of gabapentin.[112] All patients underwent continuous recording of scratching activity for a minimum of 24 hours at baseline and after four weeks of treatment with the study drug to which they had been randomized. (a, b) Mean baseline HSA from the patients who were subsequently randomized to (a) gabapentin and to (b) placebo. (c, d) Mean HSA from the patients randomized to (c) gabapentin and to (d) placebo after at least four weeks of treatment.

rhythms,[111] on the itch sensation, and (3) in contrast to gabapentin (Figure 6.2a and c), the placebo intervention was associated with a marked and significant decrease in scratching activity (Figure 6.2b and d) in patients with chronic liver disease in a double-blind, randomized, placebo-controlled trial.[112] These striking results highlight the importance of placebo on the pruritus and scratching, enticing investigators to explore the physiology of the placebo effect in the treatment of this symptom. In this regard, as the magnitude of the placebo effect depends on the subject's expectations for an effect, a practical lesson from this study is that the design of clinical trials of interventions to treat the pruritus of cholestasis, and probably any type of pruritus, must include the assessment of the subjects' expectations, prior to entrance into a study.[113,114] Accordingly, clinical trials for the treatment of pruritus in the twenty-first century should include behavioral methodology,[115] preferably portable,[116] the development of which is surely feasible in the current, continuously growing technological environment.

6.6 TREATMENT OF THE PRURITUS OF CHOLESTASIS

Patients with liver disease and pruritus often seek the attention of dermatologists. Internal medicine causes of pruritus including liver disease and malignancy (e.g., lymphoma) should be excluded in patients without dermatological diseases. The management of women with cholestasis of pregnancy is not discussed in this review. There is a consensus that ursodeoxycholic acid is associated with a decrease in the pruritus in the mothers, and with an improved prognosis in their infants;[8] patients with cholestasis of pregnancy should be treated in collaboration with the obstetricians in a high risk pregnancy service.

The skin of patients with pruritus from liver disease is devoid of primary pruritic skin lesions; however, lesions secondary to scratching, including excoriations, and sometimes prurigo nodularis can be found; to avoid scratching is very difficult as pain induced by vigorous scratching is often associated with relief of the pruritus.[117] The use of moisturizers, emollients, and other topical preparations have not been submitted to studies in patients with the pruritus of cholestasis; however, the skin must be protected and measures to keep it healthy applied. It has been the practice of the author to refer patients with pruritus from liver disease to a dermatologist to exclude primary skin diseases that may contribute to the pruritus.

Selected studies in patients with the pruritus of cholestasis are provided in Table 6.1, and guidelines for the treatment of patients in Table 6.2. Recommendations on the specific use of some drugs are provided below, although a rationale for the use of certain interventions to treat the pruritus is not always readily apparent.

6.6.1 PROCEDURES AIMED AT REMOVAL OF PRURITOGENS FROM THE BODY

The most commonly used drug to treat this form of pruritus is cholestyramine,[118] a resin that is not absorbed and that binds anions in the small intestine increasing their fecal excretion, including that of bile acids and cholesterol. The mechanism by which pruritus decreases in some patients with cholestasis in association with cholestyramine is unknown. The side effects of this resin tend to be minor in most

TABLE 6.1
Selected Publications on the Treatment of Itch from Liver Disease

Medication	Aim	Dose/Mode of Administration/Frequency	Type of Study/Duration	N	Study End Points	Results	Reference
Cholestyramine	Removal of pruritogen(s)	3.3–12 g/P.O./day	Single blind, open label/placebo-controlled crossover[a]/6–32 months[a]	27	Not reported	23 patients experienced relief of pruritus[a]	118
Rifampicin	Unknown	150 mg P.O./BID if serum bilirubin >3 mg/dL; 150 mg P.O./TID if serum bilirubin <3	Double-blind, randomized placebo-controlled crossover/4 weeks[b]	9	Change in VAS	Highly significant decrease in the 7-day summed VAS[b]	41
Naloxone	To decrease opioidergic tone	0.2 micrograms/kg/min/IV continuous infusions preceded by 0.4 mg IV bolus	Double-blind, placebo-controlled randomized crossover/4 consecutive days	29	Change in HAS	Geometric mean HAS 34% lower on naloxone than on placebo	65
Naltrexone	To decrease opioidergic tone	25 mg P.O./BID on day 1 followed by 50 mg P.O. daily	Randomized placebo-controlled/4 weeks[c]	16	Change in VAS	Daytime VAS down by 54% and nighttime VAS down by 44%	67

(continued)

TABLE 6.1 (Continued)
Selected Publications on the Treatment of Itch from Liver Disease

Medication	Aim	Dose/Mode of Administration/Frequency	Type of Study/Duration	N	Study End Points	Results	Reference
Sertraline	To change serotoninergic tone	75 to 100 mg P.O. daily	First phase, dose finding study/4 weeks	21	Change in VAS, decrease in skin excoriations	Decrease in mean VAS by 53% in first phase and decrease in mean VAS by 33% from baseline on sertraline	144
			Second phase, randomized, double-blind placebo-controlled, cross-over design/6 weeks	12		Excoriations decreased	

Note: VAS, mean visual analogue score; HSA, hourly scratching activity; *N*, total number of patients.

a Compared to an observational control group that did not receive cholestyramine but which received norethandrolone or no treatment.

b The use of cholestyramine was allowed to continue; the number of cholestyramine packs per day that the patients used was counted. Mean change in VAS not reported, VAS graphed per patient.

c Concomitant use of antipruritic medications were allowed.

TABLE 6.2
Treatment of Patients with the Pruritus of Cholestasis

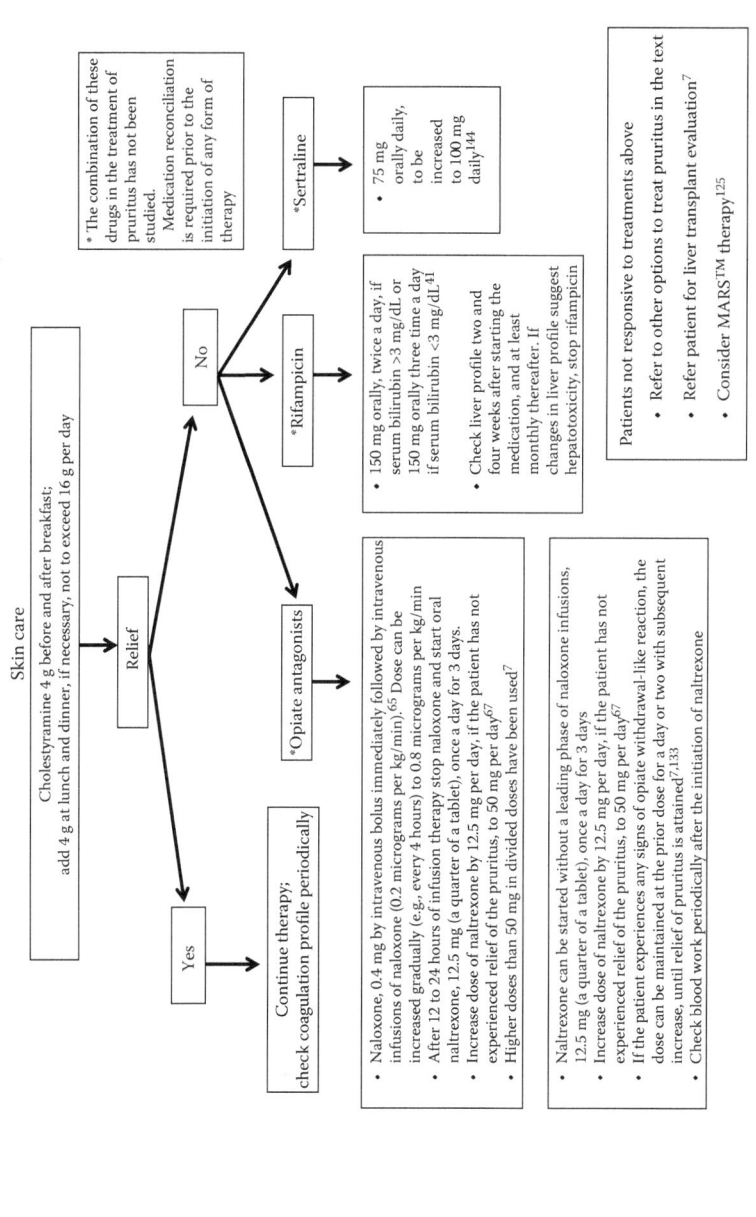

Skin care
Cholestyramine 4 g before and after breakfast; add 4 g at lunch and dinner, if necessary, not to exceed 16 g per day

Relief

Yes

Continue therapy; check coagulation profile periodically

No

*Opiate antagonists

- Naloxone, 0.4 mg by intravenous bolus immediately followed by intravenous infusions of naloxone (0.2 micrograms per kg/min).[66] Dose can be increased gradually (e.g., every 4 hours) to 0.8 micrograms per kg/min After 12 to 24 hours of infusion therapy stop naloxone and start oral naltrexone, 12.5 mg (a quarter of a tablet), once a day for 3 days.
- Increase dose of naltrexone by 12.5 mg per day, if the patient has not experienced relief of the pruritus, to 50 mg per day[67]
- Higher doses than 50 mg in divided doses have been used[7]

- Naltrexone can be started without a leading phase of naloxone infusions, 12.5 mg (a quarter of a tablet), once a day for 3 days
- Increase dose of naltrexone by 12.5 mg per day, if the patient has not experienced relief of the pruritus, to 50 mg per day[67]
- If the patient experiences any signs of opiate withdrawal-like reaction, the dose can be maintained at the prior dose for a day or two with subsequent increase, until relief of pruritus is attained[7,133]
- Check blood work periodically after the initiation of naltrexone

*Rifampicin

- 150 mg orally, twice a day, if serum bilirubin >3 mg/dL or 150 mg orally three time a day if serum bilirubin <3 mg/dL[41]
- Check liver profile two and four weeks after starting the medication, and at least monthly thereafter. If changes in liver profile suggest hepatotoxicity, stop rifampicin

*Sertraline

- 75 mg orally daily, to be increased to 100 mg daily[144]

* The combination of these drugs in the treatment of pruritus has not been studied. Medication reconciliation is required prior to the initiation of any form of therapy

Patients not responsive to treatments above
- Refer to other options to treat pruritus in the text
- Refer patient for liver transplant evaluation[7]
- Consider MARS™ therapy[125]

patients (e.g., bloating). It is recommended that cholestyramine be taken immediately before and after breakfast as the rationale for its use is to bind the pruritogen(s) that accumulates in the gallbladder during the overnight fast and that is poured into the small bowel after breaking the fast. Colesevalan, also a resin, was tested in a placebo-controlled study in which a positive effect on the amelioration of pruritus was defined, seemingly arbitrarily, as a 40% decrease in the visual analogue scores for pruritus; the effect of the drug was not better than that of placebo.[119]

A transient relief from pruritus has been reported in association with anion adsorption and plasma separation,[120] and the extracorporeal liver support systems Prometheus™ and MARS™.[121–125] The analysis of plasma removed from patients with what was defined as resistant pruritus via MARS™ was reported to reveal 60 proteins, one of which, SLURP1, was three times higher in the samples extracted from patients with cholestasis than in those from the control group.[126] These interventions may never be submitted to controlled clinical trials because of their nature; the tremendous need to provide relief to patients with severe pruritus, however, supports the use of this type of intervention in a controlled environment, although the placebo effect of these interventions may be substantial and the relief provided temporary.[122]

Reports of relief of the pruritus have also been published in association with nasobiliary drainage in PBC,[127] partial internal diversion of bile in patients with progressive familial intrahepatic cholestasis,[128] and partial external diversion of bile[129,130] and ileal diversion in children with cholestasis,[131] in whom an improvement in quality of life was also reported.

6.6.2 CHANGES IN NEUROTRANSMISSION

6.6.2.1 Opiate Antagonists

The results of several clinical studies,[43,49–52,64–67] including those that included behavioral methodology,[49,50,64,65] and the rationale on which their use is based, increase opioidergic tone,[27] support the administration of opiate antagonists to treat the pruritus of cholestasis. The opiate antagonists available for the treatment of the pruritus of cholestasis are naloxone and naltrexone (Tables 6.1 and 6.2). To decrease the probability of an opiate withdrawal-like reaction that some patients experience, treatment with opiate antagonists can be started with continuous intravenous infusions in the in-patient or out-patient setting (e.g., endoscopy suite, ambulatory surgery suite) for several hours prior to the introduction of the oral medication.[7,132,133] If at any point of the dose increase the patient exhibits signs of withdrawal, the dose can be maintained at the prior dose for a day or two with subsequent increases, until relief of the pruritus is achieved (Table 6.2). The symptoms and signs related to the opiate withdrawal-like reaction cease on the dose that triggered it; thus, it is not necessary to stop the medication. Patients should be evaluated with blood work after the initiation of naltrexone; from a large study of patients with alcoholism in which the drug was given for 1 year, however, no hepatotoxicity was reported.[134] The metabolism of naltrexone is slow in patients with compensated and decompensated cirrhosis, in comparison to that of a control group, with predominance of naltrexone over 6 β-naltrexol in the patients;[135] however, pruritus tends to cease in patients with hepatic synthetic dysfunction; thus, naltrexone is not commonly needed for patients with decompensated liver disease.[11]

Pain was reported in three patients with inflammatory conditions and with cholestasis and pruritus who had experienced relief on naltrexone.[136] This complication is not common; if it occurs, the cause of pain has to be determined and, if none found, the pain may be from the blocking the endogenous opioid system, which is a survival system that mediates endogenous analgesia.[137] The dose of the opiate antagonist can be decreased or the medication stopped as necessary if pain persists.

6.6.2.2 Serotonin Antagonists

Ondansetron was reported to relief pruritus in a group of patients with pruritus from liver disease measured by subjective methodology;[138–140] however, it did not have a therapeutic advantage over placebo in studies that applied behavioral methodology.[141,142]

6.6.2.3 Serotonin Re-Uptake Inhibitors

It was reported that the review of diaries from patients who participated in a trial for the treatment of PBC revealed an association between the use of sertraline and relief of pruritus.[143] In a subsequent randomized placebo-controlled study, sertraline was reported to be associated with relief of pruritus, as assessed by a visual analogue scale, and with an improvement in the skin appearance as evaluated on physical examination.[144] Selective serotonin reuptake inhibitors have been reported to decrease pruritus in polycythaemia vera[145] and in patients with malignancy whose pruritus was multifactorial.[146,147] Mertazapine, a noradrenergic and serotoninergic antidepressant,[148] and other antidepressants have been reported to relieve pruritus in a small heterogeneous group of patients with malignancy and uremia[149] and in patients with nocturnal itch from skin diseases.[150] These reports tend to suggest that in addition to the serotonin neurotransmitter system, noradrenaline may participate in the mediation of pruritus not limited to that associated with liver disease.

Dronabinol relieved intractable pruritus in three patients with chronic liver disease[151] and was associated with marginal ameliorating effects in a small series of patients (NVBergasa, unpublished). Impaired coordination and drowsiness of various degrees were reported as side effects (Ref. 151 and NVBergasa, unpublished). Antinociception was increased in rats with cholestasis secondary to bile duct resection in association with the administration of an agonist at a cannabinoid receptor;[152] thus, it can be hypothesized that the relief of pruritus in association with dronabinol in some patients with liver disease may be due to an increase in the threshold to experience nociception of which pruritus has been defined as its second order, the first one being pain.[153]

6.6.2.4 Antibiotics

Rifampicin was associated with relief in pruritus in patients with PBC in a study that explored the enzyme inducing capacity of this drug.[154] Rifampicin stimulates PXR, which induces drug-metabolizing enzymes and transporters.[37] Rifampicin has been studied in clinical trials where it was found to relieve pruritus in patients with liver disease,[41,155] although it is not helpful in all patients.[42] It was concluded from a meta-analysis of controlled randomized clinical trials that rifampicin was safe; however, rifampicin can be hepatotoxic.[156,157] Thus, follow-up of liver profile is necessary when patients are started on this drug and stopped if there is suggestion of hepatotoxicity.

Refractory pruritus in a group of patients with PBC was reported to be ameliorated by metronidazole.[158] The long-term use of metronidazole can be associated with peripheral neuropathy; thus, only short-term courses may be appropriate.

6.6.3 MISCELLANEOUS TYPES OF DRUGS THAT HAVE BEEN REPORTED TO RELIEF PRURITUS IN CHOLESTASIS

Phenobarbital[40,159,160] has been used to treat pruritus for some time, especially in children. The sedative effect of this drug may be associated with its ameliorating effect and also, its choleretic activity.[159] Gabapentin may be useful in a selective group of patients with high expectations for relief from the drug.[112] Pregabalin has been reported to relieve pruritus in uremia[161] and in that associated with burns.[162] This drug was associated with marked relief in the itch secondary to a generalized drug-induced rash at doses of 150 mg twice a day (NVBergasa, unpublished, 2012). Its effect on the pruritus of cholestasis merits investigation.

S-adenosylmethionine is an anticholestatic agent reported to decrease cholestasis and to improve pruritus in patients with intrahepatic cholestasis of pregnancy in clinical trials.[163,164]

Grapefruit juice has been inconsistently reported to relieve pruritus in patients with cholestasis.[165,166] Grapefruit juice causes drug interactions by inhibiting CYP3A4 in the intestine[167] and in the hepatocyte, in an exposure-dependent manner,[168] in contrast to rifampicin, which induces this enzyme in the hepatocyte.[38] It is unknown how this effect may be relevant in the relief of the pruritus; however, the effect of these compounds in some patients may be related to absorption and to metabolism of unknown pruritogens.

Controlled studies of steroids and other immunosuppressants to treat the pruritus of cholestasis have not been published. There are reports on the use of steroids to treat pruritus not associated with inflammatory skin conditions, including cholestasis;[11,169] however, long-term steroid therapy can be associated with osteopenia[170] to which patients with liver disease can be predisposed.[171] Thus, their random use is not recommended. The use of tacrolimus to treat pruritus in PBC has also been reported but not followed up in controlled studies.[172]

A reduction in the visual analogue score for pruritus was reported in a small heterogeneous group of patients with cholestasis in association with skin irradiation for a period of 8 weeks.[173] Phototherapy decreased in the number of epidermal and calcitonin gene related peptide-positive dermal nerve fibers in a group of patients with inflammatory skin disease associated with pruritus;[174] the rationale for the use of phototherapy to the skin to treat patients with the pruritus of cholestasis, however, is not apparent. The effect of bright-light phototherapy indirectly directed toward the eyes has already been discussed.[110]

The administration of anesthetics including propofol[175,176] and lidocaine[177] have been reported to relief pruritus from liver disease. The effect of lidocaine on the pruritus of cholestasis may be related to the role of transient receptor potential vanilloid receptor-1 in the transmission of the itch sensation.[178]

In summary, the pathogenesis of the pruritus of cholestasis is, most likely, multifactorial, and the perception of pruritus may be informed by the genetic composition of the patient. Increased opioidergic tone contributes to the pathogenesis of the pruritus of cholestasis, at least in part; a central mechanism has been proposed.[27] Substances that under physiological conditions are excreted in bile accumulate in body tissues as a result of cholestasis and may stimulate peripheral pruriceptive neurons, providing a peripheral component to the pruritus. In addition, the constant peripheral stimulation may also lead to central sensitization for itch.[104]

The relevance of results from cellular and from behavioral research in laboratory animals to the pruritus of cholestasis must be sought in an effort to develop effective antipruritic drugs. The incorporation of behavioral methodology in studies of pruritus offers the opportunity to learn features of scratching behavior and to define objective end points in clinical trials. In the absence of behavioral methodology, yes or no, in answer to the question of relief from patients must suffice as a measure of success.

Bile acids, which accumulate in body tissues in cholestasis, were recently reported to mediate scratching behavior in mice by mechanisms that involved gastrin releasing peptide and the mu opioid receptor, the receptor through which morphine mediates scratching behavior in non-human primates, and likely, human beings. TGR5, a G protein coupled plasma membrane receptor for bile acids, was necessary for mice to exhibit scratching behavior in association with bile acid administration. These results were interpreted to suggest a potential role of TGR5 antagonists for the treatment of the pruritus of cholestasis. Although the animal model used in these experiments is not a model of the pruritus of cholestasis, these interesting results identify the opioid receptor as mechanistically relevant in scratching behavior associated with the administration of bile acids in mice. If bile acids contribute to the pruritus of cholestasis in human beings, the mu opioid receptor may mediate the pruritogenic effect of these substances,[179] supporting further a role of the endogenous opioid system in the pruritus of cholestasis.

REFERENCES

1. Reichen, J., and Simon, F. (1988). Cholestasis. In *The Liver: Biology and Pathobilogy*, 2nd ed., Arias, I. M. et al. (Eds.). New York: Raven Press, pp. 1105–1124.
2. Ferenci, P., Zollner, G., and Trauner, M. (2002). Hepatic transport systems. *J. Gastroenterol. Hepatol.* 17(Suppl): S105–S112.
3. Bull, L. N. (2002). Hereditary forms of intrahepatic cholestasis. *Curr. Opin. Genet. Dev.* 12(3): 336–242.
4. Chia, S. C., Bergasa, N. V., Kleiner, D. E., Goodman, Z., Hoofnagle, J. H., and Di Bisceglie, A. M. (1998). Pruritus as a presenting symptom of chronic hepatitis C. *Dig. Dis. Sci.* 43(10): 2177–2183.
5. Bergasa, N. V. (2008). Pruritus in primary biliary cirrhosis: Pathogenesis and therapy. *Clin. Liver Dis.* 12(2): 385–406.

6. Elias, E. (1993). Liver transplantation. *J. R. Coll. Phys. London* 27: 224–232.

7. Neuberger, J., and Jones, E. A. (2001). Liver transplantation for intractable pruritus is contraindicated before an adequate trial of opiate antagonist therapy. *Eur. J. Gastroenterol. Hepatol.* 13(11): 1393–1394.

8. Bacq, Y., Sentilhes, L., Reyes, H. B., Glantz, A., Kondrackiene, J., Binder, T., Nicastri, P. L. et al. (2012). Efficacy of ursodeoxycholic acid in treating intrahepatic cholestasis of pregnancy: A meta-analysis. *Gastroenterology* 143(6): 1492–1501.

9. Newton, J. L., Bhala, N., Burt, J., and Jones, D. E. (2006). Characterisation of the associations and impact of symptoms in primary biliary cirrhosis using a disease specific quality of life measure. *J. Hepatol.* 44(4): 776–783.

10. Poupon, R., Chazouilleres, O., Balkau, B., and Poupon, R. E. (1999). Clinical and biochemical expression of the histopathological lesions of primary biliary cirrhosis. UDCA-PBC Group. *J. Hepatol.* 30(3): 408–412.

11. Lloyd-Thomas, H. G., and Sherlock, S. (1952). Testosterone therapy for the pruritus of obstructive jaundice. *Br. Med. J.* 2(4797): 1289–1291.

12. Carbone, M., Mells, G., Pells, G., Dawas, M. F., Newton, J. L., Heneghan, M., Neuberger, J. M. et al. (2012). Sex and age are determinants of the clinical phenotype of primary biliary cirrhosis and response to ursodeoxycholic acid. *Gastroenterology* 144(3): 560–569.

13. Rishe, E., Azarm, A., and Bergasa, N. V. (2008). Itch in primary biliary cirrhosis: A patients' perspective. *Acta Derm. Venereol.* 88(1): 34–37.

14. Heathcote, J. (1997). The clinical expression of primary biliary cirrhosis. *Semin. Liver Dis.* 17(1): 23–33.

15. Huesmann, M., Huesmann, T., Osada, N., Phan, N. Q., Kremer, A. E., and Stander, S. (2012). Cholestatic pruritus: A retrospective analysis on clinical characteristics and treatment response. *J. Dtsch. Dermatol. Ges.* 11(2): 158–168.

16. Mezey, E., Burns, C., Burdick, J. F., and Braine, H. G. (2002). A case of severe benign intrahepatic cholestasis treated with liver transplantation. *Am. J. Gastroenterol.* 97(2): 475–477.

17. Jarnagin, W. R., Burke, E., Powers, C., Fong, Y., and Blumgart, L. H. (1998). Intrahepatic biliary enteric bypass provides effective palliation in selected patients with malignant obstruction at the hepatic duct confluence. *Am. J. Surg.* 175(6): 453–460.

18. Stiehl, A. (1974). Bile acids and bile acid sulfates in the skin of patients with cholestasis and pruritus. *Z. Gastroenterol.* 12(2): 121–124.

19. Varadi, D. P. (1974). Pruritus induced by crude bile and purified bile acids. Experimental production of pruritus in human skin. *Arch. Dermatol.* 109(5): 678–681.

20. Ricci, P., Hofmann, A. F., Hagey, L. R., Jorgensen, R. A., Dickson, E. R., and Lindort, K. D. (1998). Adjuvant cholylsarcosine during ursodeoxycholic acid treatment of primary biliary cirrhosis. *Dig. Dis. Sci.* 43(6): 1292–1295.

21. Fiorucci, S., Clerici, C., Antonelli, E., Orlandi, S., Goodwin, B., Sadeghpour, B. M., Sabatino, G. et al. (2005). Protective effects of 6-ethyl chenodeoxycholic acid, a farnesoid X receptor ligand, in estrogen-induced cholestasis. *J. Pharmacol. Exp. Ther.* 313(2): 604–612.

22. Mencarelli, A., and Fiorucci, S. (2010). FXR an emerging therapeutic target for the treatment of atherosclerosis. *J. Cell Mol. Med.* 14(1–2): 79–92.

23. Gittlen, S. D., Schulman, E. S., and Maddrey, W. C. (1990). Raised histamine concentrations in chronic cholestatic liver disease. *Gut* 31(1): 96–99.

24. Ossipov, M. H., Lai, J., King, T., Vanderah, T. W., and Porreca, F. (2005). Underlying mechanisms of pronociceptive consequences of prolonged morphine exposure. *Biopolymers* 80: 319–324.

25. Kuraishi, Y., Nagasawa, T., Hayashi, K., and Satoh, M. (1995). Scratching behavior induced by pruritogenic but not algesiogenic agents in mice. *Eur. J. Pharmacol.* 275: 229–233.

26. King, T., Gardell, L., Wang, R., Vardanyan, A., Ossipov, M. H., Malan, T. P. J., Vanderah, T. W. et al. (2005). Role of NK-1 neurotransmission in opioid-induced hyperalgesia. *Pain* 116: 276–288.

27. Jones, E. A., and Bergasa, N. V. (1990). The pruritus of cholestasis: From bile acids to opiate agonists. *Hepatology* 11(5): 884–887.

28. Trivedi, M., and Bergasa, N. V. (2010). Serum concentrations of substance P in cholestasis. *Ann. Hepatol.* 9(2): 177–180.

29. Ko, M. C., and Naughton, N. N. (2009). Antinociceptive effects of nociceptin/orphanin FQ administered intrathecally in monkeys. *J. Pain* 10(5): 509–516.

30. Renback, K., Inoue, M., and Ueda, H. (1999). Lysophosphatidic acid-induced, pertussis toxin-sensitive nociception through a substance P release from peripheral nerve endings in mice. *Neurosci. Lett.* 270(1): 59–61.

31. Renback, K., Inoue, M., Yoshida, A., Nyberg, F., and Ueda, H. (2000). Vzg-1/lysophosphatidic acid-receptor involved in peripheral pain transmission. *Brain Res. Mol. Brain Res.* 75(2): 350–354.

32. Cohen, A., Sagron, R., Somech, E., Segal-Hayoun, Y., and Zilberberg, N. (2009). Pain-associated signals, acidosis and lysophosphatidic acid, modulate the neuronal K(2P)2.1 channel. *Mol. Cell Neurosci.* 40(3): 382–389.

33. Ahn, D. K., Lee, S. Y., Han, S. R., Ju, J. S., Yang, G. Y., Lee, M. K., Youn, D. H., and Bae, Y. C. (2009). Intratrigeminal ganglionic injection of LPA causes neuropathic pain-like behavior and demyelination in rats. *Pain* 146(1–2): 114–120.

34. Kremer, A. E., Martens, J. J., Kulik, W., Rueff, F., Kuiper, E. M., van Buuren, H. R., van Erpecum, K. J. et al. (2012). Lysophosphatidic acid is a potential mediator of cholestatic pruritus. *Gastroenterology* 139(3): 1008–1018.

35. Kremer, A. E., van Dijk, R., Leckie, P., Schaap, F. G., Kuiper, E. M., Mettang, T., Reiners, K. S. et al. (2012). Serum autotaxin is increased in pruritus of cholestasis, but not of other origin, and responds to therapeutic interventions. *Hepatology* 56(4): 1391–1400.

36. Laleman, W., Wilmer, A., Evenepoel, P., Elst, I. V., Zeegers, M., Zaman, Z., Verslype, C., Fevery, J., and Nevens, F. (2006). Effect of the molecular adsorbent recirculating system and Prometheus devices on systemic haemodynamics and vasoactive agents in patients with acute-on-chronic alcoholic liver failure. *Crit. Care* 10(4): R108.

37. Tirona, R. G., and Kim, R. B. (2005). Nuclear receptors and drug disposition gene regulation. *J. Pharm. Sci.* 94(6): 1169–1186.

38. Timsit, Y. E., and Negishi, M. (2007). CAR and PXR: The xenobiotic-sensing receptors. *Steroids* 72(3): 231–246.

39. Bergasa, N. V. (2005). The pruritus of cholestasis. *J. Hepatol.* 43(6): 1078–1088.

40. Bachs, L., Parés, A., Elena, M., Piera, C., and Rodés, J. (1989). Comparison of rifampicin with phenobarbitone for treatment of pruritus in biliary cirrhosis. *Lancet* 1(8638): 574–576.

41. Ghent, C. N., and Carruthers, S. G. (1988). Treatment of pruritus in primary biliary cirrhosis with rifampin. Results of a double-blind, crossover, randomized trial. *Gastroenterology* 94(2): 488–493.

42. Woolf, G. M., and Reynold, T. B. (1990). Failure of rifampicin to relieve pruritus in chronic liver disease. *J. Clin. Gastroenterol.* 12: 174–177.

43. Thornton, J. R., and Losowsky, M. S. (1988). Opioid peptides and primary biliary cirrhosis. *Br. Med. J.* 297(6662): 1501–1504.

44. Ballantyne, J. C., Loach, A. B., and Carr, D. B. (1988). Itching after epidural and spinal opiates. *Pain* 33(2): 149–160.

45. Ballantyne, J. C., Loach, A. B., and Carr, D. B. (1989). The incidence of pruritus after epidural morphine. *Anaesthesia* 44(10): 863.

46. Thomas, D. A., and Hammond, D. L. (1995). Microinjection of morphine into the rat medullary dorsal horn produces a dose-dependent increase in facial scratching. *Brain Res.* 695(2): 267–270.

47. Ko, M. C., Lee, H., Song, M. S., Sobczyk-Kojiro, K., Mosberg, H. I., Kishioka, S., Woods, J. H., and Naughton, N. N. (2003). Activation of kappa-opioid receptors inhibits pruritus evoked by subcutaneous or intrathecal administration of morphine in monkeys. *J. Pharmacol. Exp. Ther.* 305(1): 173–179.

48. Ko, M. C., Song, M. S., Edwards, T., Lee, H., and Naughton, N. N. (2004). The role of central mu opioid receptors in opioid-induced itch in primates. *J. Pharmacol. Exp. Ther.* 310(1): 169–176.

49. Bergasa, N. V., Schmitt, J. M., Talbot, T. L., Alling, D. W., Swain, M. G., Turner, M. L., Jenkins, J. B., and Jones, E. A. (1998). Open-label trial of oral nalmefene therapy for the pruritus of cholestasis. *Hepatology* 27(3): 679–684.

50. Bergasa, N. V., Alling, D. W., Talbot, T. L., Wells, M. C., and Jones, E. A. (1999). Oral nalmefene therapy reduces scratching activity due to the pruritus of cholestasis: A controlled study. *J. Am. Acad. Dermatol.* 41(3 Pt 1): 431–434.

51. Terg, R., Coronel, E., Sorda, J., Munoz, A. E., and Findor, J. (2002). Efficacy and safety of oral naltrexone treatment for pruritus of cholestasis, a crossover, double blind, placebo-controlled study. *J. Hepatol.* 37(6): 717–722.

52. Mansour-Ghanaei, F., Taheri, A., Froutan, H., Ghofrani, H., Nasiri-Toosi, M., Bagherzadeh, A. H., Farahvash, M. J. et al. (2006). Effect of oral naltrexone on pruritus in cholestatic patients. *World J. Gastroenterol.* 12(7): 1125–1128.

53. Bergasa, N. V., Alling, D. W., Vergalla, J., and Jones, E. A. (1994). Cholestasis in the male rat is associated with naloxone-reversible antinociception. *J. Hepatol.* 20(1): 85–90.

54. Bergasa, N. V., Rothman, R. B., Vergalla, J., Xu, H., Swain, M. G., and Jones, E. A. (1992). Central mu-opioid receptors are down-regulated in a rat model of cholestasis. *J. Hepatol.* 15(1–2): 220–224.

55. Bergasa, N. V., Sabol, S. L., Young, W. S., 3rd, Kleiner, D. E., and Jones, E. A. (1995). Cholestasis is associated with preproenkephalin mRNA expression in the adult rat liver. *Am. J. Physiol.* 268(2 Pt 1): G346–G354.

56. Bergasa, N. V., Vergalla, J., Swain, M. G., and Jones, E. A. (1996). Hepatic concentrations of proenkephalin-derived opioids are increased in a rat model of cholestasis. *Liver* 16(5): 298–302.

57. Swain, M. G., Rothman, R. B., Xu, H., Vergalla, J., Bergasa, N. V., and Jones, E. A. (1992). Endogenous opioids accumulate in plasma in a rat model of acute cholestasis. *Gastroenterology* 103(2): 630–635.

58. Spivey, J., Jorgensen, R., Gores, G., and Lindor, K. (1994). Methionine-enkephalin concentrations correlate with stage of disease but not pruritus in patients with primary biliary cirrhosis. *Am. J. Gastroenterol.* 89(11): 2018–2032.

59. Bergasa, N. V., Liau, S., Homel, P., and Ghali, V. (2002). Hepatic Met-enkephalin immunoreactivity is enhanced in primary biliary cirrhosis. *Liver* 22(2): 107–113.

60. Boyella, V. D., Nicastri, A. D., and Bergasa, N. V. (2008). Human hepatic met-enkephalin and delta opioid receptor-1 immunoreactivities in viral and autoimmune hepatitis. *Ann. Hepatol.* 7(3): 221–216.

61. Thornton, J. R., and Losowsky, M. S. (1989). Methionine enkephalin is increased in plasma in acute liver disease and is present in bile and urine. *J. Hepatol.* 8(1): 53–59.

62. Dagenais, C., Ducharme, J., and Pollack, G. M. (2001). Uptake and efflux of the peptidic delta-opioid receptor agonist. *Neurosci. Lett.* 301(3): 155–158.

63. Banks, W. A., and Kastin, A. J. (1990). Peptide transport systems for opiates across the blood–brain barrier. *Am. J. Physiol.* 259(1 Pt 1): E1–E10.

64. Bergasa, N. V., Talbot, T. L., Alling, D. W., Schmitt, J. M., Walker, E. C., Baker, B. L., Korenman, J. C. et al. (1992). A controlled trial of naloxone infusions for the pruritus of chronic cholestasis. *Gastroenterology* 102(2): 544–549.

65. Bergasa, N. V., Alling, D. W., Talbot, T. L., Swain, M. G., Yurdaydin, C., Turner, M. L., Schmitt, J. M., Walker, E. C., and Jones, E. A. (1995). Effects of naloxone infusions in patients with the pruritus of cholestasis. A double-blind, randomized, controlled trial. *Ann. Intern. Med.* 123(3): 161–167.

66. Carson, K. L., Tran, T. T., Cotton, P., Sharara, A. I., and Hunt, C. M. (1996). Pilot study of the use of naltrexone to treat the severe pruritus of cholestatic liver disease. *Am. J. Gastroenterol.* 91: 1022–1023.

67. Wolfhagen, F. H. J., Sternieri, E., Hop, W. C. J., Vitale, G., Bertolotti, M., and van Buuren, H. R. (1997). Oral naltrexone treatment for cholestatic pruritus: A double-blind, placebo-controlled study. *Gastroenterology* 113(4): 1264–1269.

68. Zellos, A., Roy, A., and Schwarz, K. B. (2010). Use of oral naltrexone for severe pruritus due to cholestatic liver disease in children. *J. Pediatr. Gastroenterol. Nutr.* 51(6): 787–789.

69. Tandon, P., Rowe, B. H., Vandermeer, B., and Bain, V. G. (2007). The efficacy and safety of bile acid binding agents, opioid antagonists, or rifampin in the treatment of cholestasis-associated pruritus. *Am. J. Gastroenterol.* 102(7): 1528–1536.

70. Lindor, K. D., Gershwin, M. E., Poupon, R., Kaplan, M., Bergasa, N. V., and Heathcote, E. J. (2009). Primary biliary cirrhosis. *Hepatology* 50(1): 291–308.

71. Dietrich, C. G., Geier, A., and Oude Elferink, R. P. (2003). ABC of oral bioavailability: Transporters as gatekeepers in the gut. *Gut* 52(12): 1788–1795.

72. Dombrowski, S. M., Desai, S. Y., Marroni, M., Cucullo, L., Goodrich, K., Bingaman, W., Mayberg, M. R., Bengez, L., and Janigro, D. (2001). Overexpression of multiple drug resistance genes in endothelial cells from patients with refractory epilepsy. *Epilepsia* 42(12): 1501–1506.

73. Keppler, D., Leier, I., and Jedlitschky, G. (1997). Transport of glutathione conjugates and glucuronides by the multidrug resistance proteins MRP1 and MRP2. *Biol. Chem.* 378(8): 787–791.

74. Hoffmaster, K. A., Zamek-Gliszczynski, M. J., Pollack, G. M., and Brouwer, K. L. (2005). Multiple transport systems mediate the hepatic uptake and biliary excretion of the metabolically stable opioid peptide [D-penicillamine2,5]enkephalin. *Drug Metab. Dispos.* 33(2): 287–293.

75. Floreani, A., Carderi, I., Variola, A., Rizzotto, E. R., Nicol, J., and Bergasa, N. V. (2006). A novel multidrug-resistance protein 2 gene mutation identifies a subgroup of patients with primary biliary cirrhosis and pruritus. *Hepatology* 43(5): 1152–1154.

76. Wendel, B., and Hoehe, M. (1998). The human mu opioid receptor gene: 5' regulatory and intronic sequences. *J. Mol. Med.* 76: 525–532.

77. Tan, E. C., Chong, S. A., Mahendran, R., Tan, C. H., and Teo, Y. Y. (2003). Mu opioid receptor gene polymorphism and neuroleptic-induced tardive dyskinesia in patients with schizophrenia. *Schizophr. Res.* 65(1): 61–63.

78. Oslin, D. W., Berrettini, W., Kranzler, H. R., Pettinati, H., Gelernter, J., Volpicelli, J. R., and O'Brien, C. P. (2003). A functional polymorphism of the mu-opioid receptor gene is associated with naltrexone response in alcohol-dependent patients. *Neuropsychopharmacology* 28(8): 1546–1552.

79. Bond, C., LaForge, K. S., Tian, M., Melia, D., Zhang, S., Borg, L., Gong, J. et al. (1998). Single-nucleotide polymorphism in the human mu opioid receptor gene alters beta-endorphin binding and activity: Possible implications for opiate addiction. *Proc. Natl. Acad. Sci. USA* 95(16): 9608–9613.

80. Wei, L. X., Floreani, A., Variola, A., El Younis, C., and Bergasa, N. V. (2008). A study of the mu opioid receptor gene polymorphism A118G in patients with primary biliary cirrhosis with and without pruritus. *Acta Derm. Venereol.* 88(4): 323–326.

81. Tsai, F. F., Fan, S. Z., Yang, Y. M., Chien, K. L., Su, Y. N., and Chen, L. K. (2010). Human opioid mu-receptor A118G polymorphism may protect against central pruritus by epidural morphine for post-cesarean analgesia. *Acta Anaesthesiol. Scand.* 54(10): 1265–1269.

82. Jimenez, N., Anderson, G. D., Shen, D. D., Nielsen, S. S., Farin, F. M., Seidel, K., and Lynn, A. M. (2012). Is ethnicity associated with morphine's side effects in children? Morphine pharmacokinetics, analgesic response, and side effects in children having tonsillectomy. *Paediatr. Anaesth.* 22(7): 669–675.

83. McLean, W. H., and Hull, P. R. (2007). Breach delivery: Increased solute uptake points to a defective skin barrier in atopic dermatitis. *J. Invest. Dermatol.* 127(1): 8–10.

84. Margolis, D. J., Apter, A. J., Gupta, J., Hoffstad, O., Papadopoulos, M., Campbell, L. E., Sandilands, A., McLean, W. H., and Rebbeck, T. R. (2012). The persistence of atopic dermatitis and filaggrin (FLG) mutations in a US longitudinal cohort. *J. Allergy Clin. Immunol.* 130(4): 912–917.

85. Xie, W., Samoriski, G. M., McLaughlin, J. P., Romoser, V. A., Smrcka, A., Hinkle, P. M., Bidlack, J. M., Gross, R. A., Jiang, H., and Wu, D. (1999). Genetic alteration of phospholipase C beta3 expression modulates behavioral and cellular responses to mu opioids. *Proc. Natl. Acad. Sci. USA* 96(18): 10385–10390.

86. Ko, M. C., and Naughton, N. N. (2000). An experimental itch model in monkeys: Characterization of intrathecal morphine-induced scratching and antinociception. *Anesthesiology* 92(3): 795–805.

87. Inan, S., and Cowan, A. (2006). Nalfurafine, a kappa opioid receptor agonist, inhibits scratching behavior secondary to cholestasis induced by chronic ethynylestradiol injections in rats. *Pharmacol. Biochem. Behav.* 85(1): 39–43.

88. Wikstrom, B., Gellert, R., Ladefoged, S. D., Danda, Y., Akai, M., Ide, K., Ogasawara, M. et al. (2005). Kappa-opioid system in uremic pruritus: Multicenter, randomized, double-blind, placebo-controlled clinical studies. *J. Am. Soc. Nephrol.* 16(12): 3742–3747.

89. Nakao, K., and Mochizuki, H. (2009). Nalfurafine hydrochloride: A new drug for the treatment of uremic pruritus in hemodialysis patients. *Drugs Today* 45(5): 323–329.

90. Kumagai, H., Ebata, T., Takamori, K., Muramatsu, T., Nakamoto, H., and Suzuki, H. (2010). Effect of a novel kappa-receptor agonist, nalfurafine hydrochloride, on severe itch in 337 haemodialysis patients: A Phase III, randomized, double-blind, placebo-controlled study. *Nephrol. Dial. Transplant* 25(4): 1251–1257.

91. Reisine, T., and Pasternak, G. (1996). Opioid analgesics and antagonists. In *The Pharmacological Basis of Therapeutics*, 9th ed., Molinoff, P. B., and Ruddon, R. W. (Eds.). New York: McGraw-Hill, pp. 521–555.

92. Yosipovitch, G., and Stander, S. (2006). Meeting report of the 3rd International Workshop for the Study of Itch. *J. Invest. Dermatol.* 126(9): 1928–1930.

93. Dawn, A. G., and Yosipovitch, G. (2006). Butorphanol for treatment of intractable pruritus. *J. Am. Acad. Dermatol.* 54(3): 527–531.

94. Stehling, L. C., and Zauder, H. L. (1978). Double-blind comparison of butorphanol tartrate and meperidine hydrochloride in balanced anaesthesia. *J. Int. Med. Res.* 6(5): 384–387.

95. Fisher, M. A., and Glass, S. (1997). Butorphanol (Stadol): A study in problems of current drug information and control. *Neurology* 48(5): 1156–1160.

96. Zylicz, Z., and Krajnik, M. (1999). Codeine for pruritus in primary billiary cirrhosis. *Lancet* 353(9155): 813.

97. Lewis, J. W., and Husbands, S. M. (2004). The orvinols and related opioids—High affinity ligands with diverse efficacy profiles. *Curr. Pharm. Des.* 10(7): 717–732.

98. Sun, Y. G., and Chen, Z. F. (2007). A gastrin-releasing peptide receptor mediates the itch sensation in the spinal cord. *Nature* 448: 700–703.

99. Liu, X. Y., Liu, Z. C., Sun, Y. G., Ross, M., Kim, S., Tsai, F. F., Li, Q. F. et al. (2012). Unidirectional cross-activation of GRPR by MOR1D uncouples itch and analgesia induced by opioids. *Cell* 147(2): 447–458.

100. Jordan, B. A., Cvejic, S., and Devi, L. A. (2000). Opioids and their complicated receptor complexes. *Neuropsychopharmacology* 23: S5–S18.

101. Massey, E. (1984). Unilateral neurogenic pruritus following stroke. *Stroke* 15: 901–903.

102. Shapiro, P., and Braun, C. (1987). Unilateral pruritus after a stroke. *Arch Dermatol.* 123: 1521–1530.

103. Osterman, P. (1976). Paroxysmal itching in multiple sclerosis. *Br. J. Dermatol.* 95: 555–558.

104. Ikoma, A., Fartasch, M., Heyer, G., Miyachi, Y., Handwerker, H., and Schmelz, M. (2004). Painful stimuli evoke itch in patients with chronic pruritus: Central sensitization for itch. *Neurology* 62(2): 212–217.

105. Schmelz, M., Hilliges, M., Schmidt, R., Orstavik, K., Vahlquist, C., Weidner, C., Handwerker, H. O., and Torebjörk, H. E. (2003). Active "itch fibers" in chronic pruritus. *Neurology* 61(4): 564–566.

106. Barnes, L. B., Devous, M. D., Harris, T. S., and Mayo, M. J. (2009). The central nervous system activity profile of cholestatic pruritus. *Hepatolgy* 50(S4): 375 Abstract 153.

107. Talbot, T. L., Schmitt, J. M., Bergasa, N. V., Jones, E. A., and Walker, E. C. (1991). Application of piezo film technology for the quantitative assessment of pruritus. *Biomed. Instrum. Technol.* 25(5): 400–403.

108. Savin, J., Paterson, W., and Oswald, I. (1973). Scratching during sleep. *Lancet* ii: 296–297.

109. Savin, J., Paterson, W., Oswald, I., and Adam, K. (1975). Further studies of scratching during sleep. *Br. J. Dermatol.* 93: 297–302.

110. Bergasa, N. V., Link, M. J., Keogh, M., Yaroslavsky, G., Rosenthal, R. N., and McGee, M. (2001). Pilot study of bright-light therapy reflected toward the eyes for the pruritus of chronic liver disease. *Am. J. Gastroenterol.* 96(5): 1563–1570.

111. Moore, R. Y. (1993). Organization of the mammalian circadian system. In *Circadian Clocks and Their Readjustments*, Chadwick, D. J., and Ackrill, K. (Eds.). London: John Wiley, pp. 88–106.

112. Bergasa, N. V., McGee, M., Ginsburg, I. H., and Engler, D. (2006). Gabapentin in patients with the pruritus of cholestasis: A double-blind, randomized, placebo-controlled trial. *Hepatology* 44(5): 1317–1323.

113. de la Fuente-Fernandez, R., Ruth, T. J., Sossi, V., Schulzer, M., Calne, D. B., and Stoessl, A. J. (2001). Expectation and dopamine release: Mechanism of the placebo effect in Parkinson's disease. *Science* 293: 1164–1166.

114. Wager, T. D., Rilling, J. K., Smith, E. E., Sokolik, A., Casey, K. L., Davidson, R. J., Kosslyn, S. M., Rose, R. M., and Cohen, J. D. (2004). Placebo-induced changes in FMRI in the anticipation and experience of pain. *Science* 303(5661): 1162–1167.

115. Jones, E. A. (2002). Trials of opiate antagonists for the pruritus of cholestasis: Primary efficacy endpoints and opioid withdrawal-like reactions. *J. Hepatol.* 37(6): 835–863.

116. Molenaar, H. A., Oosting, J., and Jones, E. A. (1998). Improved device for measuring scratching activity in patients with pruritus. *Med. Biol. Eng. Comput.* 36(2): 220–224.

117. Aoki, T. (2003). Pleasure of "scratch" is a complex sensation of itch an pain. In *Second International Workshop for the Study of Itch*. Toyama, p. 11.

118. Datta, D. V., and Sherlock, S. (1966). Cholestyramine for long term relief of the pruritus complicating intrahepatic cholestasis. *Gastroenterology* 50(3): 323–332.

119. Kuiper, E. M., van Erpecum, K. J., Beuers, U., Hansen, B. E., Thio, H. B., de Man, R. A., Janssen, H. L., and van Buuren, H. R. (2010). The potent bile acid sequestrant colesevelam is not effective in cholestatic pruritus: Results of a double-blind, randomized, placebo-controlled trial. *Hepatology* 52(4): 1334–1340.

120. Pusl, T., Denk, G. U., Parhofer, K. G., and Beuers, U. (2006). Plasma separation and anion adsorption transiently relieve intractable pruritus in primary biliary cirrhosis. *J. Hepatol.* 45(6): 887–891.

121. Sturm, E., Franssen, C. F., Gouw, A. B. S., Boverhof, R., De Knegt, R. J., Stellaard, F., Bijleveld, C. M., and Kuipers, F. (2002). Extracorporal albumin dialysis (MARS) improves cholestasis and normalizes low apo A-I levels in a patient with benign recurrent intrahepatic cholestasis (BRIC). *Liver* 22(Suppl 2): 72–75.

122. Doria, C., Mandala, L., Smith, J., Vitale, C. H., Lauro, A., Gruttadauria, S., Marino, I. R., Foglieni, C. S., Magnone, M., and Scott, V. L. (2003). Effect of molecular adsorbent recirculating system in hepatitis C virus-related intractable pruritus. *Liver Transpl.* 9(4): 437–443.

123. Pares, A., Cisneros, L., Salmeron, J. M., Caballeria, L., Mas, A., Torras, A., and Rodés, J. (2004). Extracorporeal albumin dialysis: A procedure for prolonged relief of intractable pruritus in patients with primary biliary cirrhosis. *Am. J. Gastroenterol.* 99(6): 1105–1110.

124. Acevedo Ribo, M., Moreno Planas, J. M., Sanz Moreno, C., Rubio Gonzalez, E. E., Rubio Gonzalez, E., Boullosa Grana, E., Sanchez-Turrión, V., Sanz Guajardo, D., and Cuervas-Mons, V. (2005). Therapy of intractable pruritus with MARS. *Transplant Proc.* 37(3): 1480–1481.

125. Pares, A., Herrera, M., Aviles, J., Sanz, M., and Mas, A. (2010). Treatment of resistant pruritus from cholestasis with albumin dialysis: Combined analysis of patients from three centers. *J. Hepatol.* 53(2): 307–312.

126. Gay, M., Pares, A., Carrascal, M., Bosch-i-Crespo, P., Gorga, M., Mas, A., and Abian, J. (2011). Proteomic analysis of polypeptides captured from blood during extracorporeal albumin dialysis in patients with cholestasis and resistant pruritus. *PLoS One* 6(7): e21850.

127. Beuers, U., Gerken, G., and Pusl, T. (2006). Biliary drainage transiently relieves intractable pruritus in primary biliary cirrhosis. *Hepatology* 44(1): 280–281.

128. Gun, F., Erginel, B., Durmaz, O., Sokucu, S., Salman, T., and Celik, A. (2010). An outstanding non-transplant surgical intervention in progressive familial intrahepatic cholestasis: Partial internal biliary diversion. *Pediatr. Surg. Int.* 26(8): 831–834.

129. Whitington, P., and Whitington, G. (1988). Partial external diversion of bile for the treatment of intractable pruritus associated with intrahepatic cholestasis. *Gastroenterology* 95: 130–136.

130. Schukfeh, N., Metzelder, M. L., Petersen, C., Reismann, M., Pfister, E. D., Ure, B. M., and Kuebler, J. F. (2012). Normalization of serum bile acids after partial external biliary diversion indicates an excellent long-term outcome in children with progressive familial intrahepatic cholestasis. *J. Pediatr. Surg.* 47(3): 501–505.

131. Ng, V. L., Ryckman, F. C., Porta, G., Miura, I. K., de Carvalho, E., Servidoni, M. F. et al. (2000). Long-term outcome after partial external biliary diversion for intractable pruritus in patients with intrahepatic cholestasis. *J. Pediatric Gastroenterol. Nutr.* 30(2): 152–156.

132. Jones, E. A., and Dekker, L. R. (2000). Florid opioid withdrawal-like reaction precipitated by naltrexone in a patient with chronic cholestasis. *Gastroenterology* 118(2): 431–432.

133. Jones, E. A., Neuberger, J., and Bergasa, N. V. (2002). Opiate antagonist therapy for the pruritus of cholestasis: The avoidance of opioid withdrawal-like reactions. *QJM* 95(8): 547–552.

134. Krystal, J. H., Cramer, J. A., Krol, W. F., Kirk, G. F., and Rosenheck, R. A. (2001). Naltrexone in the treatment of alcohol dependence. *New Engl. J. Med.* 345: 1734–1739.

135. Bertolotti, M., Ferrari, A., Vitale, G., Stefani, M., Trenti, T., Loria, P., Carubbi, F., Carulli, N., and Sternieri, E. (1997). Effect of liver cirrhosis on the systemic availability of naltrexone in humans. *J. Hepatol.* 27: 505–511.

136. McRae, C. A., Prince, M. I., Hudson, M., Day, C. P., James, O. F., and Jones, D. E. (2003). Pain as a complication of use of opiate antagonists for symptom control in cholestasis. *Gastroenterology* 125(2): 591–596.

137. Willer, J. C., Dehen, H., and Cambier, J. (1981). Stress-induced analgesia in humans: Endogenous opioids and naloxone-reversible depression of pain reflexes. *Science* 212(4495): 689–691.

138. Schworer, H., and Ramadori, G. (1993). Improvement of cholestatic pruritus by ondansetron. *Lancet* 341: 1277.

139. Muller, C., Pongratz, S., Pidlich, J., Penner, E., Kaider, A., Schemper, M., Raderer, M., Scheithauer, W., and Ferenci, P. (1998). Treatment of pruritus in chronic liver disease with the 5-hydroxytryptamine receptor type 3 antagonist ondansetron: A randomized, placebo-controlled, double-blind cross-over trial. *Eur. J. Gastroenterol. Hepatol.* 10: 865–870.

140. Schworer, H., Hartmann, H., and Ramadori, G. (1995). Relief of cholestatic pruritus by a novel class of drugs: 5-hydroxytryptamine type 3 (5-HT3) receptor antagonists: Effectiveness of Ondansetron. *Pain* 61(1): 33–37.

141. O'Donohue, J. W., Pereira, S. P., Ashdown, A. C., Haigh, C. G., Wilkinson, J. R., and Williams, R. A. (2005). Controlled trial of ondansetron in the pruritus of cholestasis. *Aliment. Pharmacol. Ther.* 21(8): 1041–1045.

142. Jones, E. A., Molenaar, H. A., and Oosting, J. (2007). Ondansetron and pruritus in chronic liver disease: A controlled study. *Hepatogastroenterology* 54(76): 1196–1199.

143. Browning, J., Combes, B., and Mayo, M. J. (2003). Long-term efficacy of sertraline as a treatment for cholestatic pruritus in patients with primary biliary cirrhosis. *Am. J. Gastroenterol.* 98(12): 2736–2741.

144. Mayo, M. J., Handem, I., Saldana, S., Jacobe, H., Getachew, Y., and Rush, A. J. (2007). Sertraline as a first-line treatment for cholestatic pruritus. *Hepatology* 45(3): 666–674.

145. Diehn, F., and Tefferi, A. (2001). Pruritus in polycythaemia vera: Prevalence, laboratory correlates and management. *Br. J. Haematol.* 115: 619–621.

146. Zylicz, Z., Krajnik, M., Sorge, A. A., and Costantini, M. (2003). Paroxetine in the treatment of severe non-dermatological pruritus: A randomized, controlled trial. *J. Pain Symptom Manage.* 26(6): 1105–1112.

147. Zylicz, Z., Smits, C., and Krajnik, M. (1998). Paroxetine for pruritus in advanced cancer. *J. Pain Symptom Manage.* 16(2): 121–124.

148. Nutt, D. (1997). Mirtazapine: Pharmacology in relation to adverse effects. *Acta Psychiatr. Scand. Suppl.* 391: 31–37.

149. Davis, M. P., Frandsen, J. L., Walsh, D., Andresen, S., and Taylor, S. (2003). Mirtazapine for pruritus. *J. Pain Symptom Manage.* 25(3): 288–291.

150. Hundley, J. L., and Yosipovich, Y. (2004). Mirtazapine for reducing nocturnal itch in patients with chronic pruritus: A pilot study. *J. Am. Acad. Dermatol.* 50: 889–891.

151. Neff, G. W., O'Brien, C. B., Reddy, K. R., Bergasa, N. V., Regev, A., Molina, E., Amaro, R. et al. (2002). Preliminary observation with dronabinol in patients with intractable pruritus secondary to cholestatic liver disease. *Am. J. Gastroenterol.* 97(8): 2117–2119.

152. Gingold, A. R., and Bergasa, N. V. (2003). The cannabinoid agonist WIN 55, 212-2 increases nociception threshold in cholestatic rats: Implications for the treatment of the pruritus of cholestasis. *Life Sci.* 73(21): 2741–2747.

153. Magerl, W. (1996). Neural mechanisms of itch sensation. *IASP Newl.* Sections 4–7.

154. Hoensch, H. P., Balzer, K., Dylewizc, P., Kirch, W., Goebell, H., and Ohnhaus, E. E. (1985). Effect of rifampicin treatment on hepatic drug metabolism and serum bile acids in patients with primary biliary cirrhosis. *Eur. J. Clin. Pharmacol.* 28(4): 475–477.

155. Bachs, L., Parés, A., Elena, M., Piera, C., and Rodés, J. (1992). Effects of long-term rifampicin administration in primary biliary cirrhosis. *Gastroenterology* 102(6): 2077–2080.

156. Prince, M. I., Burt, A. D., and Jones, D. E. (2002). Hepatitis and liver dysfunction with rifampicin therapy for pruritus in primary biliary cirrhosis. *Gut* 50(3): 436–439.

157. Heathcote, E. J. (2007). Is rifampin a safe and effective treatment for pruritus caused by chronic cholestasis? *Nat. Clin. Pract. Gastroenterol. Hepatol.* 4(4): 200–201.

158. Berg, C. L., and Gollan, J. L. (1992). Primary biliary cirrhosis: New therapeutic directions. *Scand. J. Gastroenterol. Suppl.* 192: 43–49.

159. Bloomer, J. R., and Boyer, J. L. (1975). Phenobarbital effects in cholestatic liver diseases. *Ann. Intern. Med.* 82(3): 310–307.

160. Ghent, C. N., Bloomer, J. R., and Hsia, Y. E. (1978). Efficacy and safety of long-term phenobarbital therapy of familial cholestasis. *J. Pediat.* 93(1): 127–132.

161. Shavit, L., Grenader, T., Lifschitz, M., and Slotki, I. (2012). Use of pregabalin in the management of chronic uremic pruritus. *J. Pain Symptom Manage.* 45(4): 776–781.

162. Ahuja, R. B., and Gupta, G. K. (2013). A four arm, double blind, randomized and placebo controlled study of pregabalin in the management of post-burn pruritus. *Burns* 39(1): 24–29.

163. Frezza, M., Surrenti, C., Manzillo, G., Fiaccadori, F., Bortolini, M., and Di Padova, C. (1990). Oral S-adenosylmethionine in the symptomatic treatment of intrahepatic cholestasis. A double-blind, placebo-controlled study. *Gastroenterology* 99(1): 211–215.

164. Frezza, M., Centini, G., Cammareri, G., Le Grazie, C., and Di Padova, C. (1990). S-adenosylmethionine for the treatment of intrahepatic cholestasis of pregnancy. Results of a controlled clinical trial. *Hepatogastroenterology* 37(Suppl 2): 122–125.

165. Horsmans, Y., and Geubel, A. P. (1996). Pruritus associated with cholestatic liver disease. *Ann. Intern. Med.* 125(8): 701.

166. Cadranel, J. F., Di Martino, V., and Devergie, B. (1997). Grapefruit juice for the pruritus of cholestatic liver disease. *Ann. Intern. Med.* 126(11): 920–921.

167. Dolton, M. J., Roufogalis, B. D., and McLachlan, A. J. (2012). Fruit juices as perpetrators of drug interactions: The role of organic anion-transporting polypeptides. *Clin. Pharmacol. Ther.* 92(5): 622–630.

168. Veronese, M. L., Gillen, L. P., Burke, J. P., Dorval, E. P., Hauck, W. W., Pequignot, E., Waldman, S. A., and Greenberg, H. E. (2003). Exposure-dependent inhibition of intestinal and hepatic CYP3A4 in vivo by grapefruit juice. *J. Clin. Pharmacol.* 43(8): 831–839.

169. Walt, R., Daneshmend, T., and Fellows, I. (1988). Effect of stanozolol on itching in primary biliary cirrhosis. *Br. Med. J.* 296: 607.

170. Mitchison, H. C., Bassendine, M. F., Malcolm, A. J., Watson, A. J., Record, C. O., and James, O. F. (1989). A pilot, double-blind, controlled 1-year trial of prednisolone treatment in primary biliary cirrhosis: Hepatic improvement but greater bone loss. *Hepatology* 10(4): 420–429.

171. Guanabens, N., Pares, A., Ros, I., Caballeria, L., Pons, F., Vidal, S., Monegal, A., Peris, P., and Rodés, J. (2005). Severity of cholestasis and advanced histological stage but not menopausal status are the major risk factors for osteoporosis in primary biliary cirrhosis. *J. Hepatol.* 42(4): 573–577.

172. Aguilar-Bernier, M., Bassas-Vila, J., Sanz-Munoz, C., and Miranda-Romero, A. (2005). Successful treatment of pruritus with topical tacrolimus in a patient with primary biliary cirrhosis. *Br. J. Dermatol.* 152(4): 808–809.

173. Decock, S., Roelandts, R., Steenbergen, W. V., Laleman, W., Cassiman, D., Verslype, C., Fevery, J., Pelt, J. V., and Nevens, F. (2012). Cholestasis-induced pruritus treated with ultraviolet B phototherapy: An observational case series study. *J. Hepatol.* 57(3): 637–641.

174. Wallengren, J., and Sundler, F. (2004). Phototherapy reduces the number of epidermal and CGRP-positive dermal nerve fibres. *Acta Derm. Venereol.* 84(2): 111–115.

175. Borgeat, A., Wilder-Smith, O. H. G., and Mentha, G. (1993). Subhypnotic doses of propofol relieve pruritus associated with liver disease. *Gastroenterlogy* 104: 244–247.

176. Borgeat, A., Wilder-Smith, O., Saiah, M., and Rifat, K. (1992). Subhypnotic doses of propofol relieve pruritus induced by epidural and intrathecal morphine. *Anesthesiology* 76: 510–512.

177. Villamil, A. G., Bandi, J. C., Galdame, O. A., Gerona, S., and Gadano, A. C. (2005). Efficacy of lidocaine in the treatment of pruritus in patients with chronic cholestatic liver diseases. *Am. J. Med.* 118(10): 1160–1163.

178. Shim, W. S., Tak, M. H., Lee, M. H., Kim, M., Kim, M., Koo, J. Y., Lee, C. H., and Oh, U. (2007). TRPV1 mediates histamine-induced itching via the activation of phospholipase A2 and 12-lipoxygenase. *J. Neurosci.* 27(9): 2331–2337.

179. Alemi, F., Kwon, E., Poole, D. P., Lieu, T., Lyo, V., Cattaruzza, F. et al. The TGR5 receptor mediates bile acid-induced itch and analgesia. *J. Clin. Invest.* 123(4): 1513–1530.

7 Neuropathic Itch

Anne Louise Oaklander

CONTENTS

7.1   BACKGROUND AND INTRODUCTION

Dermatologists are increasingly aware that some cases of unexplained chronic itch are caused by underlying neurological disease, and some now seek neurological care for these patients, but few neurologists feel prepared to help. This overview is intended both to guide clinicians and to suggest research topics. A PubMed search referencing the terms "neuropathic itch" plus "neuropathic pruritus" on December 22, 2012, yielded only 101 primary and review citations. Other relevant papers are not retrieved by these search terms as they may not even contain the words "itch" or "pruritus," but instead refer to "distressing sensory symptoms" or to self-injurious scratching (often by other names, e.g., "trigeminal tropic syndrome"). The fragmented literature attests to the current rudimentary state of awareness about neuropathic itch.

7.1.1   DEFINITIONS OF NEUROPATHIC ITCH

The epistemology of neuropathic itch (NI) is based on that of neuropathic pain, which has so far worked well, no doubt because most peripheral itch neurons are a subset of peripheral pain neurons. NI is thus defined as perception of itch in the absence of pruritogenic stimuli (formerly known to dermatologists as *pruritus sine materia*). In other words, there is uncoupling of the stimulus-response curve for itch sensation since dysfunction or disease of itch-signaling neurons is causing them to fire without cause. At a practical level, NI can only be diagnosed after dermatological and systemic causes have been excluded, so dermatological evaluation should usually precede referral to a neurologist.

The International Forum for the Study of Itch (IFSI) recognizes four types of itch causality—dermatologic, systemic, neurologic, and psychiatric (Ständer et al. 2007) plus mixed and unknown causes. Etiology Category III (NI) is defined as "arising from diseases or disorders of the central or peripheral nervous system, e.g. nerve damage, nerve compression, nerve irritation" (Ständer et al. 2007, p. 87). From the neurological perspective, NI is primarily a modality-specific sensory hallucination, and the literature on other types of sensory hallucinations, e.g., tinnitus, the Charles Bonnet syndrome, phantom pain (recently reviewed in Sacks [2012]) can help inform understanding.

At the time of this writing, pain specialists are digesting 2012 revisions by the International Association for the Study of Pain (IASP) to their definition of neuropathic pain, which restricts it to "pain caused by a lesion or disease of the somatosensory system" (http://www.iasp-pain.org/Content/NavigationMenu/GeneralResource Links/PainDefinitions/default.htm#Neuropathicpain). This replaces the definition in the IASP's Classification of Chronic Pain (Merskey and Bogduk 1994, p. 212) that defined neuropathic pain more broadly, specifically as "pain initiated or caused by a primary lesion or dysfunction in the nervous system." In the 2012 version, "dysfunction" has been removed and a lesion or disease affecting the nervous system is now required. The requirement for a structural abnormality is problematic because in many patients, the causal lesions can often be hard to localize, meaning that they therefore no longer qualify as having neuropathic pain (Jensen et al. 2011, 2012; Oaklander et al.

2012). Regarding pruritus, the causal pathologies are even less defined than for neuropathic pain, so the author favors the less restrictive IFSI definition of neuropathic itch that permits "disorders" as well as "diseases" of the nervous system.

7.1.2 NEUROANATOMY AND CELL BIOLOGY

As is typical in neurology, discovering the anatomical pathways for normal itch sensation has been aided by studying patients with neurological diseases or injuries that caused their itch. The relevant methods include localization by neurological examination, imaging (Chapter 24), and rarely, anatomical or surgical pathology. Studying patients has demonstrated that NI is caused by lesions that specifically affect itch neurons, just as other neurological symptoms reflect the functions of the specific neurons damaged. As always, the cause of the lesion is less important than its location. A stroke, tumor, or multiple sclerosis plaque in the same brain region will generate similar signs and symptoms. As imaging methods advance, it should improve our understanding of the pathways and mechanisms of NI. Better visualization of white matter tracts is particularly relevant for network disorders such as NI. For instance, applying 7T magnetic resonance imaging (MRI) to the spinal cord of a patient with NI after cavernous hemangioma (see below) suggested the hypothesis that his late-onset central itch might be triggered by delayed white matter degeneration (Cohen-Adad et al. 2012).

Peripheral nervous system (PNS) lesions appear to be the most common causes of NI, most likely because these are more common than injuries to the well-protected central nervous system (CNS). Neuropathological study shows clearly that injury to specific neurons is necessary to produce itch, namely, to the subset of sensory small fibers that encode itch signals, which includes C-fiber and A-delta primary afferents with various transduction patterns (Chapter 9). Neuropathic itch is caused by the same type of neurological injuries and diseases that cause neuropathic pain as well. Many patients present with pain alone, or pain plus itch, but some report only itch, for unknown reasons. Lesions that primarily affect motor neurons, amyotrophic lateral sclerosis, for instance, are not associated with NI. Lesions anywhere in the PNS appear capable of causing NI, whether distally (axonopathy) or centered on the sensory ganglion (e.g., shingles) or the proximal nerve roots. Even the most distal PNS lesion can affect the spinal cord and then the brain to cause NI. While cutting a nerve reduces incoming sensory signals (after the injury barrage) and thus would be expected to reduce sensation, in some individuals later compensatory mechanisms paradoxically augment sensations coming from nerve-injured areas. The causes include abnormal firing of nearby preserved primary afferents, loss of tonic dorsal-horn inhibitory circuits (Basbaum and Wall 1976; Woolf and Wall 1982), and alterations in thalamic and cortical circuits (Chapters 23 and 26).

NI involves nonneural cells and mechanisms as well. Some diseases that cause NI also cause itch through other mechanisms (e.g., cutaneous inflammation as during shingles) or the mixed itch of uremic neuropathy. Also, scratching NI-affected skin can trigger additional pruritogenic cutaneous inflammation, irritation, ulceration, or scarring. Even under normal conditions, nonneural cells participate in itch signaling. Small fibers release neuropeptides such as calcitonin gene-related peptide (CGRP) and substance P that have paracrine effects on nearby nonneural cells that augment itch. Keratinocytes express the heat-sensing vanilloid type 3 (TRPV3) receptor (Peier et al. 2002) and several neuronal sodium channels (Zhao et al. 2008) so they may contribute to or modulate normal and/or neuropathic itch. Secretions of injured small fibers stimulate mast cell degranulation, which not only initiates itch-signaling action potentials through histamine, serotonin, and tryptase-activated proteinase-activated receptor-2 receptors on primary afferent nerve endings, but additionally stimulates release of proinflammatory cytokines from T-cells to further augment inflammation and neuronal activation (Chapters 19 and 20).

7.1.3   What Is the Relationship between NI and Psychogenic Itch?

In the past, physicians routinely interpreted itch syndromes without evident cause (*pruritus sine materia*) as psychogenic. Recognition of NI and of how little we know about itch mechanisms is curbing this habit of blaming the patients rather than our own uncertainty. In practice, psychogenic itch diagnoses (Etiology Category IV itch; [Ständer et al. 2007]) are considered in two types of patients, those with or without major psychiatric diagnoses. For patients without psychiatric diagnoses, it is probably inappropriate for nonpsychiatrists to diagnose psychogenic itch. The psychiatric diagnoses associated with unexplained chronic itch—somatization disorder, obsessive compulsive disorder, and thought disorders—are serious illnesses with specific diagnostic criteria. If they are suspected, a psychiatrist should be consulted to diagnose and treat the patient. Even in patients with established major psychiatric diagnoses, new itch complaints will more often have dermatologic or medical causes. My few patients with obsessive-compulsive disorder plus postherpetic itch (PHI) have successfully avoided picking or scratching their itchy areas, recognizing the danger.

Psychotic patients with itch are the most difficult to evaluate because sensory hallucinations are a central symptom of psychosis. It is unclear whether or not psychiatric illness itself can trigger delusions of itch, although considering the prevalence of auditory and other hallucinations in some of the psychoses, this has be presumed possible. When psychotic patients develop unexplained itch, it is often attributed to their psychosis, despite the lack of any known specific association. Advanced imaging is now identifying abnormal brain structure and activity associated with somatic delusions and hallucinations in such patients, blurring the distinction between psychogenic and central neuropathic itch. MRI has identified reduced left fronto-insular gray matter volume in schizophrenic patients with somatic delusions as compared with schizophrenics without hallucinations and healthy controls (Spalletta et al. 2012). The thalamus, particularly the right ventral-anterior thalamic nucleus that projects to prefrontal-temporal cortex and the inferior frontal lobe, is significantly larger in hallucinating patients as well (Spalletta et al. 2012).

We all harbor illogical explanations and expectations regarding health symptoms, and some are widely accepted, e.g., health benefits of expensive organic food or colon cleanses. The most common irrational explanation for unexplained chronic itch is undiagnosed parasite infestation. It is difficult to determine in any specific patient how illogical (delusional) this is, given that the sensation of itch is thought to have evolved to signal parasite contact. Our brains have likely been hard wired to interpret itchy stimuli as insect contact until proven otherwise. However, some patients with unexplained chronic itch cling to the parasite explanation despite repeated testing showing no evidence of this, and sometimes even after other causes of itch are diagnosed.

Names for this syndrome include primary delusional parasitosis, Ekbom syndrome, and Morgellon's syndrome. Some of these individuals have related convictions that there are tiny threads or fibers in or on their skin triggering their itch. They are a mixed bunch. Some have delusional explanation for itch that is caused by other identifiable causes (Fleury et al. 2008), whereas in others the causes of itch remain unknown, making it hard to exclude implausible explanations. Intellectual impairment and cultural and religious traditions can also make it hard for patients to evaluate the relative validity of their physicians' explanations versus those suggested by the Internet, friends, and spiritual leaders. Low-level psychoses and schizophrenia are more prevalent than appreciated, and patients with delusional thinking and other mild psychotic symptoms often remain unrecognized even by physicians. Illogical beliefs and paranoid or conspiracy theories are part of all cultures, with no distinct border between "normal" and "delusional" thinking. Typically, such individuals remain in the community without exhibiting flagrant enough behavior to trigger psychiatric evaluation and diagnosis.

In my limited experience caring for patients harboring delusional explanations for obscure sensory symptoms, I have usually identified a neurological cause (although the patient often disagrees), a psychiatrist has confirmed the presence of mild thought disorder, and antipsychotic medications have lessened the delusions but not the sensory symptoms. Such patients are ill served by our fragmented health care system, which enables them to reject and conceal diagnoses that displease them—such as schizophreniform disorder plus atypical trigeminal neuropathy—and to decline treatment for such rejected diagnoses. Instead, they self-refer to new physicians who unwittingly repeat the cycle of unnecessary diagnostic testing and ineffectual treatments for their delusional diagnoses. Tragically, some patients impoverish themselves to "treat" their delusions, investing thousands of dollars on mold remediation, insect extermination, or unnecessary surgeries or medications.

Regardless of a patient's own interpretation of the cause of their itch, it is their physician's responsibility to look for "actionable" causes, including not only dermatological and systemic causes but also mastocytosis, restless leg syndrome, and neuropathic itch.

7.1.4 ANIMAL MODELS OF NI

Because itch is subjective, itch assessment in animals traditionally relied on measuring hair loss or wounds caused by scratching or other forms of self-injury directed toward skin that was experimentally irritated or denervated. Video systems now permit quantitation of even noninjurious scratch bouts. Although it can never be proven that scratching and similar behaviors (rubbing and biting) in animals are motivated by itch, the evidence for this overwhelming because a wide range of nonhuman animals scratch in response to the same provocations (dermatologic, systemic, and neuropathic) that cause human itching and scratching.

Until recently, human NI and associated lesions were usually attributed to psychiatric causes or pain (see trigeminal trophic syndrome below), so it is not surprising that self-injury to denervated skin in nerve-injured experimental animals was and still is often attributed to neuropathic pain rather than itch. Although neuropathic self-injury was noted from the very first nerve-injury experiments, it was first systematically studied in rodents and called "autotomy" by pain scientists Pat Wall, Marshall Devor, and colleagues, who developed a grading system for it (Wall et al. 1979). They and others initially interpreted autotomy as modeling *anesthesia dolorosa* type neuropathic pain, but from the beginning there was disagreement, especially from neurosurgeons who know that patients with neuropathic pain including *anesthesia dolorosa* protect rather than injure their painful areas (Levitt et al. 1992). Neurosurgical pioneer Will Sweet also pointed out that although dorsal rhizotomy was one of the main triggers for experimental autotomy in animals, it does not cause pain in humans (Sweet 1981). By the end of his career, Dr. Wall had come to believe that neuropathic autotomy in rodents more likely reflected neuropathic itch rather than pain (P. Wall, pers. comm.). He was additionally influenced by later reports of "autotomy" in nerve-injured humans who reported that their self-injurious scratching was motivated by itch and not by pain (Oaklander et al. 2002). Other scientists use the term "overgrooming" for pathological scratching at experimentally nerve-injured skin. These behaviors develop in some but not all rodents after experimental lesions throughout the PNS and in the spinal cord that damage pain/itch pathways (Yezierski et al. 1998). Genetic (Mogil et al. 1999) and dietary factors (Shir et al. 1997) influence whether or not rats develop autotomy after nerve injuries and demonstrate that NI is a complex disorder with both environmental and genetic causes.

Widespread NI caused by polyneuropathy has been modeled by treating neonatal rodents with systemic capsaicin, which permanently ablates TRPV1+ primary afferents and triggers 80% to 90% of rats to engage in self-injurious overgrooming (Maggi et al. 1987). Interestingly, this is largely restricted to the areas around the eyes, ears, and snout, consistent with the idea that these locations may be enriched in itch receptors, perhaps because their moistness and lack of keratinization attracts insects (Maggi et al. 1987). It is unclear whether or not neuropathies that damage axons only while preserving the cell bodies in the sensory ganglia also produce NI.

Focal NI caused by nerve injury has been modeled by studies of autotomy after sciatic nerve transection in rodents. Action potentials from the damaged nerve appear required to maintain autotomy, because chronic administration of local anesthetics to a sciatic nerve that completely suppresses sciatic action potentials does not trigger autotomy (Blumenkopf and Lipman 1991).

Dermatomal NI caused by ganglion lesions has been modeled by surgical extirpation of specific dorsal root ganglia. The resultant overgrooming centered on the denervated dermatome is associated with ectopic firing within the dorsal horn (Asada et al. 1996) and modulated by descending dopaminergic inhibition (Tseng and Lin 1998).

Dermatomal NI caused by radiculopathy is modeled by the autotomy that develops after lesioning rodent dorsal rootlets or roots. This seems to correspond to the rare phenomenon of self-injury after plexus injuries in humans (e.g., McCann et al. 2004). As in humans, there seems to be a rostro-caudal gradient, or at least for developing autotomy as this develops after dorsal rhizotomy of cervical and thoracic roots, but not lumbosacral nerve roots (Lombard et al. 1979).

NI caused by myelopathy has been linked by my group to a rat model of spinal-cord injury, involving micro-injection of quisqualate excitotoxin into the deep dorsal horn, which creates a small necrotic cavity (Dey et al. 2005). The dermatomal scratching and biting that some injected rats develop was interpreted as modeling neuropathic pain until we suggested it more closely modeled NI in humans with intramedullary cavernous hemangiomas (see below). We later found that injected rats that self-injure have deafferentation of their itchy dermatomes, implying that their small spinal cord injuries cause retrograde degeneration of primary afferent neurons as well as of intrinsic spinal neurons (Brewer et al. 2008). Thus, myelopathy-induced NI may not be purely central. This model also provided the information that only injected rats with lesions in the deep dorsal horn that spared the superficial lamina developed dermatomal self-injury (Yezierski et al. 1998). NI in these rats may thus be maintained by spontaneous activity in second order, lamina I, itch projection neurons responding to both loss of local inhibitory circuits (Ross et al. 2010) but also loss of normal peripheral input.

NI caused by brain lesions has not been systematically modeled in animals.

7.2   DIAGNOSIS AND MEASUREMENT

Itch, a subjective sensation, is diagnosed and measured based mostly on subject report. The following steps are recommended for clinical practice to establish the diagnosis of NI:

- Eliminate dermatological and systemic causes including medications.
- Establish inadequacy of conventional antipruritic treatments including anti-histamines and anti-inflammatories.
- Ask about known neurological disorder or nerve injury (e.g., shingles, degenerative spine disease, and peripheral neuropathy).
- Refer to a neurologist for history, neurological examination, and testing for nuerological causality and treatment recommendations.

There are no diagnostic tests for NI, but the following types of measurement can help support a clinical diagnosis or rule out other possibilities. Clinical approaches to measurement rely on questionnaires and rating scales, which are subjective (Wahlgren 1995). One objective approach is to measure the consequences of itch, namely, the scratch response, although this is an indirect correlate modified by other variables including subject volition, attention, and motivation. Photographic or videographic documentation of scratching and associated lesions has clinical and research utility.

Neurophysiological studies, specifically nerve conduction study and electro-myography (EMG) are often used to document peripheral nerve damage. However conventional electrodiagnostic testing is insensitive to small-fiber neuropathy; surface electrode nerve conduction studies only measure large (rapidly conducting) myelinated A fiber activity in mixed peripheral nerves, not the A-delta and C-fibers primarily involved in itch. With electromyography, abnormal "neurogenic" patterns only identify axonal injuries to motor axons. Abnormal studies can be clinically useful when they identify associated polyneuropathy or a focal nerve injury that also affects large-diameter sensory axons and/or motor axons, but normal studies do not rule out the presence of small-fiber nerve damage, or of subthreshold injuries, or injuries to nonstudied nerves (e.g., small sensory branches), and thus they lack negative predictive value for excluding small-fiber predominant nerve injuries. Microneurography or other special methods (Zotova and Arezzo 2013) may be required for electrophysiological diagnosis or confirmation of ectopic firing of peripheral itch axons.

Quantitative sensory testing is a more formal method of measuring sensory thresholds (most commonly thermal, mechanical, and vibratory) than examination, but despite its numerical results, the findings remain subjective and depend on subject motivation and alertness (Freeman et al. 2003). No diagnostically useful patterns have emerged for neuropathic pain conditions (Pappagallo et al. 2000). This method is mainly a research tool at present.

Radiological studies, specifically MRI or CT are optimal for detecting foraminal stenosis or other structural lesions that can impinge on cranial or spinal nerve roots to cause brachioradial pruritus or other radiculopathies. MR neurography and ultrasound are emerging techniques for visualizing nerves, especially in entrapment neuropathies.

Pathologic diagnosis of small-fiber polyneuropathy once required surgical biopsy of a sensory nerve. Now neurodiagnostic skin biopsy specially fixed and immunolabeled with axonal markers and/or autonomic function usually suffices (England et al. 2009). Skin biopsies have linked PHI to severe loss of distal unmyelinated axons (Oaklander et al. 2002) and to increased innervation in notalgia paresthetica (Savk et al. 2002; Springall et al. 1991) attributed to secondary inflammation. Neurodiagnostic skin biopsy is not routinely performed by dermatologists, and it must be planned in advance as it requires different fixatives and processing than those used for routine dermatopathological study. It involves removing a small piece of skin under local anesthesia, vertically sectioning and immunolabeling with a pan-axonal marker (i.e., PGP9.5) to permit evaluation and quantitation of sensory axons in the dermis and epidermis. Because normative densities differ at different body locations, and there is considerable interindividual variability, these biopsies are not useful diagnostically except rarely, when comparing biopsies from patients' itch-affected and matching unaffected areas confirm focal small-fiber losses.

7.3   CLINICAL NI SYNDROMES

7.3.1   WIDESPREAD NI FROM GENERALIZED PERIPHERAL NERVE DAMAGE (POLYNEUROPATHIES)

Widespread small-fiber predominant polyneuropathies (SFPN) typically include symptoms of burning pain, allodynia, hyperalgesia, sweating, and microvascular changes. Itch is currently not adequately recognized as part of this syndrome

(Figure 7.1). SFPN has metabolic (e.g., diabetes), infectious, rheumatologic, toxic, paraneoplastic, autoimmune, and hereditary causes, some of which are curable and thus a diagnosis mandates additional testing for cause (Amato and Oaklander 2004). Most small-fiber polyneuropathies are length dependent, beginning with foot pain that can progress proximally in a stocking-and-glove pattern. Occasional patients present with patchy, proximal, or total-body sensory symptoms caused by widespread ganglionopathy/neuronopathy; described below. If large (A alpha or beta) myelinated nerve fibers are also affected, there can be colocalizing hyporeflexia, weakness, and vibratory and proprioceptive deficits.

Toxic polyneuropathy caused by infusing hydroxyethyl starch (HES), a synthetic colloid widely used to increase intravascular volume in surgical, emergency, and intensive care patients, can cause widespread NI plus kidney and coagulation disorders. These are attributed to starch deposition in tissues including kidney and peripheral nerve (Kamann et al. 2007). Neuropathy symptoms typically include diffuse itching and sensorineural hearing loss (Klemm et al. 2007). It is debated whether or not newer formulations are safer (Hartog et al. 2011).

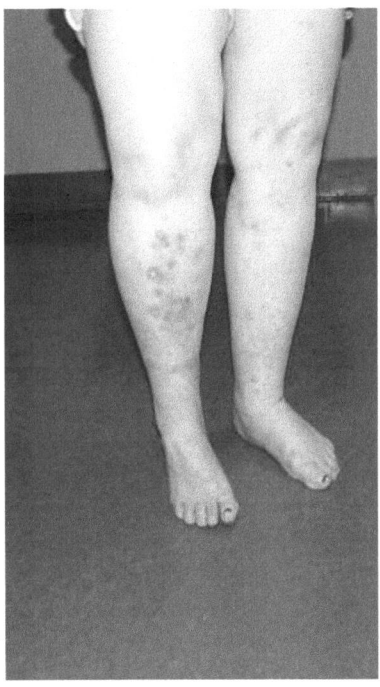

FIGURE 7.1 (See color insert.) Widespread neuropathic itch caused by small-fiber polyneuropathy. A patient with chronic itch and scratching-induced lesions on her distal legs. She had negligible sweat production from the forearm, proximal leg, distal leg, and foot sites after iontophoresis of acetylcholine, reflecting dysfunction of her cholinergic sudomotor sweat fibers, consistent with SFPN. (Reprinted from *Seminars in Cutaneous Medicine and Surgery*, 30(2), A. L. Oaklander, Neuropathic itch, 87–92, Copyright 2011, with permission from Elsevier.)

7.3.2 FOCAL NI FROM NERVE AND NERVE-ROOT LESIONS INCLUDING BRACHIORADIAL PRURITUS AND NOTALGIA PARESTHETICA

Any type of injury to itch-signaling peripheral afferent neurons can trigger itch in and near the receptive field of the injured neurons. As with neuropathic pain, it is entirely unknown why only a few patients with such injuries (which are common) develop clinically significant NI. One association is with neurofibromas, benign tumors that arise from Schwann cells. The itch is attributed to the abundant mast cells within these tumors (Johnson et al. 2000).

Radiculopathies involve damage to cranial or spinal nerve roots. These are established causes of neuropathic itch (see below) and other sensory symptoms including pain, and/or reduced sensations in the territory of affected nerve roots. If severe, patients can develop colocalizing motor deficits. Mechanical compression or irritation at the neural foramen ("pinched nerve") is the most common cause; others include inflammation or neoplasm (e.g., meningioma and schwannoma) and even Tarlov cysts (Oaklander et al. 2013).

Plexopathies refer to lesions of the intertwining neural networks near the spine where axons within nerve roots re-sort into peripheral nerves. Symptoms reflect often patchy or incomplete combinations of nerve root and nerve damage. The brachial plexus is affected more often than the lumbosacral plexus, and motor fibers are affected along with sensory fibers. Movements that irritate the affected plexus can worsen sensory symptoms. Common causes include autoimmunity (e.g., in brachial plexitis, also known as neuralgic amyotrophy or Parsonage-Turner syndrome [van Alfen and van Engelen 2006]), tumor or radiation therapy, diabetes, and tissue entrapment. Additional causes of brachial plexopathy include Pancoast lung tumors, neurogenic thoracic outlet syndrome, and SEPT9 mutations, which cause recurrent attacks (Jeannet et al. 2001). In children, traction injuries during birth have been associated with self-mutilation of the affected arm (McCann et al. 2004). Although such children are preverbal, and others have interpreted such behaviors as a sign of psychopathology or neuropathic pain, the author continues to interpret them as painless self-injury caused by the conjunction of neuropathic itch and loss of protective nociceptive pain sensation from severe sensory-nerve injury (Brewer et al. 2008; Dey et al. 2005), as in the trigeminal trophic syndrome discussed below.

Mononeuropathy, meaning damage to a single nerve, is most often caused by trauma. While originally associated with military trauma, medical injuries are now a more common cause of traumatic nerve injury in developed countries. Less frequent causes include internal lesions such as impingement, entrapment, scarring, infection (e.g., leprosy), or inflammation. Because some causes are curable or require independent treatment (e.g., tumor and aneurysm), focal peripheral neuropathies of unclear etiology require additional evaluation to identify their cause. Symptoms can spread outside the dermatome of the affected nerve (e.g., in the complex regional pain syndrome) owing to electrical coupling in the skin between nociceptive afferents (Meyer et al. 1985), secondary irritation of nearby neurons within nerve trunks, nerve roots or the DRG, the dorsal horn, and occasionally even the sensory cortex of the brain. Furthermore, there are poorly understood links between contralateral anatomically matched primary afferents that allow some types of unilateral nerve injuries to produce "mirror-image" effects on the contralateral uninjured areas (Oaklander et al. 1998; Oaklander and Belzberg 1997; Oaklander and Brown 2004; Scott et al. 2000). Induction or exacerbation of symptoms by percussing the involved nerve (Tinel's sign) can help with localization. Diagnostic aids include electrophysiological testing and advanced imaging modalities (see below).

Brachioradial pruritus is the historic term referring to NI originating from the cervical spinal nerves. NI characteristically presents with bilateral or less often unilateral itchy patches (with or without cutaneous stigmata of scratching) on the forearms, elbows, or upper outer arms (Massey and Massey 1986). Because of the location, and because most patients have worsening in the summer and improvement with protection against the sun (Veien and Laurberg 2011), ultraviolet exposure may be a contributory factor, but one third to one half of patients have radiological evidence of degenerative osteoarthritis of the spine compressing their cervical nerve roots (Abbott 1998; Goodkin et al. 2003; Veien and Laurberg 2011). As below, this same phenotype is less often caused by spinal cord lesions (myelopathy) including tumors (Kavak and Dosoglu 2002), so cervical MRI should be considered in patients without electrophysiological or other evidence of a peripheral lesion, particularly in patients with bilateral symptoms or signs or other symptoms suggestive of cervical myelopathy. A single-case report suggests efficacy of topical injection of botulinum toxin-A (Kavanagh and Tidman 2012).

Notalgia paresthetica is another well-known focal neuropathic itch syndrome. This historical term describes usually unilateral itchy areas on the back, near and below the scapula in the mid-thoracic dermatomes (Massey and Pleet 1979). These areas often exhibit the stigmata of chronic scratching as well, and dermatopathological examination reveals postinflammatory hyperpigmentation (Savk et al. 2002). Focal entrapment or irritation has been attributed to the location where nerves turn to traverse through the muscles of the back at an angle, and indeed, electromyographic study showed that seven of nine patients with notalgia paresthetica had evidence of chronic denervation of paraspinous muscles; localizing the causal lesion to the posterior rami of the spinal nerves (Massey and Pleet 1981). Radiological study also provides evidence of degenerative spine disease in up to 60% of studied patients so this same phenotype can equally be caused by radiculopathy (Raison-Peyron et al. 1999; Savk and Savk 2004, 2005), and no doubt by *zoster sine herpete*. Additionally, given the fact that even intramedullary lesions including spinal cord tumors can present as notalgia paresthetica (Johnson et al. 2000), thoracic MRI should be considered in notalgia paresthetica patients without electrophysiological evidence of a peripheral cause, particularly in those with bilateral symptoms or other signs and symptoms suggestive of myelopathy. Regarding treatment of notalgia paresthetica, there are case reports of efficacy of various neuropathic-pain medications including topical capsaicin and oxcarbazepine (Savk et al. 2001). Exercises to alter muscle configuration and presumably reduce nerve impingement have also been recommended (Fleischer et al. 2011; Savk and Savk 2004). Prospective study of injection of botulinum toxin type A in four patients with notalgia paresthetica, one with meralgia paresthetica and one with neuropathic itch of the foot showed mean reductions of itch by 28% at 6 weeks posttreatment (Wallengren and Bartosik 2010).

7.3.3 FOCAL NI FROM SENSORY GANGLIA LESIONS (NEURONOPATHIES/ GANGLIONOPATHIES) INCLUDING SHINGLES

Attacks on the sensory ganglia (cranial and dorsal root ganglia) cause sensory symptoms centered on the peripheral distribution of these ganglia. Motor function is usually spared because motor axons bypass most sensory ganglia, although motor axons may undergo secondary bystander damage due to proximity to degenerating sensory axons within the nerve roots, plexi, or nerves. Multiple sensory ganglia can also be affected by generalized conditions to cause widespread or patchy symptoms (polyneuropathy). Because sensory ganglia lack a blood–nerve barrier, they are vulnerable to autoimmune attack such as in Sjögren's, and paraneoplastic syndromes. Given the abundance of itch-associated primary afferents within the DRG (Han et al. 2012), the strong link between ganglionopathies and NI is no surprise.

Shingles (herpes zoster) is by far the most common type of sensory ganglionitis, and it is currently the best characterized cause of focal NI (Figure 7.2). Although clinicians recognized that dermatomal itch was a common sequel of zoster, and there were scattered reports, one well-studied patient who scratched through her skull into the frontal lobe of her brain after V1 shingles, brought much wider awareness of PHI (Oaklander et al. 2002). A global epidemiologic study of 586 adults with prior shingles demonstrated that itch, usually mild or moderate, affected roughly one third of patients with acute shingles or PHN. Furthermore, shingles affecting the face, head, or neck was significantly more likely to trigger PHI than shingles affecting more caudal dermatomes (Oaklander et al. 2003).

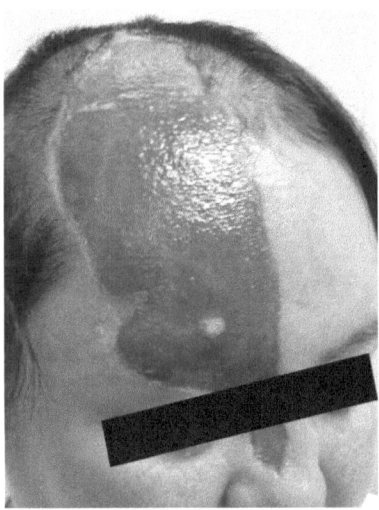

FIGURE 7.2 (**See color insert.**) Facial neuropathic itch caused by shingles (trigeminal trophic syndrome). A 21-year old woman with trigeminal trophic syndrome in the right V1 dermatome after cranial irradiation for pituitary adenoma followed by right V1 herpes zoster. Her ulceration was attributed to constant rubbing and scratching of persistent dysesthesias. (Reproduced from M. A. Nagel and D. Gilden, *Neurology* 77 (15):1499, 2011. With permission.)

7.3.4   NI Syndromes Arising from Lesions within the Spinal Cord (Myelopathy)

These are well-recognized causes of NI. It is not always clear whether the causal lesions are affecting the PNS or CNS, because primary afferent neurons extend into the spinal cord. Many peripheral sensory neurons synapse segmentally on projection neurons (e.g., spinothalamic tract neurons; see Chapter 22), but the central axons of some unmyelinated fibers as well as of myelinated fibers ascend in the dorsal columns (Briner et al. 1988). Given the small size of the spinal cord, many lesions likely affect all these compartments (Brewer et al. 2008).

It is important to recognize that NI can be the presenting sign of spinal cord injury because this requires neurological evaluation and treatment if possible. Scattered reports associate NI with various causes of myelopathy including tumors (Johnson et al. 2000; Kavak and Dosoglu 2002), multiple sclerosis (Yamamoto et al. 1981), syringomyelia (Kinsella et al. 1992), and traumatic injury causing the Brown-Séquard syndrome (Thielen et al. 2008). These authors attributed their patient's NI to axotomy of itch neurons crossing the ventral commissure en route to the contralateral anterolateral spinothalamic tract. Paramedian lesions may similarly downregulate the existing tonic inhibitory control of itch in the spinal cord, e.g., by bhlhb5 expressing neuron (Ross et al. 2010). There are more case reports of rostral than caudal lesions. One cervical ependymoma tumor extending between C4-C7 presented with brachioradial pruritus on both arms, elbows, and forearms as the only symptom (Kavak and Dosoglu 2002).

The best documented association is with inflammatory myelopathies. Among a series of 377 MS patients, 4.5% reported NI (Matthews 2005). Itch was even more common in myelopathy from neuromyelitis optica (NMO), a relapsing autoimmune/inflammatory demyelinating disorder of the spinal cord and brain (including the optic nerves) associated with aquaporin-4 autoantibody (AQP4-Ab) (Elsone et al. 2012). Twenty-seven percent among 44 patients with AQP4-Ab-positive NMO myelitis had NI attributed to their spinal cord lesion. Itch affected the trunk in 67%, the limbs in 75%, and the occiput in 25% of patients, most often accompanied by other colocalizing sensory symptoms (Elsone et al. 2012). About half had continuous pruritus whereas half had intermittent symptoms; median itch intensity was 6 of 10 (Elsone et al. 2012). The authors postulated that itch was more common in NMO than in MS because MS affects mostly white-matter tracts, whereas NMO primarily damages gray-matter neuronal cell bodies, including the lamina 1 second-order dorsal-horn neurons that process itch signals.

The second best documented association between spinal-cord injury and itch is with cavernous hemangiomas of the spinal cord (Cohen-Adad et al. 2012; Dey et al. 2005; Sandroni 2002; Vuadens et al. 1994). The author suggested that features of these rare lesions make them pruritogenic, specially their preferred location in the upper rather than the lower spinal cord and perhaps in the dorsal rather than ventral cord (although dorsal cavernomas may simply be those that are more operable and come to neurosurgical attention), plus the tendency of these lesions to trigger gliosis and bleeding. Cavernomas are probably pruritogenic for the same reasons that they are epileptogenic when in the brain; gliosis and hemosiderin deposition both excite ectopic action potentials (Dey et al. 2005). Of note, our rat model of NI from spinal cord cavernoma (see above) includes both these features, and the cervical syrinx associated with NI contained hemosiderin consistent with prior bleeding as well (Kinsella et al. 1992).

7.3.5   NI SYNDROMES AFFECTING THE HEAD AND NECK (CRANIAL NERVES AND GANGLIA)

Lesions of any cranial nerve that contain somatosensory representation, its ganglion, or its central projections can cause NI and/or neuropathic pain that localizes to the region innervated by that nerve's sensory axons. In the head and neck, this includes VI (nervus intermedius) and IX in addition to V. At least one case of neuropathic self-injury to a tonsil has been associated with damage to IX (Stefaniu et al. 1973). Although any type of lesion or injury can trigger neuropathic sensory symptoms, the most common causes are vascular compression (e.g., tic douloureux) or zoster, which can affect VI and IX as well as all three branches of V (of which involvement of V_1 or herpes zoster ophthalmicus is most common).

Cranial neuropathies are disproportionately likely to trigger NI as compared to lesions affecting nerves innervating the lower torso, legs, and the sacrally innervated dermatomes. The reasons are unknown, but it has been hypothesized that they might be related to protecting the facial mucosa, a prime target of insect attack (Oaklander et al. 2003). In PHI, after shingles, data from several independent groups confirm that higher proportions of patients with shingles of the head or neck develop PHI than patients with zoster on the torso (Oaklander et al. 2003). A similar gradient in susceptibility to autotomy after dorsal rhizotomy is described in rats (Lombard et al. 1979). Despite its proximity to the brain, neuropathic itch of the face, head, and neck is more likely to be caused by lesions of peripheral rather than central neurons. The discussion above pertaining to ganglionopathies and radiculopathies applies equally to the cranial nerve ganglia.

7.3.6   TRIGEMINAL TROPHIC SYNDROME AND OTHER FORMS OF SELF-INJURIOUS BEHAVIOR FROM NI

Group III itch according to the IFSI classification (Ständer et al. 2007) includes self-injurious scratching or mechanical stimulation of the receptive field innervated by injured sensory nerves. Such self-injury was for many years attributed by most to an aberrant response to neuropathic pain (Mailis 1996; Rapin and Ruben 1976) plus or minus coexisting psychopathology. Few patients were ever actually asked if they felt itch. Some authors mention vague descriptors of itch (e.g., aversive sensations and formication). Some patients had itch and pain, making it difficult to be certain which motivated their self-stimulation. However, given that pain triggers protective behaviors such as withdrawal, whereas itch triggers scratching, pain seems implausible as a motivation for self-injury. Publication of the case of a woman who scratched through her skull because of intractable PHI (without postherpetic pain) (Oaklander et al. 2002), helped demonstrate the self-injurious scratching was a marker of neuropathic itch colocalizing with loss of protective pain sensation. Also, see the discussion above of animal models regarding autotomy.

A contributing factor to such self-injurious behavior is many patients' (and even their physicians') unawareness that their cutaneous lesions are the consequence of their itch and not the cause of it. Teaching how to break the itch–scratch cycle is imperative. The confusing nomenclature for the skin changes caused by excessive scratching (e.g., prurigo nodularis, lichen planus, and macular amyloidosis) can obscure the true cause, which, in many cases, is chronic itch. Because scratching can occur while patients are sleeping or inattentive (Savin et al. 1973, 1990; Savin 1975), barriers to scratching (e.g., wearing mittens during sleep), and behavioral modification can be effective. Topical local anesthetics can also be helpful, testifying to the importance of remaining primary afferent signaling in maintaining this condition.

My explanation is that severe damage to primary afferent pathways can leave isolated remaining itch axons firing ectopically (without provocation) but in insufficient numbers to generate the powerful surround fields of inhibition that normally laterally inhibit itch in the dorsal horn (Oaklander et al. 2002). This is consistent with the fact that scratching, which augments nociceptive afferent firing in and around the receptive field, provides transient relief, and with reports that administering minute amounts of pruritogens to the skin using cowhage spicules produces itch disproportionate in intensity to the dose of pruritogen or to the size of the stimulated area (Sikand et al. 2011).

Trigeminal trophic syndrome (TTS) is a historical term referring to self-induced traumatic lesions of the head and face caused by the most severe forms of NI. The worst cases are literally disfiguring and even potentially fatal. Other historical names include trigeminal neuropathy with nasal ulceration, trigeminal neurotrophic ulceration, ulceration en arc, and trophic ulceration of the *ala nasi* (Jaeger 1950; Ziccardi et al. 1996). Attributed first to loss of a "trophic" substance normally supplied by axons, then to psychopathology during the Freudian era, TTS is now attributed to the disastrous conjunction of intractable neuropathic itch plus colocalizing loss of protective pain sensation caused by severe nerve injury. Some but not all patients have colocalizing neuropathic pain as well.

TTS was reported and studied in numerous publications including large case series by early twentieth century neurosurgeons including Harvey Cushing. It occurred then as an adverse effect of trigeminal rhizotomy, the only logical treatment for trigeminal neuralgia in the premedication era (Becker 1925; Cliff and Demis 1967; Cushing 1920; Harper 1985; Henderson 1967; Howell 1962; Jefferson and Schorstein 1955; Karnosh and Scherb 1940; Kavanagh et al. 1996; Knight 1954; Loveman 1933; McKenzie 1933; Philpott 1941; Schorstein 1943). TTS was also reported after other iatrogenic injuries to the trigeminal ganglion and root including injection of neurotoxins or thermal injury (Harris 1940; Walton and Keczkes 1985). Rare causes include infections such as syphilis, leishmaniasis, and leprosy (Thomas et al. 1991). Shingles currently appears to be the most common peripheral cause (Figure 7.2). The TTS syndrome can also be produced by brain lesions, most often stroke (see below). There are scattered reports of self-injurious scratching inside the mouth or nose. One patient whose "burning mouth syndrome" of unknown etiology apparently included itch, although itch was not explicitly mentioned, used "mechanical manipulation" inside her mouth to "obtain relief" to the point of causing superficial lesions. Her symptoms (and erosions) resolved after gabapentin treatment (Meiss et al. 2002).

In addition to the treatments discussed below for NI, patients with self-injury from NI require additional treatment. Most important is to make sure that patients understand that their lesions are self-induced. Behavioral modification including barriers to protect the affected area and wearing mittens during sleep can impede scratching. TTS injuries may require treatment of infections, wound management, and surgical reconstruction of defects. This must be done with innervated flaps as grafts of denervated tissues from the affected areas are less viable and patients continue to damage them (McLean and Watson 1982). Other strategies to modulate peripheral activity in itch neurons, such as transcutaneous electrical stimulation (Westerhof and Bos 1983) are worth considering.

7.3.7 Neuropathic Itch Syndromes Arising from the Brain

Brain lesions that cause NI validate the concept of "central itch"; that lesions of second- and third-order CNS itch neurons are also capable of triggering it. They demonstrate that NI is a systems or network disorder caused by imbalances between excitatory and inhibitory transneuronal signaling, rather than by consequence of lesions in one specific location. Case reports demonstrate that any type of brain lesion that affects the brain's itch neurons can cause central neuropathic itch, including TTS (King et al. 1982). Central lesions were said to cause about one fifth of cases of TTS (Spillane and Wells 1959).

Stroke is the most common brain lesion associated with NI, particularly infarctions of the lateral medulla (Wallenberg's syndrome; Figure 7.3) or slightly more rostral strokes in the lateral pons where itch signals presumably ascend (Curtis et al. 2012; Dick and Gonyea 1990; Fitzek et al. 2006; Massey 1984; Savitsky and Elpern 1948; Shapiro and Braun 1987; Wallenberg 1901). Even today, some stroke specialists are not aware of these associations (Oaklander et al. 2009). Strokes that only cause weakened chewing or reduced facial touch with preserved pain sensation do not cause NI and TTS, only those that damage nociceptive pathways and cause profound sensory loss (Fitzek et al. 2006). Such lesions identify the location of ascending itch pathways (see Chapter 22). Neurophysiological study of nonhuman primates has established that itch-signaling spinothalamic tract neurons project to the ventral posterior lateral nucleus, ventrobasal complex, and the posterior thalamic nuclei (Davidson et al. 2012; Simone et al. 2004). Correspondingly, NI has also been associated with Dejerine-Roussy syndrome (also known as the posterior thalamic syndrome and retrolenticular syndrome), caused by lesions of the ventral posterior thalamus and internal capsule (King et al. 1982). Better known symptoms include hemianesthesia and hemi-pain contralateral to the lesion. One of few studies that correlated sensory disturbances with anatomical localization of infarctions of the posterior cerebral artery (it supplies the ventrolateral thalamic sensory nuclei and white-matter tracts to the somatosensory cortex) found that all patients with sensory findings had thalamic damage, and that among patients with sensory findings, 11 of 15 had infarctions in the ventrolateral thalamus in the territory of the

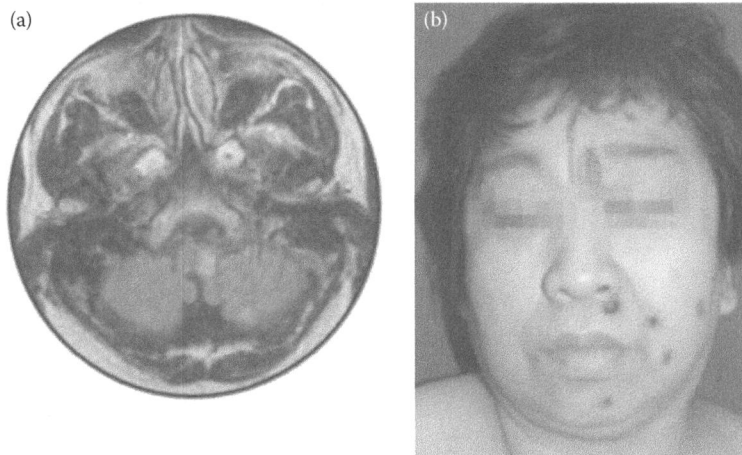

FIGURE 7.3 **(See color insert.)** Focal neuropathic itch due stroke (trigeminal trophic syndrome). A woman with trigeminal trophic syndrome (TTS) as part of Wallenberg syndrome: (a) Location of her lateral medullary infarction on imaging and (b) excoriations of the itchy areas of her face and contralateral trunk (not shown) caused by scratching. Three months poststroke, the pruritus reportedly resolved after treatment with gabapentin and topical moisturizers. (Reproduced from W. K. Seo et al., *Neurology*, 72(7), 676, 2009. With permission.)

thalamogeniculate or lateral posterior choroidal arteries (Georgiadis et al. 1999). However, as in most such studies, itch was not specifically studied.

Interestingly, Wallenberg's original patient with TTS had shingles in his stroke affected area as well, so the exact cause of that patient's NI remains uncertain (Wallenberg 1901; Wolf 1971). Less frequent brain lesions associated with NI include multiple sclerosis (Matthews 2005; Ostermann 1976; Yamamoto et al. 1981), tumors within or near the brain (Andreev and Petkov 1975; Summers and MacDonald 1988), and infections (Sullivan and Drake 1984). NI or scratch marks that spread beyond the dermatome of one individual trigeminal division suggest a causal lesion in the ganglion or brain rather than in the PNS.

Severe pruritus unresponsive to conventional treatments has been associated with several prion diseases including Creutzfeldt-Jakob disease (CJD) in humans. Scrapie (Petrie et al. 1989), which develops in sheep and goats in western Europe, is named for the characteristic injurious self-directed scratching and biting. This most often affects the rump, unlike the rostral predominance typical of other human and animal forms of NI. NI appears to occur only rarely in sporadic CJD (Shabtai et al. 1996) but was reported by 19% among 31 patients with familial CJD (Cohen et al. 2011). The itch was generalized in three patients, regional in two and localized in one patient. It was transient in one patient and continued throughout the disease in five patients. Diffusion weighted imaging associated the itch with reduced diffusion in several brain areas known to be affected by CJD, but most significantly with the midbrain periaqueductal gray. The authors postulated damage to inhibitory gating mechanism for itch however, because prions are deposited in the skin and in the peripheral nerves in CJD and other human prion diseases (Lee et al. 2005), the itch signals might also have conceivably been generated in the PNS.

In addition to the brain stem and posterior thalamus, higher brain areas participate in itch perception and processing (see Chapters 23 and 24). Positron emission tomography in normal adults implicates primary but not secondary somatosensory cortex (Brodmann 24) as well as the periaqueductal gray and cingulate gyrus in histamine-induced itch (Drzezga et al. 2001; Hsieh et al. 1994; Mochizuki et al. 2003). However, the secondary somatosensory cortex and insula have been implicated in electrically induced acute itch (Mochizuki et al. 2009). There are very rare reports of patients with small somatosensory-cortex lesions and somatotopically appropriate NI, usually associated with seizure (Maciejewski and Drop 2002) or subarachnoid hemorrhage (Canavero et al. 1997), consistent with cortical hyperactivity rather than degeneration.

7.3.8 Phantom Itch in Amputated Body Parts

Phantom itch is well described and even known to the public. It can occur after amputation of any innervated body part. Among the roughly 60% of women with phantom sensations after mastectomy, itch was the most common (Lierman 1988). The prevalence of phantom itch demonstrates the critical role of denervation in generating neuropathic itch; phantom itch is an extreme endophenotype of peripheral NI, which arise in the context of sensory deafferentation. Because the itchy area is missing and thus cannot be scratched, phantom itch is not associated with self-injury. It remains uncertain exactly which neurons fire excessively to generate phantom itch sensation. On the basis of the more plentiful studies of postamputation pain, possibilities include spontaneous activity of isolated peripheral nociceptors, perhaps those that are regenerating or entrapped in scar tissue (as with postburn itch) or neuromas. These can generate "stump" sensations perceived in remaining tissue as well as sensations perceived in prior receptive fields that have been amputated. Edge effects at the border of innervated and noninnervated tissues may also contribute. However, central mechanisms, whether originating in the spinal cord or brain, are paramount in phantom sensations including the Charles Bonnet syndrome after visual loss and tinnitus after hearing loss (Jastreboff 1990; Schultz and Melzack 1991). Phantom itch can only be studied in humans who can report it, but there are few or no studies of patients with phantom itch independent of phantom pain. It is likely that thalamic plasticity, including expansion of the representation of the proximal limb into thalamic region that used to represent the amputated part contributes, as well as unmasking of latent receptive fields (Dostrovsky 1999).

Treatment options primarily involve medications shown effective for phantom pain and other sensations, such as the gabapentinoids and tricyclics. Several methods involve increasing remaining afferent input. This includes behavioral therapies such as wearing a prosthesis to provide visual and functional input (Lierman 1988), mechanically stimulating the area or the contralateral body part and "mirror therapy" (Chan et al. 2007). Neuromodulation may have similar effects by triggering action potentials at various parts of the itch network independently of afferent input (Ahmed et al. 2011). One patient reportedly found relief from phantom itch and pain by scratching or massaging his prosthesis or the leg of another person (Weeks and

Tsao 2010). The authors hypothesized that sensory mirror neurons permitted integration of this feedback into his own sensory circuits.

7.4   MANAGING AND TREATING NEUROPATHIC ITCH

The reflex and volitional scratching triggered by NI bring fleeting relief at best, and treatments effective for conventional itch such as antihistamines and anti-inflammatories are usually ineffective, so treatment is a challenge in most cases. Pharmacotherapy is worth considering for patients with disabling NI, particularly when scratching remains uncontrolled even after its origin is explained and behavioral modification is attempted. No medication has a U.S. Food and Drug Administration indication for neuropathic itch and no high-quality clinical trials have been conducted. No studies of NI are considered Class I according to evidence-based medical guidelines, such as those of the American Academy of Neurology. The information below comes from Class IV studies. Insurance companies can deny reimbursement as there is no good evidence for most treatments. It is self-evident that the best treatments are disease modifying, such as treatment of the cause of an itch-inducing peripheral nerve or root lesion. The remarks below refer to management of symptoms for patients who do not have this option. The efficacy of medications prescribed for neuropathic itch is variable.

7.4.1   CONSERVATIVE TREATMENTS

Options to consider in all patients include those that minimize secondary itch and inflammation caused by scratching, such as maintaining good skin hydration, as well as barriers to reduce scratching. Barriers should be used in most patients with self-injury from scratching NI, particularly while patients sleep and scratch (Savin et al. 1973). Options include a lockable helmet to protect the scalp in patients with V1 TTS, rigid surgical braces to protect the limbs and torso, and wearing mittens at night. Video recordings document that patients with severe itch engage in nocturnal scratching while sleeping and are unaware of their actions. Cognitive behavioral therapy may be required to help break the itch–scratch cycle or at least to substitute less destructive methods for scratching. Other behavioral changes to consider include protecting affecting areas from itch-triggering stimuli (e.g., contact with clothing) and instituting low levels of tonic mechanical stimulation. Wearing a close-fitting binder or wrap can be helpful in several ways. Exercises to reduce nerve impingement in notalgia paresthetica have also been recommended (Fleischer et al. 2011; Savk and Savk 2004). Consideration should also be given to ways to reducing barriers to axonal regeneration and healing, among which smoking cessation and regular exercise to improve tissue perfusion are paramount. However, many patients will require drug therapies as well.

7.4.2   TOPICAL APPLICATION OF MEDICATIONS

Topical treatments that have no systemic absorption and hence no systemic adverse effects are appealing. Among them, the local anesthetics are paramount. These penetrate mucous membranes well (with some potential for systemic absorption) and thus should be considered for chronic itch affecting the mucosa. They are not administered to the eye because of the potential for suppressing protective pain sensation. On the skin, local anesthetics require an occlusive wrap or patch (e.g., 5% lidocaine patch) for penetration. Rare patients with significant self-injury require daily subcutaneous injection of long-active local anesthetics. Rarely, have caregivers or patients with V1 PHI been taught to perform supraorbital nerve blocks at home.

The utility of topical capsaicin (8-methyl-N-vanillyl-6 nonenamide) is uncertain. Punctate or miniscule experimental doses can cause itch (Sikand et al. 2011). When applied to larger areas, topical capsaicin induces predominantly pain, which inhibits itch. Over-the-counter products containing low-concentration capsaicin (usually 0.025%–0.075% w/w) have marginal efficacy against various pain conditions in clinical studies, but they require dosing three to five times daily. Systematic review found no convincing evidence for the use of low-dose topical capsaicin to treat pruritus in any medical condition, although the quality of most of the trials reviewed was poor (Gooding et al. 2010). Practically speaking, most patients find capsaicin application painful and few continue it. Scientifically speaking, capsaicin causes degeneration of distal TRPV1[+] axon terminals (Simone et al. 1998), producing the same kind of axonal injuries that are associated with neuropathic pain and itch so there is at least potential for delaying or impeding axonal regeneration. A single-dose high-concentration patch containing 8% w/w is currently licensed in the European Union for treating peripheral neuropathic pain in nondiabetic adults and in the United States for treating PHN, but PHI was not studied. Case reports support utility of other topical treatments such as topical tacrolimus and gabapentin (Nakamizo et al. 2010).

7.4.3 INTRADERMAL INJECTION OF BOTULINUM TOXIN TYPE A

BTX-A administered by local subcutaneous injection is increasingly used to treat peripheral neuropathic pain. Anti-itch effects are independent of its better known ability to weaken cholinergic neuromuscular transmission. The antipruritic benefits are based on inhibition of cutaneous mediators of neurogenic inflammation including substance P, CGRP, and glutamate, as well as inhibition of vanilloid receptor activity (Ranoux et al. 2008). For this reason, BTX-A is being explored in itch, and there is prospective evidence of efficacy in histamine-induced experimental itch (Gazerani et al. 2009). Isolated case reports (Kavanagh and Tidman 2012) and a very small study of four patients with notalgia paresthetica, one with meralgia paresthetica and one with neuropathic itch on the upper foot that showed a mean reduction of VAS by 28% at 6 weeks posttreatment (Wallengren and Bartosik 2010), suggest efficacy so this is worth considering. As of 2012, an industry-sponsored phase III trial of BTX-A (Xeomin) was registered with the US NIH's ClinicalTrials.gov Web site.

7.4.4 SYSTEMIC PHARMACOLOGIC TREATMENT

There are no controlled trials (Class I–III) of treatments for neuropathic itch; hence the recommendations below are based on Class IV evidence such as case reports or expert opinion. General recommendations include beginning with one of the medications discussed below, then raising the dose until relief is obtained or persistent adverse effects ensue. During a medication trial the dose should usually be raised to the maximum tolerated before discontinuing that medication and trying another; preferably from a different class. If partial relief is achieved from a well-tolerated medication, the second one added should be from a different class. Too often, patients are treated with low doses of multiple medications making it difficult to assess each one's efficacy or to attribute adverse effects.

Current options include systemic administration of medications that reduce ectopic neuronal firing, such as cation-channel antagonists (e.g., anti-epileptic medications), or those that augment inhibitory circuits (e.g., tricyclics). While many of these are also effective for neuropathic pain, not all are. Opioids—which are effective for neuropathic pain—trigger pruritus, particularly when administered near the spinal cord (Jinks and Carstens 2000). Isolated case reports, mostly in TTS patients, support use of carbamazepine (Bhushan et al. 1999), gabapentin (Macovei et al. 1991), amitriptyline (Finlay 1979), and pimozide (Duke 1983; Mayer and Smith 1993). One patient with intractable trigeminal itching and scratching after stroke plus shingles found efficacy of bupivicaine and clonidine administered intrathecally through a high thoracic catheter (Elkersh et al. 2003). As in neuropathic pain (Challapalli et al. 2005; Ferrini and Paice 2004), continuous systemic administration of low doses of local anesthetics by continuous subcutaneous administration at home, is a powerful treatment option that is underutilized. Brief courses of intravenous local anesthetics can be used for an in-hospital trial with cardiac monitoring to determine if there is a dose that is effective without producing adverse effects (Watson 1973).

7.4.5   NEUROSURGICAL TREATMENT OPTIONS

Prior to the mid-twentieth century development of effective drugs, surgery was the only treatment for morphine-unresponsive neuropathic pain (Warren 1828). The first neurosurgeon, Harvey Cushing, established that ablative neurosurgery (cutting nerves carrying sensation from painful areas) is usually ineffective and in some cases triggers intractable *anesthesia dolorosa* pain (Cushing 1920). In contrast, surgical decompression or removal of structural lesions causing sensory symptoms including NI is potentially curative, even in medically refractory cases. Occasional rare causes of focal NI such as tumors or vascular malformations may require nerve surgery for definitive treatment. These surgeries are currently underutilized, in part because they requires skilled diagnosis and lesion localization by history, examination, electrophysiological study, or imaging before surgery. Itch is not yet adequately recognized as a potential localizing sign for structural nerve and root injuries.

Augmentive neurostimulation using implanted bipolar electrodes has been proven effective for various neurological symptoms including neuropathic pain, and has been used for decades. Its major advantage compared to medications is that its tiny electrical currents only affect nearby cells, unlike drugs that that can spread through the body to cause diverse adverse effects. Neurostimulation has few if any systemic adverse effects. To date, there has been little or no research on potential benefits for NI. Stimulation locations used for neuropathic pain and that should be considered for refractory NI, include proximal to a focal nerve injury (Campbell and Long 1976), the dorsal column of the spinal cord, the motor cortex, and the thalamus or periaqueductal gray (Nguyen et al. 2011). The major disadvantage of neurostimulation has been the requirement for surgical implantation. However, neurons can increasingly be activated from outside the body by electromagnetic induction or direct current stimulation. Noninvasive transcranial magnetic stimulation (TMS) of the brain is an innovative alternative with much lower risk and cost. Repetitive TMS was FDA approved for treating major depression in 2008, and several small clinical trials suggest efficacy for chronic neuropathic pain (Leung et al. 2009), so TMS should be considered for NI.

7.4.6 CONTRAINDICATED THERAPIES

It is perhaps equally more important in avoiding ineffective or potentially injurious treatments as in finding effective ones. Oral treatments for normal itch (e.g., H1 blockers) are usually ineffective for NI, although worth trying due to their low cost and safety, plus some patients may have a component of conventional itch that can be ameliorated by such treatment, e.g., due to substance P release from ectopically firing injured nociceptive neurons (Hagermark et al. 1978). The uncertain utility of capsaicin as a treatment for NI is discussed above. Although local anesthetics are often effective when applied topically or given systemically, there is no evidence that nerve or nerve root blocks containing local anesthetics (and/or corticosteroids) have long-term benefits that justify their risk and cost, despite transient efficacy (Eisenberg et al. 1997; Goulden et al. 1998).

7.5 CONCLUSIONS AND FUTURE DIRECTIONS

In summary, neuropathic itch, at the watershed between dermatology and neurology, has received inadequate medical and scientific attention considering how frequent and disabling it is. However, maturation of the vibrant underlying neuroscience is beginning to attract notice from neurologists. Research priorities and opportunities in the field of neuropathic itch include clarifying whether or not autotomy after experimental nerve injury is a model of neuropathic itch or of pain. If this models itch, such animal models might be used to screen potential treatments for neuropathic itch including medications (Duckrow and Taub 1977) and neurosurgical treatment options (Rossitch et al. 1993; Sarkis et al. 1984).

There are even more urgent clinical needs, since many if not most physicians are still unaware of the very existence of neuropathic itch, and neither dermatologists nor neurologists are taught how to evaluate and treat such patients. There has never been a US Food and Drug Administration approved treatment for neuropathic itch. Given the large overlap between neuropathic itch and pain, itch should be incorporated as a secondary outcome measure in clinical trials for new treatments for neuropathic pain.

ACKNOWLEDGMENT

Supported in part by the Public Health Service (NINDS K24-NS059892).

REFERENCES

Abbott, L. G. (1998). Neuropathic pruritus. *Australas J. Dermatol.* 39(3): 198–199.

Ahmed, M. A., Mohamed, S. A., and Sayed, D. (2011). Long-term antalgic effects of repetitive transcranial magnetic stimulation of motor cortex and serum beta-endorphin in patients with phantom pain. *Neurol. Res.* 33(9): 953–958.

Amato, A. A., and Oaklander, A. L. (2004). Case records of the Massachusetts General Hospital. Case 16-2004. A 76-year-old woman with pain and numbness in the legs and feet. *New Engl. J. Med.* 350(21): 2181–2189.

Andreev, V. C., and Petkov, I. (1975). Skin manifestations associated with tumours of the brain. *Br. J. Dermatol.* 92(6): 675–678.

Asada, H., Yamaguchi, Y., Tsunoda, S., and Fukuda, Y. (1996). Relation of abnormal burst activity of spinal neurons to the recurrence of autotomy in rats. *Neurosci. Lett.* 213(2): 99–102.

Basbaum, A. I., and Wall, P. D. (1976). Chronic changes in the response of cells in adult cat dorsal horn following partial deafferentation: The appearance of responding cells in a previously non-responsive region. *Brain Res.* 116(2): 181–204.

Becker, S. W. (1925). Dermatitis in association with disease or injury of the peripheral nerves. *Arch. Derm. Syphilol.* 12: 235.

Bhushan, M., Parry, E. J., and Telfer, N. R. (1999). Trigeminal trophic syndrome: Successful treatment with carbamazepine. *Br. J. Dermatol.* 141(4): 758–759.

Blumenkopf, B., and Lipman, J. J. (1991). Studies in autotomy: Its pathophysiology and usefulness as a model of chronic pain [see comments]. *Pain* 45(2): 203–209.

Brewer, K. L., Lee, J. W., Downs, H., Oaklander, A. L., and Yezierski, R. P. (2008). Dermatomal scratching after intramedullary quisqualate injection: Correlation with cutaneous denervation. *J. Pain* 9(11): 999–1005.

Briner, R. P., Carlton, S. M., Coggeshall, R. E., and Chung, K. S. (1988). Evidence for unmyelinated sensory fibres in the posterior columns in man. *Brain* 111(Pt 5): 999–1007.

Campbell, J. N., and Long, D. M. (1976). Peripheral nerve stimulation in the treatment of intractable pain. *J. Neurosurg.* 45(6): 692–699.

Canavero, S., Bonicalzi, V., and Massa-Micon, B. (1997). Central neurogenic pruritus: A literature review. *Acta Neurol. Belg.* 97(4): 244–247.

Challapalli, V., Tremont-Lukats, I. W., McNicol, E. D., Lau, J., and Carr, D. B. (2005). Systemic administration of local anesthetic agents to relieve neuropathic pain. *Cochrane Database Syst. Rev.* (4) CD003345.

Chan, B. L., Witt, R., Charrow, A. P., Magee, A., Howard, R., Pasquina, P. F., Heilman, K. M., and Tsao, J. W. (2007). Mirror therapy for phantom limb pain. *New Engl. J. Med.* 357(21): 2206–2207.

Chandler, R. L. (1961). Encephalopathy in mice produced by inoculation with scrapie brain material. *Lancet* 1(7191): 1378–1379.

Cliff, I. S., and Demis, D. J. (1967). Giant ulcer of the face following surgery for trigeminal neuralgia. *Arch. Intern. Med.* 119(2): 218–222.

Cohen, O. S., Chapman, J., Lee, H., Nitsan, Z., Appel, S., Hoffman, C., Rosenmann, H., Korczyn, A. D., and Prohovnik, I. (2011). Pruritus in familial Creutzfeldt-Jakob disease: A common symptom associated with central nervous system pathology. *J. Neurol.* 258(1): 89–95.

Cohen-Adad, J., Zhao, W., Wald, L. L., and Oaklander, A. L. (2012). 7T MRI of spinal cord injury. *Neurology* 79(22): 2217.

Curtis, A. R., Oaklander, A. L., Johnson, A., and Yosipovitch, G. (2012). Trigeminal trophic syndrome from stroke: An under-recognized central neuropathic itch syndrome. *Am. J. Clin. Dermatol.* 13(2): 125–128.

Cushing, H. (1920). The major trigeminal neuralgias and their surgical treatment based on experiences with 332 Gasserian operations. *Am. J. Med. Sci.* 160: 157–184.

Davidson, S., Zhang, X., Khasabov, S. G., Moser, H. R., Honda, C. N., Simone, D. A., and Giesler, G. J., Jr. (2012). Pruriceptive spinothalamic tract neurons: Physiological properties and projection targets in the primate. *J. Neurophysiol.* 108(6): 1711–1723.

Dey, D. D., Landrum, O., and Oaklander, A. L. (2005). Central neuropathic itch from spinalcord cavernous hemangioma: A human case, a possible animal model, and hypotheses about pathogenesis. *Pain* 113(1–2): 233–237.

Dick, M. T., and Gonyea, E. (1990). Trigeminal neurotrophic ulceration with Wallenberg's syndrome. *Neurology* 40(10): 1634–1635.

Dostrovsky, J. O. (1999). Immediate and long-term plasticity in human somatosensory thalamus and its involvement in phantom limbs. *Pain Suppl.* 6: S37–S43.

Drzezga, A., Darsow, U., Treede, R. D., Siebner, H., Frisch, M., Munz, F., Weilke, F., Ring, J., Schwaiger, M., and Bartenstein, P. (2001). Central activation by histamine-induced itch: Analogies to pain processing: A correlational analysis of O-15 H2O positron emission tomography studies. *Pain* 92(1–2): 295–305.

Duckrow, R. B., and Taub, A. (1977). The effect of diphenylhydantoin on self-mutilation in rats produced by unilateral multiple dorsal rhizotomy. *Exp. Neurol.* 54(1): 33–41.

Duke, E. E. (1983). Clinical experience with pimozide: Emphasis on its use in postherpetic neuralgia. *J. Am. Acad. Dermatol.* 8(6): 845–850.

Eisenberg, E., Barmeir, E., and Bergman, R. (1997). Notalgia paresthetica associated with nerve root impingement. *J. Am. Acad. Dermatol.* 37(6): 998–1000.

Elkersh, M. A., Simopoulos, T. T., Malik, A. B., Cho, E. H., and Bajwa, Z. H. (2003). Epidural clonidine relieves intractable neuropathic itch associated with herpes zoster-related pain. *Reg. Anesth. Pain Med.* 28(4): 344–346.

Elsone, L., Townsend, T., Mutch, K., Das, K., Boggild, M., Nurmikko, T., and Jacob, A. (2012). Neuropathic pruritus (itch) in neuromyelitis optica. *Mult. Scler.* 19(4): 475–479.

England, J. D., Gronseth, G. S., Franklin, G., Carter, G. T., Kinsella, L. J., Cohen, J. A., Asbury, A. K. et al. (2009). Practice parameter: Evaluation of distal symmetric polyneuropathy: Role of autonomic testing, nerve biopsy, and skin biopsy (an evidence-based review). Report of the American Academy of Neurology, American Association of Neuromuscular and Electrodiagnostic Medicine, and American Academy of Physical Medicine and Rehabilitation. *Neurology* 72(2): 177–184.

Ferrini, R., and Paice, J. (2004). How to initiate and monitor infusional lidocaine for severe and/or neuropathic pain. *J. Support Oncol.* 2: 90–94.

Finlay, A. Y. (1979). Trigeminal trophic syndrome. *Arch. Dermatol.* 115(9): 1118.

Fitzek, S., Baumgartner, U., Marx, J., Joachimski, F., Axer, H., Witte, O. W., and Fitzek, C. (2006). Pain and itch in Wallenberg's syndrome: Anatomical-functional correlations. *Suppl. Clin. Neurophysiol.* 58: 187–194.

Fleischer, A. B., Meade, T. J., and Fleischer, A. B. (2011). Notalgia paresthetica: Successful treatment with exercises. *Acta Derm. Venereol.* 91(3): 356–357.

Fleury, V., Wayte, J., and Kiley, M. (2008). Topiramate-induced delusional parasitosis. *J. Clin. Neurosci.* 15(5): 597–599.

Freeman, R., Chase, K. P., and Risk, M. R. (2003). Quantitative sensory testing cannot differentiate simulated sensory loss from sensory neuropathy. *Neurology* 60(3): 465–470.

Gazerani, P., Pedersen, N. S., Drewes, A. M., and Arendt-Nielsen, L. (2009). Botulinum toxin type A reduces histamine-induced itch and vasomotor responses in human skin. *Br. J. Dermatol.* 161(4): 737–745, doi:10.1111/j.1365-2133.2009.09305.x.

Georgiadis, A. L., Yamamoto, Y., Kwan, E. S., Pessin, M. S., and Caplan, L. R. (1999). Anatomy of sensory findings in patients with posterior cerebral artery territory infarction. *Arch. Neurol.* 56: 835–838.

Gooding, S. M. D., Canter, P. H., Coelho, H. F., Boddy, K., and Ernst, E. (2010). Systematic review of topical capsaicin in the treatment of pruritus. *Int. J. Dermatol.* 49(8): 858–865.

Goodkin, R., Wingard, E., and Bernhard, J. D. (2003). Brachioradial pruritus: Cervical spine disease and neurogenic/neuropathic pruritus. *J. Am. Acad. Dermatol.* 48(4): 521–524.

Goulden, V., Toomey, P. J., and Highet, A. S. (1998). Successful treatment of notalgia paresthetica with a paravertebral local anesthetic block. *J. Am. Acad. Dermatol.* 38(1): 114–116.

Hagermark, O., Hokfelt, T., and Pernow, B. (1978). Flare and itch induced by substance P in human skin. *J. Invest. Dermatol.* 71(4): 233–235.

Han, L., Ma, C., Liu, Q., Weng, H. J., Cui, Y., Tang, Z., Kim, Y. et al. (2012). A subpopulation of nociceptors specifically linked to itch. *Nat. Neurosci.* 16(2): 174–182.

Harper, N. (1985). Trigeminal trophic syndrome. *Postgrad. Med. J.* 61(715): 449–451.

Harris, W. (1940). An analysis of 1,433 cases of paroxysmal trigeminal neuralgia (Trigeminal-Tic) and the end-results of gasserian alcohol injection. *Brain* 63(3): 209–224.

Hartog, C. S., Kohl, M., and Reinhart, K. (2011). A systematic review of third-generation hydroxyethyl starch (HES 130/0.4) in resuscitation: Safety not adequately addressed. *Anesth. Analg.* 112(3): 635–645.

Henderson, W. R. (1967). Trigeminal neuralgia: The pain and its treatment. *Br. Med. J.* 1: 7–13.

Howell, J. B. (1962). Neurotrophic changes in the trigeminal territory. *Arch. Dermatol.* 86: 442–449.

Hsieh, J. C., Hagermark, O., Stahle-Backdahl, M., Ericson, K., Eriksson, L., Stone-Elander, S., and Ingvar, M. (1994). Urge to scratch represented in the human cerebral cortex during itch. *J. Neurophysiol.* 72(6): 3004–3008.

Jaeger, H. (1950). Un type noveau d'ulcere neurotrophique de l'aile du nez apres neurotomie retrogasserienne. *Dermatologica* 100(4/6): 201–206.

Jastreboff, P. J. (1990). Phantom auditory perception (tinnitus): Mechanisms of generation and perception. *Neurosci. Res.* 8: 221–254.

Jeannet, P. Y., Watts, G. D., Bird, T. D., and Chance, P. F. (2001). Craniofacial and cutaneous findings expand the phenotype of hereditary neuralgic amyotrophy. *Neurology* 57(11): 1963–1968.

Jefferson, G., and Schorstein, J. (1955). Injuries of the trigeminal nerve, its ganglion and its divisions. *Br. J. Surg.* 42(176): 561–581.

Jensen, T. S., Baron, R., Haanpää, M., Kalso, E., Loeser, J. D., Rice, A. S. C., and Treede, R. D. (2011). A new definition of neuropathic pain. *Pain* 152(10): 2204–2205.

Jensen, T. S., Baron, R., Haanpää, M., Kalso, E., Loeser, J. D., Rice, A. S. C., and Treede, R. D. (2012). Response to letters by Oaklander et al. and Horowitz. *Pain* 153(4): 935–936.

Jinks, S. L., and Carstens, E. (2000). Superficial dorsal horn neurons identified by intracutaneous histamine: Chemonociceptive responses and modulation by morphine. *J. Neurophysiol.* 84(2): 616–627.

Johnson, R. E., Kanigsberg, N. D., and Jimenez, C. L. (2000). Localized pruritus: A presenting symptom of a spinal cord tumor in a child with features of neurofibromatosis. *J. Am. Acad. Dermatol.* 43(5 Pt 2): 958–961.

Kamann, S., Flaig, M. J., and Korting, H. C. (2007). Hydroxyethyl starch-induced itch: Relevance of light microscopic analysis of semi-thin sections and electron microscopy. *J. Dtsch. Dermatol. Ges.* 5(3): 204–208.

Karnosh, U., and Scherb, R. E. (1940). Trophic lesions in the distribution of the trigeminal nerve. *JAMA* 115: 2144–2147.

Kavak, A., and Dosoglu, M. (2002). Can a spinal cord tumor cause brachioradial pruritus? *J. Am. Acad. Dermatol.* 46(3): 437–440.

Kavanagh, G. M., and Tidman, M. J. (2012). Botulinum A toxin and brachioradial pruritus. *Br. J. Dermatol.* 166(5): 1147.

Kavanagh, G. M., Tidman, M. J., McLaren, K. M., Goldberg, A., and Benton, E. C. (1996). The trigeminal trophic syndrome: An under-recognized complication. *Clin. Exp. Dermatol.* 21(4): 299–301.

King, C. A., Huff, F. J., and Jorizzo, J. L. (1982). Unilateral neurogenic pruritus: Paroxysmal itching associated with central nervous system lesions. *Ann. Intern. Med.* 97(2): 222–223.

Kinsella, L. J., Carney-Godley, K., and Feldmann, E. (1992). Lichen simplex chronicus as the initial manifestation of intramedullary neoplasm and syringomyelia. *Neurosurgery* 30(3): 418–421.

Klemm, E., Bepperling, F., Burschka, M. A., and Mosges, R. (2007). Hemodilution therapy with hydroxyethyl starch solution (130/0.4) in unilateral idiopathic sudden sensorineural hearing loss: A dose-finding, double-blind, placebo-controlled, international multicenter trial with 210 patients. *Otol. Neurotol.* 28(2): 157–170.

Knight, G. C. (1954). Herpes simplex and trigeminal neuralgia. *Proc. R. Soc. Med.* 47: 788–790.

Lee, C. C., Kuo, L. T., Wang, C. H., Scaravilli, F., and An, S. F. (2005). Accumulation of prion protein in the peripheral nervous system in human prion diseases. *J. Neuropathol. Exp. Neurol.* 64(8): 716–721.

Leung, A., Donohue, M., Xu, R., Lee, R., Lefaucheur, J. P., Khedr, E. M., Saitoh, Y. et al. (2009). rTMS for suppressing neuropathic pain: A meta-analysis. *J. Pain* 10(12): 1205–1216.

Levitt, M., Ovelmen-Levitt, J., Rossitch, E., Jr., and Nashold, B. S., Jr. (1992). On the controversy of autotomy: A response to L. Kruger. *Pain* 51(1): 120–121.

Lierman, L. M. (1988). Phantom breast experiences after mastectomy. *Oncol. Nurs. Forum* 15(1): 41–44.

Lombard, M. C., Nashold, B. S., Jr., Albe-Fessard, D., Salman, N., and Sakr, C. (1979). Deafferentation hypersensitivity in the rat after dorsal rhizotomy: A possible animal model of chronic pain. *Pain* 6(2): 163–174.

Loveman, A. B. (1933). An unusual dermatosis following section of the fifth cranial nerve. *Arch. Derm. Syphilol.* 28: 369–375.

Maciejewski, R., and Drop, A. (2002). Neurological symptoms as the result of enlarged dimensions and non-typical course of inferior superficial temporal vein. *Folia Morphol.* 61(1): 57–60.

Macovei, M., Popa, C., Alexianu, M. E., and Vulcan, P. (1991). Neurotrophic trigeminal syndrome after pontine stroke. *Rom. J. Neurol. Psychiatry* 29(3–4): 153–158.

Maggi, C. A., Borsini, F., Santicioli, P., Geppetti, P., Abelli, L., Evangelista, S., Manzini, S., Theodorsson-Norheim, E., Somma, V., and Amenta, F. (1987). Cutaneous lesions in capsaicin-pretreated rats. A trophic role of capsaicin-sensitive afferents? *Naunyn Schmiedebergs Arch. Pharmacol.* 336(5): 538–545.

Mailis, A. (1996). Compulsive targeted self-injurious behaviour in humans with neuropathic pain: A counterpart of animal autotomy? Four case reports and literature review. *Pain* 64(3): 569–578.

Massey, E. W. (1984). Unilateral neurogenic pruritus following stroke. *Stroke* 15(5): 901–903.

Massey, E. W., and Massey, J. M. (1986). Forearm neuropathy and pruritus. *South Med. J.* 79(10): 1259–1260.

Massey, E. W., and Pleet, A. B. (1979). Localized pruritus-notalgia paresthetica. *Arch. Dermatol.* 115(8): 982–983.

Massey, E. W., and Pleet, A. B. (1981). Electromyographic evaluation of notalgia paresthetica. *Neurology* 31(5): 642.

Matthews, W. (2005). The symptoms and signs of multiple sclerosis. In *McAlpine's Multiple Sclerosis*, 4th ed. Philadelphia, PA: Churchill Livingstone, p. 326.

Mayer, R. D., and Smith, N. P. (1993). Improvement of trigeminal neurotrophic ulceration with pimozide in a cognitively impaired elderly woman—A case report. *Clin. Exp. Dermatol.* 18(2): 171–173.

McCann, M. E., Waters, P., Goumnerova, L. C., and Berde, C. (2004). Self-mutilation in young children following brachial plexus birth injury. *Pain* 110(1–2): 123–129.

McKenzie, K. G. (1933). Observations on the results of the operative treatment of trigeminal neuralgia. *Can. Med. Assoc. J.* 22: 492–496.

McLean, N. R., and Watson, A. C. (1982). Reconstruction of a defect of the ala nasi following trigeminal anaesthesia with an innervated forehead flap. *Br. J. Plast. Surg.* 35(2): 201–203.

Meiss, F., Boerner, D., Marsch, W. C., and Fischer, M. (2002). Gabapentin—A promising treatment in glossodynia. *Clin. Exp. Dermatol.* 27(6): 525–526.

Merskey, H., and Bogduk, N. (1994). *Classification of Chronic Pain: Descriptions of Chronic Pain Syndromes and Definitions of Pain Terms,* 2nd ed. Seattle: IASP Press.

Meyer, R. A., Raja, S. N., and Campbell, J. N. (1985). Coupling of action potential activity between unmyelinated fibers in the peripheral nerve of monkey. *Science* 227(4683): 184–187.

Mochizuki, H., Inui, K., Tanabe, H. C., Akiyama, L. F., Otsuru, N., Yamashiro, K., Sasaki, A., Nakata, H., Sadato, N., and Kakigi, R. (2009). Time course of activity in itch-related brain regions: A combined MEG-MRI study. *J. Neurophysiol.* 102(5): 2657–2666.

Mochizuki, H., Tashiro, M., Kano, M., Sakurada, Y., Itoh, M., and Yanai, K. (2003). Imaging of central itch modulation in the human brain using positron emission tomography. *Pain* 105(1–2): 339–346.

Mogil, J. S., Wilson, S. G., Bon, K., Lee, S. E., Chung, K., Raber, P., Pieper, J. O. et al. (1999). Heritability of nociception I: Responses of 11 inbred mouse strains on 12 measures of nociception. *Pain* 80(1–2): 67–82.

Nakamizo, S., Miyachi, Y., and Kabashima, K. (2010). Treatment of neuropathic itch possibly due to trigeminal trophic syndrome with 0.1% topical tacrolimus and gabapentin. *Acta Derm. Venereol.* 90(6): 654–655.

Nguyen, J. P., Nizard, J., Keravel, Y., and Lefaucheur, J. P. (2011). Invasive brain stimulation for the treatment of neuropathic pain. *Nat. Rev. Neurol.* 7(12): 699–709.

Oaklander, A. L., and Belzberg, A. J. (1997). Unilateral nerve injury down-regulates mRNA for Na⁺ channel SCN10A bilaterally in rat dorsal root ganglia. *Brain Res. Mol. Brain Res.* 52(1): 162–165.

Oaklander, A. L., Bowsher, D., Galer, B. S., Haanpää, M. L., and Jensen, M. P. (2003). Herpes zoster itch: Preliminary epidemiologic data. *J. Pain* 4(6): 338–343.

Oaklander, A. L., and Brown, J. M. (2004). Unilateral nerve injury produces bilateral loss of distal innervation. *Ann. Neurol.* 55(5): 639–644.

Oaklander, A. L., Cohen, S. P., and Raju, S. V. (2002). Intractable postherpetic itch and cutaneous deafferentation after facial shingles. *Pain* 96(1–2): 9–12.

Oaklander, A. L., Dey, D., and Landrum, O. (2005). Response to letter by Dr Yezierski and Vierck. *Pain* 115(3): 420–421.

Oaklander, A. L., Long, D. M., Larvie, M., and Davidson, C. J. (2013). Case Records of the Massachusetts General Hospital: A 77-year-old woman with long-standing unilateral thoracic pain and incontinence. *New Engl. J. Med.* in press. 36(9): 853–861.

Oaklander, A. L., Romans, K., Horasek, S., Stocks, A., Hauer, P., and Meyer, R. A. (1998). Unilateral postherpetic neuralgia is associated with bilateral sensory neuron damage. *Ann. Neurol.* 44(5): 789–795.

Oaklander, A. L., Seo, W. K., and Ho Park, M. (2009). Neuropathic pruritus following Wallenberg syndrome. *Neurology* 73(19): 1605–1606.

Oaklander, A. L., Wilson, P. R., Moskovitz, P. A., Manning, D. C., Lubenow, T., Levine, J. D., Harden, R. N. et al. (2012). Response to "A new definition of neuropathic pain." *Pain* 153(4): 934–935.

Ostermann, P. O. (1976). Paroxysmal itching in multiple sclerosis. *Br. J. Dermatol.* 95(5): 555–558.

Pappagallo, M., Oaklander, A. L., Quatrano-Piacentini, A. L., Clark, M. R., and Raja, S. N. (2000). Heterogenous patterns of sensory dysfunction in postherpetic neuralgia suggest multiple pathophysiologic mechanisms. *Anesthesiology* 92(3): 691–698.

Peier, A. M., Reeve, A. J., Andersson, D. A., Moqrich, A., Earley, T. J., Hergarden, A. C., Story, G. M. et al. (2002). A heat-sensitive TRP channel expressed in keratinocytes. *Science* 296(5575): 2046–2049.

Petrie, L., Heath, B., and Harold, D. (1989). Scrapie: Report of an outbreak and brief review. *Can. Vet. J.* 30(4): 321–327.

Philpott, O. S. (1941). Trophic ulcer complicating operative procedures for the relief of trigeminal neuralgia. *Rocky Mountain Med. J.* 38: 626–629.

Raison-Peyron, N., Meunier, L., Acevedo, M., and Meynadier, J. (1999). Notalgia paresthetica: Clinical, physiopathological and therapeutic aspects. A study of 12 cases. *J. Eur. Acad. Dermatol. Venereol.* 12(3): 215–221.

Ranoux, D., Attal, N., Morain, F., and Bouhassira, D. (2008). Botulinum toxin type A induces direct analgesic effects in chronic neuropathic pain. *Ann. Neurol.* 64(3): 274–283.

Rapin, I., and Ruben, R. J. (1976). Patterns of anomalies in children with malformed ears. *Laryngoscope* 86(10): 1469–1502.

Ross, S. E., Mardinly, A. R., McCord, A. E., Zurawski, J., Cohen, S., Jung, C., Hu, L. et al. (2010). Loss of inhibitory interneurons in the dorsal spinal cord and elevated itch in Bhlhb5 mutant mice. *Neuron* 65(6): 886–898.

Rossitch, E., Jr., Abdulhak, M., Ovelmen-Levitt, J., Levitt, M., and Nashold, B. S., Jr. (1993). The expression of deafferentation dysesthesias reduced by dorsal root entry zone lesions in the rat. *J. Neurosurg.* 78(4): 598–602.

Sacks, O. (2012). *Hallucinations.* New York: Alfred A. Knopf.

Sandroni, P. (2002). Central neuropathic itch: a new treatment option? *Neurology* 59(5): 778–779.

Sarkis, D., Souteyrand, J. P., and Albe-Fessard, D. (1984). Self-stimulation in the ventral tegmental area suppresses self-mutilation in rats with forelimb deafferentiation. *Neurosci. Lett.* 44(2): 199–204.

Savin, J. A. (1975). Studies of scratching during sleep. *Proc. R. Soc. Med.* 68(8): 529.

Savin, J. A., Adam, K., Oswald, I., and Paterson, W. D. (1990). Pruritus and nocturnal wakenings. *J. Am. Acad. Dermatol.* 23(4 Pt 1): 767–768.

Savin, J. A., Paterson, W. D., and Oswald, I. (1973). Scratching during sleep. *Lancet* 2(7824): 296–297.

Savitsky, N., and Elpern, S. P. (1948). Gangrene of face following occlusion of posterior inferior cerebellar artery. *Arch. Neurol. Psychiatry* 60: 388–391.

Savk, E., Bolukbasi, O., Akyol, A., and Karaman, G. (2001). Open pilot study on oxcarbazepine for the treatment of notalgia paresthetica. *J. Am. Acad. Dermatol.* 45(4): 630–632.

Savk, E., Dikicioglu, E., Culhaci, N., Karaman, G., and Sendur, N. (2002). Immunohistochemical findings in notalgia paresthetica. *Dermatology* 204(2): 88–93.

Savk, E., and Savk, S. O. (2004). On brachioradial pruritus and notalgia paresthetica. *J. Am. Acad. Dermatol.* 50(5): 800–801.

Savk, O., and Savk, E. (2005). Investigation of spinal pathology in notalgia paresthetica. *J. Am. Acad. Dermatol.* 52(6): 1085–1087.

Schorstein, J. (1943). Erosion of the ala nasi following trigeminal denervation. *J. Neurol. Psychiatry* 6: 46–51.

Schultz, G., and Melzack, R. (1991). The Charles Bonnet syndrome: "Phantom visual images." *Perception* 20: 809–825.

Scott, C., Perry, M. J., Raven, P. E., Massey, E. J., and Lisney, S. J. (2000). Capsaicin-sensitive afferents are involved in signalling transneuronal effects between cutaneous sensory nerves. *Neuroscience* 95(2): 535–541.

Shabtai, H., Nisipeanu, P., Chapman, J., and Korczyn, A. D. (1996). Pruritus in Creutzfeldt-Jakob disease. *Neurology* 46(4): 940–941.

Shapiro, P. E., and Braun, C. W. (1987). Unilateral pruritus after a stroke. *Arch. Dermatol.* 123(11): 1527–1530.

Shir, Y., Ratner, A., and Seltzer, Z. (1997). Diet can modify autotomy behavior in rats following peripheral neurectomy. *Neurosci. Lett.* 236: 71–74.

Sikand, P., Shimada, S. G., Green, B. G., and LaMotte, R. H. (2011). Sensory responses to injection and punctate application of capsaicin and histamine to the skin. *Pain* 152(11): 2485–2494.

Simone, D. A., Nolano, M., Johnson, T., Wendelschafer-Crabb, G., and Kennedy, W. R. (1998). Intradermal injection of capsaicin in humans produces degeneration and subsequent reinnervation of epidermal nerve fibers: Correlation with sensory function. *J. Neurosci.* 18(21): 8947–8959.

Simone, D. A., Zhang, X., Li, J., Zhang, J. M., Honda, C. N., LaMotte, R. H., and Giesler, G. J., Jr. (2004). Comparison of responses of primate spinothalamic tract neurons to pruritic and algogenic stimuli. *J. Neurophysiol.* 91(1): 213–222.

Spalletta, G., Piras, F., Alex, R. I., Caltagirone, C., and Fagioli, S. (2012). Fronto-thalamic volumetry markers of somatic delusions and hallucinations in schizophrenia. *Psychiatry Res.* 202(1): 54–64.

Spillane, J. D., and Wells, C. E. C. (1959). Isolated trigeminal neuropathy: A report of 16 cases. *Brain* 82: 391–426.

Springall, D. R., Karanth, S. S., Kirkham, N., Darley, C. R., and Polak, J. M. (1991). Symptoms of notalgia paresthetica may be explained by increased dermal innervation. *J. Invest. Dermatol.* 97(3): 555–561.

Ständer, S., Weisshaar, E., Mettang, T., Szepietowski, J. C., Carstens, E., Ikoma, A., Bergasa, N. V. et al. (2007). Clinical classification of itch: A position paper of the International Forum for the Study of Itch. *Acta Derm. Venereol.* 87(4): 291–294.

Stefaniu, A., Romascanu, G., and Calarasu, R. (1973). Glossopharyngeal neuralgia causing tonsillar autotraumatism. *Otorinolaringologie* 18(6): 453–456.

Sullivan, M. J., and Drake, M. E., Jr. (1984). Unilateral pruritus and Nocardia brain abscess. *Neurology* 34(6): 828–829.

Summers, C. G., and MacDonald, J. T. (1988). Paroxysmal facial itch: A presenting sign of childhood brainstem glioma. *J. Child Neurol.* 3(3): 189–192.

Sweet, W. H. (1981). Animal models of chronic pain: their possible validation from human experience with posterior rhizotomy and congenital analgesia (Part I of the second John J. Bonica lecture). *Pain* 10(3): 275–295.

Thielen, A. M., Vokatch, N., and Borradori, L. (2008). Chronic hemicorporal prurigo related to a posttraumatic Brown-Séquard syndrome. *Dermatology* 217(1): 45–47.

Thomas, J., Parimalam, S., Selvi, G. T., Augustine, S. M., and Muthuswami, T. C. (1991). Trigeminal trophic syndrome in Hansen's disease. *Int. J. Lepr. Other Mycobact. Dis.* 59(3): 479–480.

Tseng, S. H., and Lin, S. M. (1998). Substantia nigra lesion suppresses the antagonistic effects of N-methyl-D-aspartate receptor antagonist (MK-801) on the autotomy in the rat. *Neurosci. Lett.* 255(3): 167–171.

van Alfen, N., and van Engelen, B. G. M. (2006). The clinical spectrum of neuralgic amyotrophy in 246 cases. *Brain* 129(2): 438–450.

Veien, N. K., and Laurberg, G. (2011). Brachioradial pruritus: A follow-up of 76 patients. *Acta Derm. Venereol.* 91(2): 183–185.

Vuadens, P., Regli, F., Dolivo, M., and Uske, A. (1994). Segmental pruritus and intramedullary vascular malformation. *Schweiz. Arch. Neurol. Psychiatr.* 145(3): 13–16.

Wahlgren, C. F. (1995). Measurement of itch. *Semin. Dermatol.* 14(4): 277–284.

Wall, P. D., Devor, M., Inbal, R., Scadding, J. W., Schonfeld, D., Seltzer, Z., and Tomkiewicz, M. M. (1979). Autotomy following peripheral nerve lesions: Experimental anaesthesia dolorosa. *Pain* 7(2): 103–111.

Wallenberg, A. (1901). Klinische Beiträge zur Diagnostik acuter Herderkrankungen des verlängerten Markes und der Brücke. *Dtsch. Zeitsch. Nervenheil.* 19: 227–248.

Wallengren, J., and Bartosik, J. (2010). Botulinum toxin type A for neuropathic itch. *Br. J. Dermatol.* 163(2): 424–426.

Walton, S., and Keczkes, K. (1985). Trigeminal neurotrophic ulceration—A report of four patients. *Clin. Exp. Dermatol.* 10(5): 485–490.

Warren, J. C. (1828). Cases of neuralgia, or painful affections of nerves. *Boston Med. Surg. J.* 1(1): 1–6.

Watson, W. C. (1973). Intravenous lignocaine for relief of intractable itch. *Lancet* 1(7796): 211.

Weeks, S. R., and Tsao, J. W. (2010). Incorporation of another person's limb into body image relieves phantom limb pain: A case study. *Neurocase* 16(6): 461–465.

Westerhof, W., and Bos, J. D. (1983). Trigeminal trophic syndrome: A successful treatment with transcutaneous electrical stimulation. *Br. J. Dermatol.* 108(5): 601–604.

Wolf, J. K. (1971). *The Classical Brain Stem Syndromes: Translations of the Original Papers with Notes on the Evolution of Clinical Neuroanatomy.* Springfield, IL: Charles C. Thomas.

Woolf, C. J., and Wall, P. D. (1982). Chronic peripheral nerve section diminishes the primary afferent A-fibre mediated inhibition of rat dorsal horn neurones. *Brain Res.* 242: 77–85.

Yamamoto, M., Yabuki, S., Hayabara, T., and Otsuki, S. (1981). Paroxysmal itching in multiple sclerosis: A report of three cases. *J. Neurol. Neurosurg. Psychiatry* 44(1): 19–22.

Yezierski, R. P., Liu, S., Ruenes, G. L., Kajander, K. J., and Brewer, K. L. (1998). Excitotoxic spinal cord injury: Behavioral and morphological characteristics of a central pain model. *Pain* 75(1): 141–155.

Zhao, P., Barr, T. P., Hou, Q., Dib-Hajj, S. D., Black, J. A., Albrecht, P. J., Petersen, K. et al. (2008). Voltage-gated sodium channel expression in rat and human epidermal keratinocytes: Evidence for a role in pain. *Pain* 139(1): 90–105.

Ziccardi, V. B., Rosenthal, M. S., and Ochs, M. W. (1996). Trigeminal trophic syndrome: A case of maxillofacial self-mutilation. *J. Oral Maxillofac. Surg.* 54(3): 347–350.

Zotova, E. G., and Arezzo, J. C. (2013). Non-invasive evaluation of nerve conduction in small diameter fibers in the rat. *Physiol. J.* 2013: 254789.

8 Pruritus in Cutaneous T-Cell Lymphomas

Laurent Misery

CONTENTS

8.1 CUTANEOUS T-CELL LYMPHOMAS

Cutaneous T-cell lymphomas (CTCL) are a group of lymphoproliferative disorders characterized by a clonal accumulation of neoplastic memory T-lymphocytes in the skin (Girardi et al. 2004; Meyer et al. 2010). CTCL are the most common primary cutaneous lymphomas. Their incidence has increased from 5 cases per 1,000,000 in the 1980s to 12.7 per 1,000,000 in the 2000s (Bradford et al. 2009). Mycosis fungoides (MF) is the most common variant of CTCL. Typically, patients initially present with erythematous patches in sun-protected areas, which may progress to plaques and possibly skin tumors. Sézary syndrome (SS) is a more aggressive form of CTCL, characterized by the association of an exfoliative erythroderma with the presence of atypical mononuclear cells in the skin and in the peripheral blood (Sezary cells). Patients with early stage CTCL have a good prognosis, whereas the prognosis is poorer for patients with late-stage disease (Klemke et al. 2005; Willemze et al. 2005). CTCL can cause significant morbidity and adversely affect patients' quality of life (Demierre et al. 2003, 2005, 2006), in part owing to the fact that CTCL are highly associated with pruritus (Demierre and Taverna 2006). CTCL have multiple subtypes (Willemze et al. 2005).

8.2 PRURITUS IN CTCL

Folliculotropic form of CTCL, Sézary syndrome, and erythrodermic form of mycosis fungoides are extremely itchy as compared to patch stages of mycosis fungoides: in a case-control study of 43 patients with folliculotropic form of CTCL, 68% had severe pruritus requiring separate treatment aside from the lymphoma (Gerami 2008). In a phase 2 trial of oral vorinostat in 33 patients with refractory cutaneous T-cell lymphoma, 93% had symptomatic pruritus (Duvic et al. 2007). A recent study confirmed the high frequency of pruritus in patients with CTCL and showed that mean pruritus values were 3.5/10 in MF patients and 7.7 in SS patients (Vij and Duvic 2012).

In patients with CTCL, pruritus is frequent (Moses 2003; Bradford et al. 2009), severe, and unrelieved by emollients, topical steroids, or oral antihistamines. In CTCL, pruritus is permanent and aggravated in the evening or by heat. Most patients report diffuse pruritus, but pruritus may also be localized on specific lesions. As in systemic hematologic malignancies, CTCL-related pruritus may be worsened by water. It interferes with sleep. In the advanced stages, patients commonly report an ill-defined, severe and diffuse pruritus that may turn in "burning pain." This pruritus interferes with sleep and functioning and displays similarities with neuropathic pain (Namaka et al. 2004). In a large survey from the United States evaluating 309 patients with CTCL, Demierre et al. (2006) reported that a large majority of them (88%) declared to be bothered by itching, when 41% of them reported some pain. Pruritus may be a presenting feature in patients with CTCL and can occur in without skin lesions, making the diagnosis of CTCL difficult (Elmer and George 1999; Bradford et al. 2009). Pruritus is the most frequent and the earliest symptom or CTCL (D'Incan and Souteyran 2008). In the early stages, pruritus may be relieved by topical corticosteroids. This seemingly well-controlled dermatitis may delay the correct diagnosis (Hope et al. 1990). Moreover, for several patients the diagnosis of CTCL is difficult to obtain because histological findings and molecular evidences of the TCR clonality may take years to be found (D'Incan and Souteyran 2008).

In case of atypical pruritus (severe pruritus and/or chronic and/or permanent pruritus without systemic explanation, and/or pruritus that interferes with sleep, and or topical steroid-resistant pruritus, and/or pruritus associated with atypical dermatitis), at least three pathologic examinations of the skin and one assessment of T-cell clonality in the skin by molecular biology should be performed within one year (Meyer et al. 2010).

8.3 PATHOPHYSIOLOGY

Despite great advances in recent pruritus research, it is still unknown how T-cell lymphoma mediates the itch sensation and especially, the involved molecular mechanisms remain to be elucidated (Görge and Schiller 2010; Meyer et al. 2010).

It has been reported that mast cells and histamine may play a role in MF particularly in advanced stages of the disease (Yamamoto et al. 1997). Histamine, which is released by mast cells, is classically associated with pruritus. Other mast cell mediators appear to be of clinical importance: serotonin is a neuromediator involved in the pruritus associated with several diseases, including lymphomas (Krajnik and Zylicz 2003). Serotonin and histamine may have tumor growth-inhibition and immunomodulating properties: the use of serotonin antagonists inhibits *in vitro* the mitogenesis of normal lymphocytes and promotes the proliferation of transformed lymphocytes (Brandes et al. 1991). Vermeer and Willemze (1996) have reported two patients with MF who had an exacerbation of their disease shortly after starting fluoxetine. Antihistamines were reported to be associated with the development of atypical lymphoid infiltrates in the skin (Magro and Crowson 1995). In any case, the itch of patients with CTCL often fails to or only partially responds to antihistamines (Ahern et al. 2012).

Several data are supporting the fact that CTCL produce high amounts of cytokines. SS T-cells clones produce IL-3, IL-4, IL-5, IL-6, and IL-10, thus displaying a Th2 cytokine-secretion profile) (Dummer et al. 1996). IL-2 and IL-6 have been proposed as histamine-independent itch mediators. Single dose of IL-2 (10 MU/mL) delivered to the epidermis of healthy volunteers triggers itching. Intradermal injection of 20 µg of IL-2 in the dermis of atopic dermatitis patients or healthy subjects was associated with a low-intensity local itch and erythema (Wahlgren et al. 1995; Darsow et al. 1997). In patients with uremic pruritus, serum levels of IL-6 were found to be significantly increased as compared with patients without uremic pruritus (Kimmel et al. 2006). IL-7 induces lymphoid inflammation in mice (Rich et al. 1993), and it is produced by human keratinocytes (Heufler et al. 1993). IL-7 induces proliferation of human T-cell clones with a synergic effect on IL-2 in T-cell clones of CTCL (Dalloul et al. 1992).

The role of IL-31 (Sonkoly et al. 2006; Misery and Ständer 2010; Ahern et al. 2012), as histamine-independent itch mediator, could have a very important role in the pathogenesis of pruritus in CTCL because it is known to be produced by CD4+ T-cells and to be the putative main mediator for itch in other diseases like atopic dermatitis. Nonetheless, there are no data about IL-31 in CTCL at the moment.

Proteases such as kallicreins and cathepsin contribute to the integrity and protective barrier function of the skin (Meyer-Hoffert 2009). Protease regulation is altered in inflammatory skin conditions. In atopic dermatitis, psoriasis, exfoliative dermatitis excessive protease activation was shown to contribute to skin barrier function damage (Meyer-Hoffert 2009). Investigating the activity of proteases in the skin of patients with CTCL may be also important to understand CTCL-related pruritus pathophysiology.

The role of substance P, which is a neuropeptide, in itch is observed in many diseases (Misery and Ständer 2010). Its putative role in patients with CTCL has been dramatically underlined by clinical cases with effects of aprepitant on this untractable itch (Duval and Dubertret 2009; Ständer et al. 2010; Booken et al. 2011).

8.4 THERAPEUTICS

Management of pruritus should be directed at the underlying cause. CTCL are chronic diseases, and multiple therapies are available, each with varying efficacy and side-effect profiles. The main outcome of CTCL treatment remains disease control, and few studies have been designed to evaluate the efficacy of therapeutics in the management of CTCL-related pruritus. Most of the practices are based on physician experience.

The efficacy of high potency topical corticosteroids in early-stage CTCL is well known (Zackheim et al. 1998; Zackheim 2003). Topical corticosteroids have been shown to rapidly decrease histamine-induced itch (Yosipovitch et al. 1996). In early stages of the disease, CTCL-related pruritus can be efficiently diminished by topical steroids (Elmer and George 1999). High potency topical steroids should remain the treatment of first choice at these stages (Trautinger et al. 2006). They can be associated with antihistamines to diminish pruritus in CTCL and with other specific topical or systemic drugs aimed at controlling the tumor activity. Naltrexone, an opioid-receptor antagonist, has been reported to be efficient in CTCL-related pruritus (Brune et al. 2004). However, this study was open-labeled, and the use of naltrexone for the treatment of pruritus in CTCL should be further evaluated. It has recently been reported that the use of wet dressing in association with topical steroids may be an interesting option for the management of pruritic skin diseases other than atopic dermatitis (Bingham et al. 2009). The authors report a retrospective uncontrolled study on 331 patients with pruritic skin diseases recalcitrant to previous therapies (topical steroids, antihistamines, and emollients). Among them, 12 patients with CTCL were included. All patients improved after treatment with the association of topical steroids and wet dressing. This association may be an interesting option for the management of intense flares of itching in patients with CTCL.

Although their antipruritic effect in CTCL has not been evaluated, oral corticosteroids may also be used to reduce CTCL-related pruritus. In our experience, doses from 10 to 30 mg/d may significantly improve pruritus in patients with CTCL. However, numerous side effects are associated with long-term use of oral steroids. These side effects have to be taken into account when advocating oral steroids.

Second-line treatments of CTCL-related pruritus are not well defined. CTCL is one of the major dermatologic conditions for which phototherapy continues to be a valuable treatment modality (Baron and Stevens 2003). The most documented treatment for the management of pruritus in CTCL is PUVA. It has been used for decades in the management of CTCL and is effective in early-stage CTCL (Yosipovitch et al. 1996; Zackheim 2003; Trautinger et al. 2006; Zackheim et al. 2003). In early-stage CTCL, PUVA induces significant clinical improvement and long-lasting disease-free intervals (Baron and Stevens 2003). Its efficacy can further be improved when used in conjunction with other treatment modalities such as interferon-alpha (Mostow et al. 1993) or retinoids (Gniadecki et al. 2007; Papadavid et al. 2008). The efficacy of narrowband UVB phototherapy on pruritus is also well established (Yashar et al. 2003; Rivard and Henry 2005). Patients may be treated with a starting dose of 70% of the UVB-minimal erythematous dose. The dose may be increased by 10% to 15% per session as tolerated. Sessions will be performed three times weekly until regression of the lesions, and then pursued once to twice weekly in case of successful response. In the first 2 weeks of treatment with phototherapy, itch could be worsened. Patients can be maintained on topical steroids or antihistamine treatment during phototherapy. Phototherapy should be preferred to topical steroids when the pruritus is diffuse or intense or interferes with sleep.

Different systemic drugs have been reported to have effects on CTCL-related pruritus. There is a lack of evidence of the efficacy of these drugs, and their use should be a third choice prescription for the management of pruritus.

A significant reduction of CTCL-related pruritus has been reported with the use of drugs aimed at controlling the disease. The effects of denileukin diftitox (an IL-2 receptor targeted fusion protein) on quality of life and pruritus of 71 patients with CTCL have been evaluated as part of a multicenter phase III trial (Olsen et al. 2001; Duvic et al. 2002). Assessments of pruritus severity (10-cm visual analog scale) were significantly improved after 6 months of treatment as compared with baseline evaluation. Alemtuzumab, a monoclonal anti-CD52 antibody has been evaluated in the treatment of CTCL (phase II study, 22 patients) (Lundin et al. 2003). In this study, a reduction of six points in the median pruritus score (self-assessed visual analog scale) at the end of the treatment was reported. Most recently, a significant reduction of the pruritus has been reported with the use of SAHA (Suberoylanilide hydroxamic acid, vorinostat) for refractory CTCL: 45% of the patients had a reduction of at least three points for at least 4 weeks as compared with baseline evaluation of pruritus (Duvic et al. 2007). Brightman and Demierre (2005) reported on the use of thalidomide in one patient with MF with a reduction in pruritus assessment score. Thalidomide is known to be effective on several pruritic skin diseases (Summey and Yosipovitch 2005; Yosipovitch and Fleischer 2003) and may be an interesting option for the management of patients with refractory pruritic CTCL. Low-dose methotrexate has been used to treat CTCL for many years (Zackheim et al. 1996, 2003). However, the effects of low-dose methotrexate on CTCL-related pruritus are poorly documented. In our experience, patients treated with methotrexate for a CTCL may have significant pruritus improvement. Further studies should be conducted to evaluate antipruritic effects of methotrexate in CTCL.

Demierre et al. (2006) have reported their experience on the use of gabapentin—an anticonvulsant—and mirtazapine—an antidepressant—to manage severe pruritus associated with advanced stages CTCL. Both gabapentin and mirtazapine have been reported as effective in the treatment of pruritus (Bigata 2005; Yesudian and Wilson 2005). The authors recommended a starting dose of gabapentin 300 mg at night and titrate upward, without exceeding 2400 mg daily. For patients who did not respond, the authors recommend to progressively substitute gabapentin with a low dose of mirtazapine (7.5–15 mg daily in the evening).

Recently, Sepmeyer et al. (2007) reported an open-label study of combination therapy with rosiglitazone and bexarotene for CTCL, with interesting effects on pruritus: three of the four patients had relieved pruritus after 16 weeks of a regimen comprising bexarotene (190–425 mg/m² daily) and rosiglitazone (4–8 mg daily). Rosiglitazone is a PPAR (peroxisome proliferator-activated receptor) agonist. PPAR are transcription factors that heterodimerize with retinoid X receptors and bind to peroxisome proliferator response elements in the promoter region of target genes. Interestingly, retinoid X receptors are the targets of bexarotene. Bexarotene transactivates PPAR-alpha target genes in a PPAR-alpha dependent manner (Martin et al. 2005). PPAR agonists may improve the therapeutic response to retinoid X receptor agonists and may help to reduce CTCL-related pruritus through a better control of the disease. The role of PPAR agonists in CTCL should be further explored.

For the particular case of Sézary syndrome, extracorporeal photochemotherapy (ECP) can be considered as an interesting therapeutic option to reduce both disease activity and pruritus. ECP has been reported to induce long-term remissions of Sézary syndrome (Zic 2003; Scarisbrick et al. 2008). ECP has been reported to reduce pruritus, not only in Sézary syndrome, but also in atopic dermatitis (Bouwhuis et al. 2002). Sessions have to be performed twice monthly until relapse of the disease.

Finally, a dramatic improvement was noted in three patients with Sézary syndrome suffering from a severe and uncontrolled pruritus after a treatment by aprepitant, which an antagonist of NK1 receptor for substance P (Duval and Dubertret

2009). This success was confirmed by other publications (Ständer et al. 2010; Booker et al. 2011).

8.5 CONCLUSION

In patients with CTCL, pruritus is frequent and severe. This symptom negatively affects the quality of life of the patients. Management of severe pruritus is not supported by clinical evidence and is based on expert opinion. First-line treatment recommendation should be topical steroids or phototherapy, sometimes in combination with disease-modifying treatment such as interferon or retinoids. In case of insufficient response, anticonvulsants or antidepressants can be used. A hopeful alternative is given by NK1 antagonists.

REFERENCES

Ahern, K., Gilmore, E. S., and Poligone, B. 2012. Pruritus in cutaneous T-cell lymphoma: A review. *J Am Acad Dermatol* 67: 760–768.

Baron, E. D., and Stevens, S. R. 2003. Phototherapy for cutaneous T-cell Lymphoma. *Dermatol Ther* 16: 303–310.

Bigata, X., Sais, G., and Soler, F. 2005. Severe chronic urticaria: Response to mirtazapine. *J Am Acad Dermatol* 53: 916–917.

Bingham, L. G., Noble, J. W., and Davis, M. D. 2009. Wet dressing used with topical corticosteroids for pruritic dermatoses: A retrospective study. *J Am Acad Dermatol* 60: 792–800.

Booken, N., Heck, M., Nicolay, J. P., Klemke, C. D., Goerdt, S., and Utikal, J. 2011. Oral apepritant in the therapy of refractory pruritus in erythrodermic cutaneous T-cell lymphomas. *Br J Dermatol* 164: 665–667.

Bouwhuis, S. A., McEvoy, M. T., and Davis, M. D. 2002. Sustained remission of Sezary syndrome. *Eur J Dermatol* 12: 287–290.

Bradford, P. T., Devesa, S. S., Anderson, W. F., and Toro, J. R. 2009. Cutaneous lymphoma incidence patterns in the United States: A population-based study of 3884 cases. *Blood* 113: 5064–5073.

Brandes, L. J., La Bella, F. S., and Warrington, F. C. 1991. Increased therapeutic index of antineoplastic drugs in combination with intracellular histamine antagonist. *J Natl Cancer Inst* 83: 1329–1336.

Brightman, L., and Demierre, M. F. 2005. Thalidomide in mycosis fungoides. *J Am Acad Dermatol* 52: 1100–1101.

Brune, A., Metze, D., Luger, T. A., and Ständer, S. 2004. Antipruritic therapy with the oral opioid receptor antagonist naltrexone. Open, non-placebo controlled administration in 133 patients. *Hautarzt* 55: 1130–1136.

Dalloul, A., Laroche, L., Bagot, M. et al. 1992. Interleukin-7 is a growth factor for Sezary lymphoma cells. *J Clin Invest* 90: 1054.

Darsow, U., Scharein, E., Bromm, B., and Ring, J. 1997. Skin testing of the pruritogenic activity of histamine and cytokines (interleukin 2 and tumour necrosis factor-alpha) at the dermal epidermal junction. *Br J Dermatol* 137: 415–417.

Demierre, M. F., Gan, S., Jones, J., and Miller, D. R. 2006. Significant impact of cutaneous T-cell lymphoma on patients' quality of life. *Cancer* 107: 2504–2511.

Demierre, M. F., Kim, Y. H., and Zackheim, H. S. 2003. Prognosis, clinical outcomes and quality of life issues in cutaneous T-cell lymphoma. *Hematol Oncol Clin North Am* 17: 1485–1507.

Demierre, M. F., and Taverna, J. 2006. Mirtazapine and gabapentin for reducing pruritus in cutaneous T-cell lymphoma. *J Am Acad Dermatol* 55: 543–544.

Demierre, M. F., Tien, A., and Miller, D. 2005. Health-related quality of life assessment in patients with cutaneous T-cell lymphoma. *Arch Dermatol* 141: 325–330.

D'Incan, M., and Souteyran, P. 2008. Mycosis fongoïde et formes apparentées. In *Dermatologie et Infections Sexuellement Transmissibles*, Saurat, J. H., Lachapelle, J. M., Lipsker, D., and Thomas, L. (Eds.). Paris: Masson.

Dummer, R., Heald, P. W., Nestle, F. O. et al. 1996. Sezary syndrome T-cells clones display T-helper2 cytokines and express the accessory factor-1 (interferon gamma receptor beta-chain). *Blood* 88: 1383–1389.

Duval, A., and Dubertret, L. 2009. Aprepitant as an antipruritic agent? *N Engl J Med* 361: 1415–1416.

Duvic, M., Kuzel, T. M., Olsen, E. A. et al. 2002. Quality of life improvements in cutaneous T-cell lymphoma patients treated with denileukin diftitox (ONTAK). *Clin Lymphoma* 43: 121–126.

Duvic, M., Talpur, R., Ni, X. et al. 2007. Phase 2 trial of oral vorinostat (suberoylanilide hydroxamic acid, SAHA) for refractory cutaneous T-cell lymphoma (CTCL). *Blood* 109: 31–39.

Elmer, K. B., and George, R. M. 1999. Cutaneous T-cell lymphoma presenting as benign dermatoses. *Am Fam Physician* 59: 2809–2813.

Gerami, P., Rosen, S., Kuzel, T., Boone, S. L., and Guitart, J. 2008. Folliculotropic mycosis fungoides. *Arch Dermatol* 144: 738–746.

Girardi, M., Heald, P. W., and Wilson, L. D. 2004. The pathogenesis of mycosis fungoides. *N Engl J Med* 350: 1978–1988.

Gniadecki, R., Assaf, C., Bagot, M. et al. 2007. The optimal use of bexarotene in cutaneous T-cell lymphoma. *Br J Dermatol* 157: 433–440.

Görge, T., and Schiller, M. 2010. Pruritus in cutaneous T-cell lymphoma. In *Pruritus*, Misery, L., and Ständer, S. (Eds.). London: Springer-Verlag, pp. 121–124.

Heufler, C., Topar, G., Grasseger, A., Stanzl, U., Koch, F., Romani, N., Namen, A. E., and Schuler, G. 1993. Interleukin 7 is produced by murine and human keratinocytes. *J Exp Med* 178: 1109–1114.

Hope, R. T., Wood, G. S., and Abel, E. A. 1990. Mycosis fungoides and the Sezary syndrome: Pathology, staging, and treatment. *Curr Probl Cancer* 14: 293–371.

Kimmel, M., Alscher, D. M., Dunst, R. et al. 2006. The role of micro-inflammation in the pathogenesis of uraemic pruritus in haemodialysis patients. *Nephrol Dial Transplant* 21: 749–755.

Klemke, C. D., Mansmann, U., Poenitz, N., Dippel, E., and Goerdt, S. 2005. Prognostic factors and prediction of prognosis by the CTCL severity index in mycosis fungoides and sezary syndrome. *Br J Dermatol* 153: 118–124.

Krajnik, M., and Zylicz, Z. 2001. Understanding pruritus in systemic disease. *J Pain Symptom Manage* 21: 151–168.

Lundin, J., Hagberg, H., Repp, R. et al. 2003. Phase 2 study of alemtuzumab (anti-CD52 monoclonal antibody) in patients with advanced mycosis fungoides/sezary syndrome. *Blood* 101: 4267–4272.

Magro, C. M., and Crowson, A. N. 1995. Drugs with antihistaminic properties as a cause of atypical cutaneous lymphoid hyperplasia. *J Am Acad Dermatol* 32: 419–428.

Martin, P. G., Lasserre, F., Calleja, C. et al. 2005. Transcriptional modulations by RXR agonists are only partially subordinated to PPARalpha signaling and attest additional, organ-specific, molecular cross-talks. *Gene Expr* 12: 177–192.

Meyer, N., Paul, C., and Misery, L. 2010. Pruritus in cutaneous T-cell lymphomas: Frequent, often severe and difficult to treat. *Acta Derm Venereol* 90: 12–17.

Meyer-Hoffert, U. 2009. Reddish, scaly, and itchy: How proteases and their inhibitors contribute to inflammatory skin diseases. *Arch Immunol Ther Exp* 57: 345–354.

Misery, L., and Ständer, S. 2010. *Pruritus*. London: Springer-Verlag.

Moses, S. 2003. Pruritus. *Am Fam Physician* 68: 1135–1142, 1145–1146.

Mostow, E., Neckel, S., Oberhelman, L., and Anderson, T. 1993. Complete remissions in psoralen and UV-A (PUVA)-refractory mycosis fungoides-type cutaneous T-cell lymphoma with combined interferon alpha and PUVA. *Arch Dermatol* 129: 747–752.

Namaka, M., Gramlich, C. R., Ruhlen, D., Melanson, M., Sutton, I., and Major, J. 2004. A treatment algorithm for neuropathic pain. *Clin Ther* 26: 951–979.

Olsen, E. A., Duvic, M., Frankel, A. et al. 2001. Pivotal phase III trial of two dose levels of denileukin diftitox for the treatment of cutaneous T-cell lymphoma. *J Clin Oncol* 19: 376–388.

Papadavid, E., Antoniou, C., Nikolaou, V. et al. 2008. Safety and efficacy of low-dose bexarotene and PUVA in the treatment of patients with mycosis fungoides. *Am J Clin Dermatol* 9: 169–173.

Rich, B. E., Campos-Torres, J., Tepper, R. I., Moreadith, R. W., and Leder, P. 1993. Cutaneous lymphoproliferation and lymphomas in interleukin 7 transgenic mice. *J Exp Med* 177: 305–316.

Rivard, J., and Henry, W. L. 2005. Ultraviolet phototherapy for pruritus. *Dermatol Ther* 18: 344–354.

Scarisbrick, J. J., Taylor, P., Holtick, U. et al. 2008. Photopheresis expert group. U.K. consensus statement on the use of extracorporeal photopheresis for treatment of cutaneous T-cell lymphoma and chronic graft-versus-host disease. *Br J Dermatol* 158: 659–678.

Sepmeyer, J. A., Greer, J. P., Koyama, T., and Zic, J. A. 2007. Open-label pilot study of combination therapy with rosiglitazone and bexarotene in the treatment of cutaneous T-cell lymphoma. *J Am Acad Dermatol* 56: 584–587.

Sonkoly, E., Muller, A., Lauerma, A. I. et al. 2006. IL-31: A new link between T cells and pruritus in atopic skin inflammation. *J Allergy Clin Immunol* 117: 411–417.

Ständer, S., Siepmann, D., Herrgott, I., Sunderkotter, C., and Luger, T. A. 2010. Targeting the neurokinin receptor 1 with apepritant: A novel antipruritic strategy. *Plos One* 5: e10968.

Summey, B. T., and Yosipovitch, G. 2005. Pharmacologic advances in the systemic treatment of itch. *Dermatol Ther* 18: 328–332.

Trautinger, F., Knobler, R., Willemze, R. et al. 2006. EORTC consensus recommendations for the treatment of mycosis fungoides/sezary syndrome. *Eur J Cancer* 42: 1014–1030.

Vermeer, M. H., and Willemze, R. 1996. Is mycosis fungoides exacerbated by fluoxetine? *J Am Acad Dermatol* 35: 635–636.

Vij, A., and Duvic, M. 2012. Prevalence and severity of pruritus in cutaneous T cell lymphoma. *Int J Dermatol* 51: 930–934.

Wahlgren, C. F., Tengvall Linder, M., Hägermark, O., and Scheynius, A. 1995. Itch and inflammation induced by intradermally injected interleukin-2 in atopic dermatitis patients and healthy subjects. *Arch Dermatol·Res* 287: 572–580.

Willemze, R., Jaffe, E. S., Burg, G. et al. 2005. WHO-EORTC classification for cutaneous lymphomas. *Blood* 105: 3768–3785.

Yamamoto, T., Katayama, I., and Nishioka, N. 1997. Role of mast cell and stem cell factor in hyperpigmented mycosis fungoides. *Blood* 90: 1338–1340.

Yashar, S. S., Gielczyk, R., Scherschun, L., and Lim, H. W. 2003. Narrow-band ultraviolet B for vitiligo, pruritus, and inflammatory dermatoses. *Photodermatol Photoimmunol Photomed* 19: 164–168.

Yesudian, P. D., and Wilson, N. J. E. 2005. Efficacy of gabapentin in the management of pruritus of unknown origin. *Arch Dermatol* 141: 1507–1509.

Yosipovitch, G., and Fleischer, A. 2003. Itch associated with skin diseases: Advances in pathophysiology and emerging therapies. *Am J Clin Dermatol* 4: 617–622.

Yosipovitch, G., Szolar, C., Hui, X. Y., and Maibach, H. 1996. High-potency topical cortico-steroid rapidly decrease histamine-induced itch but not thermal sensation and pain in human beings. *J Am Acad Dermatol* 35: 118–120.

Zackheim, H. S. 2003. Treatment of mycosis fungoides/Sezary syndrome: The University of California, San Francisco (UCSF) approach. *Int J Dermatol* 4: 53–56.

Zackheim, H. S., Kashani-Sabet, M., and Amin, S. 1998. Topical corticosteroids for mycosis fungoides. Experience in 79 patients. *Arch Dermatol* 134: 949–954.

Zackheim, H. S., Kashani-Sabet, M., and Hwang, S. T. 1996. Low-dose methotrexate to treat erythrodermic cutaneous T-cell lymphoma: Results in twenty-nine patients. *J Am Acad Dermatol* 34: 626–631.

Zackheim, H. S., Kashani-Sabet, M., and McMillan, A. 2003. Low-dose methotrexate to treat mycosis fungoides: A retrospective study in 69 patients. *J Am Acad Dermatol* 49: 873–878.

Zic, J. A. 2003. The treatment of cutaneous T-cell lymphoma with photopheresis. *Dermatol Ther* 16: 337–346.

9 Pruriceptors

Matthias Ringkamp and Richard Meyer

CONTENTS

9.1 INTRODUCTION

Although the sensations of itch and pain are quite distinct and result in different behavioral responses (i.e., scratching versus withdrawal), they share multiple important features. In human skin, both sensations can be produced by stimuli of the same modality (e.g., mechanical, thermal, chemical, and electrical), with stimuli at low intensities producing itch and, at higher intensities, inducing the sensation of pain (von Frey 1922; Lewis et al. 1927; Bishop 1943). Itch or pain can furthermore be induced from apparently identical spots in human skin (von Frey 1922; Bishop 1943), and loss of itch sensation is paralleled by the loss of pain sensitivity (Török 1907; Thöle 1912; von Frey 1922; Bickford 1938; McMurray 1950; Nolano et al. 2000).

Following the intradermal administration of histamine or capsaicin, a neurogenically mediated increase in blood flow ("flare") can be observed in the skin surrounding the application site, and mechanical stimulation of the surrounding skin produces the sensation of itch or pain (Simone et al. 1991; LaMotte et al. 1991). Capsaicin desensitization of human skin leads to loss of pain induced by noxious heat or capsaicin and the loss of histamine-induced itch sensation (Jancso et al. 1985, Tóth-Kása et al. 1986; Nolano et al. 1999; Fuchs et al. 2000; Weisshaar et al. 1998). These commonalities suggest that the sensations of itch and pain are either mediated through an identical neuronal pathway or neuronal pathways that are closely related anatomically and functionally.

Several lines of evidence support the idea that itch and pain are indeed mediated through different primary afferent pathways. Noxious counterstimuli are able to provide a marked and persistent relief from itch while producing only a modest, nonsignificant, and transient reduction of pain (Ward et al. 1996). Opioids may induce itch while they produce pronounced analgesia (e.g., Ballantyne et al. 1988), and these effects are likely mediated by opioids acting on distinct dorsal horn neurons (Liu et al. 2011). Ablation of dorsal horn neurons expressing the gastrin-releasing peptide receptor leads to pronounced inhibition of scratching behavior in mice without affecting behavior induced by noxious stimuli (Sun et al. 2009). Mice, in which the central transmission of nociceptive input is diminished, show decreased pain behaviors to noxious stimuli, while spontaneous and stimulus-evoked scratching behavior is increased (Liu et al. 2010). These findings suggest that the sensations of itch and pain are mediated through distinct afferent pathways.

In this chapter we will review the primary afferent neuronal apparatus that responds to pruritic stimuli in primate and that therefore likely mediates the sensation of itch. As it will become evident, these afferents also respond to noxious stimuli that in human produce the sensation of pain. However, this does not exclude the possibility that these afferents indeed act as "pruriceptors."

9.2   PARALLEL PATHWAYS FOR HISTAMINERGIC AND NONHISTAMINERGIC ITCH

Although itch can be produced by mechanical, heat, and electrical stimuli, histamine has been the preferred tool to study the neuronal mechanism underlying itch sensation as it reliably produces itch following intradermal injection or iontophoresis. However, in clinical practice, many chronic itch conditions are not well treated with antihistamines, indicating that histamine is likely not a major contributing mediator and that nonhistaminergic mechanisms may play a more prominent role. Therefore, studies of these nonhistaminergic mechanisms may provide novel therapeutic strategies for combating chronic itch.

Spicules from the plant *Mucuna pruriens* (also known as cowhage) have been shown to produce itch sensation upon insertion into the skin (Shelley and Arthur 1957). Upon insertion of cowhage, subjects experience the sensation of itch and also nociceptive sensation of pricking and burning (Sikand et al. 2009). The active pruritogen in cowhage spicules was identified by Shelley and Arthur (1955a,b) as a proteinase and named mucunain. This novel cysteine proteinase was found to activate the proteinase-activated receptor 2 (PAR2) and PAR4 (Reddy et al. 2008). The PAR2 receptor has been implicated in the itch associated with atopic dermatitis (Steinhoff et al. 2003). As will be shown below, cowhage-induced itch appears to be independent of histamine, and several lines of evidence suggest that cowhage- and histamine-induced itch are mediated through different afferent pathways in the peripheral nervous system.

9.2.1   FLARE TO HISTAMINE BUT NOT COWHAGE

Topical administration of histamine into human skin leads to vasodilation that is visible as a reddening ("flare") in the skin surrounding the application site (Simone et al. 1987;

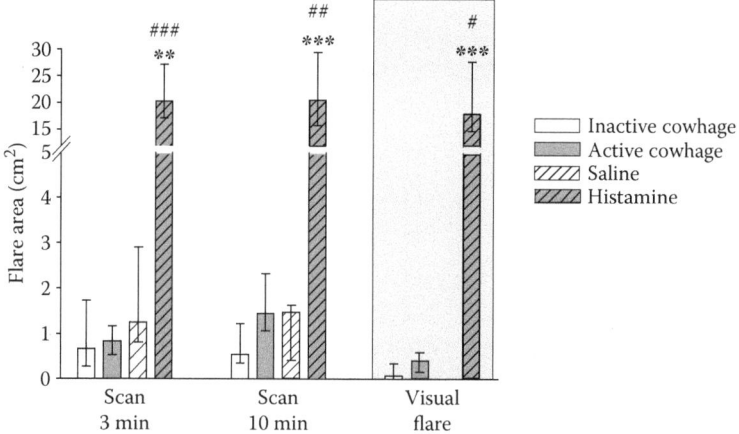

FIGURE 9.1 Histamine but not cowhage produces a flare in humans. The flare area produced by saline, inactive cowhage, active cowhage, and histamine were determined 3 and 10 min after stimulus application by laser-Doppler blood-flow measurements and also by visual inspection in a total of 12 subjects. The flare area produced by histamine was significantly larger than those produced by saline or active cowhage (*** $p < 0.001$; **$p < 0.01$ histamine versus saline; ###, $p < 0.001$; ##, $p < 0.01$; #, $p < 0.05$ histamine versus active cowhage). (Reprinted from Johanek, L. M. et al., *J Neurosci*, 27: 7490–7497, 2007. With permission.)

Magerl et al. 1990). As determined from laser-Doppler blood-flow imaging studies, the flare induced by a small amount of histamine (1 µg/10 µl) covers an area of more than 20 cm^2 (Figure 9.1). The area of flare correlated with the amount of itch; subjects that had larger areas of flare reported a greater magnitude of itch sensation (Johanek et al. 2007). Flare is absent in denervated skin; i.e., it is of neurogenic origin. It is thought to be mediated by an axon reflex where action potentials from an excited branch of a nerve fiber are antidromically propagated into other branches of that afferent resulting in the release of vasoactive substances from their peripheral terminals. The large area of flare suggests that the peripheral afferents responsible for histamine-induced itch must have large receptive fields and an extensive terminal arborization in the skin.

In contrast to histamine, insertion of cowhage spicules into the skin produces only a small area of local reddening but not a large area of flare (Figure 9.1). Because cowhage spicules produced a similar magnitude of itch, this difference is not due to less activation of peripheral afferents but rather suggests that a different population of afferents is activated.

9.2.2 TOPICAL ANTIHISTAMINE TREATMENT BLOCKS ITCH FROM HISTAMINE BUT NOT COWHAGE

When the skin is topically treated with an antihistamine cream, the itch and flare produced by intradermal injections of histamine are greatly diminished (Figure 9.2). In contrast, the itch produced by cowhage is not altered by the antihistamine treatment (Johanek et al. 2007). This is consistent with the finding that cowhage acts at the PAR receptor and that it produces itch through histamine-independent mechanisms.

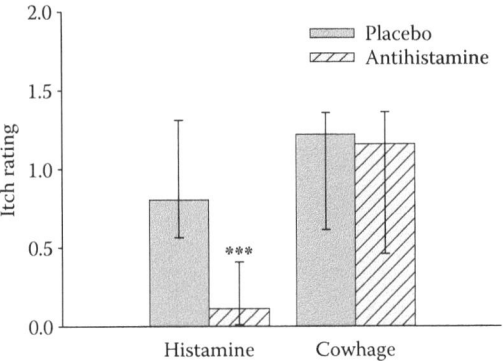

FIGURE 9.2 Histamine but not cowhage-induced itch was significantly reduced by pretreatment with a topical antihistamine. Skin sites were either pretreated with placebo cream or an antihistamine cream and subjects ($n = 15$) rated the intensity of itch sensation from cowhage and histamine applied to these sites. Itch ratings were averaged over the 0–3.5 min time period after application of the pruritic stimulus and the median ± 25th and 75th quartiles are presented (Friedman RM ANOVA, $p < 0.0001$, followed by Dunn's Multiple Comparison, ***$p < 0.001$). (Adapted from Johanek, L. M. et al., *J Neurosci*, 27: 7490–7497, 2007. With permission.)

9.2.3 TOPICAL CAPSAICIN ELIMINATES COWHAGE-INDUCED ITCH

Repeated topical applications of capsaicin lead to a desensitization of the skin to heat stimuli; i.e., nonpainful and painful heat cannot be detected at the treated skin. Cowhage-induced itch is completely abolished in capsaicin-desensitized skin (Figure 9.3); thus superficial, capsaicin-sensitive nerve terminals are likely responsible for cowhage itch. In contrast, the itch to intradermal histamine is only slightly decreased, suggesting that the capsaicin treatment regimen used in those studies

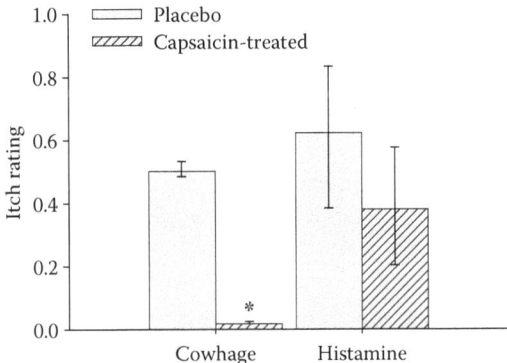

FIGURE 9.3 Capsaicin pretreatment prevented cowhage-induced itch. Subjects ($n = 6$) rated the intensity of itch sensation from cowhage and histamine at sites pretreated for 3 days with either a placebo cream or 0.1% capsaicin cream. Itch ratings were averaged over the 0–3.5 min time period. (median ± 25th and 75th quartiles; *$p < 0.05$, Wilcoxon matched pairs test). (Adapted from Johanek, L. M. et al., *J Neurosci*, 27: 7490–7497, 2007. With permission.)

likely spares the deeper cutaneous terminals that are activated by intradermally injected histamine. However, both cowhage- and histamine-induced itch are likely mediated through capsaicin sensitive nerve endings as previous studies using different capsaicin treatment regiments and different capsaicin formulations reported a complete loss or significant reduction of histamine-induced itch in capsaicin-desensitized skin (Tóth-Kása et al. 1986; Weisshaar et al. 1998).

9.2.4 ITCH PRODUCED BY HISTAMINE IS NOT CORRELATED WITH ITCH PRODUCED BY COWHAGE

The magnitude of itch to histamine and cowhage varies greatly across subjects. However, the ratings to repeated applications within a given subject are highly correlated; a subject who gave a low (or high) rating to one administration would likely give a similar rating to another administration of the same agent. In contrast, the itch ratings to cowhage are not correlated with the itch ratings to histamine; a subject who gave a low rating to histamine was not necessarily likely to give a low rating to cowhage. This supports the notion that the mechanisms of cowhage- and histamine-induced itch are independent.

9.3 ROLE OF C-FIBERS IN ITCH SENSATION

C-fiber afferents can be divided into two classes based on their response to mechanical stimuli. Mechanically insensitive C-fibers (C-MIAs) are either unresponsive to mechanical stimuli or have a very high mechanical threshold. These afferents respond to heat and various noxious chemical stimuli (e.g., capsaicin) and are often considered to be chemoreceptors. C-MIAs in humans typically have large receptive field areas ($\sim$5 cm^2). In contrast, mechanically sensitive C-fibers are polymodal and typically also respond to heat (CMHs) and noxious chemicals. In humans, CMHs have relatively small receptive field areas ($\sim$2 cm^2) (Schmidt et al. 2002).

9.3.1 C-MIAS RESPOND TO HISTAMINE BUT NOT COWHAGE

Approximately 30% of C-MIAs respond to application of histamine (Schmelz et al. 1997, 2003). Importantly, the time course of the response to histamine matches the time course of itch sensation providing indirect, but strong evidence that C-MIAs encode histamine-induced itch (Figure 9.4). In contrast, human polymodal C-fibers are only weakly activated by histamine (Handwerker et al. 1991; Schmelz et al. 1997). In addition to exhibiting a time course of histamine-induced activation that matches human itch ratings, the large receptive fields of C-MIAs provide an anatomical correlate for the large axon-reflexive flare associated with histamine administration. In contrast, cowhage does not produce a response in C-MIAs (Namer et al. 2008; Johanek et al. 2008). Thus, C-MIAs play an important role in histamine itch, but they are not involved in itch produced by cowhage spicules.

It should be noted that C-MIAs mediate the pain associated with intradermal capsaicin injection (Schmelz et al. 2000) and that histamine-sensitive C-MIAs respond to capsaicin and a variety of chemical stimuli that produce either a weak sensation in

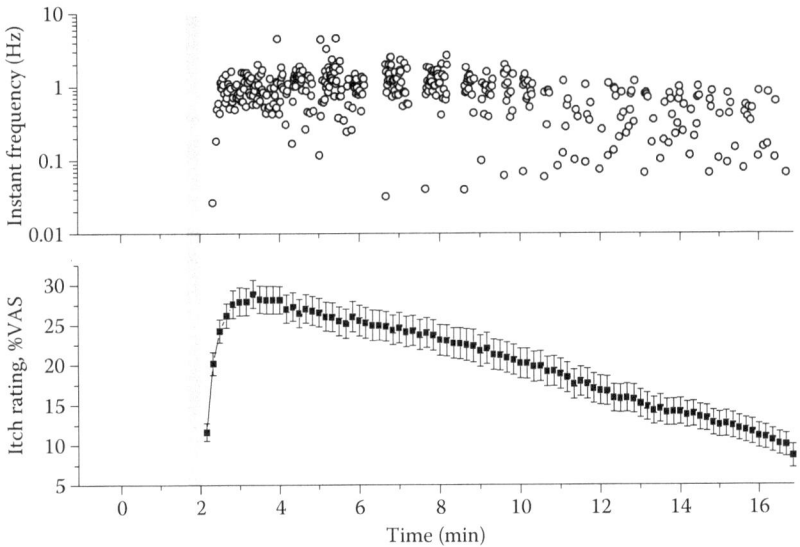

FIGURE 9.4 Histamine activates mechanically insensitive C-fibers (C-MIAs). (top) Instantaneous frequency plot showing an example of histamine induced activation in one C-MIA obtained in microneurographical recordings from the peroneal nerve in humans. For each action potential recorded, its instantaneous frequency was plotted versus the time of occurrence. Following histamine iontophoresis (gray bar), the afferent fiber becomes activated with maximal activity 1–2 min after iontophoresis. The neuronal activity then decreases over time but still can be recorded for about 15 min. (bottom) Itch ratings from 21 healthy subjects in response to an identical histamine stimulus. Itch ratings peak about 1 min after iontophoresis and then slowly decline over the next 15 min. (Adapted from Schmelz, M. et al., *J Neurosci,* 17: 8003–8008, 2007. With permission.)

humans (e.g., prostaglandin E2) or a mixed sensation of itch and pain (Schmelz et al. 2003). It is still unclear if C-MIAs exist that are specifically or selectively responsive to histamine or, alternatively, if C-MIAs exist that only respond to noxious stimuli.

9.3.2 CMHs Respond to Histamine and Cowhage

In nonhuman primates, the vast majority of CMHs respond to topical application of cowhage and intradermal injection of histamine (Johanek et al. 2008). However, the average response to cowhage was more than twice the response to histamine. It is likely that the divergent results concerning the histamine sensitivity of CMHs of human and nonhuman primates are due to differences in stimulus application (injection versus iontophoresis) and recording techniques. Thus, in microneurography experiments in humans, the signal to noise ratio is small, and therefore activity in unmyelinated fibers is tracked by the "marking" technique (Schmelz et al. 1997), which involves continuous electrical stimulation of the afferent under study at a low frequency (typically 0.25 Hz). However, this ongoing stimulation may diminish excitability of the action potential initiation site at the peripheral receptive terminal and inhibit excitation, especially when the agonist is weakly excitatory. In nonhuman primates, CMH responses to histamine and cowhage were not correlated; a fiber that gave a large response to cowhage would not necessarily give a large response to histamine, providing further evidence that both are likely mediated through different mechanisms. Cowhage-induced excitation was often bursting in nature; short, high-frequency bursts of action potentials could be followed by an interval of no activity. Even in a given fiber, responses to repetitive application of cowhage were variable, suggesting that spicules likely have to be positioned closely to the receptive terminal to induce neuronal excitation. Similar to findings in nonhuman primates, microneurographic recordings from CMHs in humans show that the vast majority of these afferents is activated by cowhage (Namer et al. 2008). In mice, *in vivo* recordings from dorsal root ganglion neurons show about 40% of polymodal C-fiber afferents responsive to cowhage. All of the cowhage responders were also sensitive to capsaicin, but only a subgroup was responsive to histamine (Ma et al. 2012).

The finding that a majority of CMHs in primates responds to active cowhage spicules poses a dilemma: these afferents respond to noxious heat in a temperature-dependent manner, and their response at different temperatures matches well with human pain ratings. Therefore, CMHs have been hypothesized to encode pain from noxious heat stimuli (Meyer and Campbell 1981). So how can neuronal input from the same afferent underlie such different sensations as "pain" or "itch"? One possibility is that the discharge pattern in primary afferents is crucial for the activity to be interpreted as pain or itch. For example, the bursting discharge that was frequently observed in polymodal nociceptors after cowhage application may be crucial for the neuronal activity to be interpreted as itch. Alternatively, pain and itch may be mediated by different subclasses of polymodal unmyelinated afferents. Indeed, CMHs can be subclassified into two groups based on their response to a stepped heat stimulus from 38°C to 49°C (Figure 9.5). Quick C-fibers (QCs) exhibit their peak discharge during the rising phase of the heat stimulus; the response quickly adapts during the plateau phase. In contrast, slow C-fibers (SCs) exhibit their peak discharge during the plateau phase of the heat stimulus. The heat threshold (41.1 ± 0.2°C) and mechanical threshold (1.23 ± 0.02 bars) of QCs is significantly lower than the corresponding thresholds of SCs (46.1 ± 0.3°C; 2.00 ± 0.10 bars; $p < 0.001$). However, the conduction velocity of QCs (0.90 ± 0.02 m/s) does not differ from that of SCs (0.90 ± 0.01 m/s). The two different fiber types exhibited a similar time course of response following the insertion of cowhage spicules into the receptive field (Figure 9.6). However, the total response of QCs was almost twice that of SCs. Furthermore, these data are consistent with the hypothesis that the QC fibers terminate close to the surface of the epidermis (where itch from superficial stimuli is thought to originate), whereas the receptive endings of SC fibers are situated deeper in the dermis. Further investigations are, however, necessary to clarify the role of QCs in itch and pain sensation.

Taken together, these results strongly suggest that CMHs play an important role in the nonhistaminergic itch produced by cowhage. In addition, both C-MIAs and CMHs may play a role in itch from histamine.

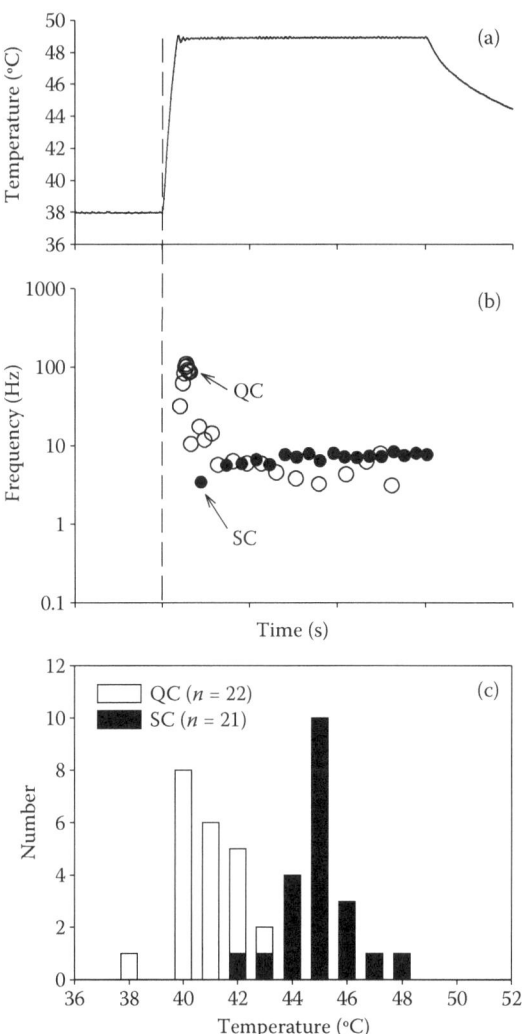

FIGURE 9.5 Two types of heat responses are observed in heat-sensitive C-fibers. (a) Stimulus waveform of the laser heat stimulus (49°C, 3 s) used to differentiate responses to heat. (b) The quickly responding C-fiber (QC, open circles) exhibits a high-frequency neuronal discharge during the rising phase of the temperature that adapts during the plateau phase of the stimulus. The slowly responding C-fiber (SC, solid circles) has a slow response onset and shows an almost constant discharge over the stimulation period. Each dot corresponds to the instantaneous frequency and time of occurrence of an action potential. (c) Histogram for the heat thresholds in QC- and SC-fibers shows two almost nonoverlapping distributions. Heat thresholds were determined with a staircase heat stimulus from 38°C to 49°C with an increment of 1°C per 1 s. (Reprinted from Johanek, L. M. et al., *J Neurosci*, 28: 7659–7669, 2008. With permission.)

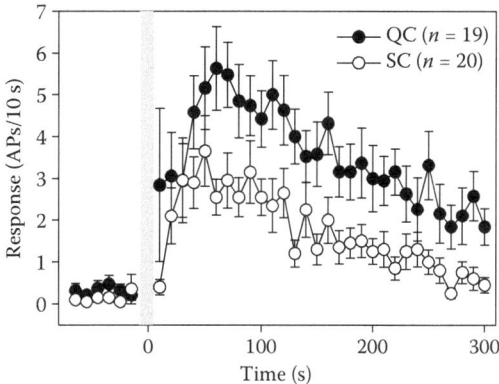

FIGURE 9.6 Time course of cowhage-induced excitation in QC- and SC-fibers looks similar, but the discharge in QC-fibers is larger (bin size = 10 s). Gray bar corresponds to time of administration of cowhage to the receptive field (Reprinted from Johanek, L. M. et al., *J Neurosci*, 28: 7659–7669, 2008. With permission.)

9.4 ROLE FOR A-FIBERS IN ITCH SENSATION

Although much attention has focused on unmyelinated fibers, small myelinated fibers may also play a role in itch sensation. To assess the role of myelinated fibers, a selective nerve block of myelinated fibers was performed in human subjects (Ringkamp et al. 2011). When conduction in A-fibers, but not C-fibers, was blocked, there was a significant reduction in the magnitude of the sensations of itch, pricking, and burning produced by cowhage. Importantly, the A-fiber block did not equally affect itch sensation across subjects. Although a successful A-fiber block was obtained in all subjects (as indicated by loss of cooling sensation and delayed detection of pricking pain to mechanical stimulation), only about half of the subjects showed a block of itch sensation. This indicates that a subgroup of subjects had an itch sensation that was mainly mediated by activity in A-fiber afferents, whereas others had itch that largely depended on activity in C-fibers. This difference may reflect an inherent difference between subjects. Alternatively, some test sites may be innervated preferentially either by A- or C-fibers. Surprisingly, the time course for itch sensation differed between subjects with A-fiber and C-fiber dominated itch (Figure 9.7); the time of peak itch sensation was earlier in the subjects with C-fiber dominated itch than those with A-fiber dominated itch (Figure 9.8b).

Electrophysiological experiments in primates reveal that about 40% of myelinated nociceptive afferents respond to cowhage (Ringkamp et al. 2011). The majority of these fibers did not respond to histamine or capsaicin and therefore responded selectively to cowhage. About 15% of A-fibers were selectively responsive to histamine. Thus, A-fibers may provide selective input for cowhage- and histamine-induced activity, and this segregation could underlie the differential response of

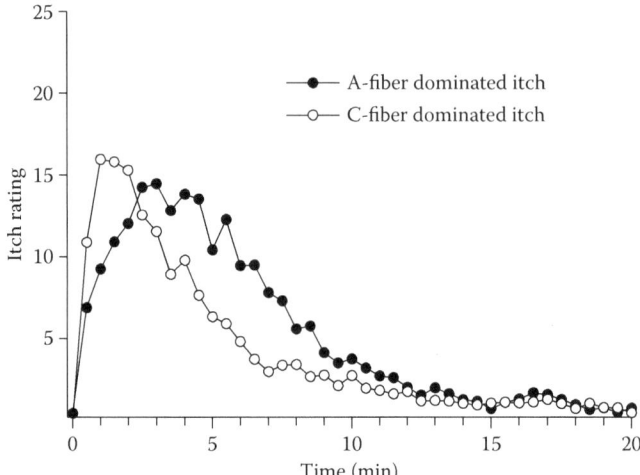

FIGURE 9.7 A-fiber and C-fiber dominated itch have a different time course. In the absence of a nerve block, the average time course of itch sensation differs in subjects whose itch responded well to the selective myelinated fiber block ($n = 6$) and those that did not ($n = 8$). In subjects with C-fiber dominated itch, the peak sensation occurred earlier and the sensation decreased more rapidly than in A-fiber dominated itch. (Reprinted from Ringkamp, M. et al., *J Neurosci,* 31: 14841–14849, 2011. With permission.)

spinothalamic tract neurons to histamine and cowhage observed in nonhuman primates (Davidson et al. 2007). The magnitude of the A-fiber response to cowhage was more than three times greater than the response in C-fibers. In addition, the time course for the response to cowhage is different between the unmyelinated and myelinated afferents (Figure 9.8a); the C-fiber response started earlier and reached a peak earlier than the A-fiber response. Interestingly, the time for the peak itch sensation for subjects with A-fiber dominated itch matched the time for peak discharge in the A-fibers (Figure 9.8b). Similarly, the time for the peak itch sensation for subjects with C-fiber dominated itch matched the time for peak discharge in the C-fiber afferents.

As cowhage-induced itch is abolished in capsaicin-desensitized skin and dramatically reduced by an A-fiber block, one has to conclude that A-fibers activated by cowhage are capsaicin sensitive. This is consistent with the observation that many mechanosensitive A-fibers became desensitized after intradermal capsaicin injection, although only a minority of cowhage-sensitive A-fibers was excited by capsaicin in electrophysiological experiments. In a previous study, capsaicin-insensitive A-fiber afferents have been shown to account for secondary hyperalgesia to pinprick stimuli (Magerl et al. 2001). Therefore, it appears that capsaicin-insensitive and capsaicin-sensitive mechanosensitive A-fibers exist in human skin and that only the latter group is involved in mediating cowhage-induced itch.

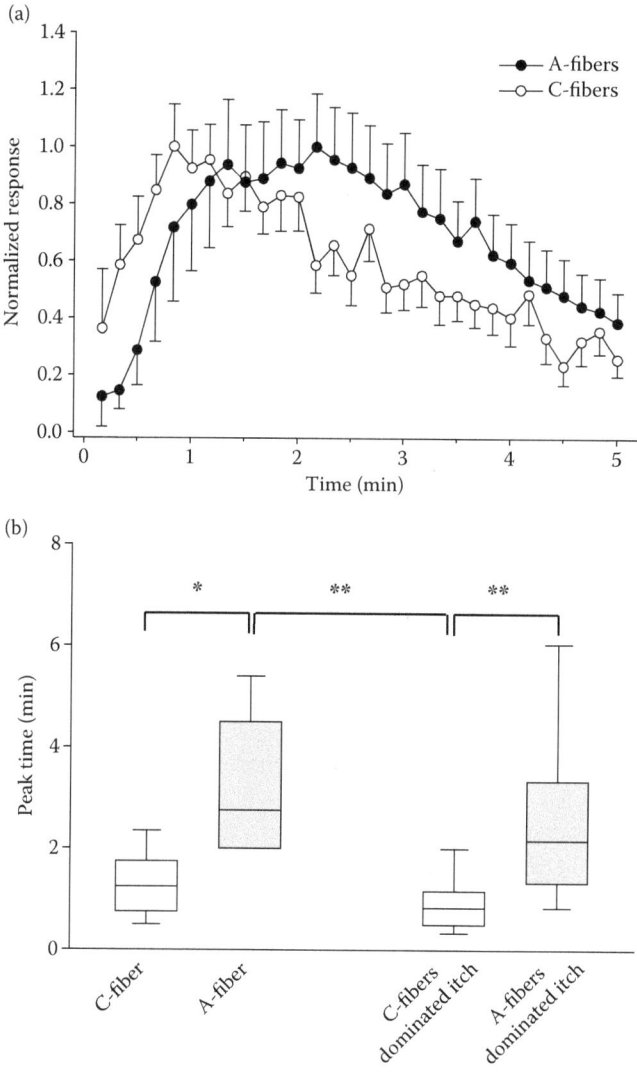

FIGURE 9.8 Comparison of the time course of responses to cowhage in A- and C-fibers. (a) For unmyelinated fibers (open circles) the peak response to cowhage occurred earlier than in myelinated fibers (solid circles). For each fiber, the response at a given time point was normalized by dividing by that fiber's maximum response. (b) The peak itch sensation in subjects with A-fiber dominated itch occurred significantly later than in subjects with C-fiber dominated itch. Similarly, the peak response to cowhage occurred later in myelinated fibers than in unmyelinated fibers (Kruskal-Wallis ANOVA, $p < 0.001$ followed by multiple Mann-Whitney U tests corrected for multiple comparisons, $*p < 0.05$, $**p < 0.01$). Medians, quartiles, and 5/95% percentiles are plotted. (Reprinted from Ringkamp, M. et al., *J Neurosci,* 31: 14841–14849, 2011. With permission.)

9.5   CONCLUSIONS

The results from the psychophysical and electrophysiological studies indicate that multiple primary afferent pathways in primate are involved in itch sensation: mechano-insensitive C-fibers (C-MIA), mechanoheat-sensitive C-fibers (CMH) and mechano-sensitive A-fiber nociceptors. Histamine appears to activate all three pathways but likely produces a stronger activation in CMIAs. In contrast, cowhage selectively activates mechanosensitive C- and A-fiber afferents. On the basis of these findings, one may be tempted to hypothesize that histamine-dependent forms of itch are preferentially mediated through C-MIAs, whereas histamine-independent forms are mediated through mechanosensitive C- and A-fibers. Consistent with this hypothesis is the observation that the itch produced by bovine adrenal medulla 8-22 (BAM8-22) peptide is not blocked by antihistamine treatment and is not accompanied by a neurogenic flare response (Sikand et al. 2011). Similarly, a flare response is not observed when itch is produced by intradermal injection of ß-alanine, which activates a subclass of murine dorsal root ganglion neurons that are unresponsive to histamine (Liu et al. 2012). However, results showing the effects of BAM8-22 and ß-alanine on primary afferent fibers in primates are still lacking. In mice, several pruritic agents (i.e., ß-alanine, BAM8-22, and chloroquine) exert their effects through activation of mas related G-protein coupled receptors (Mrgpr) such as MrgprD, MrgprC11, and MrgprA3 (Liu et al. 2009, 2012). It is currently unclear what nerve fibers in primates express MrgprD or hMrgprX1, the homologue receptor to murine MrgprC11 and MrgprA3. Further studies in primates are therefore necessary to investigate the responses to pruritic agents that activate Mrgprs and to see if responses to such agents are restricted to functionally distinct classes of afferents. Results from such studies are expected to increase our understanding underlying the neuronal mechanisms of itch and are necessary to improve the treatment strategies for patients suffering from chronic itch.

REFERENCES

Ballantyne, J. C., Loach, A. B., and Carr, D. B. (1988) Itching after epidural and spinal opiates. *Pain* 33: 149–160.

Bickford, R. G. (1938) Experiments relating to the itch sensation, its peripheral mechanism and central pathways. *Clin Sci* 3: 377–386.

Bishop, G. H. (1943) Responses to electrical stimulation of single sensory units of skin. *J Neurophysiol* 6: 361–382.

Davidson, S., Zhang, X., Yoon, C. H. et al. (2007) The itch-producing agents histamine and cowhage activate separate populations of primate spinothalamic tract neurons. *J Neurosci* 27: 10007–10014.

Fuchs, P. N., Campbell, J. N., and Meyer, R. A. (2000) Secondary hyperalgesia persists in capsaicin desensitized skin. *Pain* 84: 141–149.

Handwerker, H. O., Forster, C., and Kirchhoff, C. (1991) Discharge patterns of human C-fibers induced by itching and burning stimuli. *J Neurophysiol* 66: 307–315.

Jancso, G., Obal, F., Jr., Toth, K. I. et al. (1985) The modulation of cutaneous inflammatory reactions by peptide-containing sensory nerves. *Int J Tissue React* 7: 449–457.

Johanek, L. M., Meyer, R. A., Friedman, R. M. et al. (2008) A role for polymodal C-fiber afferents in nonhistaminergic itch. *J Neurosci* 28: 7659–7669.

Johanek, L. M., Meyer, R. A., Hartke, T. et al. (2007) Psychophysical and physiological evidence for parallel afferent pathways mediating the sensation of itch. *J Neurosci* 27: 7490–7497.

LaMotte, R. H., Shain, C. N., Simone, D. A. et al. (1991) Neurogenic hyperalgesia: Psychophysical studies of underlying mechanisms. *J Neurophysiol* 66: 190–211.

Lewis, T., Grant, R. T., and Marvin, H. M. (1927) Vascular reactions of the skin to injury. Part X.—The intervention of a chemical stimulus illustrated especially by the flare. The response to faradism. *Heart* 29: 139–160.

Liu, Q., Sikand, P., Ma, C. et al. (2012) Mechanisms of itch evoked by beta-alanine. *J Neurosci* 32: 14532–14537.

Liu, Q., Tang, Z., Surdenikova, L. et al. (2009) Sensory neuron-specific GPCR Mrgprs are itch receptors mediating chloroquine-induced pruritus. *Cell* 139: 1353–1365.

Liu, X. Y., Liu, Z. C., Sun, Y. G. et al. (2011) Unidirectional cross-activation of GRPR by MOR1D uncouples itch and analgesia induced by opioids. *Cell* 147: 447–458.

Liu, Y., Abdel, S. O., Zhang, L. et al. (2010) VGLUT2-dependent glutamate release from nociceptors is required to sense pain and suppress itch. *Neuron* 68: 543–556.

Ma, C., Nie, H., Gu, Q. et al. (2012) In vivo responses of cutaneous C-mechanosensitive neurons in mice to punctate chemical stimuli that elicit itch and nociceptive sensations in humans. *J Neurophysiol* 107: 357–363.

Magerl, W., Fuchs, P. N., Meyer, R. A. et al. (2001) Roles of capsaicin-insensitive nociceptors in cutaneous pain and secondary hyperalgesia. *Brain* 124: 1754–1764.

Magerl, W., Westerman, R. A., Möhner, B. et al. (1990) Properties of transdermal histamine iontophoresis: Differential effects of season, gender, and body region. *J Invest Dermatol* 94: 347–352.

McMurray, G. A. (1950) Experimental study of a case of insensitivity to pain. *Arch Neurol Psychiat* 64: 650.

Meyer, R. A., and Campbell, J. N. (1981) Peripheral neural coding of pain sensation. *Johns Hopkins APL Tech Dig* 2: 164–171.

Namer, B., Carr, R., Johanek, L. M. et al. (2008) Separate peripheral pathways for pruritus in man. *J Neurophysiol* 100: 2062–2069.

Nolano, M., Crisci, C., Santoro, L. et al. (2000) Absent innervation of skin and sweat glands in congenital insensitivity to pain with anhidrosis. *Clin Neurophysiol* 111: 1596–1601.

Nolano, M., Simone, D. A., Wendelschafer-Crabb, G. et al. (1999) Topical capsaicin in humans: Parallel loss of epidermal nerve fibers and pain sensation. *Pain* 81: 135–145.

Reddy, V. B., Iuga, A. O., Shimada, S. G. et al. (2008) Cowhage-evoked itch is mediated by a novel cysteine protease: A ligand of protease-activated receptors. *J Neurosci* 28: 4331–4335.

Ringkamp, M., Schepers, R. J., Shimada, S. G. et al. (2011) A role for nociceptive, myelinated nerve fibers in itch sensation. *J Neurosci* 31: 14841–14849.

Schmelz, M., Schmidt, R., Bickel, A. et al. (1997) Specific C-receptors for itch in human skin. *J Neurosci* 17: 8003–8008.

Schmelz, M., Schmidt, R., Handwerker, H. O. et al. (2000) Encoding of burning pain from capsaicin-treated human skin in two categories of unmyelinated nerve fibres. *Brain* 123(Pt 3): 560–571.

Schmelz, M., Schmidt, R., Weidner, C. et al. (2003) Chemical response pattern of different classes of C-nociceptors to pruritogens and algogens. *J Neurophysiol* 89: 2441–2448.

Schmidt, R., Schmelz, M., Weidner, C. et al. (2002) Innervation territories of mechano-insensitive C nociceptors in human skin. *J Neurophysiol* 88: 1859–1866.

Shelley, W. B., and Arthur, R. P. (1955a) Mucunain, the active pruritogenic proteinase of cowhage. *Science* 122: 469–470.

Shelley, W. B., and Arthur, R. P. (1955b) Studies on cowhage (*mucuna pruriens*) and its pruritogenic proteinase, mucunain. *Arch Dermatol* 72: 399–406.

Shelley, W. B., and Arthur, R. P. (1957) The neurohistology and neurophysiology of the itch sensation in man. *Arch Dermatol* 76: 296–323.

Sikand, P., Dong, X., and LaMotte, R. H. (2011) BAM8-22 peptide produces itch and nociceptive sensations in humans independent of histamine release. *J Neurosci* 31: 7563–7567.

Sikand, P., Shimada, S. G., Green, B. G. et al. (2009) Similar itch and nociceptive sensations evoked by punctate cutaneous application of capsaicin, histamine and cowhage. *Pain* 144: 66–75.

Simone, D. A., Alreja, M., and LaMotte, R. H. (1991) Psychophysical studies of the itch sensation and itchy skin ("Alloknesis") produced by intracutaneous injection of histamine. *Somatosens Mot Res* 8: 271–279.

Simone, D. A., Ngeow, J. Y., Whitehouse, J. et al. (1987) The magnitude and duration of itch produced by intracutaneous injections of histamine. *Somatosens Res* 5: 81–92.

Steinhoff, M., Neisius, U., Ikoma, A. et al. (2003) Proteinase-activated receptor-2 mediates itch: A novel pathway for pruritus in human skin. *J Neurosci* 23: 6176–6180.

Sun, Y. G., Zhao, Z. Q., Meng, X. L. et al. (2009) Cellular basis of itch sensation. *Science* 325: 1531–1534.

Thöle (1912) Über Jucken und Kitzeln in Beziehung zu Schmerzgefühl und Tastempfindung. *Neurolog Centralbl* 31: 610–617.

Török, L. (1907) Über das Wesen der Juckempfindung. *Zeitsch Psychol* 46: 23–35.

Tóth-Kása, I., Jancso, G., Bognar, A. et al. (1986) Capsaicin prevents histamine-induced itching. *Int J Clin Pharmacol Res* 6: 163–169.

von Frey, M. (1922) Zur Physiologie der Juckempfindung. *Arc Neerl Physiol* 7: 142–145.

Ward, L., Wright, E., and McMahon, S. B. (1996) A comparison of the effects of noxious and innocuous counterstimuli on experimentally induced itch and pain. *Pain* 64: 129–138.

Weisshaar, E., Heyer, G., Forster, C. et al. (1998) Effect of topical capsaicin on the cutaneous reactions and itching to histamine in atopic eczema compared to healthy skin. *Arch Dermatol Res* 290: 306–311.

10 Peripheral Neuronal Mechanism of Itch
Histamine and Itch

*Robin L. Thurmond, Kayvan Kazerouni,
Sandra R. Chaplan, and Andrew J. Greenspan*

CONTENTS

10.1 INTRODUCTION

The study of the physiological effects of histamine has a long history, spanning more than a century. Drugs that target histamine have been very successful, and even those not in the medical profession recognize that "antihistamines" are effective treatments for the symptoms of allergies and rhinitis. Histamine research has also spawned a wealth of scientific literature, and a search of the term "histamine" in PubMed yields almost 80,000 references. Even back in 1953, Sir Henry Dale himself said in his memoirs that histamine "…is by now almost too familiar" (Dale 1953, p. 119).

Yet new functions are still being uncovered, and this is prompting the development of new drugs that target histamine.

Histamine results from the decarboxylation of histidine, and was first described synthetically by Windaus and Vogt in 1907 (Windaus and Vogt 1907). Shortly thereafter, Dale and colleagues found that it was a natural constituent of ergot, which was potent in inducing contraction of certain muscles (Barger and Dale 1910; Dale and Laidlaw 1911). Further work suggested that the effects of histamine were similar to those seen with anaphylaxis (Dale and Laidlaw 1911). Interestingly, Dale did not use the obvious name, histamine, as he explained in his memoirs (Dale 1953), because it was claimed to be too similar to a trademark; Dale instead used the chemical name β-iminazolylethylamine. In addition to this, Dale took great care in his early work not to suggest that histamine was actually involved in normal physiological processes: "...a possibility that was almost clamouring to be recognized," in his own words (Dale 1953, p. 337). The statement that histamine played a physiological role *in vivo* had to await the proof of its existence as a natural component of tissue. This was hinted at by early work (Barger and Dale 1911; Abel and Kubota 1919) but was not shown conclusively until 1927 (Best et al. 1927). By 1929, in his Croonain Lectures, Dale readily acknowledged that histamine was a normal constituent of many tissues (Dale 1929).

We now know that histamine is produced by many cell types and can be stored by cells such as mast cells, basophils, enterochromaffin-like cells, and neurons. Release from these cells mediates many of the functions that we now intimately associate with histamine, i.e., allergy, gastric acid secretion, neuron transmission, anaphylaxis, etc. Early work linked histamine with responses to trauma in the skin. Injury, firm stroking of the skin, or many other stimuli cause a triple response of local vasodilation, local edema, and flare, which were described in work by Sir Thomas Lewis and coworkers and summarized in his book, *The Blood Vessels of the Human Skin* (Lewis 1927). Lewis recognized that injury to the skin liberated a substance that resembled, at least in its action, histamine. He carefully termed this the H-substance because there was no proof that histamine existed in skin, although it was clear he suspected the agent was histamine. He states, "...it is difficult to refrain from stating without reserve the simple conclusion that the vasodilator substance considered and the H-substance are one and the same, and that this substance is histamine" (Lewis 1927, p. 235). Others were not as cautious. In his 1929 Croonain Lectures, Dale discussed this in depth and concluded that "...the suggestion was obvious that... Lewis' H-substance, was histamine itself, not newly formed...but already existing...", although he acknowledged that final proof required the substance to be isolated (Dale 1929, p. 1235).

10.2 ANTIHISTAMINES AND HISTAMINE RECEPTORS

The emerging role for histamine in a variety of physiological functions led to the idea that blocking its action could have beneficial therapeutic effects. Bovet and Staub were the first to show that synthetic compounds could block the actions of histamine, thus setting the stage for the development of the antihistamine drugs known today (Bovet and Staub 1937; Staub and Bovet 1937). A few years later, the first clinical antihistamine, 2339 RP/phenbenzamine/Antergan, was characterized (Halpern 1942). This compound was shown to block the effects of intradermal histamine injection and be beneficial in the treatment of urticaria and serum sickness (Celice et al. 1942; Decourt 1942; Parrot 1942). In the United States, data on this new drug were hard to come by because of World War II in Europe (see Code 1945) and another antihistamine, diphenhydramine/Benadryl®, was being developed (Loew and Kaiser 1945; Loew et al. 1945). The first clinical description of this compound appears to be that of Curtis and Owens (Curtis and Owens 1945), who showed that 14 of 18 patients with urticaria had at least a partial response to diphenhydramine. They suggested that this was further evidence that Lewis' H-substance was indeed histamine and that it was a major factor causing the symptoms of urticaria. Further early work showed a clear benefit of diphenhydramine in urticaria, hay fever, and dermatographism but less so in asthma (Code 1945; Feinberg and Friedlaender 1945; Koelsche et al. 1945; McElin and Horton 1945; O'Leary and Farber 1945; McGavack et al. 1946). Diphenhydramine is still in use, and in the United States is available over the counter.

Very early, it was recognized that these antihistamines acted by competing with histamine at its site of action, and this was consistent with the notion of antagonism by competition for binding to receptors (Wells et al. 1945). However, while the early antihistamines were very effective in blocking some of the actions of histamine, other actions were not affected. In particular, gastric acid secretion and the positive chronotropic response in isolated atria were not inhibited, and it was noted that these effects were most likely due to the activation of a different receptor (Ashford et al. 1949; Trendelenburg and Hobbs 1960; Ash and Schild 1966). Ash and Schild (1966) proposed that the receptor blocked by the known antihistamines be called the H_1 receptor (H_1R) but stated that the classification of the other receptors needed the development of selective antagonists. This breakthrough came when Sir James Black and colleagues found ligands specific for the receptor that mediated gastric acid secretion and other effects and termed this the H_2 receptor (H_2R) (Black et al. 1972). This was borne out in the clinic, where the H_2R antagonist burimamide completely inhibited gastric acid secretion in humans (Black et al. 1972; Wyllie et al. 1972), in contrast to H_1R antagonists, which previously had been shown to be ineffective (Ashford et al. 1949). Later, it was recognized that a third receptor existed that acted as an autoreceptor in the central nervous system (Arrang et al. 1983). In this case, the relative potency of known agonists and the dissociation constants for antagonists suggested a new receptor, which was termed the H_3 receptor (H_3R). Most recently the histamine H_4 receptor (H_4R) was identified using molecular cloning techniques (Oda et al. 2000; Hough 2001). In retrospect, this receptor was also predicted by pharmacological evidence, as were the other three receptors. In particular, a novel histamine receptor was identified on human eosinophils whose pharmacology did not match the three known receptors (Raible et al. 1994). It is now known that this receptor on eosinophils is the H_4R (Buckland et al. 2003; Ling et al. 2004).

All of the receptors for histamine are G-protein receptors, and thus signal through activating specific G-proteins (or a recent review, see Bongers et al. [2010]). These interactions, and the signaling pathways activated, can vary in different cell types or under different physiological conditions. However, in general, the H_1R induces phospholipase C activation, inositol phosphate production, and calcium mobilization, via coupling to $G\alpha_q$ proteins. In the brain, stimulation of cyclic AMP production can also be seen (Marley et al. 1991). The H_1R is expressed on multiple cell types, including neurons in the central nervous system, leading to effects on sleep/wake cycles that likely result in the sedation effects seen with first-generation antihistamines that cross the blood–brain barrier. In addition, the receptor is found on smooth muscle cells, where it mediates contraction, and endothelial cells, where it is involved in vascular permeability. Like the H_1R, the H_2R is also expressed on many cell types, and, in particular, it is expressed on gastric parietal cells, where it mediates gastric acid release. Activation of the H_2R activates $G\alpha_s$, leading to cyclic AMP formation.

The H_3R has a more restrictive expression pattern, and is mainly found in the central nervous system, where it acts as a presynaptic autoreceptor with a role in central and peripheral neurotransmission. The H_3R activates $G\alpha_{i/o}$ proteins, leading to inhibition of cyclic AMP formation, enhanced calcium mobilization, and activation of mitogen-activated protein kinases and ion channels. H_3R antagonists are in clinical studies for a number of indications, but there have been recent failures in attention deficit hyperactivity disorder (Kuhne et al. 2011; Herring et al. 2012; Weisler et al. 2012).

The H_4R is expressed on many of the cells involved in immune and inflammatory responses. It appears to be mainly coupled to $G\alpha_{i/o}$ proteins, and signals through increases in intracellular calcium (Hofstra et al. 2003). Activation of MAP kinases and the transcription factor AP-1 have also been reported (Morse et al. 2001; Gutzmer et al. 2005; Desai and Thurmond 2011). There are no marketed H_4R antagonists, although preclinical data point to a possible role in inflammation and pruritus (Thurmond et al. 2008).

Selective agonists and antagonists for the four histamine receptors were crucial in the discovery of the individual receptors, and their use clinically and preclinically has helped in defining their relative roles. However, one must understand the pharmacology and relative potencies of these ligands at the different receptors before conclusions can be drawn. This is not always straightforward, because some of this information is not readily available, and the general classification of the different ligands has changed as new histamine receptors were discovered. This is clearly seen for impromidine, which was originally described as an H_2R agonist, but is now known to be an H_4R partial agonist as well as an H_3R antagonist, and in fact was used for the initial characterization of the H_3R (Arrang et al. 1983; Lim et al. 2005). This makes interpreting older literature complicated, because some of the conclusions may need to be revisited in light of the other receptors that may not have been known at the time. Even today, the selectivity of all of the histamine receptor ligands is not completely described, making the task more challenging. Furthermore, one has to be very careful as to the characterization of agonist versus antagonist. Some ligands, such as betahistine, are agonists at one receptor (H_1R), but antagonists at other receptors (H_3R) (Arrang et al. 1985). Moreover, there are examples of ligands acting as agonists of a receptor in one system, but antagonists of the same receptor in another (Galandrin et al. 2007; Kojetin et al. 2008). The selectivity of some of the commonly used histamine receptor ligands is given in Tables 10.1 and 10.2. Finally, the word "antihistamine" typically is used to refer to drugs that are antagonists (and also inverse agonists) of the H_1R. Here, we will use the more general definition of a drug that blocks the action of histamine at any of the receptors, and thus those that inhibit the H_1R are specified as H_1R antihistamines.

10.3 HISTAMINE-INDUCED SKIN PRURITUS

Since almost the time of its discovery, histamine has been injected into the skin to probe its function (Eppinger 1913). Injection of histamine causes the same triple response described by Lewis, and is the reason that many, including Lewis himself, suspected the H-substance was histamine (Lewis 1927). In the skin, application of histamine causes a wheal along with reddening that appear within a minute, increase for approximately 5 min and are largely gone in 30 min (Sollmann and Pilcher 1917). The fact that these reactions can be observed even in healthy subjects makes it an ideal tool for exploring the functions of drugs that target the inhibition of histamine. The early reports on Antergan and diphenhydramine showed that these drugs could block the wheal and flare response to intradermal injection of histamine (Celice et al. 1942; Parrot 1942; McElin and Horton 1945; McGavack et al. 1946). In 1949, Bain proposed that the shift in a histamine intradermal dose response could help compare various drugs, using the early drugs mepyramine and promethazine as examples (Bain et al. 1949; Bain 1951).

Itch is also produced when histamine is delivered to the skin. The first mention of histamine inducing itch was by Eppinger (1913), and this was further expanded by Sollmann and Pilcher (1917). In both cases, a drop of histamine solution applied onto a scratch in the skin producing reddening, wheal, and itch (Eppinger 1913; Sollmann and Pilcher 1917). The itch occurs about 20 to 30 s after application of histamine, rises to a maximum in a couple of minutes, and then declines until it disappears by about 10 min (Bickford 1937). However, Lewis' classic description of the triple response hardly mentions itch, although at the beginning of Chapter 4 he states "...followed by swelling of the skin, a local edema popularly termed a wheal...or urticaria because it is associated with itching" (Lewis 1927, p. 46). This is perhaps due to his focus on the vasculature. Several years later in his book *Pain*, Lewis deals with itch extensively and states that any kind of injury that produces the triple response will also produce itch (Lewis 1942).

Certain stimuli can inhibit the itch response, including scratching the site of irritation. In addition, noxious painful stimuli such as a pin prick, heat, or cold can inhibit itch sensations (Bickford 1937; Graham et al. 1951), and it has been postulated that scratching causes pain and thereby relieves itch (Rothman 1941). Rothman suggested that itch and pain are different degrees of the same sensation (Rothman 1941), whereas Lewis stated that they are separate phenomena (Lewis 1942). Nevertheless, there appears to be a relationship between the sensations of itch and pain. Under the right conditions, histamine injection itself has been reported to produce pain. Rosenthal and Minard showed that if the superficial layers of the skin were removed, the application of histamine caused a tingling and burning sensation at lower doses and then a stinging sharp pain at high concentrations (Rosenthal and Minard 1939). The depth of histamine injection may be important and injection into deep skin layers may cause pain (Broadbent 1955; Lindahl 1961; Darsow et al. 1996). Indeed, it was noted early on that subcutaneous injection of histamine itself could relieve pruritus (Ernstene and Banks 1933) and that this may be due to the induction of pain responses. However, Lewis commented that the sensation of pain is only occasional with intradermal histamine injection and that the predominant sensation is itch (Lewis 1942). The predominance of itch has been confirmed by others, although higher concentrations of histamine can yield a weak burning sensation (Cormia 1952; Sikand et al. 2009).

The fact that other stimuli can inhibit itch sensations even if given distal to the site of itch (Graham et al. 1951; Yosipovitch et al. 2005, 2007), and the fact that itch sensations can be blocked by local anesthetics (Shelley and Melton 1950), point to a neuronal pathway mediating itch. It has been shown that histamine can activate a subset of mechanically insensitive C-fibers that transmit the itch response to the spinal cord (Schmelz et al. 1997). These fibers also appear to be responsible for the flare response (Schmelz et al. 2000). While these fibers respond strongly to histamine, they also respond to other nociceptive compounds like capsaicin and bradykinin (Schmelz et al. 2003). Other polymodal C-fibers can weakly respond to histamine and may also play a role (Handwerker et al. 1991; Schmelz et al. 1997; Johanek et al. 2008). From the spinal cord, the signals are transmitted to the brain via specific spinothalamic tract (STT) neurons, and therefore the itch signal appears to have specific neuronal pathways in both the periphery and the central nervous system (Andrew and Craig 2001). However, the STT neurons are polymodal, responding to capsaicin as well as mechanical and thermal stimuli, and may provide a mechanism for the inhibition of itch by other stimuli (Schmelz 2001; Simone et al. 2004; Davidson et al. 2007). Indeed, it has been shown that scratching can reduce the histamine-induced activity of these neurons (Davidson et al. 2009).

Functional brain imaging studies confirm that the pattern of regional neuronal activation resulting from experimentally induced itch is complex, and contains elements of sensation, emotion, and motivation (for a comparatively recent review, see Paus et al. [2006]). A functional MRI study by Mochizuki et al. (2007) attempted direct comparisons between itch (histamine) and pain (cold pack) stimuli. While both had major features in common and showed considerable anatomical overlap, the activation patterns of itch and cold pain appeared nonetheless to be clearly distinguishable. Areas that were activated for both included somatosensory cortex (sensory localization), pre- and orbitofrontal cortex (decision making, reward/aversiveness), anterior cingulate and insula (affective aspects), and primary and secondary motor cortical areas. However, thalamic activation was considerably more prominent in pain. While Drzezga et al. (2001) also observed thalamic activation to be absent in itch and asserted that this likely represents a true difference between pain and itch, not all investigators have found similarity (Mochizuki et al. 2003; Herde et al. 2007; Schneider et al. 2008; Papoiu et al. 2012). Substantial differences in methods of regional cerebral blood flow imaging and itch stimulation likely account for a large proportion of the discrepancies.

A number of studies have highlighted differences in the areas of motor cortex activation during the experience of itch, as compared to pain. Hsieh et al. (1994), the first to conduct a functional brain imaging study of itch, reported that application of histamine to the right arm resulted in left somatosensory cortical activation, consistent with the somatotopical localization of itch sensation for the right arm, as well as a notable degree of activation of right-sided cortical structures implicated in motor planning or execution (inferior parietal lobule, supplemental motor area, and premotor area). Similar observations were made by Darsow et al. and to some extent Drzezga et al. (Darsow et al. 2000; Drzezga et al. 2001). These authors have interpreted the findings as substantiating the early definition of itching as an unpleasant sensation associated with the desire to scratch (Rothman 1941). Activation of contralateral extremity motor initiation areas appears to represent the intent to scratch the affected area with the other arm. In contrast, most studies of neuronal activation due to painful stimuli have observed activation of the motor and premotor cortex on the same side of the brain as the sensory activation, interpreted as indicative of the reflex intent to withdraw the affected limb itself from the stimulus.

Of note, different pruritus stimuli may result in a number of distinguishable patterns of neuronal pathway activation. Cowhage (a legume that can cause itch) and allergens such as house dust mite and grass pollen extracts have been shown to give somewhat different regional activation patterns compared to histamine (Leknes et al. 2007; Papoiu et al. 2012). Additionally, central nervous system processing of chronic pruritus may be altered compared to acute experimental pruritus: subjects with chronic pruritus due to atopic dermatitis activated more brain regions than healthy individuals when exposed to a skin histamine challenge (Schneider et al. 2008). Thus, central nervous system pruritus pathways appear to reflect complex primary afferent input as well as cognitive, emotional, and behavioral integration of the itch experience, and may also show plasticity over time.

Early clinical studies with the first antihistamines showed that the pruritus associated with urticaria (Curtis and Owens 1945; Feinberg and Friedlaender 1945; O'Leary and Farber 1945) and at least one case of atopic dermatitis (Feinberg and Friedlaender 1945) could be reduced. Clinical studies showed that the pruritus, along with the wheal and flare associated with injection of histamine, could be blocked by the first clinical antihistamine, Antergan (Decourt 1942; Parrot 1942). Similar to the use of the histamine dose response for wheal and flare proposed by Bain (Bain et al. 1949; Bain 1951), Cormia and Kuykendall used changes in the threshold of histamine concentrations required to induce itch as a way to compare the antipruritic activity of drugs, and used this to compare the efficacy of antihistamine drugs with several other classes (Cormia and Kuykendall 1954). When the paper was presented at the Annual Meeting of the American Dermatological Association, Walter Shelley commented that the presentation was the first controlled study to his knowledge on the effects of pharmacological agents on experimentally induced pruritus and that it provided a method for rapidly assessing new antipruritics in man (Cormia and Kuykendall 1954). Since then, many antihistamines have been shown to be effective in inhibiting histamine-induced itch, as well as wheal and flare reactions. These drugs include cetirizine, mepyramine, chlorpheniramine, chlorcyclizine, levomepromazine, acrivastine, levocetirizine, loratadine, epinastine, olopatadine, hydroxyzine, desloratadine, fexofenadine, and bepotastine (Hagermark 1973; Davies et al. 1979; Hagermark et al. 1979; Davies and Greaves 1980; Levander et al. 1985, 1991; Coulie et al. 1991; Lahti and Haapaniemi 1993; Weisshaar et al. 1997; Clough et al. 2001; Furue et al. 2001; Denham et al. 2003; Morita et al. 2005; Kupczyk et al. 2007; Tanizaki et al. 2012).

Antihistamines can also block the itch induced by the intradermal injection of other pruritogens. Some agents, such as compound 48/80, can directly liberate histamine from mast cells. This causes a wheal, flare, and itch response that can be inhibited by antihistamines (Fjellner and Haegermark 1981; Rukwied et al. 2000). Other substances may work in a similar fashion. Substance P injected intradermally induces wheal, flare, and itch, and all of these responses are inhibited by antihistamines, although in one case the itch induced by a high dose of substance P was not completely inhibited (Hagermark et al. 1978; Fjellner and Haegermark 1981; Hosogi et al. 2006). Therefore, it appears that the predominant effect of substance P injection in the skin may be to release histamine. This same mechanism probably also applies to other neuropeptides that induce skin reactions, including vasoactive peptide, neurotensin, and secretin. These all produced wheal, flare, and itch that were blocked by mepyramine (Fjellner and Haegermark 1981). Platelet activating factor and prostaglandin E2 also induced itch that was blocked by antihistamines or skin histamine depletion (Hagermark et al. 1978; Fjellner and Hagermark 1985). It cannot be ruled out, however, that itch induced by many of these factors may be driven by direct activation of afferent fibers and then histamine is involved downstream of this activation. Nevertheless, the fact that these agents induce a wheal response that responds to antihistamines strongly suggests that histamine is released at the site of injection.

However, antihistamines do not block the itch induced by all pruritogens. Serotonin induced a weak itch that was not significantly inhibited by antihistamines (Weisshaar et al. 1997; Hosogi et al. 2006). While bradykinin induced itch in the skin along with a mild flare, and both responses were reduced with an antihistamine (Cormia and Dougherty 1960; Hagermark 1974; Hosogi et al. 2006), the protease responsible for the formation of bradykinin, kallikrein, caused itch that differs from that of bradykinin and was not inhibited by antihistamines (Cormia and Dougherty 1960; Hagermark 1974). Other proteases can also induce itch when injected into the skin (Shelley and Arthur 1955). Some, such as trypsin and chymase, may act by releasing histamine, because they caused a wheal, flare, and itch response that was sensitive to antihistamines or histamine depletion (Hagermark et al. 1972; Hagermark 1973). This may not be true for another protease, papain, which only produced an itch response without wheal and flare (Shelley and Arthur 1955; Hagermark et al. 1972; Hagermark 1973; Davies et al. 1979). H_1R antihistamines have a small, but insignificant, inhibition of the itch; however, the combination of chlorpheniramine (H_1R antagonist) and the H_2R antagonist, cimetidine, is effective (Hagermark 1973; Davies et al. 1979). The effects of papain may be similar to those of cowhage. Cowhage is a legume that is covered in spicules. These spicules can invoke an intense itch that, owing to its limited wheal and flare, appears to be distinct from histamine (Sollmann and Pilcher 1917; Broadbent 1953; Johanek et al. 2007; Sikand et al. 2009). The active ingredient appears to be a cysteine protease similar to papain called mucunain, which is thought to induce itch by activating protease activated receptors (Shelley and Arthur 1955; Reddy et al. 2008). The itch effects induced by cowhage do not appear to be sensitive to a topical antihistamine

(Johanek et al. 2007), but an early study did show that oral mepyramine led to a slight reduction in itch (Broadbent 1953).

10.4 WHICH HISTAMINE RECEPTORS MEDIATE PRURITUS?

Clinical data show that itch induced by intradermal injection of histamine is mediated via the activation of the H_1R, because selective antagonists of the H_1R can suppress this response. Cetirizine and its active enantiomer, levocetirizine, are highly selective for the H_1R (Table 10.1). Pretreatment with cetirizine or levocetirizine reduced the pruritus score and itch duration after histamine skin prick by greater than 90% (Coulie et al. 1991; Levander et al. 1991; Lahti and Haapaniemi 1993;

TABLE 10.1
Binding Affinities for Selected H_1R Ligands

Ligand	Typically Ascribed Function	hH_1R pK_i[a]	hH_2R pK_i[b]	hH_3R pK_i[c]	hH_4R pK_i[d]
Histamine	HR agonist	4.4	5.1	8.2	8.1
2-Methylhistamine	H_1R agonist	4.4[e]	3.6[c]	4.4	6.3
Betahistine	H_1R agonist	4.5[e]		5.2[f]	4.1
Acrivastine	H_1R antagonist	7.5			
Alcaftadine	H_1R antagonist	8.5	7.2	<5	5.4
Azelastine	H_1R antagonist	6.8[g]		<5[g]	<5
Bromodiphenhydramine	H_1R antagonist	7.9[f]			
Brompheniramine	H_1R antagonist	8.3[f]			
Carbinoxamine	H_1R antagonist	8.2		<4	
Cetirizine	H_1R antagonist	8.2	<6	<6	<5
Chlorpheniramine	H_1R antagonist	8.6	5.1	5.5	4.6
Clemastine	H_1R antagonist	9.8[g]			4.8
Cyproheptadine	H_1R antagonist	8.5[f]	7.9[e]	<5	4.8
Desloratadine	H_1R antagonist	8.4	6.4[e]	<5[g]	5.1[g]
Dexbrompheniramine	H_1R antagonist	8.7[g]		5.6[g]	
Diphenhydramine	H_1R antagonist	7.9	5.8	4.6	4.4
Ebastine	H_1R antagonist	9.0			<5
Fexofenadine	H_1R antagonist	7.5			<5
Hydroxyzine	H_1R antagonist	8.7			<5
Ketotifen	H_1R antagonist	9.3	6.0	5.6	4.3
Levocetirizine	H_1R antagonist	8.2	<6	<6	<4
Loratadine	H_1R antagonist	7.2		<5[g]	4.7
Mepyramine	H_1R antagonist	8.8	4.6	<5	<5
Olopatadine	H_1R antagonist	7.5	4.0	4.1	<4[g]
Phenindamine	H_1R antagonist	7.7[f]			
Pheniramine	H_1R antagonist	7.2	4.9	5.1	
Promethazine	H_1R antagonist	8.8	6.5	4.5	4.1

(continued)

TABLE 10.1 (Continued)
Binding Affinities for Selected H$_1$R Ligands

Ligand	Typically Ascribed Function	hH$_1$R pK$_i$[a]	hH$_2$R pK$_i$[b]	hH$_3$R pK$_i$[c]	hH$_4$R pK$_i$[d]
Pyrilamine	H$_1$R antagonist	9.1	5.0	5.3	<4
Rupatadine	H$_1$R antagonist	7.0[e]			
Terfenadine	H$_1$R antagonist	7.9		<5[g]	<5
Triprolidine	H$_1$R antagonist	8.6	5.5	5.3	4.1

[a] Data averaged from the following references: Tran et al. (1978); Chang et al. (1979); Arrang et al. (1985); De Backer et al. (1993); Leurs et al. (1994b); Sharif et al. (1996); Kato et al. (1997); Merlos et al. (1997); Bakker et al. (2001); Gillard et al. (2002); Esbenshade et al. (2003); Govoni et al. (2003); Seifert et al. (2003); Thurmond et al. (2004); Bielory et al. (2005); Ligneau et al. (2007); Yu et al. (2010); Rossbach et al. (2011); Appl et al. (2012); and Gallois-Bernos and Thurmond (2012).

[b] Data averaged from the following references: Harada et al. (1983); Gantz et al. (1991); Eriks et al. (1993); Leurs et al. (1994a, 1995); Sharif et al. (1996); Kreutner et al. (2000); Saitoh et al. (2002); Esbenshade et al. (2003); Thurmond et al. (2004); Bielory et al. (2005); Lim et al. (2005); Preuss et al. (2007); Yu et al. (2010); Rossbach et al. (2011); Appl et al. (2012); Gallois-Bernos and Thurmond (2012); Levocetirizine New Drug Application #22-064; and Loratadine New Drug Application #21-165.

[c] Data averaged from the following references: Arrang et al. (1985); Sharif et al. (1996); Kato et al. (1997); Lovenberg et al. (1999); Coge et al. (2001); Liu et al. (2001a); Wieland et al. (2001); O'Reilly et al. (2002); Wellendorph et al. (2002); Wulff et al. (2002); Esbenshade et al. (2003); Thurmond et al. (2004); Bielory et al. (2005); Lim et al. (2005); Gbahou et al. (2006); Ligneau et al. (2007); Yu et al. (2010); Rossbach et al. (2011); Appl et al. (2012); Gallois-Bernos and Thurmond (2012); and Levocetirizine New Drug Application #22-064.

[d] Data averaged from the following references: Liu et al. (2001a,b); Morse et al. (2001); Zhu et al. (2001); O'Reilly et al. (2002); Esbenshade et al. (2003); Thurmond et al. (2004); Lim et al. (2005); Gbahou et al. (2006); Ligneau et al. (2007); Deml et al. (2009); Yu et al. (2010); Rossbach et al. (2011); Appl et al. (2012); and Gallois-Bernos and Thurmond (2012).

[e] Data from guinea pig.

[f] Data from rat.

[g] Unpublished data.

Clough et al. 2001; Denham et al. 2003). In addition, both compounds inhibited the histamine-induced wheal and flare by 80%–90%. Fexofenadine, another highly selective H$_1$R antagonist (Table 10.1), also completely inhibited histamine-induced itch responses (Tanizaki et al. 2012). To further support the role of the H$_1$R in mediating itch, intraepidermal injection of the H$_1$R agonist, 2-methylhistamine, caused itch, and the threshold for itch induction could be shifted by the H$_1$R antagonist chlorpheniramine (Davies and Greaves 1980). These data point to a major role for the H$_1$R in mediating histamine-induced dermal itch responses.

The H$_2$R has also been investigated for a role in dermal responses to histamine. The selective H$_2$R antagonists cimetidine and ranitidine (Table 10.2) have been shown

TABLE 10.2
Binding Affinities for Selected H_2R, H_3R and H_4R Ligands

Ligand	Typically Ascribed Function	$hH_1R\ pK_i$[a]	$hH_2R\ pK_i$[b]	$hH_3R\ pK_i$[c]	$hH_4R\ pK_i$[d]
Histamine	HR agonist	4.4	5.1	8.2	8.1
Dimaprit	H_2R agonist	<5[g]	4.6	6.1	6.4[g]
Impromidine	H_2R agonist		7.6	7.0	7.8
4-Methylhistamine	H_2R/H_4R agonist	<5[e]	5.1	4.7	7.7
Burimamide	H_2R antagonist	4.3[f]	5.4	7.4	7.0
Cimetidine	H_2R antagonist	4.8	5.8	4.7	5.0
Famotidine	H_2R antagonist	<5[g]	7.6		<5
Ranitidine	H_2R antagonist	4.5	6.7	4.9	<5
Immethridine	H_3R agonist			9.1	6.6
R-α-Methylhistamine	H_3R/H_4R agonist	<5[e]	<4	8.7	6.8
N-Methylhistamine	H_3R agonist	4.2	4.3	9.0	7.1
Clobenpropit	H_3R antagonist	5.5	5.5	9.0	8.0
Iodophenpropit	H_3R antagonist		5.6	8.1	7.8
Pitolisant	H_3R antagonist	5.8	5.0	8.1	<4
Thioperamide	H_3R antagonist	3.9	4.3	7.6	7.0
JNJ 28610244	H_4R agonist	<5	<6	<5	7.3
JNJ 7777120	H_4R antagonist	<5	<6	5.3	8.1

[a] Data averaged from the following references: Tran et al. (1978); Chang et al. (1979); Arrang et al. (1985); De Backer et al. (1993); Leurs et al. (1994b); Sharif et al. (1996); Kato et al. (1997); Merlos et al. (1997); Bakker et al. (2001); Gillard et al. (2002); Esbenshade et al. (2003); Govoni et al. (2003); Seifert et al. (2003); Thurmond et al. (2004); Bielory et al. (2005); Ligneau et al. (2007); Yu et al. (2010); Rossbach et al. (2011); Appl et al. (2012); and Gallois-Bernos and Thurmond (2012).

[b] Data averaged from the following references: Harada et al. (1983); Gantz et al. (1991); Eriks et al. (1993); Leurs et al. (1994a, 1995); Sharif et al. (1996); Kreutner et al. (2000); Saitoh et al. (2002); Esbenshade et al. (2003); Thurmond et al. (2004); Bielory et al. (2005); Lim et al. (2005); Preuss et al. (2007); Yu et al. (2010); Rossbach et al. (2011); Appl et al. (2012); Gallois-Bernos and Thurmond (2012); Levocetirizine New Drug Application #22-064; and Loratadine New Drug Application #21-165.

[c] Data averaged from the following references: Arrang et al. (1985); Sharif et al. (1996); Kato et al. (1997); Lovenberg et al. (1999); Coge et al. (2001); Liu et al. (2001a); Wieland et al. (2001); O'Reilly et al. (2002); Wellendorph et al. (2002); Wulff et al. (2002); Esbenshade et al. (2003); Thurmond et al. (2004); Bielory et al. (2005); Lim et al. (2005); Gbahou et al. (2006); Ligneau et al. (2007); Yu et al. (2010); Rossbach et al. (2011); Appl et al. (2012); Gallois-Bernos and Thurmond (2012); and Levocetirizine New Drug Application #22-064.

[d] Data averaged from the following references: Liu et al. (2001a,b); Morse et al. (2001); Zhu et al. (2001); O'Reilly et al. (2002); Esbenshade et al. (2003); Thurmond et al. (2004); Lim et al. (2005); Gbahou et al. (2006); Ligneau et al. (2007); Deml et al. (2009); Yu et al. (2010); Rossbach et al. (2011); Appl et al. (2012); and Gallois-Bernos and Thurmond (2012).

[e] Unpublished data.

[f] Data from rat.

[g] Data from guinea pig.

to reduce histamine-induced wheal and flare responses, but the effects are variable and modest, especially compared to H_1R antagonists (Marks and Greaves 1977; Hagermark et al. 1979; Meyrick-Thomas et al. 1985a,b; Kupczyk et al. 2007). There are also some limited and unconvincing data for cimetidine and ranitidine having minor effects on histamine-induced itch (Davies et al. 1979; Kupczyk et al. 2007). Robertson and Greaves (1978) mention that intradermal injection of 4-methylhistamine, an H_2R agonist, could cause itch or pain that was not modified by an H_2R or H_1R antagonist, but no data were given. However, in a later paper the same group showed that neither 4-methylhistamine nor dimaprit, both H_2R agonists, caused itch (Davies and Greaves 1980). In addition, preclinically, H_2R antagonists do not appear to modify histamine-induced itch responses (Bell et al. 2004; Dunford et al. 2007). Therefore, the current data suggest that it is unlikely that the H_2R mediates histamine-induced dermal reactions.

Currently, there are limited clinical data on the role of the H_3R and H_4R in mediating histamine-induced itch. However, preclinically, it has been shown that the H_4R is involved in pruritic responses. In mice, histamine-induced itch was inhibited by H_4R antagonists (Dunford et al. 2007; Yamaura et al. 2009; Shin et al. 2012). Furthermore, in H_4R-deficient mice, histamine-induced scratching was reduced, but not completely abolished, and this residual itch could be completely suppressed by an H_1R antagonist (Dunford et al. 2007). Compounds with H_4R agonist activity have been shown to induce itch upon intradermal injection in mice (Bell et al. 2004; Dunford et al. 2007; Yu et al. 2010). In particular, injection of 4-methylhistamine and JNJ 28610244 led to scratching, but this response was not present in H_4R-deficent mice and was inhibited by antagonists of the H_4R (Dunford et al. 2007; Yu et al. 2010). Therefore, in mice, it appears that histamine induces itch via H_1R and H_4R. H_4R antagonists can also inhibit scratching induced by substance P, haptens, and in models of dermatitis (Rossbach et al. 2009; Yamaura et al. 2009; Cowden et al. 2010; Suwa et al. 2011; Ohsawa and Hirasawa 2012). However, one should be cautious when interpreting scratching in mice as reflective of an itch sensation because depending on the site on the body it can also reflect pain (Shimada and LaMotte 2008; LaMotte et al. 2011).

There is some evidence that it is the H_4R on neurons that mediates its role in pruritus. Histamine-induced scratching still occurred in mast cell deficient mice, and the lack of a response in H_4R-deficient mice could not be recovered by reconstitution with wild-type hematopoietic cells that express the H_4R (Dunford et al. 2007). Compound 48/80 is a mast cell degranulator that also has the ability to activate sensory neurons directly (Eglezos et al. 1992). It can induce itch even in the absence of mast cells by directly activating these sensory neurons, and this could be inhibited by an H_4R antagonist or in H_4R-deficient mice indicating that the H_4R is downstream of sensory fiber activation (Dunford et al. 2007). The H_4R has been detected in the central nervous system, including the brain, spinal cord, and dorsal root ganglia (Connelly et al. 2009; Strakhova et al. 2009). Skin-specific sensory neurons have been shown to express the H_4R, and activation of the receptor led to an increase in calcium (Rossbach et al. 2011). Therefore, the preclinical data suggest that the H_4R could be involved in histamine-induced itch responses via activation of specific neuronal pathways. As noted earlier, 4-methylhistamine, a dual H_2R/H_4R agonist

(Lim et al. 2005), did not appear to cause itch in humans (Davies and Greaves 1980). While this may call into question the role of the H_4R in mediating itch, the interpretation may be complicated by the dose used and the hypothesis that it is the H_4R in the central nervous system that mediated pruritus. Therefore, definitive evidence for the role, or lack thereof, of the H_4R in pruritus in humans will have to await further clinical data.

Intradermal injection of the H_3R agonist, R-α-methylhistamine, in humans caused a wheal and flare response that was much weaker than histamine (Kavanagh et al. 1998). The flare, but not the wheal, was blocked by an H_1R antagonist, terfenadine, whereas the H_2R antagonist cimetidine had no effect (Kavanagh et al. 1998). Unfortunately, itch effects were not reported. However, preclinical data also suggest that *blocking* the H_3R can cause itch. Intradermal injection of iodophenpropit or clobenpropit, H_3R antagonists that also have H_4R activity (Table 10.2), induced scratching in mice that was not inhibited by H_1R antagonists (Hossen et al. 2003). These ligands also induced itch in H_1R-deficient mice (Hossen et al. 2006). Thioperamide, another dual H_3R/H_4R antagonist, also causes scratching in mice that can be inhibited by the dual H_3R/H_4R agonist R-α-methylhistamine (Sugimoto et al. 2004). Rossbach et al. (2011) tested the effects of pitolisant, a highly selective H_3R antagonist (Table 10.2), in mice. Pitolisant injected intradermally induced scratching that could be completely blocked with the H_3R agonist, immethridine. An H_3R agonist on its own did not induce itch. The H_3R antagonist-induced scratching could be partially blocked by cetirizine and with the H_4R antagonist, JNJ 7777120. The combination completely inhibited scratching, while an H_2R antagonist, ranitidine, had no effect. The authors suggest that the H_3R acts to suppress itch responses that are mediated via the H_1R and H_4R (Rossbach et al. 2011).

10.5 OCULAR AND NASAL PRURITUS

Histamine can also induce itch in the eye and nose. Nasal challenge with histamine leads to itching, sneezing, rhinorrhea, and congestion that are very similar to the symptoms of allergic rhinitis (Mygind 1982; Miadonna et al. 1987; Shelton and Eiser 1994). Exposure of allergic subjects to nasal allergens produces the same symptom with a delay relative to the histamine effects, suggesting that allergens degranulate mast cells leading to histamine release (Mygind 1982). Antihistamines that target the H_1R completely block histamine-induced itching, sneezing, and rhinorrhea, providing a mechanistic rationale for their effects in allergic rhinitis and indicates that these effects are due to activation of the H_1R (Secher et al. 1982; Kirkegaard et al. 1983; Mygind et al. 1983; Naclerio and Togias 1991; Hilberg 1995; Wood-Baker et al. 1996; Wang et al. 2001; Taylor-Clark et al. 2005). In addition, an intranasally administered H_1R agonist, betahistine, induced itching, sneezing, and rhinorrhea similar to histamine (Shelton and Eiser 1994). Inhibition by H_1R antagonists was only seen when they were given nasally in the same nostril as intranasal histamine, but not when they were given contralaterally, indicating that the effect is a local one (Kirkegaard et al. 1983; Mygind et al. 1983). The H_2R does not appear to be involved, because the H_2R antagonist, ranitidine, does not appear to have any effect on these symptoms (Secher et al. 1982; Wood-Baker et al. 1996).

While the itching, sneezing, and rhinorrhea appear to be completely mediated by the H_1R, nasal congestion appears to differ. Antagonism of the H_1R receptor appears to have only a partial effect on histamine-induced congestion (Secher et al. 1982; Havas et al. 1986; Naclerio and Togias 1991; Hilberg et al. 1995; Wood-Baker et al. 1996; Wang et al. 2001). When given intranasally, the H_2R agonists dimaprit and impromidine caused congestion similar to histamine, but not itching, sneezing, or rhinorrhea, and H_2R antagonists like ranitidine partially inhibited histamine-induced congestion (Secher et al. 1982; Taylor-Clark et al. 2005). The combination of H_1R and H_2R antagonist provided further, but still incomplete, congestion relief (Secher et al. 1982; Havas et al. 1986; Wood-Baker et al. 1996). This has led to speculation that another histamine receptor mediates congestion. This speculation has focused on the H_3R. The H_3R agonist R-α-methylhistamine induced nasal blockage if given intranasally and this was inhibited by thioperamide, but not by an H_1R or H_2R antagonist (Taylor-Clark et al. 2005). However, one clinical study did not find a benefit of a combined H_1R/H_3R antagonist over cetirizine on the total nasal symptom score or nasal blockage after allergen challenge (Daley-Yates et al. 2012) although another study did show a trend toward improvement of congestion when an H_3R antagonist was given in combination with fexofenadine (Stokes et al. 2012).

No clinical data exist for H_4R antagonists for nasal itch or other symptoms. However, in a mouse model of allergic rhinitis, the H_4R antagonist JNJ 7777120 significantly inhibited sneezing and nasal rubbing to a level comparable to a H_1R antagonist (Takahashi et al. 2009).

Ocular administration of histamine also causes itch as well as redness (Abelson et al. 1980). Both of these effects appear to be mediated by the H_1R; itch, in particular, is almost completely eliminated by H_1R antagonists (Abelson et al. 1980; Kirkegaard et al. 1982; Abelson and Smith 1988). The H_2R does not appear to have a role in inducing itch in the eye as local application of the H_2R agonist dimaprit did not lead to itch, although it did cause redness that could be blocked by cimetidine (Abelson and Udell 1981). There is no evidence that the H_3R is expressed in the conjunctiva, but the H_4R is present and increases in inflammatory conditions (Leonardi et al. 2011). Preclinically, the H_4R agonist 4-methylhistamine induced scratching when given ocularly, and this effect, as well as histamine-induce scratching, was inhibited by the H_4R antagonist JNJ 7777120 (Nakano et al. 2009). The topical antihistamine alcaftadine has weak H_4R antagonist activity that has been suggested to contribute to its efficacy (Namdar and Valdez 2011).

10.6 ANTIHISTAMINES IN CLINICAL PRACTICE

Pruritus is a symptom associated with a wide range of systemic and dermatological diseases. It is a primary symptom in urticaria, atopic dermatitis, and allergic rhino-conjunctivitis, and to a lesser extent in psoriasis. Pruritus can also be an indicator of an underlying systemic disorder, such as cholestasis, renal insufficiency, or malignancy. These diseases often differ greatly in their pathophysiology and yet share the common manifestation of chronic itch. Treatment with H_1R antihistamines has been proven to be of benefit in several of these conditions, but not others, where these drugs are nevertheless still used owing to the lack of other treatment options.

The itch in many dermatological conditions (e.g., atopic dermatitis) and allergic diseases (e.g., allergic rhinitis) is chiefly generated peripherally by localized mediators in the skin or upper respiratory system and is classified as *pruritoceptive*. In contrast, the pathophysiology of pruritus in systemic diseases is more variable and does not necessarily resemble that of the dermal and allergic diseases. This itch is triggered in the central nervous system as a reaction to circulating pruritogens and is classified as *neurogenic*, such as in cholestasis. Finally, some pruritic conditions are considered *neuropathic*, e.g., due to nerve entrapment, or *psychogenic*, e.g., delusional parasitosis (Twycross et al. 2003; Yosipovitch et al. 2003).

The currently marketed H_1R antihistamines are divided into two categories (Table 10.3). First-generation H_1R antihistamines were first introduced in the 1930s. They are characterized by poor selectivity for the H_1R and their ability to easily cross into the blood–brain barrier. This ability to cross the blood brain barrier allows them to interact with the H_1R in the central nervous system, leading to sedation and other neuropsychiatric adverse effects (Shamsi and Hindmarch 2000; Casale et al. 2003). Many practitioners have exploited these side effects in the treatment of primary insomnia and sleep disturbance from allergic symptoms. The first-generation H_1R antihistamines vary in their selectivity for the H_1R and can also have antimuscarinic, antiserotonin, and anti-α-adrenergic effects (Haas et al. 2008; Church et al. 2010). Many first-generation H_1R antihistamines were introduced to the market before the Kefauver Harris Amendment, or "Drug Efficacy Amendment," was added to the US Federal Food, Drug, and Cosmetic Act in 1962, requiring drug manufacturers to provide proof of efficacy before drug approval. While many of these drugs have proven to have acceptable safety profiles, with millions of patient exposures, and have been demonstrated to have comparable efficacy to more recently approved drugs (Salo et al. 1989; Monroe 1992; Breneman 1996), they have never been evaluated for efficacy and safety in rigorous regulatory quality clinical trials. These drugs are metabolized by liver enzymes to different degrees, and changes in metabolism can impact drug exposure with implications for their efficacy and safety profiles (del Cuvillo et al. 2006). Some of the "antipruritic" effects of first-generation H_1R antihistamines for some conditions, such as atopic dermatitis, may be attributable to their sedative effects (O'Donoghue and Tharp 2005).

Second-generation antihistamines were first developed in the 1980s and were often derived from metabolites of the older drugs, such as fexofenadine from terfenadine. They are more selective for the H_1R, thus reducing the side effect burden, and are mildly sedating or nonsedating, because they do not penetrate into the brain as readily. In general, the longer half-life of second-generation H_1R antihistamines permits more convenient once daily dosing (del Cuvillo et al. 2006). The second-generation H_1R antihistamines have been rigorously studied and approved for allergic and dermatologic indications (Table 10.3). In addition to antagonizing the effects of histamine, some second-generation H_1R antihistamines are also thought to have independent anti-inflammatory properties, blocking the release of various inflammatory mediators from mast cells and other inflammatory cell types (Eda et al. 1994; Abdelaziz et al. 1998; Nori et al. 2003; Okamoto et al. 2009), but the clinical relevance of these effects is difficult to differentiate from the H_1R effects. Overall, there are probably no dramatic differences in terms of efficacy between first- and

TABLE 10.3
Selected Oral H$_1$R Antihistamines

Generic Name	Generation	Indications
Acrivastine	Second	Allergic rhinitis
		Chronic urticaria
Bilastine	Second	Allergic rhinitis
		Chronic urticaria
Bepotastine	Second	Allergic rhinitis
		Chronic urticaria
Carbinoxamine	First	OTC for symptomatic relief (hives, rhinitis)
Cetirizine	Second	Allergic rhinitis
		Chronic urticaria
Chlorpheniramine	First	OTC for symptomatic relief (hives, rhinitis)
Clemastine	First	Allergic rhinitis
		Chronic urticaria
Cyproheptadine	First	OTC for symptomatic relief (hives, rhinitis)
Desloratadine	Second	Allergic rhinitis
		Chronic urticaria
Dexbrompheniramine	First	OTC for symptomatic relief (hives, rhinitis)
Diphenhydramine	First	OTC for symptomatic relief (hives, rhinitis)
Doxylamine	First	OTC for symptomatic relief (hives, rhinitis)
Ebastine	Second	Allergic rhinitis
		Chronic urticaria
Fexofenadine	Second	Allergic rhinitis
		Chronic urticaria
Hydroxyzine	First	OTC for symptomatic relief (hives, rhinitis)
Levocetirizine	Second	Allergic rhinitis
		Chronic urticaria
Loratadine	Second	Allergic rhinitis
		Chronic urticaria
Mizolastine	Second	Allergic rhinitis
		Chronic urticaria
Promethazine	First	OTC for symptomatic relief (hives, rhinitis)
Rupatadine	Second	Allergic rhinitis
		Chronic urticaria
Triprolidine	First	OTC for symptomatic relief (hives, rhinitis)

Note: Certain recent H$_1$R antihistamines, such as levocetirizine and desloratadine, have been classified as third generation, as they were developed as metabolites of second-generation drugs. OTC, over the counter.

second-generation H$_1$R antihistamines, but rather their advantages come from dosing convenience and safety.

Both the first- and second-generation H$_1$R antihistamines are efficacious first-line antipruritic treatments in a number of histamine-mediated diseases including acute and chronic urticaria, allergic conjunctivitis, and allergic rhinitis. They are also used to treat other pruritic skin diseases such as atopic dermatitis and psoriasis. The H_1R antihistamines are not as efficacious in reducing pruritus in these diseases; yet they are still often used as a result of their sedative effects, some limited efficacy, and because more effective antipruritic treatment options do not exist (other than treatments for the primary skin disorder).

Historically, efficacy of H_1R antihistamines has been used to define whether histamine is involved in a particular pruritic disease. For example, the itch in urticaria and allergic rhinitis is considered to be histamine mediated, because it is very effectively blocked by H_1R antihistamines. Conversely, the itch in atopic dermatitis and psoriasis is not considered histamine mediated, because these drugs are not effective in those conditions. The conclusion that histamine is involved if H_1R antihistamines are effective is valid. However, we should reconsider the conclusion that histamine is not relevant if H_1R or H_2R antihistamines do not work, given that there are two other histamine receptors, the H_3R and H_4R, for which there is recent evidence of a possible role in itch perception in preclinical studies (Dunford et al. 2007; Rossbach et al. 2011).

There are several factors that should be considered when using an antihistamine to treat pruritus. First, the choice of an ocular, intranasal, or systemic antihistamine should depend on the location, distribution, and extent of the particular condition. It is important to consider that the more recent oral drugs are dosed once daily, while topical antihistamines are often dosed every 8 to 12 hours because of the washout of drug by tears and nasal secretions. Second, some conditions require regular daily antihistamine administration, while for other conditions, antihistamines can be taken as needed. Third, first-generation sedating H_1R antihistamines may be better suited to treat nocturnal pruritus, while second-generation mildly or nonsedating H_1R antihistamines are more appropriate to treat daytime itch. However, the central nervous system effects should be carefully evaluated as they may negatively affect quality of life and job performance. Fourth, the entire side effect profiles of these agents should be considered, especially as many of the older drugs have safety concerns (see General Safety Considerations section) that may be underappreciated in the community and by physicians (Church et al. 2010). Finally, many of the antihistamines are also now available as fixed dose combination products that extend relief of symptoms, such as allergic rhinitis treatments that contain pseudoephedrine for decongestion (with the possibility of side effects, such as insomnia) or corticosteroids in antihistamine nasal sprays for added anti-inflammatory effects (Hampel et al. 2010).

The rest of this section will review diseases and conditions in which antihistamines are used clinically. Diseases in which histamine is thought to play a role and where antihistamines have been demonstrated to have a therapeutic benefit will be reviewed, followed by conditions in which H_1R antihistamines are sometimes used, but show negligible efficacy.

10.6.1 Urticaria

Urticaria is a common dermatological condition manifested as highly pruritic wheals, which when persisting for more than 6 weeks or longer is termed chronic urticaria. Many triggers for chronic physical urticaria have been identified, such as heat, cold, sun, pressure, vibration, water, exercise, or contact, but in many patients with chronic urticaria a trigger cannot be identified (Greaves 1995; Zuberbier and Maurer 2007). The underlying cause of the urticaria may not be appreciated, as it is dependent on the extent of the patient evaluation (Ring et al. 1999), and there has been recent discussion on updating the classification of urticaria (Maurer et al. 2013). Excluding those with physical urticaria, about half of the subjects test positive for IgG antibodies to IgE or to the IgE receptor α-subunit and thus has an autoimmune character (Kaplan and Greaves 2009). In these subjects, the antibodies are pathogenic and activate mast cells (Kaplan and Greaves 2009). Wheals present as smooth, blanching, erythematous, and highly pruritic skin elevations appearing most frequently on the arms, back, and legs, while the scalp is the least affected (Yosipovitch et al. 2002). The circumscribed lesions can last from a few minutes to several days and vary in size from a millimeter to several centimeters in diameter or larger.

Both increased numbers of mast cells (Natbony et al. 1983) and levels of histamine (Kaplan et al. 1978) have been observed in the affected skin of subjects with urticaria. Wheals emerge as a result of the release of histamine from mast cells and basophils, and perhaps other mediators such as leukotrienes involved in the urticarial response. These mediators induce pruritus via sensory nerve stimulation, as well as vasodilation, leading to erythema and increased permeability of capillaries and small veins leading to localized edema.

The approach to management begins with eliminating the triggers, which includes minimizing physical stimuli, reducing stress, and avoiding the implicated foods, food additives or drugs, which may only represent pseudoallergens (Michaelsson and Juhlin 1973; Doeglas 1975; Supramaniam and Warner 1986; Zuberbier et al. 1995). Regardless, pharmaceutical intervention is often necessary. The effective use of the H_1R antihistamine Benadryl® (diphenhydramine) in treating urticaria has been known since as early as 1945 (Curtis and Owens 1945). In one early study, 25 of 35 patients with urticaria experienced immediate and complete relief of symptoms, while the final seven patients saw definite improvement, and only three out of 35 patients experienced no response (O'Leary and Farber 1945).

Today, H_1R antihistamines continue to be the mainstay of treatment for urticaria, and these patients are generally treated by primary care doctors (Zazzali et al. 2012). Several clinical guidelines and expert opinions have been published that recommend second-generation H_1R antihistamines as the standard of care for patients with urticaria, because they achieve a significant reduction in chronic urticaria symptoms and exhibit a superior safety profile compared to the older drugs (Kaplan 2002; Powell et al. 2007; Khan 2008; Zuberbier et al. 2009; Ortonne 2011). Numerous clinical trials have been conducted assessing second-generation H_1R antihistamines versus placebo in subjects with chronic urticaria lasting 4 to 6 weeks evaluating reduction in pruritus, as well as wheal number, size, and frequency of occurrence. These include studies of bilastine (Zuberbier and Maurer 2010), loratadine (Monroe 1992; Dubertret et al. 1999), cetirizine (Breneman 1996), desloratadine (Monroe et al. 2003; Ortonne et al. 2007), fexofenadine (Kaplan et al. 2005), levocetirizine (Nettis et al. 2006), mizolastine (Brostoff et al. 1996), and rupatadine (Gimenez-Arnau et al. 2007). A few trials comparing two H_1R antihistamines have been conducted, and while one 4-week study of cetirizine versus fexofenadine versus placebo favored cetirizine (Handa et al. 2004), and another cetirizine study with loratadine and placebo favored loratadine (Guerra et al. 1994), these studies overall are too limited in number, generally show comparability across the drugs, and lack reproducibility to conclude that one drug will be consistently more effective than another (Monroe 1992; Breneman 1996; Kalis 1996; Dubertret et al. 1999).

While the majority of chronic urticaria patients experience symptomatic relief with second-generation H_1R antihistamines (Kozel and Sabroe 2004), there are patients refractory to first-line treatment who are prescribed second-line treatments that have not been as well studied and are somewhat controversial (Ortonne 2011). The two most frequent second-line treatment choices are sedating first-generation H_1R antihistamines, especially when the symptoms interfere with sleep at night, or to increase the dose (updosing) of the second-generation H_1R antihistamine up to four times the standard dose (Zuberbier 2012). While the evidence of comparable efficacy for first-generation H_1R antihistamines as compared to the newer drugs is supported by clinical trials (Salo et al. 1989; Kalivas et al. 1990; Monroe 1992), the central nervous system effects (e.g., sleep disturbance, memory, and learning impairment) and other off target side effects (especially anticholinergic) can have a significant impact on quality of life and work performance, even when used solely at bedtime (Church et al. 2010). Despite limited clinical studies for support, "updosing" or increasing the dose of second-generation H_1R antihistamines is recommended in clinical guidelines. Data to support this approach for pruritus symptoms come from greater efficacy in the higher dosage range of levocetirizine, desloratadine, and rupatadine (Kameyoshi et al. 2007; Gimenez-Arnau et al. 2009; Siebenhaar et al. 2009; Staevska et al. 2010). It has been hypothesized that some of the enhanced efficacy for urticaria at above labeled dosages may be due to non-H_1R anti-inflammatory properties, such as inhibiting other important mediators involved in urticaria (Wang et al. 2005; Weller and Maurer 2009).

Another possible treatment option for refractory patients is to block the effects of the H_2R, which, in addition to its role in suppressing gastric acid, has been shown to be expressed in the skin as well as mast cells (Lippert et al. 2004). Several H_2R antihistamines are approved for use in treating gastric ulcers (Table 10.4). There are no data on H_2R antagonists, such as ranitidine or cimetidine, as effective monotherapy treatments for urticaria. However, a handful of small studies, including one combining ranitidine with diphenhydramine in subjects with acute allergic conditions (including a subset with urticaria), suggest an added benefit on wheal reduction (Lin et al. 2000). A recent Cochrane Review on H_2R antihistamines for urticaria concluded that the data were too weak and unreliable to support their use (Fedorowicz et al.

TABLE 10.4
Selected Oral H_2R Antihistamines

Generic Name	Approved Indications	Off-Label Use
Cimetidine	Gastric ulcers	Urticaria
Famotidine	Gastric ulcers	Urticaria
Ranitidine	Gastric ulcers	Urticaria

2012). Regardless, in clinical practice, these drugs are sometimes reserved for refractory chronic urticaria.

Finally, with the exception of several studies of cold urticaria (Bonadonna et al. 2003; Juhlin 2004; Magerl et al. 2007; Siebenhaar et al. 2009), which show clear efficacy with various second-generation H_1R antihistamines, the benefits of treatment for other physical urticarias (e.g., heat and solar) have not been well characterized.

10.6.2 ALLERGIC RHINITIS

Allergic rhinitis is a very common allergic disease that includes two disorders, seasonal allergic rhinitis and perennial allergic rhinitis, and is characterized by the inflammation of the nasal mucous membrane in response to inhaled allergens. This inflammation leads to nasal and pharyngeal pruritus, as well as sneezing, rhinorrhea, coughing, and congestion. Many patients also experience allergic conjunctivitis concurrently, characterized by ocular pruritus, conjunctival edema, tearing, and hyperemia. There are a variety of known triggers for seasonal (outdoor) and perennial (indoor) allergic rhinitis including house dust mites, pollens, animal dander, and mold.

While histamine is a primary mediator in the pathogenesis of allergic rhinitis, the symptoms of allergic rhinitis are the result of complex interactions among numerous inflammatory mediators produced by an array of different circulating, mucosal, and lymph node-resident inflammatory cells. These mediators, including histamine, tryptase, chymase, prostaglandins, cysteinyl leukotrienes, and Th2 cytokines (e.g., IL-3, IL-4, IL-5, and IL-13), and the inflammatory cells, including mast cells, basophils, eosinophils, Th2 cells, and macrophages, are responsible for many of the responses in allergic rhinitis (Wagenmann et al. 2005; Kramer et al. 2006; Okano et al. 2006).

Treatment of allergic diseases begins with patient education on allergen avoidance. This approach can prove to be effective in cases in which elimination of the allergen is feasible. However, patients who react to more ubiquitous allergens or who have multiple allergens may not realistically be able to avoid allergen contact. Fortunately, these patients have a variety of treatment options available to them, including oral and intranasal H_1R antihistamines (Table 10.5) as well as leukotriene receptor antagonists, intranasal corticosteroids, intranasal cromolyns, and decongestants.

The efficacy of H_1R antihistamines in the treatment of allergic rhinitis has been clinically acknowledged since 1945, when Koelsche et al. (1945) tested the efficacy of diphenhydramine in patients presenting with "hay fever." In this preliminary study, 57 of the 83 patients experienced at least 50% relief of symptoms. Since then, there have been dozens of well-controlled studies on the efficacy and safety of topical or oral H_1R antihistamines (for a review, see Benninger et al. [2010]). These studies were placebo controlled, lasting generally 2 to 4 weeks, and were usually conducted in hundreds of patients with seasonal allergic rhinitis. Fewer studies have been conducted in subjects with perennial allergic rhinitis. The outcome measure most frequently employed was the Total Nasal Symptom Score, which includes pruritus. In addition to these studies, numerous head-to-head studies and reviews of one H_1R antihistamine (oral or intranasal) versus another or versus intranasal corticosteroids

TABLE 10.5

Selected Topical Intranasal and Ocular H₁R Antihistamines

Generic Name	Formulation	Approved Indications
Azelastine	Intranasal	Allergic rhinitis
	Ocular	Allergic conjunctivitis
Olopatadine	Intranasal	Allergic rhinitis
	Ocular	Allergic conjunctivitis
Alcaftadine	Ocular	Allergic conjunctivitis
Bepotastine	Ocular	Allergic conjunctivitis
Emedastine	Ocular	Allergic conjunctivitis
Epinastine	Ocular	Allergic conjunctivitis
Ketotifen	Ocular	Allergic conjunctivitis
Levocabastine	Ocular	Allergic conjunctivitis

have been conducted to determine the optimal first-line therapy (Bernstein et al. 1996; Gehanno and Desfougeres 1997; Howarth et al. 1999; Yanez and Rodrigo 2002; Hampel et al. 2003; Berger et al. 2006).

Recent clinical guidelines have been published that provide comprehensive reviews of these available treatment options (Scadding et al. 2008; Wallace et al. 2008). Well-conducted placebo-controlled studies of oral second-generation H_1R antihistamines, including bilastine (Kuna et al. 2009), cetirizine (Howarth et al. 1999; Murray et al. 2002; Noonan et al. 2003), desloratadine (Berger et al. 2002; Kim et al. 2006), fexofenadine (Bernstein et al. 1997; Wilson et al. 2002), levocetirizine (Potter 2003), and loratadine (Oei 1988; Storms et al. 1989), and the intranasal H_1R antihistamines, including azelastine (LaForce et al. 2004; Lumry et al. 2007) and olopatadine (Meltzer et al. 2005; Ratner et al. 2005) demonstrate efficacy against a range of allergic rhinitis symptoms, including nasal pruritus. The intranasal H_1R antihistamines have proven to be as effective, or occasionally superior, to the oral drugs, and consistently have a more rapid onset of action.

Overall, there is lack of consistent and substantial evidence that one oral H_1R antihistamine is more effective than another, and the addition of an oral H_1R antihistamine to a topical H_1R antihistamine probably does not provide significant added benefit for nasal pruritus (LaForce et al. 2004). The first-generation H_1R antihistamines, such as hydroxyzine and chlorpheniramine (Brooks et al. 1981; Wong et al. 1981) among others, are also effective for nasal pruritus, but are associated with more side effects. While oral and intranasal H_1R antihistamines are efficacious treatments for allergic rhinitis, the intranasal corticosteroids have been found to be somewhat more effective at reducing total symptoms, including nasal itch (Simpson 1994; Weiner et al. 1998; Yanez and Rodrigo 2002; Wallace et al. 2008; Benninger et al. 2010), and there is some mixed evidence of additive benefit on nasal itch in the combination of either oral or intranasal H_1R antihistamines and intranasal corticosteroids (Simpson 1994; Anolik 2008; Ratner et al. 2008).

The approach to treatment for nasal pruritus associated with allergic rhinitis depends on the severity of the itch as well as of other nasal symptoms. Allergic rhinitis presenting with only mild nasal symptoms can be treated with a topical H_1R antihistamine, such as azelastine or olopatadine. Mild nasal symptoms with concurrent ocular symptoms or other extra-nasal symptoms may be better treated with an oral H_1R antihistamine. Moderate to severe rhinitis symptoms or failure to respond sufficiently to topical and oral H_1R antihistamines warrants the administration of a topical intranasal corticosteroid. A fixed dose combination nasal spray of fluticasone and azelastine (Dymista™) was studied in large trials in patients with seasonal allergic rhinitis with efficacy on the Total Nasal Symptom Score (including nasal pruritus) that was superior to either drug alone (Carr et al. 2012; Meltzer et al. 2012).

In this chapter, there are three other histamine receptors (H_2R, H_3R, and H_4R) that may be relevant in allergic rhinitis symptoms. In the clinical trials presented above, nasal congestion in particular is considered to be incompletely treated with H_1R antihistamines, which is one reason why H_1R antihistamines have been developed in fixed dose combination therapies with decongestants. Because there are suggestions from preclinical and clinical studies that H_3R or H_4R may play a role in the congestion and inflammation associated with allergic rhinitis (Takahashi et al. 2009; Stokes et al. 2012), more well-conducted clinical studies with antihistamines that target these receptors in patients with allergic rhinitis may further clarify the role of histamine in this disease.

10.6.3 Allergic Conjunctivitis

Allergic conjunctivitis is a common ocular allergy that includes acute, seasonal, and perennial allergic conjunctivitis, as well as the more rare and serious vernal keratoconjunctivitis and atopic keratoconjunctivitis, which require specialist care. Upon exposure of the conjunctiva to environmental allergens, a cascade of inflammatory events is triggered that resembles those of other allergic disorders, including cross linkage of membrane-bound IgE, triggering mast cell degranulation and the release of histamine and other inflammatory mediators. As a result, symptoms can include tearing, lid and conjunctival edema and erythema, photophobia, and most saliently, pruritus. Allergen drainage to the nose via the nasolacrimal duct can cause the release of histamine and other inflammatory mediators leading to allergic rhinoconjunctivitis (Ono and Abelson 2005).

If the initial recommended approach of allergen avoidance fails to prevent symptoms or relieve pruritus (Bielory et al. 2005), then several topical second-generation H_1R antihistamines are effective and can be used as first-line therapy (Table 10.5). Alcaftadine (Greiner et al. 2011; Torkildsen and Shedden 2011), azelastine (Ciprandi et al. 1997; Horak et al. 1998), bepotastine (Abelson et al. 2009; Macejko et al. 2010), emedastine (Abelson and Kaplan 2002; D'Arienzo et al. 2002; Borazan et al. 2009), epinastine (Lanier et al. 2004; Whitcup et al. 2004), ketotifen (Crampton 2002; Greiner and Minno 2003), levocabastine (Zuber and Pecoud 1988; Abelson et al. 1995), and olopatadine (Abelson 1998; Spangler et al. 2001; Mah et al. 2007) demonstrated superiority over placebo (with active comparator often included) in conjunctival allergen challenge studies. In these studies, susceptible subjects are presented with an appropriate allergen and dilution, and upon the development of symptoms, are randomly treated with active H_1R antihistamine eye drops or placebo into one eye or the other. Several dose strengths are usually evaluated for onset and duration of action, assessing patient symptom scores, including pruritus as the usual primary endpoint, and visual examination with slit lamp (Abelson et al. 1990). There are some head-to-head studies conducted comparing two different ophthalmic H_1R antihistamines, suggesting the benefit of one treatment over the other (Ganz et al. 2003). These drugs are also evaluated in longer real-world environmental studies (Bende and Pipkorn 1987; Aguilar 2000; Giede et al. 2000; Katelaris et al. 2002; Ganz et al. 2003; James et al. 2003; Borazan et al. 2009).

In head-to-head studies of ocular versus oral H_1R antihistamines, ocular agents demonstrated both faster onset of action as well as superior clinical efficacy (Abelson and Welch 2000; Spangler et al. 2003) and are the first-line treatment, especially for ocular pruritus. The combination of ocular and oral H_1R antihistamines is superior to either treatment as monotherapy and is useful when symptoms involve the nose and pharynx as well (Abelson and Lanier 1999; Crampton 2003).

While topical mast cell stabilizers (Katelaris et al. 2002; James et al. 2003; Liu et al. 2011) and topical NSAIDS (Schechter 2008) can be effective for allergic conjunctivitis, head-to-head studies of topical H_1R antihistamines versus mast cell stabilizers, such as cromolyn or nedocromil (Orfeo et al. 2002; Greiner and Minno 2003), or topical NSAIDS, such as ketorolac (Discepola et al. 1999; Yaylali et al. 2003), have demonstrated the superiority of the H_1R antihistamines in the treatment of ocular pruritus.

10.6.4 INSECT BITES

Mosquito bites cause pruritic wheals mediated in part by the local release of histamine (Horsmanheimo et al. 1996), and clinical trials suggest clear efficacy of pretreatment with H_1R antihistamines. Studies have reported that ebastine and cetirizine decreased bite pruritus by up to 70% (Reunala et al. 1993; Karppinen et al. 2000). In a comparative study of the efficacies of cetirizine, ebastine, and loratadine, cetirizine and ebastine were statistically superior to loratadine and placebo at relieving bite lesion size and pruritus (Karppinen et al. 2002). Levocetirizine has also been proven to reduce bite lesion size and pruritus by up to 60% (Karppinen et al. 2006).

10.6.5 ATOPIC DERMATITIS

Atopic dermatitis is a common chronic or relapsing inflammatory pruritic skin disease, characterized by eczematous skin lesions and xerosis, although the clinical patterns of atopic dermatitis vary with age. Pruritus, one of the most common and characteristic symptoms of atopic dermatitis, presents a vexing treatment challenge across all phases of the disease from infancy to adulthood (Williams 2005). The exact mechanisms of pruritus in atopic dermatitis and the possible role of histamine are complex and even paradoxical. While studies have shown elevated histamine and dermal mast cells in pruritic skin of atopic dermatitis patients (Johnson et al. 1960; Juhlin 1967; Ikoma 2009), the pruritus from mast cell degranulation from an intradermal injection of compound 48/80 could not be blocked by cetirizine, suggesting that other mast cell mediators are involved (Rukwied et al. 2000). However, it has to be taken into account that compound 48/80 can also directly activate sensory fibers (Eglezos et al. 1992). Conversely, there are reports that intracutaneous injections of histamine in atopic dermatitis patients actually reduce local reactions and pruritus (Uehara 1982; Heyer et al. 1998). There remains some thought that H_1R antihistamines may still be relevant in the pruritus of atopic dermatitis through either non-H_1R mediated anti-inflammatory properties or through effects on central H_1R (Buddenkotte et al. 2010). As the question of the role of histamine in atopic dermatitis continues, other itch mediators including cytokines such as IL-31, neuropeptides, proteases, eicosanoids, and eosinophil-derived proteins have been suggested as key mediators contributing to pruritus in this common disease (Dillon et al. 2004; Takaoka et al. 2005; Bilsborough et al. 2006). The efficacy of topical corticosteroids and calcineurin inhibitors provides support for a role for inflammatory cells in pruritus.

Several clinical trials have been conducted to evaluate the efficacy of first- or second-generation antihistamines for the pruritus of atopic dermatitis, with limited evidence for efficacy. Reviews of these studies (Klein and Clark 1999; Akdis et al. 2006; Saeki et al. 2009) have concluded that any efficacy observed is possibly due to skin barrier (Amano et al. 2007) or anti-inflammatory effects (Tamura et al. 2008), but more likely chiefly due to central sedative effects, although studies of sedating antihistamines have not been as successful as some nonsedating drug studies (Diepgen 2002; Munday et al. 2002; Kawashima et al. 2003). Interestingly, doxepin, is a potent H_1R and H_2R antihistamine with anticholinergic, antiserotonergic, and antiadrenergic properties, which is primarily used as a tricyclic antidepressant; topically, it can have antipruritic benefits in atopic dermatitis (Drake et al. 1994), though its use is fairly limited.

All in all, atopic dermatitis may be the best example of a condition in which it is believed that histamine does not play a large role in pruritus, because H_1R antihistamines are generally considered ineffective (Klein and Clark 1999). However, it may be that a different histamine receptor mediates these effects. The H_4R may play a role in the pathogenesis of both the itch and skin lesions of atopic dermatitis, which could help account for the lack of efficacy seen in H_1R antihistamine treatment. Earlier in this chapter, in preclinical models the H_4R can mediate acute itch induced by histamine or other pruritogens (Dunford et al. 2007; Rossbach et al. 2009; Yamaura et al. 2009). In addition, scratching can be inhibited by H_4R antagonists in models of dermatitis (Cowden et al. 2010; Suwa et al. 2011; Ohsawa and Hirasawa 2012). Preclinical data also support a role for the H_4R in driving dermal inflammation. In several different mouse models, treatment with an H_4R antagonist reduced the lesion, the infiltration of inflammatory cells, and the production of inflammatory cytokines (Cowden et al. 2010; Seike et al. 2010; Suwa et al. 2011; Matsushita et al. 2012; Ohsawa and Hirasawa 2012). In humans, there is evidence to show that the H_4R is expressed on cells associated with atopic dermatitis such as fibroblasts and inflammatory dendritic epidermal cells (Bäumer et al. 2008; Dijkstra et al. 2008; Ikawa et al. 2008; Gschwandtner et al. 2011a,b). In addition, the receptor is expressed on Th2 cells thought to be pathogenic in the disease (Gutzmer et al. 2009). In these cells the receptor can regulate the production of IL-31, which is thought to be involved in both inflammation and pruritus in atopic dermatitis (Sonkoly et al. 2006; Gutzmer et al. 2009). Future clinical testing with H_4R antihistamines may provide benefits to patients and clarify further the role of histamine in atopic dermatitis.

10.6.6 PSORIASIS

Psoriasis is a common chronic skin disease with inflamed, dry, and scaling skin lesions that are pruritic in a majority of patients (Yosipovitch et al. 2000). While it has long been appreciated that mast cell numbers and histamine levels are increased in psoriatic skin (Krogstad et al. 1997) and increased numbers of degranulating mast cells can be observed in acute guttate psoriasis lesions (Brody 1984), H_1R antihistamines are not thought to be effective, and controlled trials for pruritus of psoriasis have not been performed. In the 1980s and 1990s, there was interest in the possible role of the H_2R in psoriasis, with some promising results from open label studies with H_2R antihistamines (Kristensen et al. 1995; Nielsen et al. 1997). However, a large randomized, placebo-controlled study conducted to test the efficacy of ranitidine for psoriasis failed to demonstrate efficacy in treating the disease (Zonneveld et al. 1997).

Similar to the situation with histamine and atopic dermatitis, the notion that "antihistamines" do not work in psoriasis, because H_1R and H_2R antihistamines are ineffective, is starting to be challenged, because there is evidence that the H_4R may be involved. In particular, it has been shown that the H_4R is expressed on plasmacytoid dendritic cells and Th17 cells from psoriasis patients, and it modulates the cytokine production and chemotaxis of these cells (Gschwandtner et al. 2011a; Mommert et al. 2012). Future clinical research may yield interesting results in this respect. However, at present, first-line treatment for the pruritus of psoriasis should focus on the underlying disease and consist of topical agents including moisturizers, emollients (Elmariah and Lerner 2011), and corticosteroids, as well as systemic treatments, such as methotrexate, anti-TNF, and anti-12/23 monoclonal antibody treatments for severely symptomatic patients (Dawn and Yosipovitch 2006; Gottlieb et al. 2009; Smith et al. 2009; Hsu et al. 2012).

10.6.7 CONTACT DERMATITIS

Contact dermatitis is characterized by pruritic vesicular skin lesions arising as a result of contact with an allergen to which the host is sensitized. Common allergic triggers include nickel, chromium, poison ivy, and poison oak (Peiser et al. 2012). Identification and avoidance of the causal allergen is usually all that is required to resolve the dermatitis. If this strategy is unsatisfactory, potent topical corticosteroids are the first line of treatment for inflammation. First-generation H_1R antihistamines have been administered to treat pruritus, especially in the evening to facilitate sleep, despite lack of support from well-controlled clinical trials. Second-generation H_1R antihistamines such as fexofenadine have efficacy in relieving the pruritus associated with contact dermatitis in pilot studies (Katagiri et al. 2006). There may be a potential in the future for therapeutic benefit in contact dermatitis from H_4R antihistamines, since they have been shown to be effective in animal models (Seike et al. 2010; Suwa et al. 2011; Matsushita et al. 2012).

10.6.8 LYMPHOMA

Cutaneous T-cell lymphomas (CTCL), including mycosis fungoides and Sézary syndrome, are often associated with chronic and severe pruritus with an incidence of pruritus of 88% (Demierre et al. 2006). Pruritus is often the presenting feature of patients with CTCL (Meyer et al. 2010). The pruritus associated with CTCL is for the most part unresponsive to oral H_1R antihistamine treatment (Meyer et al. 2010; Ahern et al. 2012). There is limited research in treating the pruritus of CTCL. In addition to treating the underlying lymphoma, the initial treatment used to target itch is corticosteroids, which have been shown to reduce pruritus, but not consistently (Zackheim et al. 1998; Ahern et al. 2012).

10.6.9 CHOLESTATIC ITCH

Pruritus is a common symptom in patients with chronic cholestatic liver disease. Up to 70% of patients with either primary biliary cirrhosis or primary sclerosing cholangitis report pruritus as a frequent symptom (Wiesner et al. 1985; Mela et al. 2003). The pruritus associated with cholestasis can be generalized or localized and either persistent or intermittent (Mela et al. 2003).

The mechanism of pruritus in cholestatic disease has not yet been fully elucidated; however, there are a number of theories. Traditionally, it has been believed that pruritus arises from cholestatic diseases as a result of the accumulation of pruritogens in the skin as a consequence of impaired secretion of bile. As such, cholestyramine is often the first-line treatment. Controlled clinical trials have shown antipruritic activity with rifampin (Ghent and Carruthers 1988), although the mechanism of action is unknown. The obstruction of the biliary tree leads to increased levels of opioid peptides, which can act on opioid receptors in the spinal cord and brain to cause itch. This hypothesis is supported by the efficacy of naloxone in patients with cholestasis-induced itch (Bergasa et al. 1992).

The contribution of histamine to the pathogenesis of pruritus in cholestasis is not clear, but it is thought not to play a large role. One study demonstrated increased plasma histamine levels in cholestatic subjects versus normal controls and that pruritic subjects with cholestasis had higher levels of histamine than those without itch (Gittlen et al. 1990). While H_1R antihistamine treatment is generally ineffective in relieving pruritus in these patients, first-generation H_1R antihistamines are prescribed solely for their use as sleep aids (Herndon 1975; Bergasa and Jones 1991). In contrast, controlled studies support the use of rifampin (Ghent and Carruthers 1988; Bachs et al. 1989; Cynamon et al. 1990; Khurana and Singh 2006) and naltrexone or naloxone (Wolfhagen et al. 1997; Mansour-Ghanaei et al. 2006).

10.6.10 UREMIC PRURITUS

Pruritus has been estimated to affect approximately 50% of patients with end-stage renal disease (Pisoni et al. 2006), but this prevalence has been declining, possibly because of improvements in dialysis. Uremic pruritus can be generalized or localized, with the back, abdomen, head, and arms being common sites of itch (Zucker et al. 2003). While the pathophysiology of uremic pruritus is still largely unknown, numerous theories have been proposed and reviewed (Schwartz and Iaina 1999; Manenti et al. 2009), and it is reasonable to consider that in these subjects with multiple medical problems that the pruritus is multifactorial.

While it is clear that histamine is not the driving force in uremic pruritus, the levels of histamine in the plasma are higher than in nonuremic patients and, also, those uremic patients with itch had higher plasma histamine levels than those without itch. However, there is no correlation between severity of pruritus and plasma histamine levels (Manenti et al. 2009).

Generally, trials of oral H_1R antihistamines have not demonstrated efficacy in relieving uremic pruritus, with a few exceptions. Oral doxepin, a first-generation H_1R antihistamine with activity at many other receptors, has been demonstrated to relieve uremic pruritus at low dosages (Pour-Reza-Gholi et al. 2007). Ketotifen and terfenadine have also shown some efficacy in relieving uremic pruritus in small studies (Russo et al. 1986; Francos et al. 1991), although terfenadine has since been removed from the market because of its risk of prolonging the QTc interval. In clinical settings, first-generation H_1R antihistamines continue to be prescribed occasionally as a first-line therapy for sedative effects despite the lack of clinical evidence for efficacy.

10.6.11 GENERAL SAFETY CONSIDERATIONS

H_1R antihistamines are considered by physicians and the general population to be a safe class of medications, as evidenced by millions of patient-years of use and their approvals for over-the-counter use. However, the side effects, particularly of the older first-generation H_1R antihistamines, should not be underestimated. Some first-generation H_1R antihistamines have antimuscarinic, anti-α-adrenergic, and antiserotonin side effects. Urinary retention, constipation, erectile dysfunction, mydriasis, blurred vision, dry eyes, and dry mouth can result from antimuscarinic effects. Appetite stimulation and weight gain can occur as a result of antiserotonin effects. Finally, dizziness from orthostatic hypotension can represent anti-α-adrenergic effects (Simons and Simons 2011). In contrast, second-generation H_1R antihistamines have almost no significant off-target effects. In fact, a recent position paper by the GA²LEN group proposes that the older first-generation H_1R antihistamines should no longer be available over the counter, given the superior safety profile and comparable efficacy of second-generation H_1R antihistamines (Church et al. 2010).

Histamine functions as an important neurotransmitter where it contributes to the regulation of attention, alertness, cognition, memory, learning, and the circadian sleep–wake cycle (Simons 2004; Haas et al. 2008; Thakkar 2011). A central feature of all first-generation H_1R antihistamines is their ability to readily cross the blood–brain barrier and interfere with neurotransmission by histamine at central nervous system H_1R. Positron emission tomography studies have shown that first-generation H_1R antihistamines occupy a majority of central nervous system H_1R at standard doses (Okamura et al. 2000). In the more vulnerable elderly, this central antihistamine blockade is associated with an increased risk of inattention, disorganized speech, altered consciousness and impaired function or alertness (McEvoy et al. 2006). First-generation H_1R antihistamines also have been reported to alter the circadian sleep/wake cycle. At night, first-generation H_1R antihistamines postpone the onset of rapid eye movement (REM) sleep as well as reduce the duration of REM sleep (Monti 1993; Boyle et al. 2006; Rojas-Zamorano et al. 2009). REM sleep is associated with the consolidation of memories, and its regular cycles are a hallmark of restful sleep. The effects of bedtime first-generation H_1R antihistamines on the onset and duration of REM sleep, as well has the long terminal half-life of some of these drugs, may explain the residual effects, or hangover, experienced the next morning. Such effects include impairment in divided attention, vigilance, working memory, and sensory-motor performance (Kay et al. 1997; Boyle et al. 2006). In contrast, the second-generation antihistamines are mildly or nonsedating with incidences of somnolence and impairment of central nervous system effects similar to those of placebo in clinical trials (Hindmarch et al. 1999; Kay and Harris 1999).

In addition to sedation, cardiac conduction effects have been associated with both first- and second-generation H_1R antagonists and have led to the withdrawal from the market of two second-generation H_1R antihistamines, astemizole and terfenadine, in most countries. These drugs were particularly dangerous when used concomitantly with potent cytochrome P450 inhibiting drugs, such as ketoconazole, which dramatically increased their plasma concentrations (Woosley 1996). This effect is not related to the H_1R but rather to blockade of potassium rectifier channels in the myocardium slowing cardiac repolarization as manifested by prolongation of the electrocardiographic QT interval. Prolongation of cardiac repolarization can lead to significant cardiac arrhythmias, such as torsades de pointes in particular (Craft 1986; Monahan et al. 1990). Many drugs, such as diphenhydramine, do not have QT prolonging effects at standard dosages, but do when overdosages occur (Zareba et al. 1997). Many of the more recent second-generation H_1R antihistamines, such as cetirizine, levocetirizine, ebastine, mizolastine, or fexofenadine, do not prolong the QT interval *in vivo* and therefore do not carry cardiac conduction risks (Chaufour et al. 1999; DuBuske 1999; Gillen et al. 2001; Hulhoven et al. 2007).

10.7   CONCLUSIONS AND FUTURE DIRECTIONS

Since they were initially introduced over 70 years ago, H_1R antihistamines have been one of the most commonly used classes of drugs, which serves a testament to their efficacy, convenient dosing regimens and formulations. Their overall safety profiles are generally favorable, with the exception of sedation associated with the earlier drugs and cardiac toxicity of a few drugs that block the cardiac potassium channels. They have been one of the most intensively studied class of drugs with hundreds of trials across several different very common diseases.

One of their primary uses has been for the relief of histamine-related pruritus. Much of this use has been justified, because histamine is a primary mediator of itch in many dermatologic and allergic diseases including urticaria, allergic rhinitis, and allergic conjunctivitis. Yet, in other diseases including atopic dermatitis and psoriasis, H_1R antihistamines are used often for their sedating properties and, in many cases, because more effective alternative antipruritic treatments do not presently exist. The lack of efficacy in these conditions may be in part due to other mediators playing a more significant role in the pathway of pruritus or, intriguingly, because H_1R or H_2R are not the appropriate drug targets. Rather, promising preclinical research suggests that H_3R or H_4R are responsible for histamine-mediated itch in these and other conditions, and trials in patients in future years may help bring forth promising new treatments for this unmet medical need.

REFERENCES

Abdelaziz MM, Devalia JL, Khair OA, Bayram H, Prior AJ and Davies RJ (1998) Effect of fexofenadine on eosinophil-induced changes in epithelial permeability and cytokine release from nasal epithelial cells of patients with seasonal allergic rhinitis. *J. Allergy Clin. Immunol.* 101:410–420.

Abel JJ and Kubota S (1919) Presence of histamine (4-imidozoleethylamine) in the hypophysis cerebri and other tissues of the body and its occurrence among the hydrolytic decomposition products of proteins. *J. Pharmacol.* 13:243–300.

Abelson MB (1998) Evaluation of olopatadine, a new ophthalmic antiallergic agent with dual activity, using the conjunctival allergen challenge model. *Ann. Allergy Asthma Immunol.* 81:211–218.

Abelson MB, Allansmith MR and Friedlaender MH (1980) Effects of topically applied occular decongestant and antihistamine. *Am. J. Ophthalmol.* 90:254–257.

Abelson MB, Chambers WA and Smith LM (1990) Conjunctival allergen challenge. A clinical approach to studying allergic conjunctivitis. *Arch. Ophthalmol.* 108:84–88.

Abelson MB, George MA and Smith LM (1995) Evaluation of 0.05% levocabastine versus 4% sodium cromolyn in the allergen challenge model. *Ophthalmology* 102:310–316.

Abelson MB and Kaplan AP (2002) A randomized, double-blind, placebo-controlled comparison of emedastine 0.05% ophthalmic solution with loratadine 10 mg and their combination in the human conjunctival allergen challenge model. *Clin. Ther.* 24:445–456.

Abelson MB and Lanier RQ (1999) The added benefit of local Patanol therapy when combined with systemic Claritin for the inhibition of ocular itching in the conjunctival antigen challenge model. *Acta Ophthalmol. Scand. Suppl.* 228:53–56.

Abelson MB and Smith LM (1988) Levocabastine. Evaluation in the histamine and compound 48/80 models of ocular allergy in humans. *Ophthalmology* 95:1494–1497.

Abelson MB, Torkildsen GL, Williams JI, Gow JA, Gomes PJ and McNamara TR (2009) Time to onset and duration of action of the antihistamine bepotastine besilate ophthalmic solutions 1.0% and 1.5% in allergic conjunctivitis: A phase III, single-center, prospective, randomized, double-masked, placebo-controlled, conjunctival allergen challenge assessment in adults and children. *Clin. Ther.* 31:1908–1921.

Abelson MB and Udell IJ (1981) H2-receptors in the human ocular surface. *Arch. Ophthalmol.* 99:302–304.

Abelson MB and Welch DL (2000) An evaluation of onset and duration of action of patanol (olopatadine hydrochloride ophthalmic solution 0.1%) compared to Claritin (loratadine 10 mg) tablets in acute allergic conjunctivitis in the conjunctival allergen challenge model. *Acta Ophthalmol. Scand. Suppl.* 230:60–63.

Aguilar AJ (2000) Comparative study of clinical efficacy and tolerance in seasonal allergic conjunctivitis management with 0.1% olopatadine hydrochloride versus 0.05% ketotifen fumarate. *Acta Ophthalmol. Scand. Suppl.* 230:52–55.

Ahern K, Gilmore ES and Poligone B (2012) Pruritus in cutaneous T-cell lymphoma: A review. *J. Am. Acad. Dermatol.* 67:760–768.

Akdis CA, Akdis M, Bieber T, Bindslev-Jensen C, Boguniewicz M, Eigenmann P, Hamid Q et al. (2006) Diagnosis and treatment of atopic dermatitis in children and adults: European Academy of Allergology and Clinical Immunology/American Academy of Allergy, Asthma and Immunology/PRACTALL Consensus Report. *J. Allergy Clin. Immunol.* 118:152–169.

Amano T, Takeda T, Yano H and Tamura T (2007) Olopatadine hydrochloride accelerates the recovery of skin barrier function in mice. *Br. J. Dermatol.* 156:906–912.

Andrew D and Craig AD (2001) Spinothalamic lamina I neurons selectively sensitive to histamine: A central neural pathway for itch. *Nat. Neurosci.* 4:72–77.

Anolik R (2008) Clinical benefits of combination treatment with mometasone furoate nasal spray and loratadine vs monotherapy with mometasone furoate in the treatment of seasonal allergic rhinitis. *Ann. Allergy Asthma Immunol.* 100:264–271.

Appl H, Holzammer T, Dove S, Haen E, Strasser A and Seifert R (2012) Interactions of recombinant human histamine H1, H2, H3, and H4 receptors with 34 antidepressants and antipsychotics. *Naunyn-Schmiedeberg's Arch. Pharmacol.* 385:145–170.

Arrang JM, Garbarg M and Schwartz JC (1983) Autoinhibition of brain histamine release mediated by a novel class (H3) of histamine receptor. *Nature* 302:832–837.

Arrang JM, Garbarg M, Tam TQ, My DTT, Yeramian E and Schwartz JC (1985) Actions of betahistine at histamine receptors in the brain. *Eur. J. Pharmacol.* 111:73–84.

Ash ASF and Schild JO (1966) Receptors mediating some action of histamine. *Br. J. Pharmacol.* 27:427–439.

Ashford CA, Heller H and Smart GA (1949) Effect of antihistamine substances on gastric secretion in man. *Br. J. Pharmacol. Chemother.* 4:157–161.

Bachs L, Pares A, Elena M, Piera C and Rodes J (1989) Comparison of rifampicin with phenobarbitone for treatment of pruritus in biliary cirrhosis. *Lancet* 1:574–576.

Bain WA (1951) Evaluation of drugs in man, with special reference to antihistamines. *Analyst* 76:573–579.

Bain WA, Broadbent JL and Warin RP (1949) Comparison of anthisan (mepyramine maleate) and phenergan as histamine antagonists. *Lancet* 257:47–52.

Bakker RA, Schoonus SBJ, Smit MJ, Timmerman H and Leurs R (2001) Histamine H1-receptor activation of nuclear factor-kB: Roles for Gbg- and Gaq/11-subunits in constitutive and agonist-mediated signaling. *Mol. Pharmacol.* 60:1133–1142.

Barger G and Dale HH (1910) 4-β-Aminoethylglyoxaline (β-Aminazolylethylamine) and the other active principles of ergot. *J. Chem. Soc., Trans.* 97:2592–2595.

Barger G and Dale HH (1911) 4-β-Aminoethylglyoxaline (β-aminazolylethylamine) and the other active principles of wrgot. *Proc. Chem. Soc., London* 26:327.

Bäumer W, Wendorff S, Gutzmer R, Werfel T, Dijkstra D, Chazot P, Stark H and Kietzmann M (2008) Histamine H4 receptors modulate dendritic cell migration through skin—immunomodulatory role of histamine. *Allergy* 63:1387–1394.

Bell JK, McQueen DS and Rees JL (2004) Involvement of histamine H_4 and H_1 receptors in scratching induced by histamine receptor agonists in BalbC mice. *Br. J. Pharmacol.* 142:374–380.

Bende M and Pipkorn U (1987) Topical levocabastine, a selective H1 antagonist, in seasonal allergic rhinoconjunctivitis. *Allergy* 42:512–515.

Benninger M, Farrar JR, Blaiss M, Chipps B, Ferguson B, Krouse J, Marple B, Storms W and Kaliner M (2010) Evaluating approved medications to treat allergic rhinitis in the United States: An evidence-based review of efficacy for nasal symptoms by class. *Ann. Allergy Asthma Immunol.* 104:13–29.

Bergasa NV and Jones EA (1991) Management of the pruritus of cholestasis: Potential role of opiate antagonists. *Am. J. Gastroenterol.* 86:1404–1412.

Bergasa NV, Talbot TL, Alling DW, Schmitt JM, Walker EC, Baker BL, Korenman JC, Park Y, Hoofnagle JH and Jones EA (1992) A controlled trial of naloxone infusions for the pruritus of chronic cholestasis. *Gastroenterology* 102:544–549.

Berger W, Hampel F, Jr., Bernstein J, Shah S, Sacks H and Meltzer EO (2006) Impact of azelastine nasal spray on symptoms and quality of life compared with cetirizine oral tablets in patients with seasonal allergic rhinitis. *Ann. Allergy Asthma Immunol.* 97:375–381.

Berger WE, Schenkel EJ and Mansfield LE (2002) Safety and efficacy of desloratadine 5 mg in asthma patients with seasonal allergic rhinitis and nasal congestion. *Ann. Allergy Asthma Immunol.* 89:485–491.

Bernstein DI, Creticos PS, Busse WW, Cohen R, Graft DF, Howland WC, Lumry WR et al. (1996) Comparison of triamcinolone acetonide nasal inhaler with astemizole in the treatment of ragweed-induced allergic rhinitis. *J. Allergy Clin. Immunol.* 97:749–755.

Bernstein DI, Schoenwetter WF, Nathan RA, Storms W, Ahlbrandt R and Mason J (1997) Efficacy and safety of fexofenadine hydrochloride for treatment of seasonal allergic rhinitis. *Ann. Allergy Asthma Immunol.* 79:443–448.

Best CH, Dale HH, Dudley HW and Thorpe WV (1927) The nature of the vaso-dilator constituents of certain tissue extracts. *J. Physiol.* 62:397–417.

Bickford RG (1937) Experiments relating to the itch sensation, it's peripheral mechanism, and central pathways. *Clin. Sci.* 3:377–386.

Bielory L, Lien KW and Bigelsen S (2005) Efficacy and tolerability of newer antihistamines in the treatment of allergic conjunctivitis. *Drugs* 65:215–228.

Bilsborough J, Leung DY, Maurer M, Howell M, Boguniewicz M, Yao L, Storey H, LeCiel C, Harder B and Gross JA (2006) IL-31 is associated with cutaneous lymphocyte antigen-positive skin homing T cells in patients with atopic dermatitis. *J. Allergy Clin. Immunol.* 117:418–425.

Black JW, Duncan WAM, Durant CJ, Ganellin CR and Parsons EM (1972) Definition and antagonism of histamine H2-receptors. *Nature* 236:385–390.

Bonadonna P, Lombardi C, Senna G, Canonica GW, Passalacqua G, Breneman D, Bronsky EA et al. (2003) Treatment of acquired cold urticaria with cetirizine and zafirlukast in combination. *J. Am. Acad. Dermatol.* 49:714–716.

Bongers G, de Esch I and Leurs R (2010) Molecular pharmacology of the four histamine receptors. *Adv. Exp. Med. Biol.* 709:11–19. ·

Borazan M, Karalezli A, Akova YA, Akman A, Kiyici H and Erbek SS (2009) Efficacy of olopatadine HCI 0.1%, ketotifen fumarate 0.025%, epinastine HCI 0.05%, emedastine 0.05% and fluorometholone acetate 0.1% ophthalmic solutions for seasonal allergic conjunctivitis: A placebo-controlled environmental trial. *Acta Ophthalmol.* 87:549–554.

Bovet D and Staub AM (1937) Protective action of some phenolic ethers in histamine intoxication. *C. R. Seances Soc. Biol. Fil.* 124:547–549.

Boyle J, Eriksson M, Stanley N, Fujita T and Kumagi Y (2006) Allergy medication in Japanese volunteers: Treatment effect of single doses on nocturnal sleep architecture and next day residual effects. *Curr. Med. Res. Opin.* 22:1343–1351.

Breneman DL (1996) Cetirizine versus hydroxyzine and placebo in chronic idiopathic urticaria. *Ann. Pharmacother.* 30:1075–1079.

Broadbent JL (1953) Observations on itching produced by cowhage, and on the part played by histamine as a mediator of the itch sensation. *Br. J. Pharmacol. Chemother.* 8:263–270.

Broadbent JL (1955) Observations on histamine-induced pruritus and pain. *Br. J. Pharmacol. Chemother.* 10:183–185.

Brody I (1984) Mast cell degranulation in the evolution of acute eruptive guttate psoriasis vulgaris. *J. Invest. Dermatol.* 82:460–464.

Brooks CD, Nelson A, Parzyck R and Maile MH (1981) Protective effect of hydroxyzine and phenylpropanolamine in the challenged allergic nose. *Ann. Allergy* 47:316–319.

Brostoff J, Fitzharris P, Dunmore C, Theron M and Blondin P (1996) Efficacy of mizolastine, a new antihistamine, compared with placebo in the treatment of chronic idiopathic urticaria. *Allergy* 51:320–325.

Buckland KF, Williams TJ and Conroy DM (2003) Histamine induces cytoskeletal changes in human eosinophils via the H_4 receptor. *Br. J. Pharmacol.* 140:1117–1127.

Buddenkotte J, Maurer M and Steinhoff M (2010) Histamine and antihistamines in atopic dermatitis. *Adv. Exp. Med. Biol.* 709:73–80.

Carr W, Bernstein J, Lieberman P, Meltzer E, Bachert C, Price D, Munzel U and Bousquet J (2012) A novel intranasal therapy of azelastine with fluticasone for the treatment of allergic rhinitis. *J. Allergy Clin. Immunol.* 129:1282–1289 e1210.

Casale TB, Blaiss MS, Gelfand E, Gilmore T, Harvey PD, Hindmarch I, Simons FE et al. (2003) First do no harm: Managing antihistamine impairment in patients with allergic rhinitis. *J. Allergy Clin. Immunol.* 111:S835–S842.

Celice J, Perrault M and Durel P (1942) Utilisation clinique des antihistaminiques de synthese. *Bull. Mem. Soc. Med. Hop. Paris* 21–22:284–288.

Chang RSL, Vihn TT and Snyder SH (1979) Heterogeneity of histamine H1-receptors: species variations in [3H]mepyramine binding of brain membranes. *J. Neurochem.* 32:1653–1663.

Chaufour S, Caplain H, Lilienthal N, L'Heritier C, Deschamps C, Dubruc C and Rosenzweig P (1999) Study of cardiac repolarization in healthy volunteers performed with mizolastine, a new H1-receptor antagonist. *Br. J. Clin. Pharmacol.* 47:515–520.

Church MK, Maurer M, Simons FE, Bindslev-Jensen C, van Cauwenberge P, Bousquet J, Holgate ST and Zuberbier T (2010) Risk of first-generation H(1)-antihistamines: A GA(2)LEN position paper. *Allergy* 65:459–466.

Ciprandi G, Buscaglia S, Catrullo A, Pesce G, Fiorino N, Montagna P, Bagnasco M and Canonica GW (1997) Azelastine eye drops reduce and prevent allergic conjunctival reaction and exert anti-allergic activity. *Clin. Exp. Allergy* 27:182–191.

Clough GF, Boutsiouki P and Church MK (2001) Comparison of the effects of levocetirizine and loratadine on histamine-induced wheal, flare, and itch in human skin. *Allergy* 56:985–988.

Code CF (1945) A discussion of benadryl as an antihistamine substance. *Proc. Staff Meet. Mayo Clin.* 20:439–445.

Coge F, Guenin S-P, Audinot V, Renouard-Try A, Beauverger P, Macia C, Ouvry C et al. (2001) Genomic organization and characterization of splice variants of the human histamine H3 receptor. *Biochem. J.* 355:279–288.

Connelly WM, Shenton FC, Lethbridge N, Leurs R, Waldvogel HJ, Faull RLM, Lees G and Chazot PL (2009) The histamine H4 receptor is functionally expressed on neurons in the mammalian CNS. *Br. J. Pharmacol.* 157:55–63.

Cormia FE (1952) Experimental histamine pruritis. I. Influence of physical and psychological factors on reactivity. *J. Invest. Dermatol.* 19:21–34.

Cormia FE and Dougherty JW (1960) Proteolytic activity in development of pain and itching. Cutaneous reactions to bradykinin and kallikrein. *J. Invest. Dermatol.* 35:21–26.

Cormia FE and Kuykendall V (1954) Experimental histamine pruritis. III. Influence of drugs on the itch threshold. *A.M.A. Arch. Dermatol. Syphilol.* 69:206–218.

Coulie PJ, Ghys L and Rihoux JP (1991) Histamine-induced wheal, flare, and pruritus: Inhibition by cetirizine, terfenadine, and placebo. *Drug Dev. Res.* 23:269–274.

Cowden JM, Zhang M, Dunford PJ and Thurmond RL (2010) The histamine H$_4$ receptor mediates inflammation and pruritus in Th2-dependent dermal inflammation. *J. Invest. Dermatol.* 130:1023–1033.

Craft TM (1986) Torsade de pointes after astemizole overdose. *Br. Med. J.* 292:660.

Crampton HJ (2002) A comparison of the relative clinical efficacy of a single dose of ketotifen fumarate 0.025% ophthalmic solution versus placebo in inhibiting the signs and symptoms of allergic rhinoconjunctivitis as induced by the conjunctival allergen challenge model. *Clin. Ther.* 24:1800–1808.

Crampton HJ (2003) Comparison of ketotifen fumarate ophthalmic solution alone, desloratadine alone, and their combination for inhibition of the signs and symptoms of seasonal allergic rhinoconjunctivitis in the conjunctival allergen challenge model: A double-masked, placebo- and active-controlled trial. *Clin. Ther.* 25:1975–1987.

Curtis AC and Owens BB (1945) Beta dimethylaminoethyl benzhydryl ether hydrochloride (benadryl) in treatment of urticaria. *Arch. Derm. Syphilol.* 52:239–242.

Cynamon HA, Andres JM and Iafrate RP (1990) Rifampin relieves pruritus in children with cholestatic liver disease. *Gastroenterology* 98:1013–1016.

D'Arienzo PA, Leonardi A and Bensch G (2002) Randomized, double-masked, placebo-controlled comparison of the efficacy of emedastine difumarate 0.05% ophthalmic solution and ketotifen fumarate 0.025% ophthalmic solution in the human conjunctival allergen challenge model. *Clin. Ther.* 24:409–416.

Dale HH (1929) Croonian lectures on some chemical factors in the control of the circulation. *Lancet* 213:1233–1237.

Dale HH (1953) *Adventures in Physiology.* New York: Macmillan.

Dale HH and Laidlaw PP (1911) The physiological action of β-iminazolylethylamine. *J. Physiol.* 41:318–344.

Daley-Yates P, Ambery C, Sweeney L, Watson J, Oliver A and McQuade B (2012) The efficacy and tolerability of two novel H1/H3 receptor antagonists in seasonal allergic rhinitis. *Int. Arch. Allergy Immunol.* 158:84–98.

Darsow U, Drzezga A, Frisch M, Munz F, Weilke F, Bartenstein P, Schwaiger M and Ring J (2000) Processing of histamine-induced itch in the human cerebral cortex: A correlation analysis with dermal reactions. *J. Invest. Dermatol.* 115:1029–1033.

Darsow U, Ring J, Scharein E and Bromm B (1996) Correlations between histamine-induced wheal, flare and itch. *Arch. Dermatol. Res.* 288:436–441.

Davidson S, Zhang X, Khasabov SG, Simone DA and Giesler GJ, Jr. (2009) Relief of itch by scratching: State-dependent inhibition of primate spinothalamic tract neurons. *Nat. Neurosci.* 12:544–546.

Davidson S, Zhang X, Yoon CH, Khasabov SG, Simone DA and Giesler GJ, Jr. (2007) The itch-producing agents histamine and cowhage activate separate populations of primate spinothalamic tract neurons. *J. Neurosci.* 27:10007–10014.

Davies MG and Greaves MW (1980) Sensory responses of human skin to synthetic histamine analogs and histamine. *Br. J. Clin. Pharmacol.* 9:461–465.

Davies MG, Marks R, Horton RJ and Storari FE (1979) The efficacy of histamine antagonists as antipruritics in experimentally induced pruritus. *Arch. Dermatol. Res.* 266:117–120.

Dawn A and Yosipovitch G (2006) Treating itch in psoriasis. *Dermatol. Nurs.* 18:227–233.

De Backer MD, Gommeren W, Moereels H, Nobels G, Van Gompel P, Leysen JE and Luyten WHML (1993) Genomic cloning, heterologous expression and pharmacological characterization of a human histamine H1 receptor. *Biochem. Biophys. Res. Commun.* 197:1601–1608.

Decourt P (1942) Action de bases antagonistes de l'histamine sur quelques maladies liees a des reactions d'hypersensibilite. *Bull. Mem. Soc. Med. Hop. Paris* 58:265–266.

Del Cuvillo A, Mullol J, Bartra J, Davila I, Jauregui I, Montoro J, Sastre J and Valero AL (2006) Comparative pharmacology of the H1 antihistamines. *J. Investig. Allergol. Clin. Immunol.* 16 Suppl 1:3–12.

Demierre MF, Gan S, Jones J and Miller DR (2006) Significant impact of cutaneous T-cell lymphoma on patients' quality of life: Results of a 2005 National Cutaneous Lymphoma Foundation Survey. *Cancer* 107:2504–2511.

Deml K-F, Beermann S, Neumann D, Strasser A and Seifert R (2009) Interactions of histamine H1-receptor agonists and antagonists with the human histamine H4-receptor. *Mol. Pharmacol.* 76:1019–1030.

Denham KJ, Boutsiouki P, Clough GF and Church MK (2003) Comparison of the effects of desloratadine and levocetirizine on histamine-induced wheal, flare and itch in human skin. *Inflamm. Res.* 52:424–427.

Desai P and Thurmond RL (2011) Histamine H4 receptor activation enhances LPS-induced IL-6 production in mast cells via ERK and PI3K activation. *Eur. J. Immunol.* 41:1764–1773.

Diepgen TL (2002) Long-term treatment with cetirizine of infants with atopic dermatitis: A multi-country, double-blind, randomized, placebo-controlled trial (the ETAC trial) over 18 months. *Pediatr. Allergy Immunol.* 13:278–286.

Dijkstra D, Stark H, Chazot PL, Shenton FC, Leurs R, Werfel T and Gutzmer R (2008) Human inflammatory dendritic epidermal cells express a functional histamine H_4 receptor. *J. Invest. Dermatol.* 128:1696–1703.

Dillon SR, Sprecher C, Hammond A, Bilsborough J, Rosenfeld-Franklin M, Presnell SR, Haugen HS et al. (2004) Interleukin 31, a cytokine produced by activated T cells, induces dermatitis in mice. *Nat. Immunol.* 5:752–760.

Discepola M, Deschenes J and Abelson M (1999) Comparison of the topical ocular antiallergic efficacy of emedastine 0.05% ophthalmic solution to ketorolac 0.5% ophthalmic solution in a clinical model of allergic conjunctivitis. *Acta Ophthalmol. Scand. Suppl.* 228:43–46.

Doeglas HM (1975) Reactions to aspirin and food additives in patients with chronic urticaria, including the physical urticarias. *Br. J. Dermatol.* 93:135–144.

Drake LA, Fallon JD and Sober A (1994) Relief of pruritus in patients with atopic dermatitis after treatment with topical doxepin cream. The Doxepin Study Group. *J. Am. Acad. Dermatol.* 31:613–616.

Drzezga A, Darsow U, Treede RD, Siebner H, Frisch M, Munz F, Weilke F, Ring J, Schwaiger M and Bartenstein P (2001) Central activation by histamine-induced itch: Analogies to pain processing: A correlational analysis of O-15 H2O positron emission tomography studies. *Pain* 92:295–305.

Dubertret L, Murrieta Aguttes M and Tonet J (1999) Efficacy and safety of mizolastine 10 mg in a placebo-controlled comparison with loratadine in chronic idiopathic urticaria: Results of the MILOR Study. *J. Eur. Acad. Dermatol. Venereol.* 12:16–24.

DuBuske LM (1999) Second-generation antihistamines: The risk of ventricular arrhythmias. *Clin. Ther.* 21:281–295.

Dunford PJ, Williams KN, Desai PJ, Karlsson L, McQueen D and Thurmond RL (2007) Histamine H_4 receptor antagonists are superior to traditional antihistamines in the attenuation of experimental pruritus. *J. Allergy Clin. Immunol.* 119:176–183.

Eda R, Townley RG and Hopp RJ (1994) Effect of terfenadine on human eosinophil and neutrophil chemotactic response and generation of superoxide. *Ann. Allergy* 73:154–160.

Eglezos A, Lecci A, Santicioli P, Giuliani S, Tramontana M, Del Bianco E and Maggi CA (1992) Activation of capsaicin-sensitive primary afferents in the rat urinary bladder by compound 48/80: A direct action on sensory nerves? *Arch. Int. Pharmacodyn. Ther.* 315:96–109.

Elmariah SB and Lerner EA (2011) Topical therapies for pruritus. *Semin. Cutan. Med. Surg.* 30:118–126.

Eppinger H (1913) *Wien. Med. Wochenschr.* 23:1414.

Eriks JC, Van DGH and Timmerman H (1993) New activation model for the histamine H2 receptor, explaining the activity of the different classes of histamine H2 receptor agonists. *Mol. Pharmacol.* 44:886–894.

Ernstene AC and Banks BM (1933) Use of histamine in the treatment of pruritus. *J. Am. Med. Assoc.* 100:328–330.

Esbenshade TA, Krueger KM, Miller TR, Kang CH, Denny LI, Witte DG, Yao BB et al. (2003) Two novel and selective nonimidazole histamine H3 receptor antagonists A-304121 and A-317920: I. In vitro pharmacological effects. *J. Pharmacol. Exp. Ther.* 305:887–896.

Fedorowicz Z, van ZEJ and Hu N (2012) Histamine H2-receptor antagonists for urticaria. *Cochrane Database Syst. Rev.* 3:CD008596.

Feinberg SM and Friedlaender S (1945) Relief of dermographism and other urticarias of histamine origin by a synthetic benzhydryl alkamine ether. *J. Allergy* 16:296–298.

Fjellner B and Haegermark O (1981) Studies on pruritogenic and histamine-releasing effects of some putative peptide neurotransmitters. *Acta Derm. Venereol.* 61:245–250.

Fjellner B and Hagermark O (1985) Experimental pruritus evoked by platelet activating factor (PAF-acether) in human skin. *Acta Derm. Venereol.* 65:409–412.

Francos GC, Kauh YC, Gittlen SD, Schulman ES, Besarab A, Goyal S and Burke JF, Jr. (1991) Elevated plasma histamine in chronic uremia. Effects of ketotifen on pruritus. *Int. J. Dermatol.* 30:884–889.

Furue M, Terao H and Koga T (2001) Effects of cetirizine and epinastine on the skin response to histamine iontophoresis. *J. Dermatol. Sci.* 25:59–63.

Galandrin S, Oligny-Longpré G and Bouvier M (2007) The evasive nature of drug efficacy: Implications for drug discovery. *Trends Pharmacol. Sci.* 28:423–430.

Gallois-Bernos AC and Thurmond RL (2012) Alcaftadine, a new antihistamine with combined antagonist activity at histamine H1, H2, and H4 receptors. *J. Receptor Ligand Channel Res.* 2012:9–20.

Gantz I, Munzert G, Tashiro T, Schaffer M, Wang L, DelValle J and Yamada T (1991) Molecular cloning of the human histamine H2 receptor. *Biochem. Biophys. Res. Commun.* 178:1386–1392.

Ganz M, Koll E, Gausche J, Detjen P and Orfan N (2003) Ketotifen fumarate and olopatadine hydrochloride in the treatment of allergic conjunctivitis: A real-world comparison of efficacy and ocular comfort. *Adv. Ther.* 20:79–91.

Gbahou F, Vincent L, Humbert-Claude M, Tardivel-Lacombe J, Chabret C and Arrang J-M (2006) Compared pharmacology of human histamine H3 and H4 receptors: Structure-activity relationships of histamine derivatives. *Br. J. Pharmacol.* 147:744–754.

Gehanno P and Desfougeres JL (1997) Fluticasone propionate aqueous nasal spray compared with oral loratadine in patients with seasonal allergic rhinitis. *Allergy* 52:445–450.

Ghent CN and Carruthers SG (1988) Treatment of pruritus in primary biliary cirrhosis with rifampin. Results of a double-blind, crossover, randomized trial. *Gastroenterology* 94:488–493.

Giede C, Metzenauer P, Petzold U and Ellers-Lenz B (2000) Comparison of azelastine eye drops with levocabastine eye drops in the treatment of seasonal allergic conjunctivitis. *Curr. Med. Res. Opin.* 16:153–163.

Gillard M, Van Der Perren C, Moguilevsky N, Massingham R and Chatelain P (2002) Binding characteristics of cetirizine and levocetirizine to human H(1) histamine receptors: Contribution of Lys(191) and Thr(194). *Mol. Pharmacol.* 61:391–399.

Gillen MS, Miller B, Chaikin P and Morganroth J (2001) Effects of supratherapeutic doses of ebastine and terfenadine on the QT interval. *Br. J. Clin. Pharmacol.* 52:201–204.

Gimenez-Arnau A, Izquierdo I and Maurer M (2009) The use of a responder analysis to identify clinically meaningful differences in chronic urticaria patients following placebo-controlled treatment with rupatadine 10 and 20 mg. *J. Eur. Acad. Dermatol. Venereol.* 23:1088–1091.

Gimenez-Arnau A, Pujol RM, Ianosi S, Kaszuba A, Malbran A, Poop G, Donado E, Perez I, Izquierdo I and Arnaiz E (2007) Rupatadine in the treatment of chronic idiopathic urticaria: A double-blind, randomized, placebo-controlled multicentre study. *Allergy* 62:539–546.

Gittlen SD, Schulman ES and Maddrey WC (1990) Raised histamine concentrations in chronic cholestatic liver disease. *Gut* 31:96–99.

Gottlieb A, Menter A, Mendelsohn A, Shen YK, Li S, Guzzo C, Fretzin S, Kunynetz R and Kavanaugh A (2009) Ustekinumab, a human interleukin 12/23 monoclonal antibody, for psoriatic arthritis: Randomised, double-blind, placebo-controlled, crossover trial. *Lancet* 373:633–640.

Govoni M, Bakker RA, van de Wetering I, Smit MJ, Menge WM, Timmerman H, Elz S, Schunack W and Leurs R (2003) Synthesis and pharmacological identification of neutral histamine H1-receptor antagonists. *J. Med. Chem.* 46:5812–5824.

Graham DT, Goodell H and Wolff HG (1951) Neural mechanisms involved in itch, itchy skin, and tickle sensations. *J. Clin. Invest.* 30:37–49.

Greaves MW (1995) Chronic urticaria. *N. Engl. J. Med.* 332:1767–1772.

Greiner JV, Edwards-Swanson K and Ingerman A (2011) Evaluation of alcaftadine 0.25% ophthalmic solution in acute allergic conjunctivitis at 15 minutes and 16 hours after instillation versus placebo and olopatadine 0.1%. *Clin. Ophthalmol.* 5:87–93.

Greiner JV and Minno G (2003) A placebo-controlled comparison of ketotifen fumarate and nedocromil sodium ophthalmic solutions for the prevention of ocular itching with the conjunctival allergen challenge model. *Clin. Ther.* 25:1988–2005.

Gschwandtner M, Mommert S, Koether B, Werfel T and Gutzmer R (2011a) The histamine H4 receptor is highly expressed on plasmacytoid dendritic cells in psoriasis and histamine regulates their cytokine production and migration. *J. Invest. Dermatol.* 131: 1668–1676.

Gschwandtner M, Schaekel K, Werfel T and Gutzmer R (2011b) Histamine H4 receptor activation on human slan-dendritic cells down-regulates their pro-inflammatory capacity. *Immunology* 132:49–56.

Guerra L, Vincenzi C, Marchesi E, Tosti A, Pretto E, Bassi R, Fubbian PR and De Costanza F (1994) Loratadine and cetirizine in the treatment of chronic urticaria. *J. Eur. Acad. Dermatol. Venereol.* 3:148–152.

Gutzmer R, Diestel C, Mommert S, Koether B, Stark H, Wittmann M and Werfel T (2005) Histamine H_4 receptor stimulation suppresses IL-12p70 production and mediates chemotaxis in human monocyte-derived dendritic cells. *J. Immunol.* 174:5224–5232.

Gutzmer R, Mommert S, Gschwandtner M, Zwingmann K, Stark H and Werfel T (2009) The histamine H4 receptor is functionally expressed on TH2 cells. *J. Allergy Clin. Immunol.* 123:619–625.

Haas HL, Sergeeva OA and Selbach O (2008) Histamine in the nervous system. *Physiol. Rev.* 88:1183–1241.

Hagermark O (1973) Influence of antihistamines, sedatives, and aspirin on experimental itch. *Acta Derm. Venereol.* 53:363–368.

Hagermark O (1974) Studies on experimental itch induced by kallikrein and bradykinin. *Acta Derm. Venereol.* 54:397–400.

Hagermark O, Hokfelt T and Pernow B (1978) Flare and itch induced by substance P in human skin. *J. Invest. Dermatol.* 71:233–235.

Hagermark O, Rajka G and Bergvist U (1972) Experimental itch in human skin elicited by rat mast cell chymase. *Acta Derm. Venereol.* 52:125–128.

Hagermark O, Strandberg K and Gronneberg R (1979) Effects of histamine receptor antagonists on histamine-induced responses in human skin. *Acta Derm. Venereol.* 59:297–300.

Halpern BN (1942) Synthetic antihistamine substances. *Arch. Int. Pharmacodyn. Ther.* 68:339–408.

Hampel F, Ratner P, Mansfield L, Meeves S, Liao Y and Georges G (2003) Fexofenadine hydrochloride, 180 mg, exhibits equivalent efficacy to cetirizine, 10 mg, with less drowsiness in patients with moderate-to-severe seasonal allergic rhinitis. *Ann. Allergy Asthma Immunol.* 91:354–361.

Hampel FC, Ratner PH, Van Bavel J, Amar NJ, Daftary P, Wheeler W and Sacks H (2010) Double-blind, placebo-controlled study of azelastine and fluticasone in a single nasal spray delivery device. *Ann. Allergy Asthma Immunol.* 105:168–173.

Handa S, Dogra S and Kumar B (2004) Comparative efficacy of cetirizine and fexofenadine in the treatment of chronic idiopathic urticaria. *J. Dermatolog. Treat.* 15:55–57.

Handwerker HO, Forster C and Kirchhoff C (1991) Discharge patterns of human C-fibers induced by itching and burning stimuli. *J. Neurophysiol.* 66:307–315.

Harada M, Terai M and Maeno H (1983) Effect of a new potent H2-receptor antagonist 3[[[2-[(diaminomethylene)amino]-4-thiazolyl]methyl]thio]-N2-sulfamoylpropionamidine (YM-11170) on gastric mucosal histamine-sensitive adenylate cyclase from guinea pig. *Biochem. Pharmacol.* 32:1635–1640.

Havas TE, Cole P, Parker L, Oprysk D and Ayiomamitis A (1986) The effects of combined H1 and H2 histamine antagonists on alterations in nasal airflow resistance induced by topical histamine provocation. *J. Allergy Clin. Immunol.* 78:856–860.

Herde L, Forster C, Strupf M and Handwerker HO (2007) Itch induced by a novel method leads to limbic deactivations a functional MRI study. *J. Neurophysiol.* 98:2347–2356.

Herndon JH, Jr. (1975) Itching: The pathophysiology of pruritus. *Int. J. Dermatol.* 14:465–484.

Herring WJ, Wilens TE, Adler LA, Baranak C, Liu K, Snavely DB, Lines CR and Michelson D (2012) Randomized controlled study of the histamine h3 inverse agonist mk-0249 in adult attention-deficit/hyperactivity disorder. *J. Clin. Psychiatry* 73:e891–e898.

Heyer G, Koppert W, Martus P and Handwerker HO (1998) Histamine and cutaneous nociception: histamine-induced responses in patients with atopic eczema, psoriasis and urticaria. *Acta Derm. Venereol.* 78:123–126.

Hilberg O (1995) Effect of terfenadine and budesonide on nasal symptoms, olfaction, and nasal airway patency following allergen challenge. *Allergy* 50:683–688.

Hilberg O, Grymer LF and Pedersen OF (1995) Nasal histamine challenge in nonallergic and allergic subjects evaluated by acoustic rhinometry. *Allergy* 50:166–173.

Hindmarch I, Shamsi Z, Stanley N and Fairweather DB (1999) A double-blind, placebo-controlled investigation of the effects of fexofenadine, loratadine and promethazine on cognitive and psychomotor function. *Br. J. Clin. Pharmacol.* 48:200–206.

Hofstra CL, Desai PJ, Thurmond RL and Fung-Leung W-P (2003) Histamine H_4 receptor mediates chemotaxis and calcium mobilization of mast cells. *J. Pharmacol. Exp. Ther.* 305:1212–1221.

Horak F, Berger UE, Menapace R, Toth J, Stubner PU and Marks B (1998) Dose-dependent protection by azelastine eye drops against pollen-induced allergic conjunctivitis. A double-blind, placebo-controlled study. *Arzneimittelforschung.* 48:379–384.

Horsmanheimo L, Harvima IT, Harvima RJ, Brummer-Korvenkontio H, Francois G and Reunala T (1996) Histamine and leukotriene C4 release in cutaneous mosquito-bite reactions. *J. Allergy Clin. Immunol.* 98:408–411.

Hosogi M, Schmelz M, Miyachi Y and Ikoma A (2006) Bradykinin is a potent pruritogen in atopic dermatitis: A switch from pain to itch. *Pain* 126:16–23.

Hossen MA, Inoue T, Shinmei Y, Fujii Y, Watanabe T and Kamei C (2006) Role of substance P on histamine H3 antagonist-induced scratching behavior in mice. *J. Pharmacol. Sci.* 100:297–302.

Hossen MA, Sugimoto Y, Kayasuga R and Kamei C (2003) Involvement of histamine H3 receptors in scratching behavior in mast cell-deficient mice. *Br. J. Dermatol.* 149:17–22.

Hough LB (2001) Genomics meets histamine receptors: New subtypes, new receptors. *Mol. Pharmacol.* 59:415–419.

Howarth PH, Stern MA, Roi L, Reynolds R and Bousquet J (1999) Double-blind, placebo-controlled study comparing the efficacy and safety of fexofenadine hydrochloride (120 and 180 mg once daily) and cetirizine in seasonal allergic rhinitis. *J. Allergy Clin. Immunol.* 104:927–933.

Hsieh JC, Hagermark O, Stahle-Backdahl M, Ericson K, Eriksson L, Stone-Elander S and Ingvar M (1994) Urge to scratch represented in the human cerebral cortex during itch. *J. Neurophysiol.* 72:3004–3008.

Hsu S, Papp KA, Lebwohl MG, Bagel J, Blauvelt A, Duffin KC, Crowley J et al. (2012) Consensus guidelines for the management of plaque psoriasis. *Arch. Dermatol.* 148:95–102.

Hulhoven R, Rosillon D, Letiexhe M, Meeus MA, Daoust A and Stockis A (2007) Levocetirizine does not prolong the QT/QTc interval in healthy subjects: Results from a thorough QT study. *Eur. J. Clin. Pharmacol.* 63:1011–1017.

Ikawa Y, Shiba K, Ohki E, Mutoh N, Suzuki M, Sato H and Ueno K (2008) Comparative study of histamine H4 receptor expression in human dermal fibroblasts. *J. Toxicol. Sci.* 33:503–508.

Ikoma A (2009) Analysis of the mechanism for the development of allergic skin inflammation and the application for its treatment: Mechanisms and management of itch in atopic dermatitis. *J. Pharmacol. Sci.* 110:265–269.

James IG, Campbell LM, Harrison JM, Fell PJ, Ellers-Lenz B and Petzold U (2003) Comparison of the efficacy and tolerability of topically administered azelastine, sodium cromoglycate and placebo in the treatment of seasonal allergic conjunctivitis and rhinoconjunctivitis. *Curr. Med. Res. Opin.* 19:313–320.

Johanek LM, Meyer RA, Friedman RM, Greenquist KW, Shim B, Borzan J, Hartke T, LaMotte RH and Ringkamp M (2008) A role for polymodal C-fiber afferents in nonhistaminergic itch. *J. Neurosci.* 28:7659–7669.

Johanek LM, Meyer RA, Hartke T, Hobelmann JG, Maine DN, LaMotte RH and Ringkamp M (2007) Psychophysical and physiological evidence for parallel afferent pathways mediating the sensation of itch. *J. Neurosci.* 27:7490–7497.

Johnson HH, Jr., DeOreo GA, Lascheid WP and Mitchell F (1960) Skin histamine levels in chronic atopic dermatitis. *J. Invest. Dermatol.* 34:237–238.

Juhlin L (1967) Localization and content of histamine in normal and diseased skin. *Acta Derm. Venereol.* 47:383–391.

Juhlin L (2004) Inhibition of cold urticaria by desloratadine. *J. Dermatol. Treat.* 15:51–59.

Kalis B (1996) Double-blind multicentre comparative study of ebastine, terfenadine and placebo in the treatment of chronic idiopathic urticaria in adults. *Drugs* 52 Suppl 1:30–34.

Kalivas J, Breneman D, Tharp M, Bruce S and Bigby M (1990) Urticaria: Clinical efficacy of cetirizine in comparison with hydroxyzine and placebo. *J. Allergy Clin. Immunol.* 86:1014–1018.

Kameyoshi Y, Tanaka T, Mihara S, Takahagi S, Niimi N and Hide M (2007) Increasing the dose of cetirizine may lead to better control of chronic idiopathic urticaria: An open study of 21 patients. *Br. J. Dermatol.* 157:803–804.

Kaplan AP (2002) Chronic urticaria—New concepts regarding pathogenesis and treatment. *Curr. Allergy Asthma Rep.* 2:263–264.

Kaplan AP and Greaves M (2009) Pathogenesis of chronic urticaria. *Clin. Exp. Allergy* 39:777–787.

Kaplan AP, Horakova Z and Katz SI (1978) Assessment of tissue fluid histamine levels in patients with urticaria. *J. Allergy Clin. Immunol.* 61:350–354.

Kaplan AP, Spector SL, Meeves S, Liao Y, Varghese ST and Georges G (2005) Once-daily fexofenadine treatment for chronic idiopathic urticaria: A multicenter, randomized, double-blind, placebo-controlled study. *Ann. Allergy Asthma Immunol.* 94:662–669.

Karppinen A, Brummer-Korvenkontio H, Petman L, Kautiainen H, Herve J-P and Reunala T (2006) Levocetirizine for treatment of immediate and delayed mosquito bite reactions. *Acta Derm. Venereol.* 86:329–331.

Karppinen A, Kautiainen H, Petman L, Burri P and Reunala T (2002) Comparison of cetirizine, ebastine and loratadine in the treatment of immediate mosquito-bite allergy. *Allergy* 57:534–537.

Karppinen A, Petman L, Jekunen A, Kautiainen H, Vaalasti A and Reunala T (2000) Treatment of mosquito bites with ebastine: A field trial. *Acta Derm. Venereol.* 80:114–116.

Katagiri K, Arakawa S, Hatano Y and Fujiwara S (2006) Fexofenadine, an H1-receptor antagonist, partially but rapidly inhibits the itch of contact dermatitis induced by diphenylcyclopropenone in patients with alopecia areata. *J. Dermatol.* 33:75–79.

Katelaris CH, Ciprandi G, Missotten L, Turner FD, Bertin D and Berdeaux G (2002) A comparison of the efficacy and tolerability of olopatadine hydrochloride 0.1% ophthalmic solution and cromolyn sodium 2% ophthalmic solution in seasonal allergic conjunctivitis. *Clin. Ther.* 24:1561–1575.

Kato M, Nishida A, Aga Y, Kita J, Kudo Y, Narita H and Endo T (1997) Pharmacokinetic and pharmacodynamic evaluation of central effect of the novel antiallergic agent bepotastine besilate. *Arzneim. Forsch.* 47:1116–1124.

Kavanagh GM, Sabroe RA, Greaves MW and Archer CB (1998) The intradermal effects of the H3 receptor agonist R α methylhistamine in human skin. *Br. J. Dermatol.* 138:622–626.

Kawashima M, Tango T, Noguchi T, Inagi M, Nakagawa H and Harada S (2003) Addition of fexofenadine to a topical corticosteroid reduces the pruritus associated with atopic dermatitis in a 1-week randomized, multicentre, double-blind, placebo-controlled, parallel-group study. *Br. J. Dermatol.* 148:1212–1221.

Kay GG, Berman B, Mockoviak SH, Morris CE, Reeves D, Starbuck V, Sukenik E and Harris AG (1997) Initial and steady-state effects of diphenhydramine and loratadine on sedation, cognition, mood, and psychomotor performance. *Arch. Intern. Med.* 157:2350–2356.

Kay GG and Harris AG (1999) Loratadine: A non-sedating antihistamine. Review of its effects on cognition, psychomotor performance, mood and sedation. *Clin. Exp. Allergy* 29 Suppl 3:147–150.

Khan DA (2008) Chronic urticaria: Diagnosis and management. *Allergy Asthma Proc.* 29:439–446.

Khurana S and Singh P (2006) Rifampin is safe for treatment of pruritus due to chronic cholestasis: A meta-analysis of prospective randomized-controlled trials. *Liver Int.* 26:943–948.

Kim K, Sussman G, Hebert J, Lumry W, Lutsky B and Gates D (2006) Desloratadine therapy for symptoms associated with perennial allergic rhinitis. *Ann. Allergy Asthma Immunol.* 96:460–465.

Kirkegaard J, Secher C, Borum P and Mygind N (1983) Inhibition of histamine-induced nasal symptoms by the H1 antihistamine chlorpheniramine maleate: Demonstration of topical effect. *Br. J. Dis. Chest* 77:113–122.

Kirkegaard J, Secher C and Mygind N (1982) Effect of the H1 antihistamine chlorpheniramine maleate on histamine-induced symptoms in the human conjunctiva. Indirect evidence for nervous H1 receptors. *Allergy* 37:203–208.

Klein PA and Clark RAF (1999) An evidence-based review of the efficacy of antihistamines in relieving pruritus in atopic dermatitis. *Arch. Dermatol.* 135:1522–1525.

Koelsche GA, Prickman LE and Carryer HM (1945) The symptomatic treatment of bronchial asthma and hay fever with benadryl. *Proc. Staff Meet. Mayo Clin.* 20:432.

Kojetin DJ, Burris TP, Jensen EV and Khan SA (2008) Implications of the binding of tamoxifen to the coactivator recognition site of the estrogen receptor. *Endocr.-Relat. Cancer* 15:851–870.

Kozel MMA and Sabroe RA (2004) Chronic urticaria: Etiology, management and current and future treatment options. *Drugs* 64:2515–2536.

Kramer MF, Jordan TR, Klemens C, Hilgert E, Hempel JM, Pfrogner E and Rasp G (2006) Factors contributing to nasal allergic late phase eosinophilia. *Am. J. Otolaryngol.* 27:190–199.

Kreutner W, Hey JA, Anthes J, Barnett A, Young S and Tozzi S (2000) Preclinical pharmacology of desloratadine, a selective and nonsedating histamine H1 receptor antagonist: Receptor selectivity, antihistaminic activity, and antiallergic effects. *Arzneim. Forsch.* 50:345–352.

Kristensen JK, Petersen LJ, Hansen U, Nielsen H, Skov PS and Nielsen HJ (1995) Systemic high-dose ranitidine in the treatment of psoriasis: An open prospective clinical trial. *Br. J. Dermatol.* 133:905–908.

Krogstad AL, Lonnroth P, Larson G and Wallin BG (1997) Increased interstitial histamine concentration in the psoriatic plaque. *J. Invest. Dermatol.* 109:632–635.

Kuhne S, Wijtmans M, Lim HD, Leurs R and de Esch IJP (2011) Several down, a few to go: Histamine H3 receptor ligands making the final push towards the market? *Expert Opin. Invest. Drugs* 20:1629–1648.

Kuna P, Bachert C, Nowacki Z, van Cauwenberge P, Agache I, Fouquert L, Roger A, Sologuren A and Valiente R (2009) Efficacy and safety of bilastine 20 mg compared with cetirizine 10 mg and placebo for the symptomatic treatment of seasonal allergic rhinitis: A randomized, double-blind, parallel-group study. *Clin. Exp. Allergy* 39:1338–1347.

Kupczyk M, Kuprys I, Bochenska-Marciniak M, Gorski P and Kuna P (2007) Ranitidine (150 mg daily) inhibits wheal, flare, and itching reactions in skin-prick tests. *Allergy Asthma Proc.* 28:711–715.

LaForce CF, Corren J, Wheeler WJ and Berger WE (2004) Efficacy of azelastine nasal spray in seasonal allergic rhinitis patients who remain symptomatic after treatment with fexofenadine. *Ann. Allergy Asthma Immunol.* 93:154–159.

Lahti A and Haapaniemi T (1993) Initiation of the effects of acrivastine and cetirizine on histamine-induced wheals and itch in human skin. *Acta Derm. Venereol.* 73:350–351.

LaMotte RH, Shimada SG and Sikand P (2011) Mouse models of acute, chemical itch and pain in humans. *Exp. Dermatol.* 20:778–782.

Lanier BQ, Finegold I, D'Arienzo P, Granet D, Epstein AB and Ledgerwood GL (2004) Clinical efficacy of olopatadine vs epinastine ophthalmic solution in the conjunctival allergen challenge model. *Curr. Med. Res. Opin.* 20:1227–1233.

Leknes SG, Bantick S, Willis CM, Wilkinson JD, Wise RG and Tracey I (2007) Itch and motivation to scratch: An investigation of the central and peripheral correlates of allergen- and histamine-induced itch in humans. *J. Neurophysiol.* 97:415–422.

Leonardi A, Di SA, Vicari C, Motterle L and Brun P (2011) Histamine H4 receptors in normal conjunctiva and in vernal keratoconjunctivitis. *Allergy* 66:1360–1366.

Leurs R, Smit MJ, Menge WM and Timmerman H (1994a) Pharmacological characterization of the human histamine H2 receptor stably expressed in Chinese hamster ovary cells. *Br. J. Pharmacol.* 112:847–854.

Leurs R, Smit MJ, Tensen CP, Ter Laak AM and Timmerman H (1994b) Site-directed mutagenesis of the histamine H1-receptor reveals a selective interaction of asparagine207 with subclasses of H1-receptor agonists. *Biochem. Biophys. Res. Commun.* 201:295–301.

Leurs R, Tulp MTM, Menge WMBP, Adolfs MJP, Zuiderveld OP and Timmerman H (1995) Evaluation of the receptor selectivity of the H3 receptor antagonists, iodophenpropit and thioperamide: An interaction with the 5-HT3 receptor revealed. *Br. J. Pharmacol.* 116:2315–2321.

Levander S, Haegermark O and Staahle M (1985) Peripheral antihistamine and central sedative effects of three H1-receptor antagonists. *Eur. J. Clin. Pharmacol.* 28:523–529.

Levander S, Staahle-Baeckdahl M and Haegermark O (1991) Peripheral antihistamine and central sedative effects of single and continuous oral doses of cetirizine and hydroxyzine. *Eur. J. Clin. Pharmacol.* 41:435–439.

Lewis T (1927) *The Blood Vessels of the Human Skin and Their Responses.* London: Shaw and Sons.

Lewis T (1942) *Pain.* New York: Macmillan.

Ligneau X, Perrin D, Landais L, Camelin JC, Calmels TP, Berrebi-Bertrand I, Lecomte JM et al. (2007) BF2.649 [1-{3-[3-(4-Chlorophenyl)propoxy]propyl}piperidine, hydrochloride], a nonimidazole inverse agonist/antagonist at the human histamine H3 receptor: Preclinical pharmacology. *J. Pharmacol. Exp. Ther.* 320:365–375.

Lim HD, van Rijn RM, Ling P, Bakker RA, Thurmond RL and Leurs R (2005) Evaluation of histamine H1-, H2-, and H3-receptor ligands at the human histamine H4 receptor: Identification of 4-methylhistamine as the first potent and selective H4 receptor agonist. *J. Pharmacol. Exp. Ther.* 314:1310–1321.

Lin RY, Curry A, Pesola GR, Knight RJ, Lee HS, Bakalchuk L, Tenenbaum C and Westfal RE (2000) Improved outcomes in patients with acute allergic syndromes who are treated with combined H1 and H2 antagonists. *Ann. Emerg. Med.* 36:462–468.

Lindahl O (1961) Experimental skin pain induced by injection of water-soluble substances in humans. *Acta Physiol. Scand.* 51:90 pp.

Ling P, Ngo K, Nguyen S, Thurmond RL, Edwards JP, Karlsson L and Fung-Leung W-P (2004) Histamine H_4 receptor mediates eosinophil chemotaxis with cell shape change and adhesion molecule upregulation. *Br. J. Pharmacol.* 142:161–171.

Lippert U, Artuc M, Gruetzkau A, Babina M, Guhl S, Haase I, Blaschke V et al. (2004) Human skin mast cells express H2 and H4, but not H3 receptors. *J. Invest. Dermatol.* 123:116–123.

Liu C, Ma X-J, Jiang X, Wilson SJ, Hofstra CL, Blevitt J, Pyati J, Li X, Chai W, Carruthers N and Lovenberg TW (2001a) Cloning and pharmacological characterization of a fourth histamine receptor (H_4) expressed in bone marrow. *Mol. Pharmacol.* 59:420–426.

Liu C, Wilson SJ, Kuei C and Lovenberg TW (2001b) Comparison of human, mouse, rat, and guinea pig histamine H_4 receptors reveals substantial pharmacological species variation. *J. Pharmacol. Exp. Ther.* 299:121–130.

Liu YL, Hu FR, Wang IJ, Chen WL and Hou YC (2011) A double-masked study to compare the efficacy and safety of topical cromolyn for the treatment of allergic conjunctivitis. *J. Formos Med. Assoc.* 110:690–694.

Loew ER and Kaiser ME (1945) Alleviation of anaphylactic shock in guinea pigs with synthetic benzohydryl alkamine ethers. *Proc. Soc. Exp. Biol. Med.* 58:235–237.

Loew ER, Kaiser ME and Moore V (1945) Synthetic benzohydryl alkamine ethers effective in preventing fatal experimental asthma in guinea pigs exposed to atomized histamine. *J. Pharmacol.* 83:120–129.

Lovenberg TW, Roland BL, Wilson SJ, Jiang X, Pyati J, Huvar A, Jackson MR and Erlander MG (1999) Cloning and functional expression of the human histamine H3 receptor. *Mol. Pharmacol.* 55:1101–1107.

Lumry W, Prenner B, Corren J and Wheeler W (2007) Efficacy and safety of azelastine nasal spray at a dose of 1 spray per nostril twice daily. *Ann. Allergy Asthma Immunol.* 99:267–272.

Macejko TT, Bergmann MT, Williams JI, Gow JA, Gomes PJ, McNamara TR and Abelson MB (2010) Multicenter clinical evaluation of bepotastine besilate ophthalmic solutions 1.0% and 1.5% to treat allergic conjunctivitis. *Am. J. Ophthalmol.* 150:122–127.

Magerl M, Schmolke J, Siebenhaar F, Zuberbier T, Metz M and Maurer M (2007) Acquired cold urticaria symptoms can be safely prevented by ebastine. *Allergy* 62:1465–1468.

Mah FS, Rosenwasser LJ, Townsend WD, Greiner JV and Bensch G (2007) Efficacy and comfort of olopatadine 0.2% versus epinastine 0.05% ophthalmic solution for treating itching and redness induced by conjunctival allergen challenge. *Curr. Med. Res. Opin.* 23:1445–1452.

Manenti L, Tansinda P and Vaglio A (2009) Uraemic pruritus: Clinical characteristics, pathophysiology and treatment. *Drugs* 69:251–263.

Mansour-Ghanaei F, Taheri A, Froutan H, Ghofrani H, Nasiri-Toosi M, Bagherzadeh AH, Farahvash MJ et al. (2006) Effect of oral naltrexone on pruritus in cholestatic patients. *World J. Gastroenterol.* 12:1125–1128.

Marks R and Greaves MW (1977) Vascular reactions to histamine and compound 48/80 in human skin: Suppression by a histamine H2-receptor blocking agent. *Br. J. Clin. Pharmacol.* 4:367–369.

Marley PD, Thomson KA, Jachno K and Johnston MJ (1991) Histamine-induced increases in cyclic AMP levels in bovine adrenal medullary cells. *Br. J. Pharmacol.* 104:839–846.

Matsushita A, Seike M, Okawa H, Kadawaki Y and Ohtsu H (2012) Advantages of histamine H4 receptor antagonist usage with H1 receptor antagonist for the treatment of murine allergic contact dermatitis. *Exp. Dermatol.* 21:714–715.

Maurer M, Bindslev-Jensen C, Gimenez-Arnau A, Godse K, Grattan CE, Hide M, Kaplan AP et al. (2013) Chronic Idiopathic Urticaria (CIU) is no longer idiopathic: Time for an update! *Br. J. Dermatol.* 168:455–456.

McElin TW and Horton BT (1945) Clinical observations on the use of benadryl; a new antihistamine substance. *Proc. Staff Meet. Mayo Clin.* 20:417–429.

McEvoy LK, Smith ME, Fordyce M and Gevins A (2006) Characterizing impaired functional alertness from diphenhydramine in the elderly with performance and neurophysiologic measures. *Sleep* 29:957–966.

McGavack TH, Elias H and Boyd LJ (1946) Some pharmacological and clinical experiences with dimethylaminoethyl benzhydryl ether hydrochloride (benadryl). *Bull. N. Y. Acad. Med.* 22:481.

Mela M, Mancuso A and Burroughs AK (2003) Review article: Pruritus in cholestatic and other liver diseases. *Aliment. Pharmacol. Ther.* 17:857–870.

Meltzer EO, Hampel FC, Ratner PH, Bernstein DI, Larsen LV, Berger WE, Finn AF, Jr. et al. (2005) Safety and efficacy of olopatadine hydrochloride nasal spray for the treatment of seasonal allergic rhinitis. *Ann. Allergy Asthma Immunol.* 95:600–606.

Meltzer EO, Laforce C, Ratner P, Price D, Ginsberg D and Carr W (2012) MP29-02 (a novel intranasal formulation of azelastine hydrochloride and fluticasone propionate) in the treatment of seasonal allergic rhinitis: A randomized, double-blind, placebo-controlled trial of efficacy and safety. *Allergy Asthma Proc.* 33:324–332.

Merlos M, Giral M, Balsa D, Ferrando R, Queralt M, Puigdemont A, Garcia-Rafanell J and Forn J (1997) Rupatadine, a new potent, orally active dual antagonist of histamine and platelet-activating factor (PAF). *J. Pharmacol. Exp. Ther.* 280:114–121.

Meyer N, Paul C and Misery L (2010) Pruritus in cutaneous T-cell lymphomas: Frequent, often severe and difficult to treat. *Acta Derm. Venereol.* 90:12–17.

Meyrick-Thomas RH, Browne PD and Kirby JDT (1985a) The effect of ranitidine, alone and in combination with clemastine, on allergen-induced cutaneous wheal-and-flare reactions in human skin. *J. Allergy Clin. Immunol.* 76:864–869.

Meyrick-Thomas RH, Browne PD and Kirby JDT (1985b) The influence of ranitidine, alone and in combination with clemastine, on histamine-mediated cutaneous weal and flare reactions in human skin. *Br. J. Clin. Pharmacol.* 20:377–382.

Miadonna A, Tedeschi A, Leggieri E, Lorini M, Folco G, Sala A, Qualizza R, Froldi M and Zanussi C (1987) Behavior and clinical relevance of histamine and leukotrienes C4 and B4 in grass pollen-induced rhinitis. *Am. Rev. Respir. Dis.* 136:357–362.

Michaelsson G and Juhlin L (1973) Urticaria induced by preservatives and dye additives in food and drugs. *Brit. J. Dermatol.* 88:525–532.

Mochizuki H, Sadato N, Saito DN, Toyoda H, Tashiro M, Okamura N and Yanai K (2007) Neural correlates of perceptual difference between itching and pain: A human fMRI study. *Neuroimage* 36:706–717.

Mochizuki H, Tashiro M, Kano M, Sakurada Y, Itoh M and Yanai K (2003) Imaging of central itch modulation in the human brain using positron emission tomography. *Pain* 105:339–346.

Mommert S, Gschwandtner M, Koether B, Gutzmer R and Werfel T (2012) Human memory Th17 cells express a functional histamine H4 receptor. *Am. J. Pathol.* 180:177–185.

Monahan BP, Ferguson CL, Killeavy ES, Lloyd BK, Troy J and Cantilena LR, Jr. (1990) Torsades de pointes occurring in association with terfenadine use. *J. Am. Med. Assoc.* 264:2788–2790.

Monroe E, Finn A, Patel P, Guerrero R, Ratner P and Bernstein D (2003) Efficacy and safety of desloratadine 5 mg once daily in the treatment of chronic idiopathic urticaria: A double-blind, randomized, placebo-controlled trial. *J. Am. Acad. Dermatol.* 48:535–541.

Monroe EW (1992) Relative efficacy and safety of loratadine, hydroxyzine, and placebo in chronic idiopathic urticaria and atopic dermatitis. *Clin. Ther.* 14:17–21.

Monti JM (1993) Involvement of histamine in the control of the waking state. *Life Sci.* 53:1331–1338.

Morita E, Matsuo H and Zhang Y (2005) Double-blind, crossover comparison of olopatadine and cetirizine versus placebo: Suppressive effects on skin response to histamine ionto-phoresis. *J. Dermatol.* 32:58–61.

Morse KL, Behan J, Laz TM, West RE, Jr., Greenfeder SA, Anthes JC, Umland S et al. (2001) Cloning and characterization of a novel human histamine receptor. *J. Pharmacol. Exp. Ther.* 296:1058–1066.

Munday J, Bloomfield R, Goldman M, Robey H, Kitowska GJ, Gwiezdziski Z, Wankiewicz A, Marks R, Protas-Drozd F and Mikaszewska M (2002) Chlorpheniramine is no more effective than placebo in relieving the symptoms of childhood atopic dermatitis with a nocturnal itching and scratching component. *Dermatology* 205:40–45.

Murray JJ, Nathan RA, Bronsky EA, Olufade AO, Chapman D and Kramer B (2002) Comprehensive evaluation of cetirizine in the management of seasonal allergic rhinitis: Impact on symptoms, quality of life, productivity, and activity impairment. *Allergy Asthma Proc.* 23:391–398.

Mygind N (1982) Mediators of nasal allergy. *J. Allergy Clin. Immunol.* 70:149–159.

Mygind N, Secher C and Kirkegaard J (1983) Role of histamine and antihistamines in the nose. *Eur. J. Respir. Dis. Suppl.* 128 (Pt 1):16–20.

Naclerio RM and Togias AG (1991) The nasal allergic reaction: Observations on the role of histamine. *Clin. Exp. Allergy* 21 Suppl 2:13–19.

Nakano Y, Takahashi Y, Ono R, Kurata Y, Kagawa Y and Kamei C (2009) Role of histamine H4 receptor in allergic conjunctivitis in mice. *Eur. J. Pharmacol.* 608:71–75.

Namdar R and Valdez C (2011) Alcaftadine: A topical antihistamine for use in allergic con-junctivitis. *Drugs Today* 47:883–890.

Natbony SF, Phillips ME, Elias JM, Godfrey HP and Kaplan AP (1983) Histologic studies of chronic idiopathic urticaria. *J. Allergy Clin. Immunol.* 71:177–183.

Nettis E, Colanardi MC, Barra L, Ferrannini A, Vacca A and Tursi A (2006) Levocetirizine in the treatment of chronic idiopathic urticaria: A randomized, double-blind, placebo-controlled study. *Br. J. Dermatol.* 154:533–538.

Nielsen HJ, Kristensen JK, Hansen U, Nielsen HI, Skov PS and Petersen LM (1997) Clinical effect of ranitidine in psoriasis. An open prospective study. *Ugeskr. Laeger* 159:598–600.

Noonan MJ, Raphael GD, Nayak A, Greos L, Olufade AO, Leidy NK, Champan D and Kramer B (2003) The health-related quality of life effects of once-daily cetirizine HCl in patients with seasonal allergic rhinitis: A randomized double-blind, placebo-controlled trial. *Clin. Exp. Allergy* 33:351–358.

Nori M, Iwata S, Munakata Y, Kobayashi H, Kobayashi S, Umezawa Y, Hosono O et al. (2003) Ebastine inhibits T cell migration, production of Th2-type cytokines and proinflamma-tory cytokines. *Clin. Exp. Allergy* 33:1544–1554.

O'Donoghue M and Tharp MD (2005) Antihistamines and their role as antipruritics. *Dermatol. Ther.* 18:333–340.

O'Leary PA and Farber EM (1945) Benadryl in the treatment of urticaria. *Proc. Staff Meet. Mayo Clin.* 20:429–432.

O'Reilly M, Alpert R, Jenkinson S, Gladue RP, Foo S, Trim S, Peter B, Trevethick M and Fidock M (2002) Identification of a histamine H4 receptor on human eosinophils-role in eosinophil chemotaxis. *J. Recept. Signal Transduct.* 22:431–448.

Oda T, Morikawa N, Saito Y, Masuho Y and Matsumoto S (2000) Molecular cloning and char-acterization of a novel type of histamine receptor preferentially expressed in leukocytes. *J. Biol. Chem.* 275:36781–36786.

Oei HD (1988) Double-blind comparison of loratadine (SCH 29851), astemizole, and placebo in hay fever with special regard to onset of action. *Ann. Allergy* 61:436–439.

Ohsawa Y and Hirasawa N (2012) The antagonism of histamine H1 and H4 receptors ameliorates chronic allergic dermatitis via anti-pruritic and anti-inflammatory effects in NC/Nga mice. *Allergy* 67:1014–1022.

Okamoto T, Iwata S, Ohnuma K, Dang NH and Morimoto C (2009) Histamine H1-receptor antagonists with immunomodulating activities: Potential use for modulating T helper type 1 (Th1)/Th2 cytokine imbalance and inflammatory responses in allergic diseases. *Clin. Exp. Immunol.* 157:27–34.

Okamura N, Yanai K, Higuchi M, Sakai J, Iwata R, Ido T, Sasaki H, Watanabe T and Itoh M (2000) Functional neuroimaging of cognition impaired by a classical antihistamine, d-chlorpheniramine. *Br. J. Pharmacol.* 129:115–123.

Okano M, Fujiwara T, Sugata Y, Gotoh D, Masaoka Y, Sogo M, Tanimoto W et al. (2006) Presence and characterization of prostaglandin D2-related molecules in nasal mucosa of patients with allergic rhinitis. *Am. J. Rhinol.* 20:342–348.

Ono SJ and Abelson MB (2005) Allergic conjunctivitis: Update on pathophysiology and prospects for future treatment. *J. Allergy Clin. Immunol.* 115:118–122.

Orfeo V, Vardaro A, Lena P, Mensitieri I, Tracey M and De Marco R (2002) Comparison of emedastine 0.05% or nedocromil sodium 2% eye drops and placebo in controlling local reactions in subjects with allergic conjunctivitis. *Eur. J. Ophthalmol.* 12:262–266.

Ortonne JP (2011) Chronic urticaria: A comparison of management guidelines. *Expert Opin. Pharmacother.* 12:2683–2693.

Ortonne JP, Grob JJ, Auquier P and Dreyfus I (2007) Efficacy and safety of desloratadine in adults with chronic idiopathic urticaria: A randomized, double-blind, placebo-controlled, multicenter trial. *Am. J. Clin. Dermatol.* 8:37–42.

Papoiu ADP, Coghill RC, Kraft RA, Wang H and Yosipovitch G (2012) A tale of two itches. Common features and notable differences in brain activation evoked by cowhage and histamine induced itch. *Neuroimage* 59:3611–3623.

Parrot J-J (1942) Les Modifications Aportees par un Antagoniste de L'histamine (2,339 R.P.) a la Reaction Vasculaire Locale de la Peau. *C. R. Seances Soc. Biol. Ses Fil.* 136:715–716.

Paus R, Schmelz M, Biro T and Steinhoff M (2006) Frontiers in pruritus research: Scratching the brain for more effective itch therapy. *J. Clin. Invest.* 116:1174–1185.

Peiser M, Tralau T, Heidler J, Api AM, Arts JH, Basketter DA, English J et al. (2012) Allergic contact dermatitis: Epidemiology, molecular mechanisms, in vitro methods and regulatory aspects. Current knowledge assembled at an international workshop at BfR, Germany. *Cell. Mol. Life Sci.* 69:763–781.

Pisoni RL, Wikstrom B, Elder SJ, Akizawa T, Asano Y, Keen ML, Saran R, Mendelssohn DC, Young EW and Port FK (2006) Pruritus in haemodialysis patients: International results from the Dialysis Outcomes and Practice Patterns Study (DOPPS). *Nephrol. Dial. Transplant.* 21:3495–3505.

Potter PC (2003) Levocetirizine is effective for symptom relief including nasal congestion in adolescent and adult (PAR) sensitized to house dust mites. *Allergy* 58:893–899.

Pour-Reza-Gholi F, Nasrollahi A, Firouzan A, Nasli Esfahani E and Farrokhi F (2007) Low-dose doxepin for treatment of pruritus in patients on hemodialysis. *Iran J. Kidney Dis.* 1:34–37.

Powell RJ, Du Toit GL, Siddique N, Leech SC, Dixon TA, Clark AT, Mirakian R, Walker SM, Huber PA and Nasser SM (2007) BSACI guidelines for the management of chronic urticaria and angio-oedema. *Clin. Exp. Allergy* 37:631–650.

Preuss H, Ghorai P, Kraus A, Dove S, Buschauer A and Seifert R (2007) Mutations of Cys-17 and Ala-271 in the human histamine H2 receptor determine the species selectivity of guanidine-type agonists and increase constitutive activity. *J. Pharmacol. Exp. Ther.* 321:975–982.

Raible DG, Lenahan T, Fayvilevich Y, Kosinski R and Schulman ES (1994) Pharmacologic characterization of a novel histamine receptor on human eosinophils. *Am. J. Respir. Crit. Care Med.* 149:1506–1511.

Ratner PH, Hampel F, Van Bavel J, Amar NJ, Daftary P, Wheeler W and Sacks H (2008) Combination therapy with azelastine hydrochloride nasal spray and fluticasone propionate nasal spray in the treatment of patients with seasonal allergic rhinitis. *Ann. Allergy Asthma Immunol.* 100:74–81.

Ratner PH, Hampel FC, Amar NJ, van Bavel JH, Mohar D, Marple BF, Roland PS et al. (2005) Safety and efficacy of olopatadine hydrochloride nasal spray for the treatment of seasonal allergic rhinitis to mountain cedar. *Ann. Allergy Asthma Immunol.* 95:474–479.

Reddy VB, Iuga AO, Shimada SG, LaMotte RH and Lerner EA (2008) Cowhage-evoked itch is mediated by a novel cysteine protease: A ligand of protease-activated receptors. *J. Neurosci.* 28:4331–4335.

Reunala T, Brummer-Korvenkontio H, Karppinen A, Coulie P and Palosuo T (1993) Treatment of mosquito bites with cetirizine. *Clin. Exp. Allergy* 23:72–75.

Ring J, Brockow K, Ollert M and Engst R (1999) Antihistamines in urticaria. *Clin. Exp. Allergy* 29 Suppl 1:31–37.

Robertson I and Greaves MW (1978) Responses of human skin blood vessels to synthetic histamine analogs. *Br. J. Clin. Pharmacol.* 5:319–322.

Rojas-Zamorano JA, Esqueda-Leon E, Jimenez-Anguiano A, Cintra-McGlone L, Mendoza Melendez MA and Velazquez Moctezuma J (2009) The H1 histamine receptor blocker, chlorpheniramine, completely prevents the increase in REM sleep induced by immobilization stress in rats. *Pharmacol. Biochem. Behav.* 91:291–294.

Rosenthal SR and Minard D (1939) Histamine as the chemical mediator for cutaneous pain. *J. Exp. Med.* 70:415–425.

Rossbach K, Nassenstein C, Gschwandtner M, Schnell D, Sander K, Seifert R, Stark H, Kietzmann M and Baeumer W (2011) Histamine H1, H3 and H4 receptors are involved in pruritus. *Neuroscience* 190:89–102.

Rossbach K, Wendorff S, Sander K, Stark H, Gutzmer R, Werfel T, Kietzmann M and Baeumer W (2009) Histamine H4 receptor antagonism reduces hapten-induced scratching behaviour but not inflammation. *Exp. Dermatol.* 18:57–63.

Rothman S (1941) Physiology of itching. *Physiol. Rev.* 21:357–381.

Rukwied R, Lischetzki G, McGlone F, Heyer G and Schmelz M (2000) Mast cell mediators other than histamine induce pruritus in atopic dermatitis patients. A dermal microdialysis study. *Br. J. Dermatol.* 142:1114–1120.

Russo GE, Spaziani M, Guidotti C, Scarpellini MG, Leri O, Bonini S, Crisciotti C and Carmenini G (1986) Pruritus in chronic uremic patients in periodic hemodialysis. Treatment with terfenadine (an antagonist of histamine H1 receptors). *Minerva Urol. Nefrol.* 38:443–447.

Saeki H, Furue M, Furukawa F, Hide M, Ohtsuki M, Katayama I, Sasaki R, Suto H and Takehara K (2009) Guidelines for management of atopic dermatitis. *J. Dermatol.* 36:563–577.

Saitoh T, Fukushima Y, Otsuka H, Ishikawa M, Tamai M, Takahashi H, Mori H et al. (2002) Effects of N-alpha-methyl-histamine on human H2 receptors expressed in CHO cells. *Gut* 50:786–789.

Salo OP, Harvey SG, Calthrop JG and Gibson JR (1989) A comparison of acrivastine versus hydroxyzine and placebo in the treatment of chronic idiopathic urticaria. *J. Int. Med. Res.* 17 Suppl 2:18B–21B.

Scadding GK, Durham SR, Mirakian R, Jones NS, Leech SC, Farooque S, Ryan D et al. (2008) BSACI guidelines for the management of allergic and non-allergic rhinitis. *Clin. Exp. Allergy* 38:19–42.

Schechter BA (2008) Ketorolac tromethamine 0.4% as a treatment for allergic conjuctivitis. *Expert Opin. Drug Metab. Toxicol.* 4:507–511.

Schmelz M (2001) A neural pathway for itch. *Nat. Neurosci.* 4:9–10.

Schmelz M, Michael K, Weidner C, Schmidt R, Torebjork HE and Handwerker HO (2000) Which nerve fibers mediate the axon reflex flare in human skin? *Neuroreport* 11:645–648.

Schmelz M, Schmidt R, Bickel A, Handwerker HO and Torebjork HE (1997) Specific C-receptors for Itch in human skin. *J. Neurosci.* 17:8003–8008.

Schmelz M, Schmidt R, Weidner C, Hilliges M, Torebjoerk HE and Handwerker HO (2003) Chemical response pattern of different classes of C-nociceptors to pruritogens and algogens. *J. Neurophysiol.* 89:2441–2448.

Schneider G, Staender S, Burgmer M, Driesch G, Heuft G and Weckesser M (2008) Significant differences in central imaging of histamine-induced itch between atopic dermatitis and healthy subjects. *Eur. J. Pain* 12:834–841.

Schwartz IF and Iaina A (1999) Uraemic pruritus. *Nephrol. Dial. Transplant.* 14:834–839.

Secher C, Kirkegaard J, Borum P, Maansson A, Osterhammel P and Mygind N (1982) Significance of H1 and H2 receptors in the human nose: Rationale for topical use of combined antihistamine preparations. *J. Allergy Clin. Immunol.* 70:211–218.

Seifert R, Wenzel-Seifert K, Burckstummer T, Pertz HH, Schunack W, Dove S, Buschauer A and Elz S (2003) Multiple differences in agonist and antagonist pharmacology between human and guinea pig histamine H1-receptor. *J. Pharmacol. Exp. Ther.* 305:1104–1115.

Seike M, Furuya K, Omura M, Hamada-Watanabe K, Matsushita A and Ohtsu H (2010) Histamine H4 receptor antagonist ameliorates chronic allergic contact dermatitis induced by repeated challenge. *Allergy* 65:319–326.

Shamsi Z and Hindmarch I (2000) Sedation and antihistamines: A review of inter-drug differences using proportional impairment ratios. *Human Psychopharmacol.* 15:S3–S30.

Sharif NA, Xu SX and Yanni JM (1996) Olopatadine (AL-4943A): ligand binding and functional studies on a novel, long-acting, H1-selective histamine antagonist and antiallergic agent for use in allergic conjunctivitis. *J. Ocul. Pharmacol. Ther.* 12:401–407.

Shelley WB and Arthur RP (1955) Mucunain, the active pruritogenic proteinase of cowhage. *Science* 122:469–470.

Shelley WB and Melton FM (1950) Relative effect of local anesthetics on experimental histamine pruritus in man. *J. Invest. Dermatol.* 15:299–300.

Shelton D and Eiser N (1994) Histamine receptors in the human nose. *Clin. Otolaryngol. Allied Sci.* 19:45–49.

Shimada SG and LaMotte RH (2008) Behavioral differentiation between itch and pain in mouse. *Pain* 139:681–687.

Shin N, Covington M, Bian D, Zhuo J, Bowman K, Li Y, Soloviev M et al. (2012) INCB38579, a novel and potent histamine H4 receptor small molecule antagonist with antiinflammatory pain and anti-pruritic functions. *Eur. J. Pharmacol.* 675:47–56.

Siebenhaar F, Degener F, Zuberbier T, Martus P and Maurer M (2009) High-dose desloratadine decreases wheal volume and improves cold provocation thresholds compared with standard-dose treatment in patients with acquired cold urticaria: A randomized, placebo-controlled, crossover study. *J. Allergy Clin. Immunol.* 123:672–679.

Sikand P, Shimada SG, Green BG and LaMotte RH (2009) Similar itch and nociceptive sensations evoked by punctate cutaneous application of capsaicin, histamine and cowhage. *Pain* 144:66–75.

Simone DA, Zhang X, Li J, Zhang J-M, Honda CN, LaMotte RH and Giesler GJ, Jr. (2004) Comparison of responses of primate spinothalamic tract neurons to pruritic and algogenic stimuli. *J. Neurophysiol.* 91:213–222.

Simons FE (2004) Advances in H1-antihistamines. *N. Engl. J. Med.* 351:2203–2217.

Simons FER and Simons KJ (2011) Histamine and H1-antihistamines: Celebrating a century of progress. *J. Allergy Clin. Immunol.* 128:1139–1150.e1134.

Simpson RJ (1994) Budesonide and terfenadine, separately and in combination, in the treatment of hay fever. *Ann. Allergy* 73:497–502.

Smith CH, Anstey AV, Barker JN, Burden AD, Chalmers RJ, Chandler DA, Finlay AY et al. (2009) British Association of Dermatologists' guidelines for biologic interventions for psoriasis 2009. *Br. J. Dermatol.* 161:987–1019.

Sollmann T and Pilcher JD (1917) Endermic reactions. I. *J. Pharmacol.* 9:309–340.

Sonkoly E, Muller A, Lauerma AI, Pivarcsi A, Soto H, Kemeny L, Alenius H et al. (2006) IL-31: A new link between T cells and pruritus in atopic skin inflammation. *J. Allergy Clin. Immunol.* 117:411–417.

Spangler DL, Abelson MB, Ober A and Gotnes PJ (2003) Randomized, double-masked comparison of olopatadine ophthalmic solution, mometasone furoate monohydrate nasal spray, and fexofenadine hydrochloride tablets using the conjunctival and nasal allergen challenge models. *Clin. Ther.* 25:2245–2267.

Spangler DL, Bensch G and Berdy GJ (2001) Evaluation of the efficacy of olopatadine hydrochloride 0.1% ophthalmic solution and azelastine hydrochloride 0.05% ophthalmic solution in the conjunctival allergen challenge model. *Clin. Ther.* 23:1272–1280.

Staevska M, Popov TA, Kralimarkova T, Lazarova C, Kraeva S, Popova D, Church DS, Dimitrov V and Church MK (2010) The effectiveness of levocetirizine and desloratadine in up to 4 times conventional doses in difficult-to-treat urticaria. *J. Allergy Clin. Immunol.* 125:676–682.

Staub AM and Bovet D (1937) Action of thymoxyethyldiethylamine (F 929) and other phenol ethers on anaphylactic shock in guinea pigs. *C. R. Seances Soc. Biol. Fil.* 125:818–821.

Stokes JR, Romero FA, Jr., Allan RJ, Phillips PG, Hackman F, Misfeldt J and Casale TB (2012) The effects of an H3 receptor antagonist (PF-03654746) with fexofenadine on reducing allergic rhinitis symptoms. *J. Allergy Clin. Immunol.* 129:409–412.e402.

Storms WW, Bodman SF, Nathan RA, Chervinsky P, Banov CH, Dockhorn RJ, Jarmoszuk I et al. (1989) SCH 434: A new antihistamine/decongestant for seasonal allergic rhinitis. *J. Allergy Clin. Immunol.* 83:1083–1090.

Strakhova MI, Nikkel AL, Manelli AM, Hsieh GC, Esbenshade TA, Brioni JD and Bitner RS (2009) Localization of histamine H4 receptors in the central nervous system of human and rat. *Brain Res.* 1250:41–48.

Sugimoto Y, Iba Y, Nakamura Y, Kayasuga R and Kamei C (2004) Pruritus-associated response mediated by cutaneous histamine H3 receptors. *Clin. Exp. Allergy* 34:456–459.

Supramaniam G and Warner JO (1986) Artificial food additive intolerance in patients with angio-oedema and urticaria. *Lancet* 2:907–909.

Suwa E, Yamaura K, Oda M, Namiki T and Ueno K (2011) Histamine H4 receptor antagonist reduces dermal inflammation and pruritus in a hapten-induced experimental model. *Eur. J. Pharmacol.* 667:383–388.

Takahashi Y, Kagawa Y, Izawa K, Ono R, Akagi M and Kamei C (2009) Effect of histamine H4 receptor antagonist on allergic rhinitis in mice. *Int. Immunopharmacol.* 9:734–738.

Takaoka A, Arai I, Sugimoto M, Yamaguchi A, Tanaka M and Nakaike S (2005) Expression of IL-31 gene transcripts in NC/Nga mice with atopic dermatitis. *Eur. J. Pharmacol.* 516:180–181.

Tamura T, Matsubara M, Amano T and Chida M (2008) Olopatadine ameliorates rat experimental cutaneous inflammation by improving skin barrier function. *Pharmacology* 81:118–126.

Tanizaki H, Ikoma A, Fukuoka M, Miyachi Y and Kabashima K (2012) Effects of bepotastine and fexofenadine on histamine-induced flare, wheal and itch. *Int. Arch. Allergy Immunol.* 158:191–195.

Taylor-Clark T, Sodha R, Warner B and Foreman J (2005) Histamine receptors that influence blockage of the normal human nasal airway. *Br. J. Pharmacol.* 144:867–874.

Thakkar MM (2011) Histamine in the regulation of wakefulness. *Sleep Med. Rev.* 15:65–74.

Thurmond RL, Desai PJ, Dunford PJ, Fung-Leung W-P, Hofstra CL, Jiang W, Nguyen S et al. (2004) A potent and selective histamine H_4 receptor antagonist with anti-inflammatory properties. *J. Pharmacol. Exp. Ther.* 309:404–413.

Thurmond RL, Gelfand EW and Dunford PJ (2008) The role of histamine H_1 and H_4 receptors in allergic inflammation: The search for new antihistamines. *Nat. Rev. Drug Discov.* 7:41–53.

Torkildsen G and Shedden A (2011) The safety and efficacy of alcaftadine 0.25% ophthalmic solution for the prevention of itching associated with allergic conjunctivitis. *Curr. Med. Res. Opin.* 27:623–631.

Tran VT, Chang RS and Snyder SH (1978) Histamine H1 receptors identified in mammalian brain membranes with [3H]mepyramine. *Proc. Natl. Acad. Sci. USA* 75:6290–6294.

Trendelenburg U and Hobbs RD (1960) Action of histamine and 5-hydroxytryptamine (5-HT) on isolated mammalian atria. *J. Pharmacol. Exp. Ther.* 130:450–460.

Twycross R, Greaves MW, Handwerker H, Jones EA, Libretto SE, Szepietowski JC and Zylicz Z (2003) Itch: Scratching more than the surface. *QJM* 96:7–26.

Uehara M (1982) Reduced histamine reaction in atopic dermatitis. *Arch. Dermatol.* 118:244–245.

Wagenmann M, Schumacher L and Bachert C (2005) The time course of the bilateral release of cytokines and mediators after unilateral nasal allergen challenge. *Allergy* 60:1132–1138.

Wallace DV, Dykewicz MS, Bernstein DI, Blessing-Moore J, Cox L, Khan DA, Lang DM et al. (2008) The diagnosis and management of rhinitis: An updated practice parameter. *J. Allergy Clin. Immunol.* 122:S1–S84.

Wang DY, Hanotte F, De VC and Clement P (2001) Effect of cetirizine, levocetirizine, and dextrocetirizine on histamine-induced nasal response in healthy adult volunteers. *Allergy (Cph)* 56:339–343.

Wang YH, Tache Y, Harris AG, Kreutner W, Daly AF and Wei JY (2005) Desloratadine prevents compound 48/80-induced mast cell degranulation: Visualization using a vital fluorescent dye technique. *Allergy* 60:117–124.

Weiner JM, Abramson MJ and Puy RM (1998) Intranasal corticosteroids versus oral H1 receptor antagonists in allergic rhinitis: Systematic review of randomised controlled trials. *Br. Med. J.* 317:1624–1629.

Weisler RH, Pandina GJ, Daly EJ, Cooper K and Gassmann-Mayer C (2012) Randomized clinical study of a histamine H3 receptor antagonist for the treatment of adults with attention-deficit hyperactivity disorder. *CNS Drugs* 26:421–434.

Weisshaar E, Ziethen B and Gollnick H (1997) Can a serotonin type 3 (5-HT3) receptor antagonist reduce experimentally-induced itch? *Inflamm. Res.* 46:412–416.

Wellendorph P, Goodman MW, Burstein ES, Nash NR, Brann MR and Weiner DM (2002) Molecular cloning and pharmacology of functionally distinct isoforms of the human histamine H3 receptor. *Neuropharmacology* 42:929–940.

Weller K and Maurer M (2009) Desloratadine inhibits human skin mast cell activation and histamine release. *J. Invest. Dermatol.* 129:2723–2726.

Wells JA, Bull HB, Dragstedt CA, Morris HC (1945) Observations on the nature of the antagonism of histamine by beta-dimethylaminoethyl benzhydryl ether (benadryl). *J. Pharmacol. Exp. Ther.* 85:122–128.

Whitcup SM, Bradford R, Lue J, Schiffman RM and Abelson MB (2004) Efficacy and tolerability of ophthalmic epinastine: A randomized, double-masked, parallel-group, active- and vehicle-controlled environmental trial in patients with seasonal allergic conjunctivitis. *Clin. Ther.* 26:29–34.

Wieland K, Bongers G, Yamamoto Y, Hashimoto T, Yamatodani A, Menge WMBP, Timmerman H, Lovenberg TW and Leurs R (2001) Constitutive activity of histamine H3 receptors stably expressed in SK-N-MC cells: Display of agonism and inverse agonism by H3 antagonists. *J. Pharmacol. Exp. Ther.* 299:908–914.

Wiesner RH, LaRusso NF, Ludwig J and Dickson ER (1985) Comparison of the clinico-pathologic features of primary sclerosing cholangitis and primary biliary cirrhosis. *Gastroenterology* 88:108–114.

Williams HC (2005) Clinical practice. Atopic dermatitis. *N. Engl. J. Med.* 352:2314–2324.

Wilson AM, Haggart K, Sims EJ and Lipworth BJ (2002) Effects of fexofenadine and deslo-ratadine on subjective and objective measures of nasal congestion in seasonal allergic rhinitis. *Clin. Exp. Allergy* 32:1504–1509.

Windaus A and Vogt W (1907) Synthese des Imidazolyl-athylamins. *Ber. Dtsch. Chem. Ges.* 40:3691–3695.

Wolfhagen FH, Sternieri E, Hop WC, Vitale G, Bertolotti M and Van Buuren HR (1997) Oral naltrexone treatment for cholestatic pruritus: A double-blind, placebo-controlled study. *Gastroenterology* 113:1264–1269.

Wong L, Hendeles L and Weinberger M (1981) Pharmacologic prophylaxis of allergic rhini-tis: Relative efficacy of hydroxyzine and chlorpheniramine. *J. Allergy Clin. Immunol.* 67:223–228.

Wood-Baker R, Lau L and Howarth PH (1996) Histamine and the nasal vasculature: The influence of H1 and H2-histamine receptor antagonism. *Clin. Otolaryngol. Allied Sci.* 21:348–352.

Woosley RL (1996) Cardiac actions of antihistamines. *Ann. Rev. Pharmacol. Toxicol.* 36:233–252.

Wulff BS, Hastrup S and Rimvall K (2002) Characteristics of recombinantly expressed rat and human histamine H3 receptors. *Eur. J. Pharmacol.* 453:33–41.

Wyllie JH, Hesselbo T and Black JW (1972) Effects in man of histamine H2-receptor blockade by burimamide. *Lancet* 2:1117–1120.

Yamaura K, Oda M, Suwa E, Suzuki M, Sato H and Ueno K (2009) Expression of histamine H4 receptor in human epidermal tissues and attenuation of experimental pruritus using H4 receptor antagonist. *J. Toxicol. Sci.* 34:427–431.

Yanez A and Rodrigo GJ (2002) Intranasal corticosteroids versus topical H1 receptor antago-nists for the treatment of allergic rhinitis: A systematic review with meta-analysis. *Ann. Allergy Asthma Immunol.* 89:479–484.

Yaylali V, Demirlenk I, Tatlipinar S, Ozbay D, Esme A, Yildirim C and Ozden S (2003) Comparative study of 0.1% olopatadine hydrochloride and 0.5% ketorolac trometh-amine in the treatment of seasonal allergic conjunctivitis. *Acta Ophthalmol. Scand.* 81:378–382.

Yosipovitch G, Ansari N, Goon A, Chan YH and Goh CL (2002) Clinical characteristics of pruritus in chronic idiopathic urticaria. *Br. J. Dermatol.* 147:32–36.

Yosipovitch G, Duque MI, Fast K, Dawn AG and Coghill RC (2007) Scratching and nox-ious heat stimuli inhibit itch in humans: A psychophysical study. *Br. J. Dermatol.* 156:629–634.

Yosipovitch G, Fast K and Bernhard JD (2005) Noxious heat and scratching decrease hista-mine-induced itch and skin blood flow. *J. Invest. Dermatol.* 125:1268–1272.

Yosipovitch G, Goon A, Wee J, Chan YH and Goh CL (2000) The prevalence and clinical characteristics of pruritus among patients with extensive psoriasis. *Br. J. Dermatol.* 143:969–973.

Yosipovitch G, Greaves MW and Schmelz M (2003) Itch. *Lancet* 361:690–694.

Yu F, Wolin RL, Wei J, Desai PJ, McGovern PM, Dunford PJ, Karlsson L and Thurmond RL (2010) Pharmacological characterization of oxime agonists of the histamine H4 recep-tor. *J. Receptor Ligand Channel Res.* 3:37–49.

Zackheim HS, Kashani-Sabet M and Amin S (1998) Topical corticosteroids for mycosis fun-goides. Experience in 79 patients. *Arch. Dermatol.* 134:949–954.

Zareba W, Moss AJ, Rosero SZ, Hajj-Ali R, Konecki J and Andrews M (1997) Electrocardiographic findings in patients with diphenhydramine overdose. *Am. J. Cardiol.* 80:1168–1173.

Zazzali JL, Broder MS, Chang E, Chiu MW and Hogan DJ (2012) Cost, utilization, and patterns of medication use associated with chronic idiopathic urticaria. *Ann. Allergy Asthma Immunol.* 108:98–102.

Zhu Y, Michalovich D, Wu H, Tan KB, Dytko GM, Mannan IJ, Boyce R et al. (2001) Cloning, expression, and pharmacological characterization of a novel human histamine receptor. *Mol. Pharmacol.* 59:434–441.

Zonneveld IM, Meinardi MM, Karlsmark T, Johansen UB, Kuiters GR, Hamminga L, Staberg B et al. (1997) Ranitidine does not affect psoriasis: A multicenter, double-blind, placebo-controlled study. *J. Am. Acad. Dermatol.* 36:932–934.

Zuber P and Pecoud A (1988) Effect of levocabastine, a new H1 antagonist, in a conjunctival provocation test with allergens. *J. Allergy Clin. Immunol.* 82:590–594.

Zuberbier T (2012) Pharmacological rationale for the treatment of chronic urticaria with second-generation non-sedating antihistamines at higher-than-standard doses. *J. Eur. Acad. Dermatol. Venereol.* 26:9–18.

Zuberbier T, Asero R, Bindslev-Jensen C, Walter Canonica G, Church MK, Gimenez-Arnau AM, Grattan CE et al. (2009) EAACI/GA(2)LEN/EDF/WAO guideline: Management of urticaria. *Allergy* 64:1427–1443.

Zuberbier T, Chantraine-Hess S, Hartmann K and Czarnetzki BM (1995) Pseudoallergen-free diet in the treatment of chronic urticaria. A prospective study. *Acta Derm. Venereol.* 75:484–487.

Zuberbier T and Maurer M (2007) Urticaria: Current opinions about etiology, diagnosis and therapy. *Acta Derm. Venereol.* 87:196–205.

Zuberbier T and Maurer M (2010) Antihistamines in the treatment of urticaria. *Adv. Exp. Med. Biol.* 709:67–72.

Zucker I, Yosipovitch G, David M, Gafter U and Boner G (2003) Prevalence and characterization of uremic pruritus in patients undergoing hemodialysis: Uremic pruritus is still a major problem for patients with end-stage renal disease. *J. Am. Acad. Dermatol.* 49:842–846.

11 Role of PAR-2 in Neuroimmune Communication and Itch

Cordula Kempkes, Joerg Buddenkotte,
Ferda Cevikbas, Timo Buhl, and Martin Steinhoff

CONTENTS

11.1 INTRODUCTION

A critical role of various proteases in skin homeostasis as well as pathobiology, including itch, was described decades ago (Arthur and Shelley 1955; Rajka 1967, 1969; Shelley and Arthur 1955). Later, after the cloning of protease-activated receptors-1 (PAR-1) in 1991 (Vu et al. 1991) and the characterization of PAR-2 in the skin (D'Andrea et al. 1998; Derian et al. 1997; Hou et al. 1998; Santulli et al. 1995; Schechter et al. 1998; Steinhoff et al. 1999), some of the effects of endogenous or exogenous proteases could be attributed—at least in part—to the activation of those G-protein-coupled receptors. In 2000, a role of PAR-2 in skin neurogenic inflammation and later in human pruritus was established (Steinhoff et al. 2000, 2003). This review highlights our current understanding of proteases as histamine-independent pruritogens that are mainly—but not exclusively—mediated via protease-activated receptors. For more detailed information about the cellular mechanisms of PAR function, the reader is referred to exquisitely detailed reviews (Cottrell et al. 2002; Ossovskaya and Bunnett 2004; Steinhoff et al. 2005).

Various proteases such as mucunain, trypsin, kallikreins (KLK), or tryptase have been demonstrated to exert capacities as pruritogens in rodents or humans *in vivo* (Cormia and Dougherty 1960; Costa et al. 2008; Hägermark 1974; Hagermark et al.

1972; Hansson et al. 2002; Ny and Egelrud 2003; Reddy et al. 2008; Stefansson et al. 2008; Steinhoff et al. 2003; Ui et al. 2006). Accordingly, increased tryptase serum levels have been described in hemodialysis patients, interestingly correlating with itch severity, as well as in patients with atopic dermatitis (AD) (Dugas-Breit et al. 2005; Kawakami et al. 2006).

On the basis of these findings, it is important to understand the mechanisms how protease activated receptors, in particular, PAR-2 but probably also PAR-4, regulate protease-dependent, but histamine-independent itch. This is also of major medical interest since serine proteases like KLK, matriptase, prostasin, or tryptase have been implicated in various pruritic diseases including atopic dermatitis, Netherton syndrome, psoriasis, anaphylaxis-associated itch, dry skin itch, or renal insufficiency-associated itch, for example (Zhu et al. 2009a; Akiyama et al. 2009, 2010a, 2010b, 2010c).

However, the itch symptom as observed in patients with AD, other eczematous diseases, neuropathic or systemic disease-associated itch, or tumor-associated itch is mostly resistant to therapy with oral antihistamines. In this context, it is important to note that the itch sensation can be triggered by endogenous (KLK, matriptase, trypsins, and prostasin) or exogenous (e.g., house dust mite allergens Dpt. 1, 3, 9 or certain *Staphylococcus aureus* toxins) factors that are partly proteases (Jeong et al. 2008; Kato et al. 2009; Lee et al. 2010; Lutfi et al. 2012; Frateschi et al. 2011). Unfortunately, the mechanisms by which itch is controlled in antihistamine unresponsive pruritic diseases are still poorly understood in humans, and effective therapeutic options represent a significant unmet need (Hong et al. 2011; Buddenkotte and Steinhoff 2010).

As will be pointed out later in this book chapter in greater detail, PAR-2 is widely expressed by different cell types including suprabasal keratinocytes, endothelial cells, mast cells, neutrophils, macrophages, dendritic cells, and sensory nerve fibers, for example (Böhm et al. 1996). Recent findings indicate that keratinocytes can also be seen as a "sensory forefront" of neuronal activation and signaling (Elias and Steinhoff 2008). Upon activation by trigger factors that are potentially deleterious for the body system, such as microbes, noxious heat/cold, UV radiation, chemicals, allergens, or proteases, keratinocytes are capable of releasing factors that subsequently activate peripheral nerve endings to induce "neurogenic" inflammation, increased skin sensitivity, and pain or itch (Roosterman et al. 2006; Ma 2010). The factors that mediate this epidermal–neuronal communication during chronic inflammation and itch are, however, currently poorly understood and include at least ATP, endorphin, and endothelin-1 (ET-1) (Caterina et al. 2000; Imamachi et al. 2009; Kido et al. 2011).

Several studies indicate that proteases like KLK, trypsin, or tryptase are important activators of neuronal and/or keratinocyte-derived PAR-2, which subsequently induces itch directly (direct activation of nerve endings) or indirectly via activation of keratinocyte PAR-2 by endogenous or exogenous proteases. Activation of keratinocyte-derived PAR-2 leads to the release of as yet unidentified mediators, which activate sensory nerves and lead to itch and neuroinflammation, as well as nerve sprouting and/or skin hypersensitivity (Figure 11.1) (Frateschi et al. 2011; Stefansson et al. 2008; Moormann et al. 2006; Steinhoff et al. 1999, 2000, 2003, 2005). This chapter gives a comprehensive overview how proteases act via PAR-2 to trigger itch and skin inflammation.

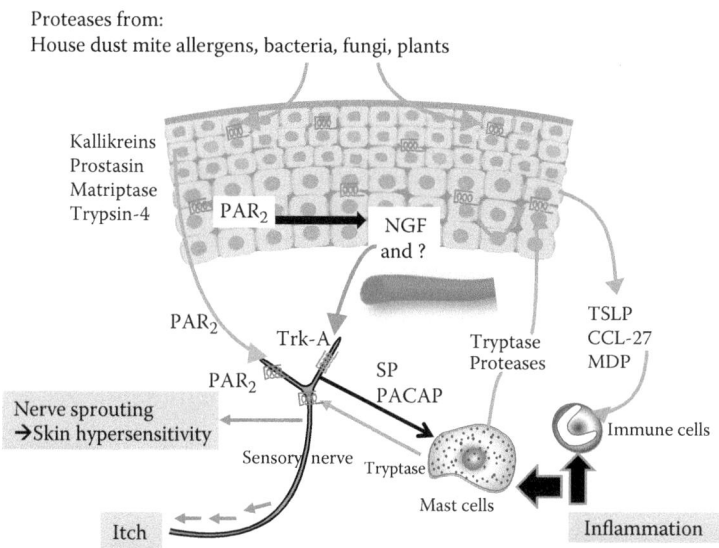

FIGURE 11.1 **(See color insert.)** Role of exogenous, keratinocyte-, and immune-cell-derived proteases on itch and inflammation in the skin. Upon stimulation by exogenous or endogenous proteases, PAR-2 becomes activated on nerve endings of primary afferent nerve fibers. Simultaneously, proteases can activate PAR-2 on keratinocytes or dermal endothelial cells resulting in release of cytokines, chemokines, neuropeptides, and reactive oxygen species. Release of certain mediators may subsequently activate high-affinity receptors on sensory nerves amplifying the pruritic induction and inflammatory response.

Our review highlights the mechanisms by which proteases act via PAR receptors to exert important biological and pathobiological functions in the skin with a focus on itch and neurogenic inflammation. Understanding these mechanisms may lead to novel therapies against acute as well as chronic, recalcitrant itch.

11.2 PROTEASE-ACTIVATED RECEPTORS AND PROTEASES

Proteases and their inhibitors have diverse functional roles to maintain body homeostasis. Proteolysis does not only occur during stress, injury, or infection but is also necessary for normal tissue function, e.g., thrombin or factor VII are involved in blood coagulation, trypsin is an important protease for food digestion, and—in the skin—members of the KLK protease family are critical for epidermal differentiation, cornification and/or pigmentation (Babiarz-Magee et al. 2004; Derian et al. 1997; Lee et al. 2010; Lin et al. 2008; Santulli et al. 1995; Seiberg et al. 2000; Seiberg 2001). Extracellular proteolysis activity is also directly sensed by cells through the unique class of G-protein-coupled receptors (GPCRs), known as PARs, which consist of four receptors with unique cellular functions. These receptors become activated by proteolytic cleavage of the N-terminal end. Thereby, a tethered ligand sequence (TLS) is unmasked that binds to the second extracellular loop of the same receptor and subsequently leads to receptor activation, signaling, and receptor internalization. It is important to understand that, in contrast to other ligand binding GPCRs, such as neurokinin 1 receptor (NK1R) or endothelin A receptor (ETAR), activation and internalization of PARs results in endosomal trafficking, ubiquitination, and degradation of the receptors in lysosomes. Thus, replenishing of the cell surface with PARs depends on transport of stored receptors in intracellular vesicular structures and *de novo* synthesis of the receptor (Böhm et al. 1996a, 1996b; Déry et al. 1999; Roosterman et al. 2003). Thus, while binding of substance P (SP) or ET-1 to NK1R or ETAR, respectively, leads to internalization and recycling of the receptors, activation of PARs by proteases results in degradation of the internalized receptors (Böhm et al. 1996a; Roosterman et al. 2003, 2004; Zhang et al. 1999).

PARs are widely expressed in different cell types and activation of PARs on these cells induces different responses, respectively. PAR-1 was the first PAR to be cloned and was originally identified and termed as the thrombin receptor (Coughlin et al. 1992; Vu et al. 1991). PAR-1 is also activated by other coagulation proteases, matrix metalloproteinase 1, and microbial proteases (Goerge et al. 2006; Schuepbach and Riewald 2010). Besides PAR-1, PAR-3 and PAR-4 were also classified as thrombin-sensitive receptors (Gandhi et al. 2011; Oikonomopoulou et al. 2006b). PAR-2 was identified as a trypsin-activated, thrombin-insensitive receptor that is upregulated by inflammatory mediators, e.g., in endothelial cells (Santulli et al. 1995). Interestingly, PAR-2 can become transactivated by thrombin-cleaved PAR-1 (Lin and Trejo 2013; O'Brien et al. 2000). However, this is the only report of PAR-induced receptor transactivation so far. Proteolysis of PAR-2 is mediated by a broad array of extracellular proteases including members of serine proteases (e.g., KLK5, KLK14, trypsin, tryptase, prostasin, and matriptase) as well as cysteine proteases (cathepsin S, house dust mite (HDM) antigen Der p1) (Bocheva et al. 2009; Cattaruzza et al. 2011; Frateschi et al. 2011; Kauffman et al. 2006; Santulli et al. 1995; Stefansson et al. 2008). PAR-2 is also known to be activated by proteases produced by microbial agents such as house dust mites, cockroaches, certain bacteria, and probably also certain parasites (Shpacovitch et al. 2007). Several reports indicate that PAR-2 activation plays an important role in inflammation (e.g., neurogenic inflammation or rheumatic arthritis), tumor progression, allergic reaction, and pain (Kempkes et al. 2012; Lam et al. 2012; Lohman et al. 2012; Nichols et al. 2012; Poole et al. 2013; Seeliger et al. 2003; Steinhoff et al. 2000). Finally, several studies suggest that activation of PAR-2 can also elicit itch, either directly by activation of the receptor on sensory nerve fibers innervating the skin, or indirectly by activating keratinocytes or immune cells (e.g., mast cells), and thereby induce a cascade leading to release of pruritogens that in turn activate sensory nerve fibers in the skin (Akiyama et al. 2009, 2010a, 2010b, 2010c, 2012, 2013; Steinhoff et al. 2003).

11.3 PAR-2 EXPRESSION IN THE SKIN

The skin is composed of three layers: epidermis, dermis, and subcutis. Several studies showed that PAR-2 expression in different cell types (keratinocytes, endothelial cells, skin-innervating nerve fibers, mast cells, neutrophils, and dendritic cells) is important for a functional skin homeostasis. The epidermis consists of several keratinocyte layers (stratum corneum, stratum granulosum, stratum spinosum, and stratum basale) that create a barrier to protect the body from the environment. Besides being a protection barrier, the epidermis can also be understood as a sensor for a broad variety of stimuli (e.g., UV radiation, temperature, pH changes, touch, chemicals, allergens, and microbial proteases). The stratum basale is formed by highly proliferating keratinocyte-specific stem cells and immature keratinocytes. These cells do not express PAR-2, but during the maturation process of keratinocytes, the expression levels of PAR-2-activating proteases (e.g., matriptase, KLK and prostasin) and PAR-2 itself become significantly upregulated. Different studies showed that PAR-2 activation of keratinocytes influences cell function by decreasing keratinocyte proliferation and inducing maturation of keratinocytes. Recent reports demonstrate the importance of PAR-2 for epidermal barrier homeostasis (Demerjian et al. 2008; Hachem et al. 2006). Keratinocytes of the stratum granulosum produce specific lipids (lamellar bodies) that prevent epidermal water loss. After acute permeability barrier disruption, e.g., by tape stripping, the pH of the stratum corneum increases from acidic (pH ~5) toward a more neutral pH. This, in turn, induces the activity of serine proteases, such as KLK5 and 14 (Brattsand et al. 2005; Hachem et al. 2003), which can activate PAR-2 leading to impairment of epidermal barrier recovery by downregulation of lamellar body secretion and induction of cornification. Jeong and colleagues observed that proteolytic allergens from cockroach and house dust mites delayed barrier recovery and lamellar body secretion *in vivo* (Jeong et al. 2008). Barrier recovery was normalized after application of a PAR-2 antagonist or protease inhibitors prior to epidermal barrier disruption. PAR-2 and the proteolytic activity of its activating proteases are tightly regulated by protease inhibitors during keratinocyte differentiation.

Different members of the KLK family have a broad spectrum of activity in the skin, and the proteolytic activity of proteases are regulated by different members of the serine protease inhibitor/lymphoepithelial-Kazal-type 5 inhibitor (Spink/LEKTI) family during keratinocyte differentiation (Deraison et al. 2007; Schechter et al. 2005). KLK5 and 14 are capable of activating PAR-2, which inhibits keratinocyte proliferation and induces the maturation of keratinocytes (Oikonomopoulou et al. 2006a, 2006b; Stefansson et al. 2008). KLK5 and KLK7 are important to initiate shedding of dead keratinocytes from the stratum granulosum by degrading corneodesmosomes (Brattsand et al. 2005). Therefore, it is important that the skin maintains a regulated proteolytic activity of KLKs during skin homeostasis. Consequently, during keratinocyte differentiation the gene expression of KLK inhibitors such as Spink5/LEKTI1, Spink9/LEKTI2, and Spink6/LEKTI3 are downregulated while KLK5, 7, and 14 as well as PAR-2 expression are upregulated. This fine-tuned interaction of proteases and protease inhibitors probably initiates keratinocyte differentiation and finally skin shedding (Deraison et al. 2007; Fischer et al. 2013; Reiss et al. 2011; Schechter et al. 2005). Vice versa, dysregulation of the LEKTI/KLK system leads to uncontrolled increased activity of KLKs, which, in turn, induces inflammatory reactions, scaly skin, and pruritus (Borgoño et al. 2007; Ishida-Yamamoto et al. 2005; Meyer-Hoffert et al. 2010). The impact of KLK dysregulation is best described by one form of a severe genetic skin disease, Netherton syndrome. Patients with this life-limiting form of ichthyosis harbor mutations in the Spink5 gene that results in the disability of LEKTI1 to inhibit KLK5 and 7 activities (Briot et al. 2009, 2010; Fortugno et al. 2011). The patients have severe skin barrier disruption defects leading to dehydration, chronic skin inflammation, itch, and a high risk of severe skin infections. Similar to patients with AD, Netherton syndrome patients and ichthyosis patients suffer from universal itch and develop a scaly, reddish skin (Briot et al. 2009, 2010; Fortugno et al. 2011). Of note, not only KLKs and protease inhibitors are important regulators of keratinocyte differentiation and skin barrier, but also PAR-2 and its activating proteases. This is of interest because the hyperkeratosis in different pruritic and inflammatory skin diseases (e.g., AD or Netherton syndrome) is accompanied by increased expression levels of PAR-2 and PAR-2-activating proteases (Briot et al. 2009, 2010; Elias and Steinhoff 2008).

Beside keratinocytes, the epidermis also contains melanocytes that produce melanin after sun exposure to protect the skin against UV radiation. Interestingly, PAR-2 has an additional important role for the melanin transport from melanocytes into keratinocytes. Upon UVB exposure, melanocytes produce and store melanin in so-called melanosomes, located in dendritic-like cell structures. Nearby keratinocytes constrict and phagocyte the melanosomes (Okazaki et al. 1976; Yamamoto and Bhawan 1994). UVB radiation enhances secretion of proteases into the extracellular matrix of the epidermis and thereby triggers PAR-2 activity on keratinocytes (Seiberg et al. 2000; Seiberg 2001). This process however leads to increased phagocytosis of melanosomes and increased skin tanning (Lin et al. 2008). Inhibition of PAR-2 activity on the other hand reduces pigment transfer after UVB exposure and also induces depigmentation of the skin (Seiberg et al. 2000; Seiberg 2001).

In the dermis, PAR-2 is expressed by dermal fibroblasts, mast cells, endothelial cells, and dermis/epidermis-innervating peripheral nerve endings. Human mast cells produce proinflammatory molecules such as histamine, proteoglycans, and PAR-2-activating proteases (e.g., tryptase and chymase). These mediators are stored in intracellular secretory granules that are released upon mast cell activation. Mast cells play an important role in several skin disorders including inflammation, hypersensitivity, and wound healing (Benoist and Mathis 2002). Although tryptase is able to stimulate PAR-2, the enzyme is a less potent activator of PAR-2 compared to trypsin, because tryptase activity seems to depend on receptor glycosylation and on the presence of sialic acid on the cell surface (Compton et al. 2002; Seymour et al. 2005). Interestingly, dermal mast cells themselves express PAR-2, suggesting a possible autocrine activation mechanism of PAR-2 on these cells. Upon allergen contact, mast cells release histamine that has several effects on the organism. Regarding neurogenic inflammation, it activates cutaneous sensory nerves and is therefore responsible for smooth muscle contraction, vasodilation, plasma extravasation, and instant itch sensations (Gazerani et al. 2009; Harvima et al. 2010; Steinhoff et al. 2003). Usually, histamine and tryptase are cosecreted upon mast cell stimulation. However, PAR-2 stimulation of mast cells did not seem to be followed by corelease of tryptase in skin mast cells (He et al. 2005; Moormann et al. 2006). In contrast, mast cells isolated from tonsils were observed to secrete tryptase upon PAR-2 stimulation but not histamine (He et al. 2005). Further investigations are needed to fully elucidate these somewhat contradictory observations. However, the PAR-2-profiling of mast cells differs between tissues; whereas human mast cells from the colon also express PAR-2; it was observed that lung mast cells were PAR-2-negative and did not respond to PAR-2 agonists (He et al. 2004, 2005).

11.4 FUNCTIONAL ROLE OF PAR-2 IN ITCH

In 1955, Arthur and Shelley published that the protease mucunain is the active compound of cowhage (*Mucuna pruriens*) and induces itch but not pain in humans (Arthur and Shelley 1955; Shelley and Arthur 1955). In particular, they observed that mucunain was responsible for a long lasting, histamine-independent itch (>30 min without wheal or flare in contrast to histamine-induced itch). Although the mechanism how mucunain induces itch was unknown, they already hypothesized that the protease may activate a "protease receptor," which is expressed by nerves or indirectly exerts its pruritic effects by inducing the release of a pruritogen from epidermal cells. Although this study was of significant importance, it took several years before Rajka and colleagues followed up on this idea and demonstrated that the proteases trypsin and chymotrypsin are capable of inducing itch in humans (Rajka et al. 1967, 1969). However, it was not before 1995 after the murine and in 1996 the human PAR-2 were cloned that the protease-mediated histamine-independent itch was on the cusp of being better understood (Boehm et al. 1996; Nystedt et al. 1995a, 1995b; Santulli et al. 1996).

In 2000, we were able to detect PAR-2 in terminal nerve endings of a subpopulation of polymodal peptidergic C nerve fibers in the skin, and we could demonstrate that rat dorsal root ganglion (DRG) neurons respond to nanomolar concentrations of trypsin or tryptase (Steinhoff et al. 2000). This was of interest because further studies confirmed the hypothesis that PAR-2 activation on DRGs induces not only neurogenic inflammation but also histamine-independent itch (Steinhoff et al. 1999, 2003). Our subsequent studies showed that PAR-2 expression in keratinocytes and skin nerve fibers were increased in patients with AD (Briot et al. 2010; Buddenkotte et al. 2005; Steinhoff et al. 2003). Interestingly, AD patients also display increased tryptase serum levels, which correlate with itch sensation (Steinhoff et al. 2003). Similarly, patients with renal insufficiency can suffer from severe pruritus because of insufficient clearance of urinary excreted and pruritogenic substances. Dugas-Breit and colleagues were able to show that the severity of itch in dialysis patients correlated well with the patients' increased serum levels of mast cell tryptase, suggesting an important role of mast cell tryptase and PAR-2 in the pathogenesis of renal pruritus (Dugas-Breit et al. 2005).

Subsequent studies in various skin mouse models helped to understand the molecular mechanisms behind PAR-2-induced itch and the role of PAR-2 in chronic pruritic diseases. In particular, intradermal injections of trypsin or tryptase were able to induce histamine-independent scratching behavior in mice (Costa et al. 2008; Tsujii et al. 2009; Ui et al. 2006). Using different protease inhibitors as well as antagonist or anti-PAR-2 antibodies, PAR-2 evoked itch sensation was significantly suppressed (Costa et al. 2008; Tsujii et al. 2009; Ui et al. 2006).

These observations can be understood by recent studies, showing that PAR-2 is expressed in a population of TRPV1-positive peptidergic nerve fibers (Amadesi et al. 2004, 2006). PAR-2-induced activation of PAR-2/TRPV1 nerve fibers induces the release of SP and calcitonin gene-related product (CGRP), which, in turn, induce secretion and activation of inflammatory mediators (as described below) and can increase itch sensitivity (Akiyama et al. 2013; Grant et al. 2007; Ständer et al. 2010;

Steinhoff et al. 2000). The expression of PAR-2 in keratinocytes seems to correlate with severity of pruritus in chronic itch diseases. Mouse models treated to develop chronic dry skin phenotype also developed an itch phenotype (Yosipovitch 2004). Similar to patients with AD or chronic eczema, these mice also developed hyper-keratosis and skin barrier dysfunction (Akiyama et al. 2010a, 2010b, 2010c, 2012; Yosipovitch 2004). Furthermore, the expression of PAR-2 and PAR-2-activating proteases was increased in the skin, and the skin was highly innervated by nerve fibers. In a skin barrier disruption mouse model, proteases from cockroach and mite allergens induced activation of PAR-2, and the mice presented a delayed epidermal barrier recovery (Jeong et al. 2008; Yosipovitch 2004). Another study used NC mice that develop an atopic-like dermatitis when housed in a nonpathogen free/conven-tional environment (Takahashi et al. 2005). In the lesional skin of these mice, trypt-ase activity was significantly increased (Tsujii et al. 2009). The intense scratching behavior of the NC mice was widely reduced by intravenous injection of the protease inhibitor nafamostat mesilate (Tsujii et al. 2009). In correlation to this study, other groups established that PAR-2 knockout mice display a reduced response in oxazo-lone and picryl chloride induced allergic dermatitis models (Kawagoe et al. 2002). These results have a highly translational value because patients with chronic eczema and AD suffer from allergen-induced itch that increases epidermal barrier disrup-tion. This can be partly explained by increased protease-activity on PAR-2 that is expressed by keratinocytes and skin nerve fibers (Ikoma et al. 2011; Lee et al. 2010; Steinhoff et al. 1999).

Different studies demonstrate that uncontrolled protease activity in the skin of mice induces atopic-dermatitis like phenotypes (Briot et al. 2009, 2010; Frateschi et al. 2011; Kim et al. 2012). It was shown in Spink5 knockout mice that increased and uncontrolled proteolytic activity of the serine proteases KLK5 and KLK7 induces a skin phenotype, which had similarity to the Netherton syndrome in humans (Briot et al. 2009). In addition, cathepsin S overexpressing mice and keratinocyte-specific prostasin transgenic mice developed atopic-like dermatitis (Kim et al. 2012). These data show that proteases are important for skin homeostasis and are tightly regu-lated to maintain a normal skin homeostasis. Dysregulation results in severe skin diseases accompanied with severe pruritus. However, a more recent study proved that the severe skin phenotype is driven by PAR-2 activity, as an epidermal PAR-2 over-expressing mouse model also results in an atopic-like dermatitis phenotype, including hyperkeratosis and skin barrier disruption (Frateschi et al. 2011). The mice display increased scratching behavior and develop skin lesion during adolescence. The tail and paw skin appears to be dry and scaly (Frateschi et al. 2011). We observed in these mice that the skin is highly innervated with nerve fibers and the mice develop hyperki-nesis and allokinesis (yet unpublished observation and part of upcoming manuscript).

In understanding PAR-2-induced itch, humans are maybe the preferred research object compared to animals as the intensity of itch sensation can be described by the object itself. Studies using cowhage showed that the protease mucunain induces PAR-2-related itch most intensely when the spicules were inserted down to the level of the basal membrane (Arthur and Shelley 1955; Shelley and Arthur 1955). Interestingly, itch was never induced when spicules were inserted into the skin area after the epidermis and upper dermis had been removed. These results are similar to histamine-evoked itch or pain. Intradermal injections of histamine induce itch sensation, but subcutaneous injection was rather painful then itchy (Rosenthal et al. 1977). Our own studies revealed that intradermal injection of a human PAR-2 agonist as well as tryptase induce pruritus in humans and that intralesional injection of tryptase enhances and prolongs itch in AD patients (Steinhoff et al. 2003). We used codeine to investigate the release of histamine and tryptase in healthy volunteers and AD patients, and thereby we were able to show that tryptase release was increased in AD patients and that it induced a histamine-independent itch in humans (Steinhoff et al. 2003). Therefore, tryptase could induce antihistamine treatment resistant itch in AD patients (Steinhoff et al. 2003).

11.5 DIFFERENCES BETWEEN PAR-2- AND MrgprC11-INDUCED ITCH

Over recent years, several studies investigating the role of murine PAR-2 in inflammation and itch utilized a peptide analog that is synthesized according to the TLS sequence of the murine PAR-2. The use of synthetic peptide sequences is frequently deployed to investigate activation and mechanism of PAR-induced cellular responses. An interesting publication of Liu and colleagues demonstrated recently that the murine-amidated peptide sequence SLIGRL-NH2 is not only capable of inducing PAR-2 activation but additionally addresses the Mas-related G-protein-coupled receptor MrgprC11 (Liu et al. 2011). This receptor is highly potent to activate Bam8-22-induced itch responses in mice (Sikand et al. 2011; Wilson et al. 2011). Liu and colleagues demonstrated that MrgprC11 could be mainly responsible for SLIGRL-NH2-induced scratching behavior in mice (Liu et al. 2011). The authors used both PAR-2 and MrgprC11 knockout mice for their studies. SLIGRL-NH2-induced itch was only abolished in MrgprC11 knockout mice but not in PAR-2 knockout mice. As an explanation, the authors pointed out that the activation of MrgprC11 is mainly induced by the RL-NH2 feature of the peptide structure. Proteases, such as trypsin, tryptase, or KLK, activate PAR-2 by cleaving the receptor upstream of the TLS sequence $\left(NH_3^+\text{-}AA_{1\text{-}37}GR/SLIGRL\text{-}EAA_{46\text{-}399}\text{-}COO^-\right)$ (Adams et al. 2011; Hollenberg et al. 2008; Ramachandran et al. 2009). Thereby, the receptor structure itself will be shortened to $NH_3^+\text{-}SLIGRL\text{-}EAA_{46\text{-}399}COO^-$, which induces activation of the PAR-2 receptor. This also means that the synthetic activating peptide sequence for MrgprC11 and PAR-2 does not occur naturally by proteolytic cleavage of the PAR-2 receptor in mice. Because there are three different synthetic PAR-2 peptide sequences commercially available (SLIGRL, SLIGRL-NH2, and fuoryl-SLIGRLO-NH2), future studies will need to furnish proof, if all three peptide sequences are potent PAR-2 and/or Mrgprc11 activators. Liu and colleagues presented evidence in their work that already the loss of the L amino acid (at position 6 after unmasking the TLS) inhibits the peptide's potential to activate MrgprC11. Interestingly, although the authors quantified an intense itch response to intradermal trypsin injection in PAR-2 knockout mice, the concentration of the used trypsin was significantly higher (8000 pmol) compared to the study of Ui and colleagues applying 75 pmol of the less potent PAR-2-activating protease tryptase (Liu et al. 2011; Ui et al. 2006). However, trypsin and tryptase are not only exclusive PAR-2 activators but also capable of accomplishing proteolytic cleavage of PAR-4 or PAR-1 as well as of other pre-proteins, which may have a pruritic effect distant from PARs (Antalis et al. 2011; Brown et al. 2006; Kahn et al. 1998). Furthermore, the pruritic response to intra-dermal injection of trypsin, tryptase, or the human synthetic agonist SLIGKV-NH2 are, as far as we know, MRGPRX1-independent (the human homolog of MrgprC11) (Lee et al. 2010; Steinhoff et al. 2003). Although Liu and colleagues describe SLIGRL-NH2 as a potent itch activator mainly functioning via MrgprC11, it is wrong to conclude that PAR-2 expressed in DRGs would be nonfunctional and not important for itch in both human and mice (Liu et al. 2011). An armada of publica-tions on this topic clearly favors the importance of proteases and neuronal PAR-2 function induced by proteases on itch, but not pain (Cormia and Dougherty 1960; Arthur and Shelley 1955; Shelley and Arthur 1955; Rajka 1967; Hägermark 1974; Steinhoff et al. 2003; Akiyama et al. 2012a, 2012b; Andoh et al. 2012a, 2012b; Kim et al. 2012; Frateschi et al. 2011; Olivry et al. 2013). Thus, the effect of a promis-cuous synthetic agonist (SLIGRL-NH2) cannot replace the observations revealed using natural proteases, PAR-2 overexpression approaches, functional studies using proteases and human studies. We also have to consider species differences, because human studies clearly indicate PAR-2 as an itch receptor in humans. Although we cannot exclude a protease-dependent but PAR-2-independent itch pathway in neurons, increased protease and PAR-2 expression in the skin of atopic dermatitis-like mouse models and AD patients suffering from severe itch sensations indeed underline the importance of PAR-2 in itch (Briot et al. 2009, 2010; Frateschi et al. 2011; Kim et al. 2012; Lee et al. 2010; Steinhoff et al. 2003). Further studies will be necessary to understand the molecular mechanism of itch and the role of PAR-2 in pruritic skin diseases.

11.6 FUNCTIONAL ROLE OF PAR-2 IN NEUROGENIC INFLAMMATION

Serine and cystein proteases, such as thrombin, kallikreins, matriptase, tryptase, trypsin, cathepsin G and S, have acute effects on the inflammatory response in the human body (Kim et al. 2012; Stefansson et al. 2008; Seitz et al. 2007; Steinhoff et al. 2005). Recruitment of downstream cellular signaling cascades of proteolytic-activated PAR-2 finally accelerates widespread inflammatory processes that are expressed by keratinocyte activation, vasodilation, extravasation of plasma proteins, and infiltration of neutrophils (Seeliger et al. 2003; Steinhoff et al. 2005; Briot et al. 2010). The precise underlying signaling and intercellular communication path-ways induced by PAR-2 in keratinocytes, dermal endothelial cells and in skin sen-sory nerves are not fully understood as of yet. The NF-kB pathway appears to play an essential role in PAR-2 mediated inflammation of skin diseases (Shpacovitch et al. 2002; Budenkotte et al. 2005; Moormann et al. 2006; Macfarlane et al. 2005; Goon et al. 2008; Dejean et al. 2012). A key component of PAR-2 induced acute and chronic inflammation appears to be a secretion of Neuropeptides such as SP and CGRP from sensory nerves in the skin (Steinhoff et al. 2000; Vergnolle et al. 2001). Histological analysis of DRG coherently reveals a high amount of PAR-2$^+$ neurons coexpressing CGRP or SP, respectively (Steinhoff et al. 2000). That serine proteases engage in neurogenic inflammation has been shown for example for tryptase. In cutaneous tissue, tryptase originated from mast cells activates PAR-2 on sensory afferents leading to secretion of CGRP and SP, both neuropeptides capable of initiating inflammatory processes through their corresponding target receptors, most likely CGRP1 receptor (CGRR2 also exists) and the NK1R, respectively (Steinhoff et al. 2000, 2003; Obreja et al. 2005; Costa et al. 2008). In a dual mechanistic mode, CGRP induces vasodilation in arteriolar vessels resulting in erythema (flare), while SP induces edema (wheal) by activating postcapillary venules, the prerequisite for immune cell infiltration into the inflammatory tissue. CGRP further boosts SP release from cutaneous nerve terminals and protects SP from degradation by neutral endopeptidases, thereby amplifying SP-induced neurogenic inflammation (Scholzen et al. 2001; Schlereth et al. 2013). Simultaneously, SP activates residing mast cells to release tryptase that not only can activate PAR-2 but inactivates CGRP to counteract the CGRP-sustained inflammatory spiral. These tryptase-releasing mast cells *are not* only found in close proximity to PAR-2 expressing cells such as C-fibers, dermal endothelial cells, or keratinocytes (D'Andrea et al. 1998, 2000; Steinhoff et al. 2003) during inflammation, but themselves express PAR-2 (also PAR-1) (Moormann et al. 2006). Therefore, mast cells may release tryptase and stimulate inflammatory or propruritic responses in mast cells in an autocrine or paracrine fashion via PAR-2. A role for the tryptase/PAR-2 ligand-receptor axis might be of importance in the pathophysiology of AD because tryptase levels in this skin disease are significantly upregulated (Steinhoff et al. 2003; Hallgren and Pejler 2006).

For decades, the preventing effect of repeatedly applied capsaicin to ameliorate neurogenic inflammation (as does denervation) is well established (Jancsó et al. 1967), but it took further decades to identify the main receptor target of that natural compound in spicy pepper, namely TRPV1 (Caterina et al. 1997). The concept of desensitization of neuronal receptors to chemical stimuli that parallels the failure of the stimuli to cause pain or an inflammatory response was later commonly applied to investigate agonist/receptor tandems that induce neurogenic inflammation. In the acute setting, capsaicin leads to nociceptive behavior as well as neurogenic inflammation implicating a close functional cross talk on receptor level between TRPV1 and PAR-2. Cotreatment experiments of PAR-2-agonists with capsaicin effectively support a partial dependency of PAR-2/TRPV1 in conducting their function. Several groups observe a PAR-2-dependent potentiation of TRPV1 activity in protease induced inflammatory pain (Dai et al. 2004), hyperalgesia (Amadesi et al. 2004), and neurogenic inflammation (Hoogerwerf et al. 2001). A recent study aiming to further solve the signaling mechanism by which PAR-2 mediates neurogenic inflammation identifies coupling of the PAR to TRPV4, PAR-2-induced generation of arachidonic acid-derived lipid mediators such as 5′-6′-EET and tyrosine phosphorylation (Tyr110) of TRPV4 as important steps in this process (Poole et al. 2013). Also, activation and translocation of PKD1 to 3 are described to engage in the process of PAR-2-initiated neurogenic inflammation, which in part appears to be further dependent on PKCε (Amadesi et al. 2009). A functional interaction of PAR-2 and TRPA1 has been described in DRG neurons where PAR-2 activation (possibly through trypsin or tryptase released in response to tissue inflammation) leads to phospholipase C (PLC) as well as phosphotidylinositol 4,5-bisphosphate (PIP2) activation. Subsequent

TRPA1 sensitization then leads to PAR-2/TRPA1-mediated inflammatory pain (Dai et al. 2007). The interaction between PAR-2 and TRPV1 or TRPA1 channels has to await further investigation.

The PAR-2-triggered inflammatory response to serine proteases certainly comprises a strong neurogenic element, but a nonneurogenic component cannot be neglected because administration of CGRP and SP receptor antagonists fail to completely ameliorate symptoms of inflammation as, e.g., formation of edema. This is supported by findings in skin tissues where PAR-2 agonists activate intracellular signaling cascades leading to activation and nuclear translocation of the proinflammatory transcription factor NF-κB (Buddenkotte et al. 2005). PAR-2-induced activation of NF-κB is crucial for the upregulation of cell adhesion molecules (e.g., ICAM-1 [Shpacovitch et al. 2002; Buddenkotte et al. 2005] and E-selectin [Seeliger et al. 2003]). Of note, both PAR-2 as well as NF-kB appear to be critical for the pathophysiology of atopic dermatitis (Pastore et al. 2000; Steinhoff et al. 2003; Macfarlane et al. 2005; Goon et al. 2008). In addition, PAR-2 induces release of cytokines (IL-6, IL-8, GRO-α) and prostaglandins (LTB4, PGE2) from keratinocytes or dermal endothelium (Hou et al. 1998; Shpacovitch et al. 2002; Takei-Taniguchi et al. 2012; Zhu et al. 2009b). Especially, cytokines may also contribute to neurogenic inflammation, pain, and pruritus as receptors bound by the IL-6 cytokine family [e.g., oncostatin M receptor (OSMR), leukemia inhibitory factor receptor (LIFR)] are expressed by cutaneous sensory neurons. Recent studies utilizing PAR-2-deficient mice further demonstrate that PAR-2 affected cutaneous inflammation *in vivo i*s probably mediated by release of the vasodilator nitric monoxide and involves SP and NF-kB (Seeliger et al. 2003).

A detailed picture of the regulation of the neurogenic inflammation elicited by PAR-2 agonists has not been fully painted yet, but the physiological control of cell responses to various inflammatory stimuli will require regulation at several levels. The first level of regulation affects the operational capability of PAR-2 agonist prior to receptor activation, which is commonly based on the balance between agonist (protease) inhibitors and agonist (protease). Dysregulation of such balance often leads to severe disease states, for example, overexpressed CAP1/Prss8 (channel-activating protease-1/protease serine S1 family member 8) causes ichthyosis and a dysbalance of the PAR-2 activating KLK5 serine protease and its inhibitor LEKTI contributes to atopic lesions in Netherton syndrome (Briot et al. 2009). A second level of regulating neurogenic inflammation processes is directed against the receptor by multiple strategies. In the way PAR-2 sensitizes TRPV4 to cause mechanical hyperalgesia in mice (Grant et al. 2007), and interacts with TRPA1 in mouse DRG neurons to sensitize for inflammatory pain (Dai et al. 2007), PAR-2 function itself on keratinocytes could be regulated through such sensitization or, in contrast, desensitization. Further possible ways to modify PAR-2 function/PAR-2 activation on the receptor level include receptor trafficking/recycling, ligand uncoupling from receptor binding site, modulation of ligand/receptor affinity or endocytosis by the extracellular microenvironment. On a third, the intracellular level, the modulation of PAR-2-induced signal transduction and their amplification/reduction mechanisms in nerves, endothelial cells and keratinocytes may significantly influence downstream signaling and regulation of transcription factors involved in PAR-2-mediated inflammation and itch. The impact of these molecular mechanisms in human disease, however, is very poorly understood.

11.7 ROLE OF PAR-4 IN ITCH

Recent studies also suggest a possible role of PAR-4 in inflammation and itch. In 2008, Tsujii and colleagues demonstrated that PAR-4 stimulation induced histamine-depending scratching behavior in mice (Tsujii et al. 2008). However, the same authors also published later, that a specific PAR-4 agonist failed to induce PAR-4-specific scratching behavior in mice (Tsujii et al. 2009). In the same year, Akiyama and colleagues presented a study that showed that PAR-4 and histamine-induced scratching behavior seem to be independent from each other, as the authors have not observed cross-tachyphylaxis between PAR-4 agonist and histamine stimulation (Akiyama et al. 2009, 2010c). We are only at the beginning of understanding the role of PAR-4 as a potential itch receptor, and further studies will need to show if PAR-4-induced itch depends on PAR-4 activation and if the expression of PAR-4 ligands and the receptor itself is modulated during inflammation and itch on sensory nerve endings or other cell types in the skin.

11.8 CONCLUSIONS AND FUTURE DIRECTIONS

Various proteases from animals, plants, and skin cells themselves are important itch inducers that mainly function via PAR-2 in the rodent and human skin. PAR-2 is upregulated in various acute and chronic pruritic diseases in mice and humans, and the increased generation and release of certain kallikreins, tryptase, or prostasin correlating with itch intensity supports the concept that proteases via PAR-2 are important histamine-independent inducers in acute and chronic itch. Although the exact molecular mechanisms of PAR-2-induced itch are still poorly understood, its clinical relevance in severe, pruritic human skin diseases is indisputable. However, we are still at the beginning of understanding PAR-2-induced itch and neurogenic inflammation in the skin and other organs including the lung and gastrointestinal tract or brain. Further studies will be necessary to demonstrate if PAR-2 activation on sensory nerves in the skin is sufficient to directly induce itch in humans, or if a complex cascade involving keratinocytes and/or immune cells in the skin is also involved, or even more important for the PAR-2-induced, histamine-independent itch.

REFERENCES

Adams MN, Ramachandran R, Yau MK et al. 2011. Structure, function and pathophysiology of protease activated receptors. *Pharmacol Ther.* 130(3):248–82.
Akiyama T, Carstens MI, Carstens E. 2010a. Enhanced scratching evoked by PAR-2 agonist and 5-HT but not histamine in a mouse model of chronic dry skin itch. *Pain.* 151(2):378–83.
Akiyama T, Carstens MI, Carstens E. 2010b. Facial injections of pruritogens and algogens excite partly overlapping populations of primary and second-order trigeminal neurons in mice. *J Neurophysiol.* 104(5):2442–50.
Akiyama T, Carstens MI, Carstens E. 2010c. Differential itch- and pain-related behavioral responses and μ-opoid modulation in mice. *Acta Derm Venereol.* 90(6):575–81.

Akiyama T, Carstens MI, Ikoma A et al. 2012. Mouse model of touch-evoked itch (alloknesis). *J Invest Dermatol.* 132(7):1886–91.

Akiyama T, Merrill AW, Zanotto K, Carstens MI, Carstens E. 2009. Scratching behavior and Fos expression in superficial dorsal horn elicited by protease-activated receptor agonists and other itch mediators in mice. *J Pharmacol Exp Ther.* 329(3):945–51.

Akiyama T, Tominaga M, Davoodi A et al. 2013. Roles for substance P and gastrin-releasing peptide as neurotransmitters released by primary afferent pruriceptors. *J Neurophysiol.* 109(3):742–8.

Amadesi S, Cottrell GS, Divino L et al. 2006. Protease-activated receptor 2 sensitizes TRPV1 by protein kinase Cepsilon- and A-dependent mechanisms in rats and mice. *J Physiol.* 575(Pt 2):555–71.

Amadesi S, Grant AD, Cottrell GS et al. 2009. Protein kinase D isoforms are expressed in rat and mouse primary sensory neurons and are activated by agonists of protease-activated receptor 2. *J Comp Neurol.* 516(2):141–56.

Amadesi S, Nie J, Vergnolle N et al. 2004. Protease-activated receptor 2 sensitizes the capsaicin receptor transient receptor potential vanilloid receptor 1 to induce hyperalgesia. *J Neurosci.* 24(18):4300–12.

Antalis TM, Bugge TH, Wu Q. 2011. Membrane-anchored serine proteases in health and disease. *Prog Mol Biol Transl Sci.* 99:1–50.

Arthur RP, Shelley WB. 1955. The role of proteolytic enzymes in the production of pruritus in man. *J Invest Dermatol.* 25(5):341–6.

Babiarz-Magee L, Chen N, Seiberg M, Lin CB. 2004. The expression and activation of protease-activated receptor-2 correlate with skin color. *Pigment Cell Res.* 17(3):241–51.

Benoist C, Mathis D. 2002. Mast cells in autoimmune disease. *Nature.* 420(6917):875–8.

Bocheva G, Rattenholl A, Kempkes C et al. 2009. Role of matriptase and proteinase-activated receptor-2 in nonmelanoma skin cancer. *J Invest Dermatol.* 129(7):1816–23. doi:10.1038/jid.2008.449.

Böhm SK, Khitin LM, Grady EF et al. 1996a. Mechanisms of desensitization and resensitization of proteinase-activated receptor-2. *J Biol Chem.* 271(36):22003–16.

Böhm SK, Kong W, Bromme D et al. 1996b. Molecular cloning, expression and potential functions of the human proteinase-activated receptor-2. *Biochem J.* 314(Pt 3):1009–16.

Borgoño CA, Michael IP, Komatsu N et al. 2007. A potential role for multiple tissue kallikrein serine proteases in epidermal desquamation. *J Biol Chem.* 282(6):3640–52.

Brattsand M, Stefansson K, Lundh C, Haasum Y, Egelrud T. 2005. A proteolytic cascade of kallikreins in the stratum corneum. *J Invest Dermatol.* 124(1):198–203.

Briot A, Deraison C, Lacroix M et al. 2009. Kallikrein 5 induces atopic dermatitis-like lesions through PAR2-mediated thymic stromal lymphopoietin expression in Netherton syndrome. *J Exp Med.* 206(5):1135–47.

Briot A, Lacroix M, Robin A et al. 2010. Par2 inactivation inhibits early production of TSLP, but not cutaneous inflammation, in Netherton syndrome adult mouse model. *J Invest Dermatol.* 130(12):2736–42.

Brown JK, Hollenberg MD, Jones CA. 2006. Tryptase activates phosphatidylinositol 3-kinases proteolytically independently from proteinase-activated receptor-2 in cultured dog airway smooth muscle cells. *Am J Physiol Lung Cell Mol Physiol.* 290(2):L259–69.

Buddenkotte J, Steinhoff M. 2010. Pathophysiology and therapy of pruritus in allergic and atopic diseases. *Allergy.* 65(7):805–21.

Buddenkotte J, Stroh C, Engels IH et al. 2005. Agonists of proteinase-activated receptor-2 stimulate upregulation of intercellular cell adhesion molecule-1 in primary human keratinocytes via activation of NF-kappa B. *J Invest Dermatol.* 124(1):38–45.

Caterina MJ, Leffler A, Malmberg AB et al. 2000. Impaired nociception and pain sensation in mice lacking the capsaicin receptor. *Science.* 288(5464):306–13.

Caterina MJ, Schumacher MA, Tominaga M et al. 1997. The capsaicin receptor: A heat-activated ion channel in the pain pathway. *Nature.* 389(6653):816–24.

Cattaruzza F, Lyo V, Jones E et al. 2011. Cathepsin S is activated during colitis and causes visceral hyperalgesia by a PAR2-dependent mechanism in mice. *Gastroenterology.* 141(5):1864–74.e1–3.

Compton SJ, McGuire JJ, Saifeddine M, Hollenberg MD. 2002. Restricted ability of human mast cell tryptase to activate proteinase-activated receptor-2 in rat aorta. *Can J Physiol Pharmacol.* 80(10):987–92.

Cormia FE, Dougherty JW. 1960. Proteolytic activity in development of pain and itching. Cutaneous reactions to bradykinin and kallikrein. *J Invest Dermatol.* 35:21–6.

Costa R, Marotta DM, Manjavachi MN et al. 2008. Evidence for the role of neurogenic inflammation components in trypsin-elicited scratching behaviour in mice. *Br J Pharmacol.* 154(5):1094–103.

Cottrell GS, Coelho AM, Bunnett NW. 2002. Protease-activated receptors: The role of cell-surface proteolysis in signalling. *Essays Biochem.* 38:169–83.

Coughlin SR, Vu TK, Hung DT, Wheaton VI. 1992. Expression cloning and characterization of a functional thrombin receptor reveals a novel proteolytic mechanism of receptor activation. *Semin Thromb Hemost.* 18(2):161–6.

Dai Y, Moriyama T, Higashi T et al. 2004. Proteinase-activated receptor 2-mediated potentiation of transient receptor potential vanilloid subfamily 1 activity reveals a mechanism for proteinase-induced inflammatory pain. *J Neurosci.* 24(18):4293–9.

Dai Y, Wang S, Tominaga M et al. 2007. Sensitization of TRPA1 by PAR2 contributes to the sensation of inflammatory pain. *J Clin Invest.* 117(7):1979–87.

D'Andrea MR, Derian CK, Leturcq D et al. 1998. Characterization of protease-activated receptor-2 immunoreactivity in normal human tissues. *J Histochem Cytochem.* 46(2):157–64.

D'Andrea MR, Rogahn CJ, Andrade-Gordon P. 2000. Localization of protease-activated receptors-1 and -2 in human mast cells: Indications for an amplified mast cell degranulation cascade. *Biotech Histochem.* 75(2):85–90.

Dejean E, Foisseau M, Lagarrigue F et al. 2012. ALK+ALCLs induce cutaneous, HMGB-1-dependent IL-8/CXCL8 production by keratinocytes through NF-κB activation. *Blood.* 119(20):4698–707.

Demerjian M, Hachem JP, Tschachler E et al. 2008. Acute modulations in permeability barrier function regulate epidermal cornification: Role of caspase-14 and the protease-activated receptor type 2. *Am J Pathol.* 172(1):86–97.

Deraison C, Bonnart C, Lopez F et al. 2007. LEKTI fragments specifically inhibit KLK5, KLK7, and KLK14 and control desquamation through a pH-dependent interaction. *Mol Biol Cell.* 18(9):3607–19.

Derian CK, Eckardt AJ, Andrade-Gordon P. 1997. Differential regulation of human keratinocyte growth and differentiation by a novel family of protease-activated receptors. *Cell Growth Differ.* 8(7):743–9.

Déry O, Thoma MS, Wong H, Grady EF, Bunnett NW. 1999. Trafficking of proteinase-activated receptor-2 and beta-arrestin-1 tagged with green fluorescent protein. Beta-Arrestin-dependent endocytosis of a proteinase receptor. *J Biol Chem.* 274(26):18524–35.

Dugas-Breit S, Schöpf P, Dugas M et al. 2005. Baseline serum levels of mast cell tryptase are raised in hemodialysis patients and associated with severity of pruritus. *J Dtsch Dermatol Ges.* 3(5):343–7.

Elias PM, Steinhoff M. 2008. "Outside-to-inside" (and now back to "outside") pathogenic mechanisms in atopic dermatitis. *J Invest Dermatol.* 128(5):1067–70.

Fischer J, Koblyakova Y, Latendorf T, Wu Z, Meyer-Hoffert U. 2013. Cross-linking of SPINK6 by transglutaminases protects from epidermal proteases. *J Invest Dermatol.* 133(5):1170–7.

Fortugno P, Bresciani A, Paolini C et al. 2011. Proteolytic activation cascade of the Netherton syndrome-defective protein, LEKTI, in the epidermis: Implications for skin homeostasis. *J Invest Dermatol.* 131(11):2223–32.

Frateschi S, Camerer E, Crisante G et al. 2011. PAR2 absence completely rescues inflammation and ichthyosis caused by altered CAP1/Prss8 expression in mouse skin. *Nat Commun.* 2:161.

Gandhi PS, Chen Z, Appelbaum E, Zapata F, Di Cera E. 2011. Structural basis of thrombin-protease-activated receptor interactions. *IUBMB Life.* 63(6):375–82.

Gazerani P, Pedersen NS, Drewes AM, Arendt-Nielsen L. 2009. Botulinum toxin type A reduces histamine-induced itch and vasomotor responses in human skin. *Br J Dermatol.* 161(4):737–45.

Goerge T, Barg A, Schnaeker EM et al. 2006. Tumor-derived matrix metalloproteinase-1 targets endothelial proteinase-activated receptor 1 promoting endothelial cell activation. *Cancer Res.* 66(15):7766–74.

Goon Goh F, Sloss CM, Cunningham MR et al. 2008. G-protein-dependent and -independent pathways regulate proteinase-activated receptor-2 mediated p65 NFkappaB serine 536 phosphorylation in human keratinocytes. *Cell Signal.* 20(7):1267–74.

Grant AD, Cottrell GS, Amadesi S et al. 2007. Protease-activated receptor 2 sensitizes the transient receptor potential vanilloid 4 ion channel to cause mechanical hyperalgesia in mice. *J Physiol.* 578(Pt 3):715–33.

Hachem JP, Crumrine D, Fluhr J et al. 2003. pH directly regulates epidermal permeability barrier homeostasis, and stratum corneum integrity/cohesion. *J Invest Dermatol.* 121(2):345–53.

Hachem JP, Houben E, Crumrine D et al. 2006. Serine protease signaling of epidermal permeability barrier homeostasis. *J Invest Dermatol.* 126(9):2074–86.

Hägermark O. 1974. Studies on experimental itch induced by kallikrein and bradykinin. *Acta Derm Venereol.* 54(5):397–400.

Hägermark D, Rajka G, Berqvist U. 1972. Experimental itch in human skin elicited by rat mast cell chymase. *Acta Derm Venereol.* 52(2):125–8.

Hallgren J, Pejler G. 2006. Biology of mast cell tryptase. An inflammatory mediator. *FEBS J.* 273(9):1871–95.

Hansson L, Bäckman A, Ny A et al. 2002. Epidermal overexpression of stratum corneum chymotryptic enzyme in mice: A model for chronic itchy dermatitis. *J Invest Dermatol.* 118(3):444–9.

Harvima IT, Nilsson G, Naukkarinen A. 2010. Role of mast cells and sensory nerves in skin inflammation. *G Ital Dermatol Venereol.* 145(2):195–204.

He SH, Chen HQ, Zheng J. 2004. Inhibition of tryptase and chymase induced nucleated cell infiltration by proteinase inhibitors. *Acta Pharmacol Sin.* 25(12):1677–84.

He SH, Xie H, Fu YL. 2005. Activation of human tonsil and skin mast cells by agonists of proteinase activated receptor-2. *Acta Pharmacol Sin.* 26(5):568–74.

Hollenberg MD, Oikonomopoulou K, Hansen KK et al. 2008. Kallikreins and proteinase-mediated signaling: Proteinase-activated receptors (PARs) and the pathophysiology of inflammatory diseases and cancer. *Biol Chem.* 389(6):643–51.

Hong J, Buddenkotte J, Berger TG, Steinhoff M. 2011. Management of itch in atopic dermatitis. *Semin Cutan Med Surg.* 30(2):71–86.

Hoogerwerf WA, Zou L, Shenoy M et al. 2001. The proteinase-activated receptor 2 is involved in nociception. *J Neurosci.* 21(22):9036–42.

Hou L, Kapas S, Cruchley AT et al. 1998. Immunolocalization of protease-activated receptor-2 in skin: Receptor activation stimulates interleukin-8 secretion by keratinocytes in vitro. *Immunology.* 94(3):356–62.

Ikoma A, Cevikbas F, Kempkes C, Steinhoff M. 2011. Anatomy and neurophysiology of pruritus. *Semin Cutan Med Surg.* 30(2):64–70.

Imamachi N, Park GH, Lee H et al. 2009. TRPV1-expressing primary afferents generate behavioral responses to pruritogens via multiple mechanisms. *Proc Natl Acad Sci USA.* 106(27):11330–5.

Ishida-Yamamoto A, Deraison C, Bonnart C et al. 2005. LEKTI is localized in lamellar granules, separated from KLK5 and KLK7, and is secreted in the extracellular spaces of the superficial stratum granulosum. *J Invest Dermatol.* 124(2):360–6.

Jancsó N, Jancsó-Gábor A, Szolcsányi J. 1967. Direct evidence for neurogenic inflammation and its prevention by denervation and by pretreatment with capsaicin. *Br J Pharmacol Chemother.* 31(1):138–51.

Jeong SK, Kim HJ, Youm JK et al. 2008. Mite and cockroach allergens activate protease-activated receptor 2 and delay epidermal permeability barrier recovery. *J Invest Dermatol.* 128(8):1930–9.

Kahn ML, Zheng YW, Huang W et al. 1998. A dual thrombin receptor system for platelet activation. *Nature.* 394(6694):690–4.

Kato T, Takai T, Fujimura T et al. 2009. Mite serine protease activates protease-activated receptor-2 and induces cytokine release in human keratinocytes. *Allergy.* 64(9):1366–74.

Kauffman HF, Tamm M, Timmerman JA, Borger P. 2006. House dust mite major allergens Der p 1 and Der p 5 activate human airway-derived epithelial cells by protease-dependent and protease-independent mechanisms. *Clin Mol Allergy.* 4:5.

Kawagoe J, Takizawa T, Matsumoto J et al. 2002. Effect of protease-activated receptor-2 deficiency on allergic dermatitis in the mouse ear. *Jpn J Pharmacol.* 88(1):77–84.

Kawakami T, Kaminishi K, Soma Y, Kushimoto T, Mizoguchi M. 2006. Oral antihistamine therapy influences plasma tryptase levels in adult atopic dermatitis. *J Dermatol Sci.* 43(2):127–34.

Kempkes C, Rattenholl A, Buddenkotte J et al. 2012. Proteinase-activated receptors 1 and 2 regulate invasive behavior of human melanoma cells via activation of protein kinase D1. *J Invest Dermatol.* 132(2):375–84.

Kido M, Buddenkotte J, Kempkes C et al. 2011. Endothelin-converting enzyme-1 modulates endothelin-1-induced pruritus by inhibiting endothelin receptor recycling and prolonging ERK1/2 signaling in murine sensory neurons. *J Invest Dermatol.* 131:S44.

Kim N, Bae KB, Kim MO et al. 2012. Overexpression of cathepsin S induces chronic atopic dermatitis in mice. *J Invest Dermatol.* 132(4):1169–76.

Lam DK, Dang D, Zhang J, Dolan JC, Schmidt BL. 2012. Novel animal models of acute and chronic cancer pain: A pivotal role for PAR2. *J Neurosci.* 32(41):14178–83.

Lee SE, Jeong SK, Lee SH. 2010. Protease and protease-activated receptor-2 signaling in the pathogenesis of atopic dermatitis. *Yonsei Med J.* 51(6):808–22.

Lin CB, Chen N, Scarpa R et al. 2008. LIGR, a protease-activated receptor-2-derived peptide, enhances skin pigmentation without inducing inflammatory processes. *Pigment Cell Melanoma Res.* 21(2):172–83.

Lin H, Trejo J. 2013. Transactivation of the PAR1-PAR2 heterodimer by thrombin elicits β-arrestin-mediated endosomal signaling. *J Biol Chem.* 288(16):11203–15.

Liu Q, Weng HJ, Patel KN et al. 2011. The distinct roles of two GPCRs, MrgprC11 and PAR2, in itch and hyperalgesia. *Sci Signal.* 4(181):ra45.

Lohman RJ, Cotterell AJ, Barry GD et al. 2012. An antagonist of human protease activated receptor-2 attenuates PAR2 signaling, macrophage activation, mast cell degranulation, and collagen-induced arthritis in rats. *FASEB J.* 26(7):2877–87.

Lutfi R, Lewkowich IP, Zhou P et al. 2012. The role of protease-activated receptor-2 on pulmonary neutrophils in the innate immune response to cockroach allergen. *J Inflamm.* 9(1):32.

Ma Q. 2010. Labeled lines meet and talk: Population coding of somatic sensations. *J Clin Invest.* 120(11):3773–8.

Macfarlane SR, Sloss CM, Cameron P, Kanke T, McKenzie RC, Plevin R. 2005. The role of intracellular Ca2$^+$ in the regulation of proteinase-activated receptor-2 mediated nuclear factor kappa B signaling in keratinocytes. *Br J Pharmacol.* 145(4):535–44.

Meyer-Hoffert U, Wu Z, Kantyka T et al. 2010. Isolation of SPINK6 in human skin: Selective inhibitor of kallikrein-related peptidases. *J Biol Chem.* 285(42):32174–81.

Moormann C, Artuc M, Pohl E et al. 2006. Functional characterization and expression analysis of the proteinase-activated receptor-2 in human cutaneous mast cells. *J Invest Dermatol.* 126(4):746–55.

Nichols HL, Saffeddine M, Theriot BS et al. 2012. β-Arrestin-2 mediates the proinflammatory effects of proteinase-activated receptor-2 in the airway. *Proc Natl Acad Sci USA.* 109(41):16660–5.

Ny A, Egelrud T. 2003. Transgenic mice over-expressing a serine protease in the skin: Evidence of interferon gamma-independent MHC II expression by epidermal keratinocytes. *Acta Derm Venereol.* 83(5):322–7.

Nystedt S, Emilsson K, Larsson AK, Strömbeck B, Sundelin J. 1995a. Molecular cloning and functional expression of the gene encoding the human proteinase-activated receptor 2. *Eur J Biochem.* 232(1):84–9.

Nystedt S, Larsson AK, Aberg H, Sundelin J. 1995b. The mouse proteinase-activated receptor-2 cDNA and gene. Molecular cloning and functional expression. *J Biol Chem.* 270(11):5950–55.

Obreja O, Biasio W, Andratsch M et al. 2005. Fast modulation of heat-activated ionic current by proinflammatory interleukin 6 in rat sensory neurons. *Brain.* 128(Pt 7):1634–41.

O'Brien PJ, Prevost N, Molino M et al. 2000. Thrombin responses in human endothelial cells. Contributions from receptors other than PAR1 include the transactivation of PAR2 by thrombin-cleaved PAR1. *J Biol Chem.* 275(18):13502–9.

Oikonomopoulou K, Hansen KK, Saifeddine M et al. 2006a. Kallikrein-mediated cell signalling: Targeting proteinase-activated receptors (PARs). *Biol Chem.* 387(6):817–24.

Oikonomopoulou K, Hansen KK, Saifeddine M et al. 2006b. Proteinase-mediated cell signalling: Targeting proteinase-activated receptors (PARs) by kallikreins and more. *Biol Chem.* 387(6):677–85.

Okazaki K, Uzuka M, Morikawa F, Toda K, Seiji M. 1976. Transfer mechanism of melanosomes in epidermal cell culture. *J Invest Dermatol.* 67(4):541–7.

Olivry T, Bizikova P, Paps JS, Dunston S, Lerner EA, Yosipovitch G. 2013. Cowhage can induce itch in the atopic dog. *Exp Dermatol.* 22(6):435–7.

Ossovskaya VS, Bunnett NW. 2004. Protease-activated receptors: Contribution to physiology and disease. *Physiol Rev.* 84(2):579–621.

Pastore S, Giustizieri ML, Mascia F et al. 2000. Dysregulated activation of activator protein 1 in keratinocytes of atopic dermatitis patients with enhanced expression of granulocyte/macrophage-colony stimulating factor. *J Invest Dermatol.* 115(6):1134–43.

Poole DP, Amadesi S, Veldhuis NA et al. 2013. Protease-activated receptor 2 (PAR2) protein and transient receptor potential vanilloid 4 (TRPV4) protein coupling is required for sustained inflammatory signaling. *J Biol Chem.* 288(8):5790–802.

Rajka G. 1967. Itch duration in the involved skin of atopic dermatitis (prurigo Besnier). *Acta Derm Venereol.* 47(3):154–7.

Rajka G. 1969. Latency and duration of pruritus elicited by trypsin in aged patients with itching eczema and psoriasis. *Acta Derm Venereol.* 49(4):401–3.

Ramachandran R, Mihara K, Mathur M et al. 2009. Agonist-biased signaling via proteinase activated receptor-2: Differential activation of calcium and mitogen-activated protein kinase pathways. *Mol Pharmacol.* 76(4):791–801.

Reddy VB, Iuga AO, Shimada SG et al. 2008. Cowhage-evoked itch is mediated by a novel cysteine protease: A ligand of protease-activated receptors. *J Neurosci.* 28(17):4331–5.

Reiss K, Meyer-Hoffert U, Fischer J et al. 2011. Expression and regulation of murine SPINK12, a potential orthologue of human LEKTI2. *Exp Dermatol.* 20(11):905–10.

Roosterman D, Cottrell GS, Schmidlin F, Steinhoff M, Bunnett NW. 2004. Recycling and resensitization of the neurokinin 1 receptor. Influence of agonist concentration and Rab GTPases. *J Biol Chem.* 279(29):30670–9.

Roosterman D, Goerge T, Schneider SW, Bunnett NW, Steinhoff M. 2006. Neuronal control of skin function: The skin as a neuroimmunoendocrine organ. *Physiol Rev.* 86(4): 1309–79.

Roosterman D, Schmidlin F, Bunnett NW. 2003. Rab5a and rab11a mediate agonist-induced trafficking of protease-activated receptor 2. *Am J Physiol Cell Physiol.* 284(5):C1319–29.

Rosenthal S, Schwartz JH, Canellos GP. 1977. Basophilic chronic granulocytic leukaemia with hyperhistaminaemia. *Br J Haematol.* 36(3):367–72.

Santulli RJ, Derian CK, Darrow AL et al. 1995. Evidence for the presence of a protease-activated receptor distinct from the thrombin receptor in human keratinocytes. *Proc Natl Acad Sci USA.* 92(20):9151–5.

Schechter NM, Brass LF, Lavker RM, Jensen PJ. 1998. Reaction of mast cell proteases tryptase and chymase with protease activated receptors (PARs) on keratinocytes and fibroblasts. *J Cell Physiol.* 176(2):365–73.

Schechter NM, Choi EJ, Wang ZM et al. 2005. Inhibition of human kallikreins 5 and 7 by the serine protease inhibitor lympho-epithelial Kazal-type inhibitor (LEKTI). *Biol Chem.* 386(11):1173–84.

Schlereth T, Breimhorst M, Werner N et al. 2013. Inhibition of neuropeptide degradation suppresses sweating but increases the area of the axon reflex flare. *Exp. Dermatol.* 22(4):299–301.

Schuepbach RA, Riewald M. 2010. Coagulation factor Xa cleaves protease-activated receptor-1 and mediates signaling dependent on binding to the endothelial protein C receptor. *J Thromb Haemost.* 8(2):379–88.

Seeliger S, Derian CK, Vergnolle N et al. 2003. Proinflammatory role of proteinase-activated receptor-2 in humans and mice during cutaneous inflammation in vivo. *FASEB J.* 17(13):1871–85.

Seiberg M. 2001. Keratinocyte-melanocyte interactions during melanosome transfer. *Pigment Cell Res.* 14(4):236–42.

Seiberg M, Paine C, Sharlow E et al. 2000. The protease-activated receptor 2 regulates pigmentation via keratinocyte-melanocyte interactions. *Exp Cell Res.* 254(1):25–32.

Seitz I, Hess S, Schulz H et al. 2007. Membrane-type serine protease-1/matriptase induces interleukin-6 and -8 in endothelial cells by activation of protease-activated receptor-2: Potential implications in atherosclerosis. *Arterioscler Thromb Vasc Biol.* 27(4):769–75.

Seymour ML, Binion DG, Compton SJ, Hollenberg MD, MacNaughton WK. 2005. Expression of proteinase-activated receptor 2 on human primary gastrointestinal myofibroblasts and stimulation of prostaglandin synthesis. *Can J Physiol Pharmacol.* 83(7):605–16.

Shelley WB, Arthur RP. 1955. Studies on cowhage (Mucuna pruriens) and its pruritogenic proteinase, mucunain. *AMA Arch Derm.* 72(5):399–406.

Shpacovitch VM, Brzoska T, Buddenkotte J et al. 2002. Agonists of proteinase-activated receptor 2 induce cytokine release and activation of nuclear transcription factor kappaB in human dermal microvascular endothelial cells. *J Invest Dermatol.* 118(2):380–5.

Shpacovitch V, Feld M, Bunnett NW, Steinhoff M. 2007. Protease-activated receptors: Novel PARtners in innate immunity. *Trends Immunol.* 28(12):541–50.

Sikand P, Dong X, LaMotte RH. 2011. BAM8-22 peptide produces itch and nociceptive sensations in humans independent of histamine release. *J Neurosci.* 31(20):7563–7.

Ständer S, Siepmann D, Herrgott I, Sunderkötter C, Luger TA. 2010. Targeting the neurokinin receptor 1 with aprepitant: A novel antipruritic strategy. *PLoS One.* 5(6):e10968.

Stefansson K, Brattsand M, Roosterman D, Kempkes C et al. 2008. Activation of proteinase-activated receptor-2 by human kallikrein-related peptidases. *J Invest Dermatol.* 128(1):18–25.

Steinhoff M, Buddenkotte J, Shpacovitch V et al. 2005. Proteinase-activated receptors: Transducers of proteinase-mediated signaling in inflammation and immune response. *Endocr Rev.* 26(1):1–43.

Steinhoff M, Corvera CU, Thoma MS et al. 1999. Proteinase-activated receptor-2 in human skin: Tissue distribution and activation of keratinocytes by mast cell tryptase. *Exp Dermatol.* 8(4):282–94.

Steinhoff M, Neisius U, Ikoma A et al. 2003. Proteinase-activated receptor-2 mediates itch: A novel pathway for pruritus in human skin. *J Neurosci.* 23(15):6176–80.

Steinhoff M, Vergnolle N, Young SH et al. 2000. Agonists of proteinase-activated receptor 2 induce inflammation by a neurogenic mechanism. *Nat Med.* 6(2):151–8.

Takahashi N, Arai I, Honma Y et al. 2005. Scratching behavior in spontaneous- or allergic contact-induced dermatitis in NC/Nga mice. *Exp Dermatol.* 14(11):830–7.

Takei-Taniguchi R, Imai Y, Ishikawa C et al. 2012. Interleukin-17- and protease-activated receptor 2-mediated production of CXCL1 and CXCL8 modulated by cyclosporine A, vitamin D3 and glucocorticoids in human keratinocytes. *J Dermatol.* 39(7):625–31.

Tsujii K, Andoh T, Lee JB, Kuraishi Y. 2008. Activation of proteinase-activated receptors induces itch-associated response through histamine-dependent and -independent pathways in mice. *J Pharmacol Sci.* 108(3):385–8.

Tsujii K, Andoh T, Ui H et al. 2009. Involvement of tryptase and proteinase-activated receptor-2 in spontaneous itch-associated response in mice with atopy-like dermatitis. *J Pharmacol Sci.* 109(3):388–95.

Ui H, Andoh T, Lee JB, Nojima H, Kuraishi Y. 2006. Potent pruritogenic action of tryptase mediated by PAR-2 receptor and its involvement in anti-pruritic effect of nafamostat mesilate in mice. *Eur J Pharmacol.* 530(1–2):172–8.

Vergnolle N, Bunnett NW, Sharkey KA et al. 2001. Proteinase-activated receptor-2 and hyperalgesia: A novel pain pathway. *Nat Med.* 7(7):821–6.

Vu TK, Hung DT, Wheaton VI, Coughlin SR. 1991. Molecular cloning of a functional thrombin receptor reveals a novel proteolytic mechanism of receptor activation. *Cell.* 64(6):1057–68.

Wilson SR, Gerhold KA, Bifolck-Fisher A et al. 2011. TRPA1 is required for histamine-independent, Mas-related G protein-coupled receptor-mediated itch. *Nat Neurosci.* 14(5):595–602.

Yamamoto O, Bhawan J. 1994. Three modes of melanosome transfers in Caucasian facial skin: Hypothesis based on an ultrastructural study. *Pigment Cell Res.* 7(3):158–69.

Yosipovitch G. 2004. Dry skin and impairment of barrier function associated with itch—New insights. *Int J Cosmet Sci.* 26(1):1–7.

Zhang J, Barak LS, Anborgh PH et al. 1999. Cellular trafficking of G protein-coupled receptor/beta-arrestin endocytic complexes. *J Biol Chem.* 274(16):10999–1006.

Zhu Y, Peng C, Xu JG et al. 2009a. Participation of proteinase-activated receptor-2 in passive cutaneous anaphylaxis-induced scratching behavior and the inhibitory effect of tacrolimus. *Biol Pharm Bull.* 32(7):1173–6.

Zhu Y, Wang XR, Peng C et al. 2009b. Induction of leukotriene B(4) and prostaglandin E(2) release from keratinocytes by protease-activated receptor-2-activating peptide in ICR mice. *Int Immunopharmacol.* 9(11):1332–6.

12 Mrgprs as Itch Receptors

Benjamin McNeil and Xinzhong Dong

CONTENTS

12.1 INTRODUCTION

The peripheral nervous system (PNS) is designed to receive inputs from the environment and transduce these into signals that are sent to the central nervous system (CNS). While most PNS neurons carry some receptors for classical neurotransmitters like glutamate, they are only weakly sensitive to these substances (or are not nearly as sensitive to them as CNS neurons are). Instead, as their primary function is to detect changes in external cues—for instance, heat and foreign chemicals—they are thought to be activated mostly through a broad range of receptors that recognize such environmental signals. These receptors can be divided into several broad groups. The groups comprising the olfactory and taste receptors represent the classic examples of families that, taken together, can detect an extraordinary range of substances from the environment. Both groups are members of the GPCR superfamily and are coupled to various G-proteins, through which they transduce their signals in second messenger mediated intracellular pathways. Each receptor is "tuned" to respond to varying degrees to chemicals with specific structures or properties. Many of these receptors exist—several dozen olfactory receptors in humans and several hundred in mice—and each is tuned differently. The circuitry underlying detection of odors is complex and knowledge of it is incomplete, though the field is moving forward rapidly, but it is thought that specific odors are detected by CNS processing of the signals sent from many different olfactory neurons that respond differently to the same odor. The ensemble input is different for every odor; in such a way, this library of receptors is capable of detecting a vast array of substances not ever made endogenously.

Among the sensations detected by the PNS, arguably the ones that induce the quickest behavioral (as opposed to motor) responses are those of pain and itch. These unpleasant sensations direct the organism to avoid a harmful situation or to remove a dangerous animal like a parasite. Like olfaction and taste, they are critical for survival, though extended experience of these sensations, often occurring in pathological states, dramatically lowers quality of life. It is known that mechanical stimulation and a wide range of chemicals can induce these sensations. While receptors have been discovered for several individual chemicals—one example is the TrpV1 channel for the neurotoxin capsaicin—other mechanisms must be discovered to account for the effects of most painful and itchy substances. One attractive hypothesis is that the neurons innervating the epithelia employ a family of receptors, analogous to the olfactory system, that serve to detect these noxious stimuli.

The Mrgpr family of receptors was discovered in 2001 and comprises 18 genes and pseudogenes in humans and 50 in mice (Dong et al. 2001) (Figure 12.1a). Many members are expressed exclusively in the dorsal root ganglia (DRG) and trigeminal ganglia (TG), which extend neurites into several layers of the skin and are responsible for most peripheral sensations, including noxious mechanical stimuli and temperature (Dong et al. 2001) (Figure 12.1b). This expression pattern raises the exciting possibility that they are specialized for somatosensation. In 2009, one of these receptors was found to be critical for the itch induced by the antimalarial drug chloroquine (Liu et al. 2009). In this and other studies, other pruritic substances were also shown to activate Mrgpr family members, further linking them to itch sensation (Liu et al. 2009, 2011). At the moment, it is unclear whether most pruritic stimuli act through these receptors, at least indirectly, but they have helped clarify the neural mechanisms underlying itch, and to a lesser extent, pain sensation. The history of research on Mrgprs is presented here, including what is known currently about this interesting family of receptors.

12.2   DISCOVERY OF THE Mrgprs

The Mrgprs were cloned around the turn of the millennium, before the human and mouse genome projects were completed, in a screen performed using a knockout mouse with a striking and fortuitous phenotype (Dong et al. 2001). In the late 1990s, a basic helix-loop-helix transcription factor called Neurogenin 1 (Ngn1) was found to be critical for the development of a subset of neurons in the DRG that express TrkA, the receptor for nerve growth factor (Ma et al. 1999). This subset includes most

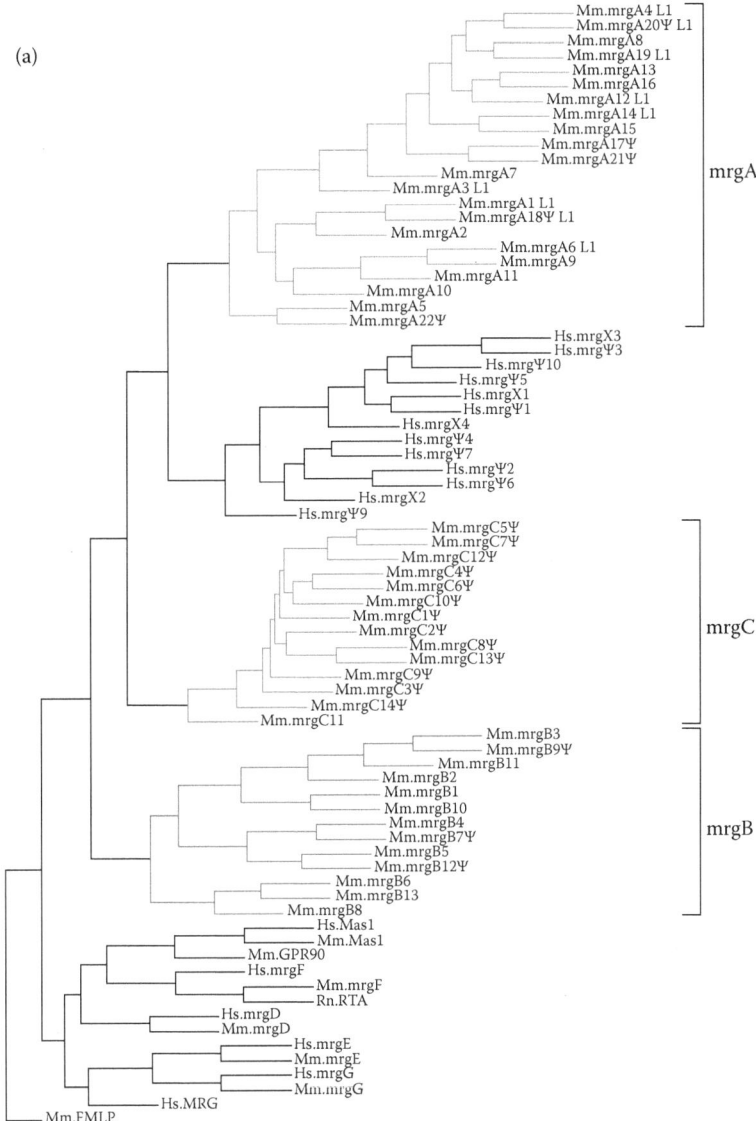

FIGURE 12.1 (See color insert.) Overview of the Mrgpr family and generation of cluster knockout mice. (a) Protein phylogenetic tree of human and mouse Mrgpr family members. Ψ denotes pseudogenes. L1 denotes genes with transposon elements within ~650 bases of the 3′ end of the gene. Note that family members outside of the MrgprA–C and MrgprX clusters have obvious orthologs, while members within clusters are more closely related to each other than they are to genes in the other species. Mouse formyl peptide receptor 1 (FMLP) was used as an outlying control. Analysis was performed using CLUSTALW, and PHYLIP software was used to generate the dendrogram, using neighbor joining with 1000 bootstrap trials. Horizontal distances are proportional to the number of different amino acids, while vertical distances are not meaningful.

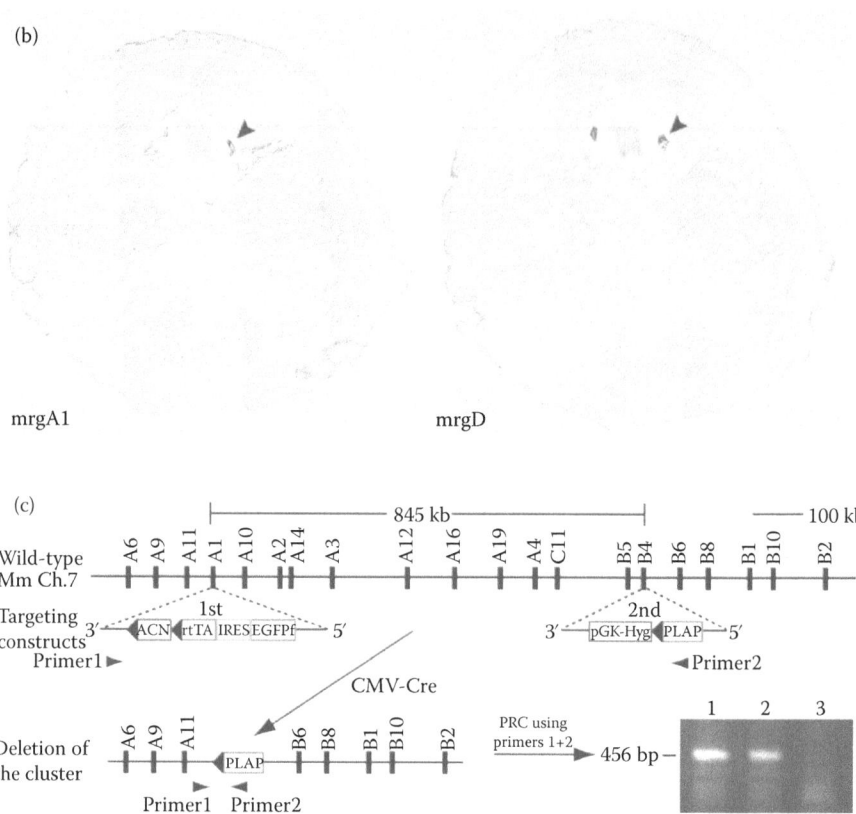

FIGURE 12.1 **(Continued)** **(See color insert.)** Overview of the Mrgpr family and generation of cluster knockout mice. (b) *In situ* localization of *MrgprA1* and *MrgprD* in representative sections of the trunk in neonatal mice. Expression of both genes is limited to the DRG.
Expression was not observed in any other region of the body in other sections. (c) Generation
of the Mrgpr-cluster $\Delta^{-/-}$ mice. Gene cassettes carrying various reporters and drug selection
genes were placed into the *MrgprA1* and *MrgprB4* open reading frames in mouse embryonic
stem cells, using conventional homologous recombination methods. Importantly, while each
cassette carried unique genes designed for other experiments, each also contained a LoxP
site. Transient expression of Cre recombinase resulted in the splicing together of these loci
and the removal of intervening DNA, generating a knockout. Primers used for genotyping
are indicated by arrows, and the PCR products from two Mrgpr-cluster $\Delta^{-/-}$ mice and a wild-
type mouse are shown at the bottom right. (Sections a and b from Dong, X. et al., *Cell*, 106,
619–632, 2001. Section c from Liu, Q. et al., *Cell*, 139, 1353–1365, 2009.)

neurons that detect nociceptive, or noxious, stimuli. Ngn1, which is expressed transiently in the DRG during embryogenesis, was shown to act as a master regulator of
neurogenesis in these neurons, controlling the expression of a host of genes including
the widely used developmental neuronal marker NeuroD (Ma et al. 1999). Without
Ngn1, most nociceptive DRG neurons failed to develop, while others involved in
types of low threshold mechanical input were still present (Ma et al. 1999). This

opened up an opportunity to identify nociceptor-specific genes, by performing subtractive hybridization of the cDNAs in newborn wild-type and Ngn1$^{-/-}$ mice. The transcripts absent from the Ngn1$^{-/-}$ mice were likely to be exclusive to nociceptors.

Several genes already known to be involved in nociceptor function were identified from the screen, verifying the approach (Dong et al. 2001). Multiple unknown transcripts were also recovered, including one that strongly resembled a GPCR (Dong et al. 2001). This was found to have ~30% to 35% identity with Mas1, the putative receptor for angiotensin (1–7), a peptide involved in blood pressure and control of osmolarity (Santos et al. 2003). Interestingly, Mas1 is not enriched in the DRG and is not known to participate in nociception, indicating that this new receptor had a different endogenous ligand. This founding member was named Mas-related gene A1 (MrgprA1). Additional screens of DRG cDNA, and BAC libraries turned up several dozen members that could be divided into three groups, the MrgprA, MrgprB, and MrgprC families (Dong et al. 2001) (Figure 12.1a). Over 50 distinct sequences were found in mice, although it is unclear how many are physiologically relevant, as many do not have open reading frames and only a few have been shown to be expressed (Dong et al. 2001). Of the originally discovered members, 21 can be annotated to current editions of the mouse genome and are not predicted to be pseudogenes (Figure 12.1a). In addition to these, a more distantly related group of receptors was discovered and named MrgprD-H (Dong et al. 2001).

Human Mrgpr family members were discovered in searches of contemporary databases (Dong et al. 2001; Lembo et al. 2002). Human orthologs of the MrgprD-H members were fairly clear, though, interestingly, only four other family members were found. These members were somewhat more closely related to the mouse MrgprA family members than to other families, though no clear orthologs existed, and these members were given the names MrgprX1-4 or SNSRs (sensory neuron specific receptors) called by different groups (Dong et al. 2001; Lembo et al. 2002). The mouse MrgprA, MrgprB, and MrgprC family genes are all found in a tight cluster on chromosome 7 (Figure 12.1c). Likewise, the human MrgprX family is found clustered on chromosome 11. Both mouse and human genomic regions are characterized by multiple repetitive transposon elements, to the point that very little in the region except for the Mrgprs is unique to the genome (Zylka et al. 2003). Four studies have reported that the human genomic region is duplicated, with up to six copies detected in some individuals (Hindson et al. 2011; Kato et al. 2008; Redon et al. 2006; Wong et al. 2007). The sequences reported for the MrgprX family are slightly different (Dong et al. 2001; Lembo et al. 2002), provoking the question of whether these represent polymorphisms between individuals, or perhaps a different situation in which each individual actually has different, highly related MrgprX genes in the duplicated DNA.

12.3 EXPRESSION OF Mrgprs

Expression of MrgprA1-8 was detected by *in situ* hybridization in a subset of DRG and trigeminal neurons, but nowhere else in the body (Dong et al. 2001) (Figure 12.1b). Among members of the MrgprA-C families expressed at birth, MrgprA1 was, by far, the most widely expressed, seen in 13.5% of all DRG neurons. MrgprD also had a fairly wide expression pattern, but most other Mrgpr family members were observed only in a few neurons (~1%) or not at all at birth (Dong et al. 2001). Notably, several of the MrgprA members colocalized in the same neurons, indicating specialization of those neurons, and a partial overlap was seen between MrgprA1 and MrgprD (Dong et al. 2001). Importantly, Mrgpr expression was observed almost exclusively in neurons that were TrkA+ and could be stained with the plant lectin IB4, a marker of nociceptors (Dong et al. 2001). Most Mrgpr+ neurons did not overlap with proinflammatory peptides Substance P or CGRP labeling, indicating that most were nonpeptidergic, though low expression of these peptides could not be ruled out (Dong et al. 2001).

Some aspects of Mrgpr family expression were notably different in adult tissues. While Mrgpr-expressing neurons were IB4+, as in newborns, Mrgpr expression patterns largely became nonoverlapping within this neuronal subset (Dong et al. 2001). Specifically, Mrgpr+ neurons could be segregated into MrgprA3+, MrgprD+, and MrgprB4+ neurons, with (~5, 30, and 1% of total DRG neurons, respectively) (Dong et al. 2001). MrgprA1 expression was strongly reduced; other MrgprA family members could be detected in MrgprA3+ neurons at lower levels, and, interestingly, MrgprC11, the only coding member of the MrgprC family, was expressed in a pattern closely overlapping the MrgprA3 signal (Dong et al. 2001; Liu et al. 2009). In addition, nearly half of the MrgprA+ neurons now expressed CGRP, while Substance P expression was still undetectable both in MrgprA+ and MrgprD+ neurons (Dong et al. 2001).

The expression patterns of the human MrgprX family and MrgprD have not been characterized at the same level of detail as in mice. Human MrgprX1, MrgprX3, and MrgprX 4 were found in a human DRG cDNA library (Lembo et al. 2002), though separate studies have reported MrgprX2 expression in the DRG, as well (Kamohara et al. 2005; Robas et al. 2003; Zhang et al. 2005). No MrgprX family members were found in a panel of human tissues that did not include the DRG or TG (Dong et al. 2001), and a more complete panel confirmed expression only in the DRG and TG (Lembo et al. 2002), indicating that expression is spatially restricted, as in mice. MrgprX2, and to a much lesser extent, MrgprX1, has been detected in mast cells, which is to date the only reported cellular localization of Mrgprs outside of neurons (Subramanian et al. 2011; Tatemoto et al. 2006).

The highly restricted tissue expression pattern—exclusive to nociceptors and divided segregated within this class—raised the possibility that the Mrgprs were involved in some aspect of somatosensation, perhaps analogous to olfactory or taste receptors. This was bolstered by a pair of studies in which genetically encoded cellular labels were knocked into the MrgprB4 and MrgprD loci (Liu et al. 2007; Zylka et al. 2005). These modifications enabled easy detection of the complex morphology of neurons that normally express these receptors. Knocking in PLAP and, separately, eGFP into the MrgprD locus revealed that neurites extending to the skin exhibited extensive branching and termination into all layers of the epidermis except the dead cell layer of the stratum corneum (Zylka et al. 2005). MrgprB4+ neurons, labeled with PLAP, were also shown to penetrate into the shallow layers of the epidermis and undergo branching (Liu et al. 2007). The morphological characteristics of these neurons poised them for detection of subtle environmental stimuli; however, the exact function(s) of these receptors and the neurons that expressed them were not obvious from morphological data alone.

12.4 GENERATION OF Mrgpr KNOCKOUT ANIMALS

A knockout approach was undertaken for members of the MrgprA-C cluster (Liu et al. 2009). Generation of knockout animals turned out not to be a trivial task, owing to the extensive repetitive sequence that blanketed the entire cluster. The repetitive elements meant that very little DNA sequence in the locus was specific to the area, a serious impediment to conventional homologous recombination techniques that require unique sequences flanking the targeted gene to align the knockout/knockin construct to the locus during mitosis. After screening more than 10,000 embryonic stem cell clones, one knockout each was obtained for MrgprA1 and MrgprB4, genes selected for their high expression and presumed importance (Liu et al. 2009).

Given the dramatically expanded number of family members in mice, it seemed plausible that some of the genes are redundant and single knockouts might not show much of a phenotype. Therefore, an unusual approach was taken to knock out the genes between and including MrgprA1 and MrgprB4 (Mrgpr-cluster$\Delta^{-/-}$ mice) (Liu et al. 2009) (Figure 12.1c). This took advantage of the remaining LoxP site in each knockout and was achieved by transient expression of Cre recombinase in stem cells harboring a copy of both knockouts to splice out the intermediate DNA (Liu et al. 2009). A total of 845 kb was deleted, containing 12 of the estimated 24 genes with complete open reading frames (*MrgprA1-4, A10, A12, A14, A16, A19, B4, B5*, and *C11*) (Liu et al. 2009) (Figure 12.1c). This includes most of the MrgprA family, and all of the genes that had been shown to be expressed at a high level in the DRG and TG. Importantly, no non-Mrgpr ORF exists in the deleted region.

12.5 ROLE OF Mrgprs IN ITCH IN MICE

The list of pruritogens used in animal studies is quite long. A landmark study in the itch field, published not long after the Mrgpr-cluster $\Delta^{-/-}$ mice were generated, demonstrated that itch sensation induced by several of these well-established pruritogens was dependent on the receptor for gastrin-releasing peptide (Sun and Chen 2007). One of the substances used in the paper, an antimalarial drug called chloroquine (CQ), stood out because it was the least expensive by a large margin. In addition, study of its itch-inducing mechanism is medically important, as CQ is still in therapeutic use. Unfortunately, oral CQ evokes strong itch sensation in black Africans, which can be so strong that it limits compliance (Sowunmi et al. 1989). Researchers in Africa have undertaken many studies on the mechanism of such a highly undesirable side effect. They found that this type of itch does not fit the profile of an allergic response and that antihistamines are largely ineffective, indicating that another way to induce itch exists alongside the classical, histamine-dependent mechanism associated with allergic responses (Ekpechi and Okoro 1964; Salako 1984).

Mrgpr-cluster $\Delta^{-/-}$ mice did not show general deficits in nociception. Multiple tests of pain sensation showed that Mrgpr-cluster $\Delta^{-/-}$ mice responded similar to wild type (Liu et al. 2009). Histamine and compound 48/80-induced itch sensation were not reduced in Mrgpr-cluster $\Delta^{-/-}$ mice, demonstrating that the ability to perceive itch was retained (Liu et al. 2009) (Figure 12.2a and b). A subcutaneous dose of CQ injected into the nape of the neck of wild-type mice induced strong scratching behavior within a couple of minutes, and which lasted for tens of minutes (Liu et al. 2009). Strikingly, CQ-induced scratching in Mrgpr-cluster $\Delta^{-/-}$ mice was reduced by over 65%, indicating that an Mrgpr family member mediated most of this effect (Liu et al. 2009) (Figure 12.2c). The residual effect may be mast cell dependent, as scratching was reduced by ~35% in mice lacking mast cells, consistent with studies showing that CQ can activate mast cells (Liu et al. 2009).

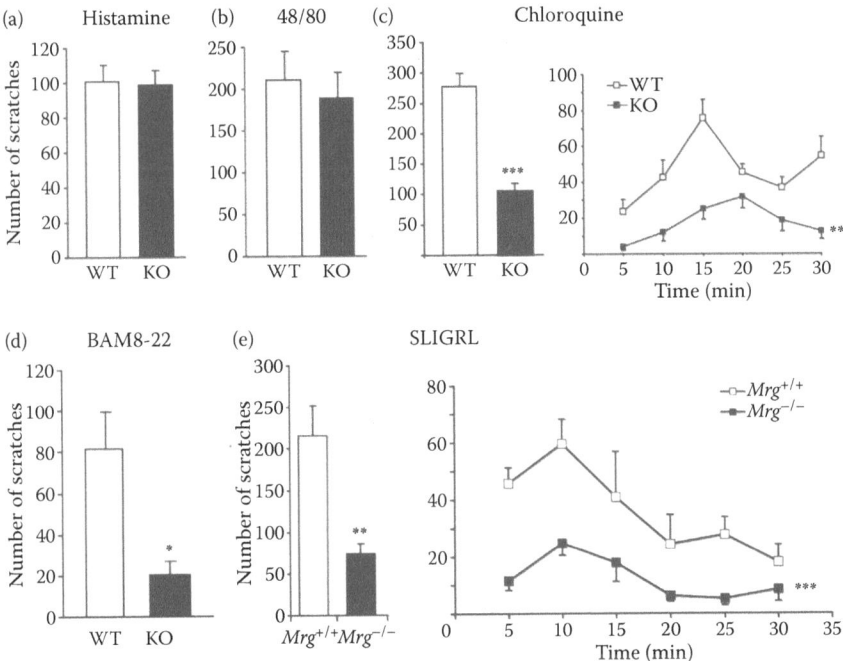

FIGURE 12.2   Scratching response to select pruritogens is drastically reduced in Mrgpr-cluster $\Delta^{-/-}$ mice. (a, b) Scratching bouts in response to histamine (10 μmol) and the mast cell degranulator Compound 48/80 (100 μg) are comparable in WT and Mrgpr-cluster $\Delta^{-/-}$ mice ($n = 12$ for each). Substances were administered subcutaneously in the nape of the neck of 2- to 3-month old mice, and scratching bouts were counted for 30 min. (c) Scratching response to CQ (200 μg) was strongly reduced in Mrgpr-cluster $\Delta^{-/-}$ mice ($n = 9$ for KO, $n = 8$ for WT). Right panel shows the number of scratches throughout the 30-min examination period. (d) Scratching induced by BAM8-22 (50 nmol) is also strongly reduced in Mrgpr-cluster $\Delta^{-/-}$ mice ($n = 8$ for each). (e) Scratching in response to SLIGRL-NH2 (100 nmol) was strongly reduced in Mrgpr-cluster $\Delta^{-/-}$ mice ($n = 7$ for KO, $n = 6$ for WT). Right panel shows the number of scratches throughout the 30-min examination period. Unlike SLIGRL, SLIGR, the PAR2 specific peptide referenced in Figure 12.4, does not evoke any scratching in neither wild-type nor Mrgpr cluster knockout mice (data not shown). (Sections a to d from Liu, Q. et al., *Cell*, 139, 1353–1365, 2009. Section e from Liu, Q. et al., *Sci. Signal*, 4, ra452011, 2011.)

Cultured DRG neurons were used to determine whether this effect correlated with DRG activation. Approximately 5% of wild-type neurons, mostly small-to-medium diameter, responded to CQ with an increase in intracellular calcium and generation of action potentials (Liu et al. 2009) (Figure 12.3a and b). This effect was completely abolished in Mrgpr-cluster $\Delta^{-/-}$ neurons (Liu et al. 2009) (Figure 12.3a and b). Because 12 possibly functioning receptors were deleted, it became imperative to find out which one(s) responded to CQ. Testing in heterologous cells revealed that MrgprA3, a highly expressed member in adult mouse DRG, had the strongest response, and that no other deleted member could be activated except for a very weak response by MrgprA1 (Figure 12.4a left panel). MrgprA3 expression in Mrgpr-cluster $\Delta^{-/-}$ mice restored sensitivity to CQ, and siRNA-mediated knockdown blocked the response in wild-type neurons, firmly establishing that MrgprA3 is the receptor for CQ in mouse DRG (Liu et al. 2009). Importantly, related chemicals were not active, establishing the specificity of the ligand–receptor relationship (Liu et al. 2009) (Figure 12.4b and c).

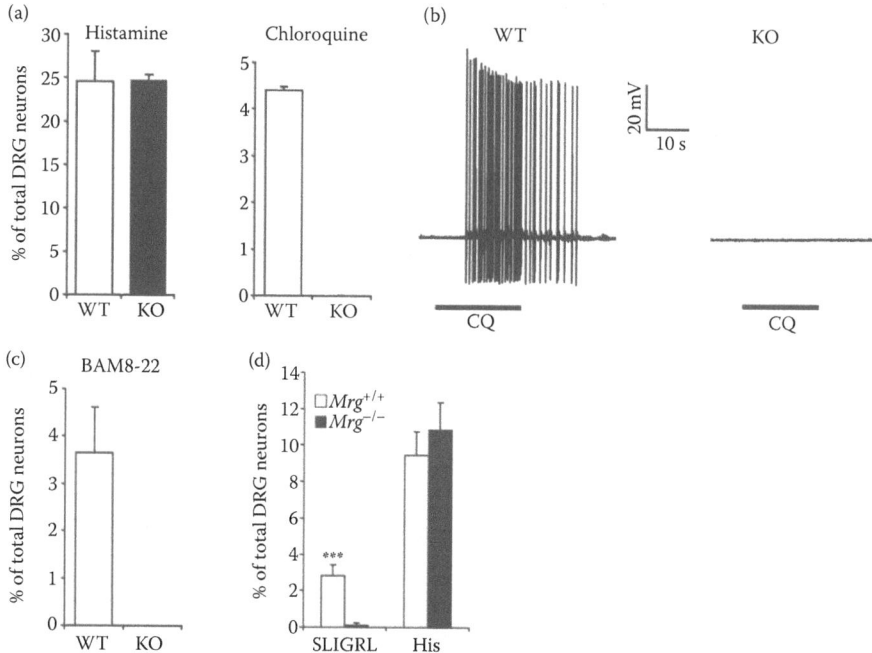

FIGURE 12.3 (**See color insert.**) DRG neurons responsive to CQ and BAM8-22 define a subset of the nociceptive population. (a) Cultured DRG neurons from young adult Mrgpr-cluster $\Delta^{-/-}$ mice show no response to 1 mM CQ but respond normally to 50 μM histamine ($n = 3$ experiments for each). (b) Representative current clamp traces demonstrating that CQ (1 mM) induces action potentials in a subset of WT but not Mrgpr-cluster $\Delta^{-/-}$ mice. (c) Cultured DRG neurons from young adult Mrgpr-cluster $\Delta^{-/-}$ mice show no response to BAM8-22, compared to WT neurons ($n = 3$ experiments for each). (d) Cultured DRG neurons from WT mice respond to histamine (50 μM) and SLIGRL-NH2 (100 μM), while Mrgpr-cluster $\Delta^{-/-}$ neurons do not respond to SLIGRL-NH2 ($n = 3$ experiments for each).

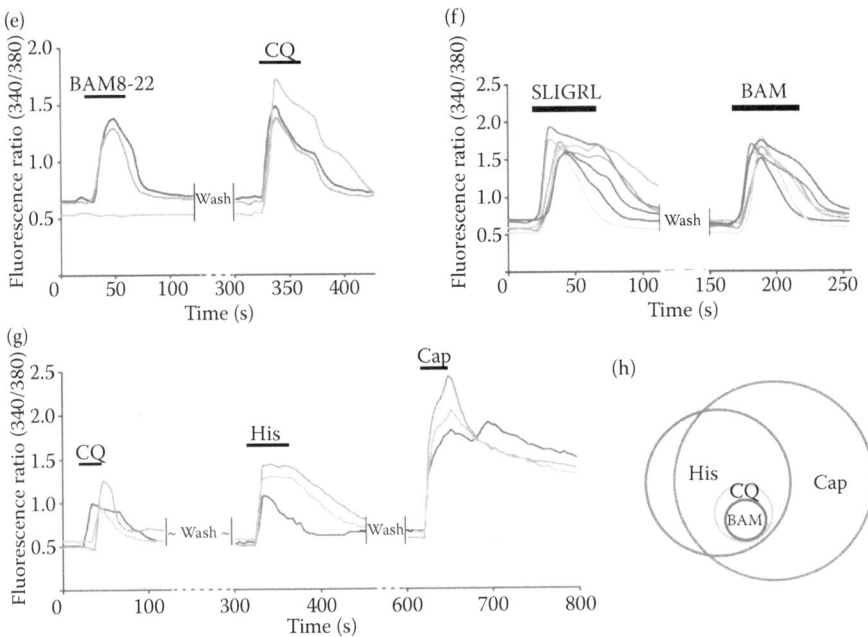

FIGURE 12.3   (Continued)   (See color insert.) (e) Representative Fura-2 calcium imaging traces showing that BAM8-22 (2 µM) responsive neurons also respond to CQ (1 mM). (f) Representative Fura-2 calcium imaging traces showing that SLIGRL-NH2 (100 µM) responsive neurons also respond to BAM8-22 (2 µM). (g) Representative Fura-2 calcium imaging traces showing that CQ (1 mM) responsive neurons define a small subset of neurons also responsive to histamine (50 µM) and capsaicin (1 µM). (h) A Venn diagram defining the overlap between neurons responsive to various substances. Circles are representative of the size of each population. Note that BAM8-22-responsive neurons all respond to CQ and that they also respond to histamine and capsaicin, though they comprise only a small subset of the latter two populations. (Sections a to c, e, g, and h from Liu, Q. et al., *Cell*, 139, 1353–1365, 2009. Sections d and f from Liu, Q. et al., *Sci. Signal*, 4, ra45, 2011.)

The discovery of CQ as an MrgprA3 agonist served as a tool to find a human ortholog, as none clearly exists from sequence homology. MrgprX1 was found to be activated by CQ, a striking finding, since MrgprX1 is also found in the DRG, and the gene rests in the same part of the human genome as MrgprA3 is in the mouse (Liu et al. 2009) (Figure 12.4a and c). A variety of peptide libraries had already been tested on MrgprX1, and the most specific ligand found was a fragment of Bovine Adrenal Medulla peptide (BAM8-22) (Lembo et al. 2002). Full length BAM is a ligand for opioid receptors and for MrgprX1; however, BAM activates the opioid receptors at its N-terminus only, and removing this to generate BAM8-22 abolished this activity but did not affect MrgprX1 activation (Lembo et al. 2002). In light of the link between Mrgprs and itch, BAM8-22 was tested for pruritogenic activity by subcutaneous injection into the nape of the neck. As with CQ, BAM8-22 induced itch in wild-type mice and a response in DRG neurons, while Mrgpr-cluster $\Delta^{-/-}$ mice

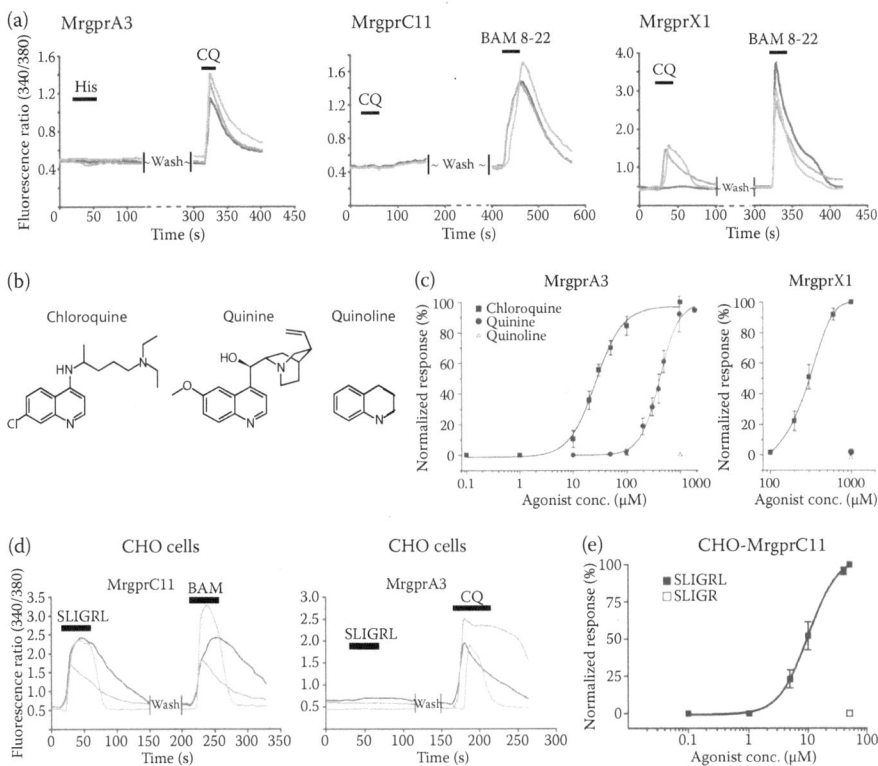

FIGURE 12.4 **(See color insert.)** Responses to various pruritogens in Mrgpr-transfected cell lines. (a) Representative traces of the rise in intracellular calcium evoked by Mrgpr agonists in transfected HEK cells. Cells were loaded with the calcium indicator Fura-2 and imaged at 340- and 380-nm excitation. The 340/380 ratio is used to establish relative calcium levels. Traces of different colors are from different cells on the same cover slips. Left panel, MrgprA3-transfected cells respond to chloroquine (CQ, 1 mM) but not to histamine (His, 50 μM). Middle panel, MrgprC11-transfected cells respond to BAM8-22 (2 μM) but not CQ (1 mM). Right panel, MrgprX1-transfected cells respond to CQ (1 mM) and BAM8-22 (2 μM). (b) Structure of chloroquine and related compounds. (c) Dose-response curves in MrgprA3 and MrgprX1 stable HEK cell lines to CQ and related compounds shown in (b). Response (%) refers to the number of cells responding to the agonist. Each data point, including standard error, is taken from ≥3 experiments and over 50 cells. (d) Representative traces in cells transfected with the indicated Mrgprs to the mouse PAR-2 agonist SLIGRL-NH2 (100 μM). Experiments were performed similarly to a, except that CHO cells were used instead of HEK cells. (e) Dose-response curve to SLIGRL-NH2 and SLIGR-NH2 in MrgprC11 stably transfected cells. Note that cells respond to SLIGRL-NH2, the prototypical PAR-2 agonist, while they do not respond to SLIGR-NH2, another PAR-2 agonist. Because all responses to SLIGRL-NH2 have been ascribed solely to PAR-2, this experiment demonstrates that SLIGR-NH2 is a more selective PAR-2 agonist and that some effects of SLIGRL-NH2 may be due to MrgprC11 activation. Each data point, including standard error, is taken from ≥3 experiments and over 100 cells. (Sections a to c from Liu, Q. et al., *Cell*, 139, 1353–1365, 2009. Sections d and e from Liu, Q. et al., *Sci. Signal*, 4, ra45, 2011.)

had almost no behavioral or neuronal response (Liu et al. 2009) (Figures 12.2d and 12.3c). Interestingly, BAM8-22 was found not to activate MrgprA3, but MrgprC11, illustrating a possible complicated relationship between Mrgprs in rodents and those in other species (Liu et al. 2009) (Figure 12.4a).

An unexpected discovery in a subsequent study perhaps hints at a much broader role for the Mrgprs in itch mediation. Over 10 years ago, a GPCR called PAR2 was linked to inflammation and itch (Steinhoff et al. 2000, 2003). PARs, or protease-activated receptors, are a unique family of GPCRs whose ligands come from their own N-termini. Upon exposure to serine proteases (e.g., trypsin and trypt-ase), mouse PAR2 is cleaved at a residue that is just N-terminal to the amino acid sequence SLIGRL, such that the unmasked sequence can now activate PAR2 itself (Ramachandran and Hollenberg 2008). For each PAR, a synthetic peptide has been designed, on the basis of the exposed sequence after protease cleavage, and these so-called activating peptides are widely used as specific agonists for their corre-sponding PARs. SLIGRL-NH2 is the activating peptide for PAR2 and induces itch in wild-type mice, but had not been tested in PAR2 $^{-/-}$ mice to confirm that it actu-ally worked through its parent receptor (Shimada et al. 2006). When this test was performed, the PAR2$^{-/-}$ mice actually scratched more than wild-type mice, ruling out a role for the receptor in inducing itch (Liu et al. 2011). Instead, when tested in the Mrgpr-cluster $\Delta^{-/-}$ mice, the scratching response was nearly abolished (Figure 12.2d and e), as was the response in DRG neurons (Figure 12.3d) (Liu et al. 2011). SLIGRL-NH2 was found to activate MrgprC11, like BAM8-22 (Liu et al. 2011) (Figure 12.4d and e). The corresponding human peptide, SLIGKV-NH2, was found to activate MrgprX2, not MrgprX1, underscoring the need to account for species differences in the ligand-binding profiles (Liu et al. 2011).

This finding invites speculation that a relationship exists between Mrgprs and proteases. If the SLIGRL/SLIGKV sequence becomes untethered by further pro-teolysis, the free peptide can activate Mrgprs in the vicinity, which would be an unprecedented form of cross-activation between receptors. Because PAR2 has been detected on small diameter DRG neurons, colocalization close enough for cross-activation is plausible. Another unexplored possibility is that, because Mrgprs can be activated by PAR ligands, they might have a similar binding pocket and might themselves be activated by proteases, though no sequences orthologous to SLIGRL/ SLIGKV exist in the extracellular regions.

In a fascinating finding, DRG neurons responding to BAM8-22, CQ, and SLIGRL almost completely overlapped, consistent with *in situ* studies showing overlap of *Mrgprc11* and *Mrgpra3* mRNA, and were also responsive to histamine and capsaicin (Liu et al. 2009; Zylka et al. 2003) (Figure 12.3e and h). This implicated this subset of neurons as specialized, itch-selective nociceptors, though the exact significance is still unclear (and may be species specific, as described in the Questions section).

12.6 Mrgpr SIGNALING PATHWAYS

Recent studies have implicated transient receptor potential (Trp) channels as intra-cellular mediators of multiple receptors involved in nociception (Bandell et al. 2004; Imamachi et al. 2009; Jordt et al. 2004). This exciting adjustment to classical modes of GPCR signaling is expanding to a diverse collection of GPCRs and is in line with experiments showing abundant expression of multiple Trp channels in nociceptive neurons (Moran et al. 2011; Wu et al. 2010). The mouse DRG neuron response to CQ was blocked by ruthenium red, a nonspecific blocker of Trp channels, and also by extracellular EGTA, a calcium chelator, implicating that calcium influx via Trp channels is a crucial part of MrgprA3 signaling (Liu et al. 2009). A further set of experiments showed that TrpA1 was required for CQ and BAM8-22 induced rises in intracellular calcium and action potentials in DRG neurons (Wilson et al. 2011). No scratching was observed in TrpA1$^{-/-}$ mice, and both MrgprA3 and MrgprC11 were shown to couple to TrpA1 in heterologous cells (Wilson et al. 2011). The pathways were not identical, however. MrgprC11 signaling was shown to require phospholipase C (PLC) as an intermediate, while MrgprA3 signaling did not require PLC but instead involved an active Gbeta/gamma complex (Wilson et al. 2011). This in itself is another interesting modification of the classical GPCR models, in which the alpha subunits are the active messenger proteins. Importantly, MrgprC11 did not require active Gbeta/gamma signaling, indicating at least partially exclusive pathways to the same mediator (Wilson et al. 2011).

A surprising finding from this series of studies is that TrpV1, an abundant channel in nociceptors and one present in MrgprA3+/MrgprC11+ neurons, was not required for signaling (Wilson et al. 2011). Very little is known about how these receptors specifically couple to TrpA1 while "avoiding" TrpV1, though there is some indication that MrgprC11 actually can couple to TrpV1 when the two are overexpressed (Wilson et al. 2011). Among the unresolved questions are whether the receptors form separate complexes with Trp channels or act on a common pool, and whether other Trp channels are involved, perhaps as heterodimers with TrpA1.

MrgprX1 may have an inhibitory role in some neurons, or at least in specific neuronal compartments. BAM8-22 strongly inhibited the high voltage activated calcium current in the somata of primary neurons heterologous expressing MrgprX1 (Chen and Ikeda 2004). This is difficult to reconcile with the observed BAM8-22 induced rise in intracellular calcium, and action potential generation, in mouse DRG neurons, but it is possible that MrgprX1 and MrgprC11 can couple to different effectors in different cell types or in different parts of the same cell. Interestingly, BAM8-22 was shown to inhibit pain responses in mouse models of several types of pathological pain (Guan et al. 2010). Perhaps this ability, which is Mrgpr dependent, is due to selective inhibition of the pain circuit.

12.7 BAM8-22, AN Mrgpr AGONIST, INDUCES ITCH IN HUMANS

Mouse studies indicated that skin injection of BAM8-22 induces itch (Liu et al. 2009). If this is the case, it could be an endogenous transmitter for itch and would make MrgprX1 a potentially useful therapeutic target for more than just CQ-induced itch. To test whether BAM8-22 can induce itch in humans, Robert LaMotte's laboratory applied BAM8-22 onto the skin of healthy volunteers by inactivated cowhage spicules loaded with the peptide (Sikand et al. 2011). Cowhage spicules penetrate the shallow layers of skin and act as an epidermal and intradermal applicator. All individuals reported strong itch sensation at a level similar to histamine application (Sikand et al.

2011). Importantly, a truncated form of BAM8-22, BAM8-18, which lacks the Mrgpr-interacting motif, failed to elicit any sensation in these people (Sikand et al. 2011). In addition, application of an antihistamine cream, which abolished histamine-evoked itch, had no effect on BAM8-22 induced itch (Sikand et al. 2011). Together these data confirm what was found in mice and demonstrate that BAM8-22 may be an endogenous itch mediator in humans and that its action likely goes through MrgprX1. Interestingly, an animal model of liver disease, which often leads to itch symptoms in humans, has elevated BAM22 levels in blood (Swain et al. 1994).

As mentioned briefly above, extensive copy number variation (CNV) in the MrgprX1 gene has been observed, first in a pair of studies of CNV across the human genome and later shown in two independent studies (Hindson et al. 2011; Kato et al. 2008; Redon et al. 2006; Wong et al. 2007). In fact, as many as six copies were found in some individuals (Hindson et al. 2011). The Mrgpr cluster contains a tremendous number of repetitive elements, and as described earlier in this chapter, duplication of a genomic region including a presumed MrgprX ancestor has been proposed to allow for the expansion of family members (Zylka et al. 2003). It seems plausible that genomic instability in this region has led to these repeats in humans. As of this writing, such extensive duplication in other Mrgprs has not been discovered, limiting studies to MrgprX1. If copy number is correlated with mRNA and protein expression levels, it may indicate a striking heterogeneity in the human population in its sensitivity to MrgprX1-recognized pruritogens.

A further possible source of variation in chloroquine sensitivity may come from polymorphisms. Numerous mouse Mrgprs have been shown to carry polymorphisms in different strains (Dong et al. 2001); some coding differences exist between the MrgprX1 genes reported by the first two labs to clone the gene (Dong et al. 2001; Lembo et al. 2002), and dbSNP at NCBI has archived many missense mutations in the human population, though how these might affect sensitivity has not been studied. Given the multiple copies of MrgprX1 in many individuals, it may be that one individual could have several different SNPs and that particular combinations may result in widely varying sensitivities. It is interesting that macaques have been shown to carry the canonical receptor genes, along with additional copies of several of the MrgprXs that differ only in a few amino acids (Zhang et al. 2005). These are slightly more divergent than single SNPs and may be the predicted additional copies on duplicated DNA. Their activation profiles actually are somewhat different from the canonical receptors, giving some validity to these possibilities (Zhang et al. 2005).

12.8 DIFFERENCES BETWEEN ITCH RECEPTORS AND OLFACTORY RECEPTORS

Olfactory neurons, like nociceptors, have numerous receptors designed to detect information from the environment. Each olfactory neuron expresses only one of these receptors, and odorous substances activate a subset of receptors, and thus a subset of neurons, to varying degrees (Gottfried 2010). It is thought that each smell activates a different subset and at different intensities, resulting in a unique signature (Gottfried 2010). The data from Mrgprs indicate that itch sensation is not as sensitive as this. Unlike olfactory neurons, a single nociceptor can express multiple Mrgprs implicated in itch, as well as histamine receptors. Thus, several different pruritogens can activate the same neurons that could carry mostly the same information. Perhaps the intensity of activation encodes additional information, but it is unclear how the signal could be more than whether something is itchy and how itchy it is.

Three findings hint that itch sensation might be more complex. First, recent data indicate that a subset of myelinated neurons, which would not include the majority of Mrgpr+ neurons, can also transmit itch signals (Ringkamp et al. 2011). Little is known about these neurons, but it opens up the possibility that different pruritogens can be perceived differently. Second, some pruritogens induce more than just itch. If some pruritogen receptors are on multiple types of neurons or other cell types involved in nociception, the ligands for those receptors might elicit a mixed sense profile. One example of such a profile, actually, is BAM8-22, which induces a stinging sensation along with itch, though it is not known why (Sikand et al. 2011). Third, BAM8-22 may be able to elicit different sensations, depending on what part of the body is exposed to it. BAM8-22 induces itch and some stinging sensation when applied to human forearms and scratching behavior when administered subcutaneously to the mouse rostral back (Liu et al. 2009; Sikand et al. 2011). However, when injected into the mouse hind paw, it induces activity more consistent with pain (and similar to the response to capsaicin) (Grazzini et al. 2004). Because no model exists for animal itch in nonhairy skin, the licking and biting observed may indeed be a response to itch, though hyperalgesia to heat was also observed, more consistent with pain (Grazzini et al. 2004). Another study showed that spinal application of BAM8-22 actually reduced pain behaviors, demonstrating a clearly complex role for MrgprC11 in mouse nociception (Guan et al. 2010). Regardless, this raises the possibility that pruritogen receptors like MrgprC11 may exist on different subtypes of neurons, depending on where they terminate in the body, or, alternatively, that the same types of neurons are part of different circuits.

12.9 OTHER Mrgprs

Much less is known about Mrgpr family members that are not orthologs of MrgprX1. The only other members linked to putative ligands are MrgprD, MrgprA9, and MrgprX2. The locations of most family members are incompletely characterized, and, except for MrgprD-G, it is not even clear whether orthologs exist across species. In this section, summaries are provided of information regarding the better characterized family members.

12.9.1 MRGPRD

MrgprD, as discussed above, is present in the majority of IB4+, nonpeptidergic DRG neurons, meaning that most nociceptors express this gene (Zylka et al. 2005). RT-PCR assays turned up evidence for lower levels of expression in the bladder, uterus, testes, and arteries in a panel of rat mRNA pools (Shinohara et al. 2004). MrgprD orthologs from humans, rats, and mice respond to beta alanine, a natural cellular metabolite, at low micromolar concentrations in heterologous cells (Shinohara et al. 2004). The most pronounced effect both in mouse and rat DRG neurons was to enhance their excitability by inhibiting potassium channel M-currents via a $G_{i/o}$ dependent pathway (Crozier et al. 2007; Rau et al. 2009). The enhancement was such that a stimulus that normally produced a single action potential now produced a short train (Crozier et al. 2007; Rau et al. 2009). Thus, the observed effect was to modulate neurons so that they were more sensitive to other stimuli. DRG neurons from MrgprD$^{-/-}$ mice were less sensitive than wild-type neurons to heat, cold, and mechanical stimulation (Rau et al. 2009). Because beta-alanine levels in the body are high enough to activate MrgprD, it may be that MrgprD is constitutively active and that nociceptors adjust their overall excitability when they sense extreme changes in beta-alanine levels.

MrgprD+ neurons do not appear to be linked to a single sensation. Mice with genetically encoded cell tracers in the MrgprD locus revealed that these neurons make connections in all parts of Lamina II in the dorsal horn of the spinal cord, the target of most nociceptors (Wang and Zylka 2009). No specific type of neuron was enriched, and no area in Lamina II was obviously more targeted. One study showed that the cells were highly responsive to ATP and almost totally unresponsive to other noxious substances (Dussor et al. 2008). It thus appears that the function of MrgprD is considerably different from MrgprA3 and MrgprC11.

12.9.2 MRGPRA9

The rat ortholog of mouse MrgprA9 was found primarily in small diameter DRG neurons, and in a separate study, mouse MrgprA9 was found in the brain and spleen, though DRG tissue was not examined (Bender et al. 2002; von Kugelgen et al. 2008). Both orthologs were activated by low nanomolar concentrations of adenine, a cell metabolite and potential transmitter, resulting primarily in $G_{i/o}$ activity (Bender et al. 2002; von Kugelgen et al. 2008). The effects of adenine in neurons were not examined, and because human MrgprX family members are unresponsive to adenine (Bender et al. 2002), it appears that this receptor is rodent specific.

12.9.3 MRGPRE AND MRGPRF

MrgprE and MrgprF appear to be linked somehow, as mouse MrgprE knockouts have strongly reduced expression of MrgprF but no other tested family member (Cox et al. 2008). MrgprE was found in multiple brain regions in mice and humans, with expression appearing to be somewhat more restricted in humans (Zhang et al. 2005). Human MrgprE also was found in the placenta (Zhang et al. 2005). Human MrgprE was detected in ~85% of DRG neurons and a subset of enteric neurons (Avula et al. 2011; Zhang et al. 2005). MrgprE knockout mice responded normally to all tested pain assays, and the function and ligands of this receptor have yet to be discovered (Cox et al. 2008).

MrgprF expression is poorly characterized, though it, too, was found in enteric neurons in mice (Avula et al. 2011). Interestingly, despite the putative link between the two receptors, MrgprE and MrgprF were only rarely detected in the same enteric neurons (Avula et al. 2011). Both receptors were downregulated in inflammatory intestinal conditions (Avula et al. 2011). The ligand(s) for MrgprF, too, awaits discovery.

12.9.4 MRGPRX2

MrgprX2 is a unique member of the family in that it is the only member shown to be expressed in a specific cell type outside of the nervous system. The mRNAs for MrgprD and MrgprE have been detected in nonnervous tissue, but the cell types are undetermined (Shinohara et al. 2004; Zhang et al. 2005). While MrgprX2 has been detected by several laboratories in the DRG (Kamohara et al. 2005; Robas et al. 2003; Zhang et al. 2005), it also is expressed in mast cells, a part of the innate immune system (Subramanian et al. 2011; Tatemoto et al. 2006). Mast cells are found in most tissues but are enriched in areas at the interface between the inside and outside of the organism, like the skin, lung, and intestine. They are thought of as "first response" cells, activated by pathogens shortly after they breach the tissue, and release immunomodulatory factors, proteases, and massive amounts of histamine and other transmitters that together help inactivate or kill the pathogens (Theoharides et al. 2012). Their classical activation pathway involves cross-linking of IgE receptors, but they also can be activated by numerous positively charged peptides, including exogenous peptides in venoms and toxins, and peptides released from neurons, like substance P and members of the bradykinin family (Ferry et al. 2002; Pundir and Kulka 2010). They occupy a prominent role in itch, because they are the major source of histamine in the skin (Yamatodani et al. 1982), and pain, because they can induce acute inflammation (Theoharides et al. 2012).

Interestingly, MrgprX2 was shown to be activated by substance P and several other mast cell activators at micromolar levels, higher than required for their canonical receptors, if known, but in line with concentrations needed for mast cell activation in mice (Tatemoto et al. 2006). These substances are generally thought to be receptor independent, as they have been shown to intercalate into lipid bilayers and possibly mimic the G-protein activating domains of GPCRs to activate G proteins directly (Ferry et al. 2002). In the absence of knockout studies, the role of MrgprX2 is unknown, especially given the high concentrations of agonists required to activate the receptor. No confirmed orthologs in rodents have been identified to facilitate study of MrgprX2. Rat MrgprB3 was shown to respond to two mast cell activators, but the concentrations required and responses of other Mrgprs was not reported (Tatemoto et al. 2006). No examination has been reported for a mouse ortholog.

12.10 DIMERIZATION

Mrgprs have been considered as though they operate independently of each other, even though multiple family members have been found in the same cells. Two studies have shown that this view may have to be reconsidered. In one study, MrgprD and MrgprE were shown to interact when coexpressed in heterologous cells (Milasta et al. 2006). The presence of MrgprE enhanced the intracellular signaling ability of MrgprD after beta-alanine stimulation and slowed down its internalization rate (Milasta et al. 2006). Local subcellular regulation of MrgprE may thus modulate MrgprD signaling so that the same ligand can have dramatically different effects, depending on where it contacts the cell. If a ligand is found for MrgprE, the same experiment can be tested on it to see if the modulation is bidirectional. This also suggests that activation of one family member may be able to influence the surface fraction and activity of others in the cell.

Another study found a possible interaction between MrgprX1 and delta opioid receptor (DOR) (Breit et al. 2006). When an agonist, BAM22, which could activate both receptors, was applied to cells overexpressing both receptors, MrgprX1 was activated at the expense of DOR signaling (Breit et al. 2006). Coapplication of agonists that activate each receptor independently resulted again in selective MrgprX1 signaling while inhibiting DOR signaling (Breit et al. 2006). The two receptors evidently could heterodimerize when overexpressed, indicating that they can form a complex and that MrgprX1 somehow dominates over DOR signaling when signals for both receptors are received (Breit et al. 2006). The physiological relevance of this is unclear, but considering that MrgprX1 agonists may inhibit pain sensation (Guan et al. 2010), a role for Mrgprs in modulation of pain pathways is an especially exciting idea. MrgprX2 can be activated directly by morphine (Akuzawa et al. 2007) MrgprX2 might be involved in morphine-induced itch.

12.11   QUESTIONS

The Mrgpr field, and indeed the entire field of itch, is exciting but immature. Even basic cell biology tools like reliable antibodies are not established, while the networks that define itch pathways in the nervous system are just beginning to be sketched out. Thus, perhaps it is not surprising that a number of fundamental questions have not been answered. In this section, we will cover some of the most outstanding issues that must be resolved to sharpen our view of the field.

12.11.1   DO MRGPR EXPRESSION PATTERNS CHANGE BETWEEN SPECIES?

In mice, studies have shown that MrgprA3, MrgprC11, and histamine receptors are colocalized in a subset of DRG neurons (Liu et al. 2009). This has led to the hypothesis that these neurons are itch selective. The story, however, becomes more complicated when data from other species are taken into account. A population of rat trigeminal and DRG neurons are responsive to chloroquine but not histamine, demonstrating that not all pruritogens activate chloroquine-sensitive neurons (Klein et al. 2011). The rat ortholog of MrgprC11, MrgprC, does not closely colocalize with other Mrgprs (Zylka et al. 2003). Humans have a mixed response to BAM8-22, reporting a stinging sensation, and thus do not experience pure itch (Sikand et al. 2011). This indicates that MrgprX1, if this is the only receptor for BAM8-22, is in a wider group of neurons than those involved in itch, though other explanations are possible.

These possible differences in expression patterns are hardly unique to Mrgprs. Expression of TrpV1 and Substance P has been reported to differ as well between rats and mice, and even between trigeminal and DRG neurons in the same species (Price and Flores 2007). It still may be the case that itch-selective neurons exist, but the markers that precisely define these neurons may not be the same between species. Also, more than one population of itch neurons may be present. In primates, different CNS neurons respond to histamine and cowhage, another pruritogen, indicating that at some point the nervous system detects them as distinct signals (Davidson et al. 2007). Myelinated neurons may also carry an itch signal, and some evidence exists in humans that neurons sensitive to pruritogens do not completely overlap (Johanek et al. 2008; Ma et al. 2012; Ringkamp et al. 2011).

12.11.2   TRULY ORTHOLOGOUS PAIRS?

The homology between putative orthologous Mrgprs in humans and mice is lower than for most GPCRs. Human MrgprX1 shares only 54% amino acid identity with the mouse MrgprC11 and less than 50% with mouse MrgprA3, its other putative ortholog. In contrast, other receptors associated with itch share much higher identity. For instance, the longest forms of the Histamine 1 receptor are 78% identical, and various splice variants of the mu opioid receptor exceed 90% identity. In fact, the relatively low sequence similarity and expansion in rodents has made pairing of orthologs between the MrgprX family and the mouse MrgprA-C cluster difficult. To date, only MrgprX1 has been shown to have mouse orthologs, and these were linked through similar ligand binding profiles and expression patterns (Han et al. 2002; Lembo et al. 2002; Liu et al. 2009). However, some divergences have been observed that complicate such a simple approach. One notable example is that MrgprX1 in humans responds to chloroquine and BAM8-22, while separate receptors are needed in mouse to detect these substances (Liu et al. 2009). SLIGRL-NH2, the canonical mouse PAR2 agonist, is detected by MrgprC11, while the comparable sequence in humans, SLIGKV-NH2, is recognized not by MrgprX1 but by its relative, MrgprX2 (Liu et al. 2011). The relevance of the differences in affinity is unknown because it is unclear which, if any, of these substances actually transmits an itch signal. It is reasonable to predict that some of these substances are unimportant to itch, that the receptors are not exposed to them, and thus that any differences in affinity are not relevant. However, it also is possible that the mouse and human receptors are not fully orthologous and may have some unique properties.

12.11.3   LIGANDS—ENDOGENOUS OR EXOGENOUS, OR BOTH?

The existence of around 50 genes and pseudogenes in mice (with 30 of them potentially encoding full reading frames), compared to ~18 in humans (10 potentially coding), is reminiscent of vomeronasal and olfactory receptors (Dong et al. 2001; Zylka et al. 2003). Similarly to the Mrgpr family, most of these sensory receptors are located in gene clusters and have substantially more members in rodents than in humans (Mombaerts 2004). It is thought that they respond to external cues and that humans have fewer because the sense of smell is much more critical for survival and communication in rodents than in humans. Can a similar inference be made for the Mrgprs? Are most of the substances detected by the Mrgpr family found in the environment? Is the difference in receptor cohort between species related to differences in environmental threats? As more ligands are discovered for the Mrgpr family members, an evaluation can be made of this and other hypotheses, but for now no explanation suffices as to why the numbers vary so widely in different species.

Ligand binding profiles actually would argue against the Mrgprs as detectors of external substances, as most identified ligands are endogenous, except for chloroquine. MrgprX2 has been shown to be activated by high concentrations of several positively charged peptides found outside the body, some of which can also activate MrgprX1, but the relevance of these low-affinity interactions is not known (Tatemoto et al. 2006). The finding that any endogenous substance can activate Mrgpr family members is curious. Currently, it is not known whether most itch and pain sensation is similar to olfaction, in which the substances directly activate neurons. It is possible that pruritogens might act indirectly by inducing other cells, like keratinocytes and endothelial cells, to release transmitters that are recognized by neurons. More studies of known endogenous Mrgpr agonists may shed some light on this highly important issue.

As a final note, receptor orthologs that have such little homology, and yet the same ligands are rarely seen. An example to be considered is that of the cytokine receptors. These are mostly not GPCRs but may yet be instructive, as they show extensive divergence between orthologs, sometimes 40% or more at the amino acid level. One such case is the broadly important receptor for the cytokine IL-4. This receptor is only 52% identical between human and mouse. Interestingly, IL-4 itself also has diverged extensively, with the precursor sharing only ~40% identity. Another example is IL-31, whose orthologs share about 30% identity, and its coreceptor IL31RA, with about 60% identity between mouse and human. It is possible, then, that receptors can diverge extensively at the amino acid level and retain their relevant ligand binding properties, because the ligands changed, as well. Thus, when considering the basic properties of the Mrgprs of large families and extensive divergence, precedents can be found in other receptor families for external and internal ligands.

ACKNOWLEDGMENT

The work was supported by grants from the NIH to X.D. (NS054791 and GM087369).

REFERENCES

Akuzawa, N., Obinata, H., Izumi, T., Takeda, S. (2007). Morphine is an exogeneous ligand for MrgX2, a G protein-coupled receptor for cortistatin. *JCAB 2*, 4–9.

Avula, L.R., Buckinx, R., Alpaerts, K., Costagliola, A., Adriaensen, D., Van Nassauw, L., and Timmermans, J.P. (2011). The effect of inflammation on the expression and distribution of the MAS-related gene receptors MrgE and MrgF in the murine ileum. *Histochem Cell Biol 136*, 569–585.

Bandell, M., Story, G.M., Hwang, S.W., Viswanath, V., Eid, S.R., Petrus, M.J., Earley, T.J., and Patapoutian, A. (2004). Noxious cold ion channel TRPA1 is activated by pungent compounds and bradykinin. *Neuron 41*, 849–857.

Bender, E., Buist, A., Jurzak, M., Langlois, X., Baggerman, G., Verhasselt, P., Ercken, M. et al. (2002). Characterization of an orphan G protein-coupled receptor localized in the dorsal root ganglia reveals adenine as a signaling molecule. *Proc Natl Acad Sci USA 99*, 8573–8578.

Breit, A., Gagnidze, K., Devi, L.A., Lagace, M., and Bouvier, M. (2006). Simultaneous activation of the delta opioid receptor (deltaOR)/sensory neuron-specific receptor-4 (SNSR-4) hetero-oligomer by the mixed bivalent agonist bovine adrenal medulla peptide 22 activates SNSR-4 but inhibits deltaOR signaling. *Mol Pharmacol 70*, 686–696.

Chen, H., and Ikeda, S.R. (2004). Modulation of ion channels and synaptic transmission by a human sensory neuron-specific G-protein-coupled receptor, SNSR4/mrgX1, heterologously expressed in cultured rat neurons. *J Neurosci 24*, 5044–5053.

Cox, P.J., Pitcher, T., Trim, S.A., Bell, C.H., Qin, W., and Kinloch, R.A. (2008). The effect of deletion of the orphan G-protein coupled receptor (GPCR) gene MrgE on pain-like behaviours in mice. *Mol Pain 4*, 2.

Crozier, R.A., Ajit, S.K., Kaftan, E.J., and Pausch, M.H. (2007). MrgD activation inhibits KCNQ/M-currents and contributes to enhanced neuronal excitability. *J Neurosci 27*, 4492–4496.

Davidson, S., Zhang, X., Yoon, C.H., Khasabov, S.G., Simone, D.A., and Giesler, G.J., Jr. (2007). The itch-producing agents histamine and cowhage activate separate populations of primate spinothalamic tract neurons. *J Neurosci 27*, 10007–10014.

Dong, X., Han, S., Zylka, M.J., Simon, M.I., and Anderson, D.J. (2001). A diverse family of GPCRs expressed in specific subsets of nociceptive sensory neurons. *Cell 106*, 619–632.

Dussor, G., Zylka, M.J., Anderson, D.J., and McCleskey, E.W. (2008). Cutaneous sensory neurons expressing the Mrgprd receptor sense extracellular ATP and are putative nociceptors. *J Neurophysiol 99*, 1581–1589.

Ekpechi, O.L., and Okoro, A.N. (1964). A pattern of pruritus due to chloroquine. *Arch Dermatol 89*, 631–632.

Ferry, X., Brehin, S., Kamel, R., and Landry, Y. (2002). G protein-dependent activation of mast cell by peptides and basic secretagogues. *Peptides 23*, 1507–1515.

Gottfried, J.A. (2010). Central mechanisms of odour object perception. *Nat Rev Neurosci 11*, 628–641.

Grazzini, E., Puma, C., Roy, M.O., Yu, X.H., O'Donnell, D., Schmidt, R., Dautrey, S. et al. (2004). Sensory neuron-specific receptor activation elicits central and peripheral nociceptive effects in rats. *Proc Natl Acad Sci USA 101*, 7175–7180.

Guan, Y., Liu, Q., Tang, Z., Raja, S.N., Anderson, D.J., and Dong, X. (2010). Mas-related G-protein-coupled receptors inhibit pathological pain in mice. *Proc Natl Acad Sci USA 107*, 15933–15938.

Han, S.K., Dong, X., Hwang, J.I., Zylka, M.J., Anderson, D.J., and Simon, M.I. (2002). Orphan G protein-coupled receptors MrgA1 and MrgC11 are distinctly activated by RF-amide-related peptides through the Galpha q/11 pathway. *Proc Natl Acad Sci USA 99*, 14740–14745.

Hindson, B.J., Ness, K.D., Masquelier, D.A., Belgrader, P., Heredia, N.J., Makarewicz, A.J., Bright, I.J. et al. (2011). High-throughput droplet digital PCR system for absolute quantitation of DNA copy number. *Anal Chem 83*, 8604–8610.

Imamachi, N., Park, G.H., Lee, H., Anderson, D.J., Simon, M.I., Basbaum, A.I., and Han, S.K. (2009). TRPV1-expressing primary afferents generate behavioral responses to pruritogens via multiple mechanisms. *Proc Natl Acad Sci USA 106*, 11330–11335.

Johanek, L.M., Meyer, R.A., Friedman, R.M., Greenquist, K.W., Shim, B., Borzan, J., Hartke, T., LaMotte, R.H., and Ringkamp, M. (2008). A role for polymodal C-fiber afferents in nonhistaminergic itch. *J Neurosci 28*, 7659–7669.

Jordt, S.E., Bautista, D.M., Chuang, H.H., McKemy, D.D., Zygmunt, P.M., Hogestatt, E.D., Meng, I.D., and Julius, D. (2004). Mustard oils and cannabinoids excite sensory nerve fibres through the TRP channel ANKTM1. *Nature 427*, 260–265.

Kamohara, M., Matsuo, A., Takasaki, J., Kohda, M., Matsumoto, M., Matsumoto, S., Soga, T., Hiyama, H., Kobori, M., and Katou, M. (2005). Identification of MrgX2 as a human G-protein-coupled receptor for proadrenomedullin N-terminal peptides. *Biochem Biophys Res Commun 330*, 1146–1152.

Kato, M., Nakamura, Y., and Tsunoda, T. (2008). An algorithm for inferring complex haplotypes in a region of copy-number variation. *Am J Human Genetics 83*, 157–169.

Klein, A., Carstens, M.I., and Carstens, E. (2011). Facial injections of pruritogens or algogens elicit distinct behavior responses in rats and excite overlapping populations of primary sensory and trigeminal subnucleus caudalis neurons. *J Neurophysiol 106*, 1078–1088.

Lembo, P.M., Grazzini, E., Groblewski, T., O'Donnell, D., Roy, M.O., Zhang, J., Hoffert, C. et al. (2002). Proenkephalin A gene products activate a new family of sensory neuron–specific GPCRs. *Nat Neurosci 5*, 201–209.

Liu, Q., Tang, Z., Surdenikova, L., Kim, S., Patel, K.N., Kim, A., Ru, F. et al. (2009). Sensory neuron-specific GPCR Mrgprs are itch receptors mediating chloroquine-induced pruritus. *Cell 139*, 1353–1365.

Liu, Q., Vrontou, S., Rice, F.L., Zylka, M.J., Dong, X., and Anderson, D.J. (2007). Molecular genetic visualization of a rare subset of unmyelinated sensory neurons that may detect gentle touch. *Nat Neurosci 10*, 946–948.

Liu, Q., Weng, H.J., Patel, K.N., Tang, Z., Bai, H., Steinhoff, M., and Dong, X. (2011). The distinct roles of two GPCRs, MrgprC11 and PAR2, in itch and hyperalgesia. *Sci Signal 4*, ra45.

Ma, C., Nie, H., Gu, Q., Sikand, P., and Lamotte, R.H. (2012). In vivo responses of cutaneous C-mechanosensitive neurons in mouse to punctate chemical stimuli that elicit itch and nociceptive sensations in humans. *J Neurophysiol 107*, 357–363.

Ma, Q., Fode, C., Guillemot, F., and Anderson, D.J. (1999). Neurogenin1 and neurogenin2 control two distinct waves of neurogenesis in developing dorsal root ganglia. *Genes Dev 13*, 1717–1728.

Milasta, S., Pediani, J., Appelbe, S., Trim, S., Wyatt, M., Cox, P., Fidock, M., and Milligan, G. (2006). Interactions between the Mas-related receptors MrgD and MrgE alter signalling and trafficking of MrgD. *Mol Pharmacol 69*, 479–491.

Mombaerts, P. (2004). Genes and ligands for odorant, vomeronasal and taste receptors. *Nat Rev Neurosci 5*, 263–278.

Moran, M.M., McAlexander, M.A., Biro, T., and Szallasi, A. (2011). Transient receptor potential channels as therapeutic targets. *Nat Rev Drug Discov 10*, 601–620.

Price, T.J., and Flores, C.M. (2007). Critical evaluation of the colocalization between calcitonin gene-related peptide, substance P, transient receptor potential vanilloid subfamily type 1 immunoreactivities, and isolectin B4 binding in primary afferent neurons of the rat and mouse. *J Pain 8*, 263–272.

Pundir, P., and Kulka, M. (2010). The role of G protein-coupled receptors in mast cell activation by antimicrobial peptides: Is there a connection? *Immunol Cell Biol 88*, 632–640.

Ramachandran, R., and Hollenberg, M.D. (2008). Proteinases and signalling: Pathophysiological and therapeutic implications via PARs and more. *Br J Pharmacol 153 Suppl 1*, S263–S282.

Rau, K.K., McIlwrath, S.L., Wang, H., Lawson, J.J., Jankowski, M.P., Zylka, M.J., Anderson, D.J., and Koerber, H.R. (2009). Mrgprd enhances excitability in specific populations of cutaneous murine polymodal nociceptors. *J Neurosci 29*, 8612–8619.

Redon, R., Ishikawa, S., Fitch, K.R., Feuk, L., Perry, G.H., Andrews, T.D., Fiegler, H. et al. (2006). Global variation in copy number in the human genome. *Nature 444*, 444–454.

Ringkamp, M., Schepers, R.J., Shimada, S.G., Johanek, L.M., Hartke, T.V., Borzan, J., Shim, B., LaMotte, R.H., and Meyer, R.A. (2011). A role for nociceptive, myelinated nerve fibers in itch sensation. *J Neurosci 31*, 14841–14849.

Robas, N., Mead, E., and Fidock, M. (2003). MrgX2 is a high potency cortistatin receptor expressed in dorsal root ganglion. *J Biol Chem 278*, 44400–44404.

Salako, L.A. (1984). Toxicity and side-effects of antimalarials in Africa: A critical review. *Bull World Health Org 62 Suppl*, 63–68.

Santos, R.A., Simoes e Silva, A.C., Maric, C., Silva, D.M., Machado, R.P., de Buhr, I., Heringer-Walther, S. et al. (2003). Angiotensin-(1-7) is an endogenous ligand for the G protein-coupled receptor Mas. *Proc Natl Acad Sci USA 100*, 8258–8263.

Shimada, S.G., Shimada, K.A., and Collins, J.G. (2006). Scratching behavior in mice induced by the proteinase-activated receptor-2 agonist, SLIGRL-NH2. *Eur J Pharmacol 530*, 281–283.

Shinohara, T., Harada, M., Ogi, K., Maruyama, M., Fujii, R., Tanaka, H., Fukusumi, S. et al. (2004). Identification of a G protein-coupled receptor specifically responsive to beta-alanine. *J Biol Chem 279*, 23559–23564.

Sikand, P., Dong, X., and LaMotte, R.H. (2011). BAM8-22 peptide produces itch and nociceptive sensations in humans independent of histamine release. *J Neurosci 31*, 7563–7567.

Sowunmi, A., Walker, O., and Salako, L.A. (1989). Pruritus and antimalarial drugs in Africans. *Lancet 2*, 213.

Steinhoff, M., Neisius, U., Ikoma, A., Fartasch, M., Heyer, G., Skov, P.S., Luger, T.A., and Schmelz, M. (2003). Proteinase-activated receptor-2 mediates itch: A novel pathway for pruritus in human skin. *J Neurosci 23*, 6176–6180.

Steinhoff, M., Vergnolle, N., Young, S.H., Tognetto, M., Amadesi, S., Ennes, H.S., Trevisani, M. et al. (2000). Agonists of proteinase-activated receptor 2 induce inflammation by a neurogenic mechanism. *Nat Med 6*, 151–158.

Subramanian, H., Kashem, S.W., Collington, S.J., Qu, H., Lambris, J.D., and Ali, H. (2011). PMX-53 as a dual CD88 antagonist and an agonist for Mas-related gene 2 (MrgX2) in human mast cells. *Mol Pharmacol 79*, 1005–1013.

Sun, Y.G., and Chen, Z.F. (2007). A gastrin-releasing peptide receptor mediates the itch sensation in the spinal cord. *Nature 448*, 700–703.

Swain, M.G., MacArthur, L., Vergalla, J., and Jones, E.A. (1994). Adrenal secretion of BAM-22P, a potent opioid peptide, is enhanced in rats with acute cholestasis. *Am J Physiol 266*, G201–G205.

Tatemoto, K., Nozaki, Y., Tsuda, R., Konno, S., Tomura, K., Furuno, M., Ogasawara, H. et al. (2006). Immunoglobulin E-independent activation of mast cell is mediated by Mrg receptors. *Biochem Biophys Res Commun 349*, 1322–1328.

Theoharides, T.C., Alysandratos, K.D., Angelidou, A., Delivanis, D.A., Sismanopoulos, N., Zhang, B., Asadi, S. et al. (2012). Mast cells and inflammation. *Biochim Biophys Acta 1822*, 21–33.

von Kugelgen, I., Schiedel, A.C., Hoffmann, K., Alsdorf, B.B., Abdelrahman, A., and Muller, C.E. (2008). Cloning and functional expression of a novel Gi protein-coupled receptor for adenine from mouse brain. *Mol Pharmacol 73*, 469–477.

Wang, H., and Zylka, M.J. (2009). Mrgprd-expressing polymodal nociceptive neurons innervate most known classes of substantia gelatinosa neurons. *J Neurosci 29*, 13202–13209.

Wilson, S.R., Gerhold, K.A., Bifolck-Fisher, A., Liu, Q., Patel, K.N., Dong, X., and Bautista, D.M. (2011). TRPA1 is required for histamine-independent, Mas-related G protein-coupled receptor-mediated itch. *Nat Neurosci 14*, 595–602.

Wong, K.K., deLeeuw, R.J., Dosanjh, N.S., Kimm, L.R., Cheng, Z., Horsman, D.E., MacAulay, C. et al. (2007). A comprehensive analysis of common copy-number variations in the human genome. *Am J Human Genetics 80*, 91–104.

Wu, L.J., Sweet, T.B., and Clapham, D.E. (2010). International Union of Basic and Clinical Pharmacology. LXXVI. Current progress in the mammalian TRP ion channel family. *Pharmacol Rev 62*, 381–404.

Yamatodani, A., Maeyama, K., Watanabe, T., Wada, H., and Kitamura, Y. (1982). Tissue distribution of histamine in a mutant mouse deficient in mast cells: Clear evidence for the presence of non-mast-cell histamine. *Biochem Pharmacol 31*, 305–309.

Zhang, L., Taylor, N., Xie, Y., Ford, R., Johnson, J., Paulsen, J.E., and Bates, B. (2005). Cloning and expression of MRG receptors in macaque, mouse, and human. *Brain Res Mol Brain Res 133*, 187–197.

Zylka, M.J., Dong, X., Southwell, A.L., and Anderson, D.J. (2003). Atypical expansion in mice of the sensory neuron-specific Mrg G protein-coupled receptor family. *Proc Natl Acad Sci USA 100*, 10043–10048.

Zylka, M.J., Rice, F.L., and Anderson, D.J. (2005). Topographically distinct epidermal nociceptive circuits revealed by axonal tracers targeted to Mrgprd. *Neuron 45*, 17–25.

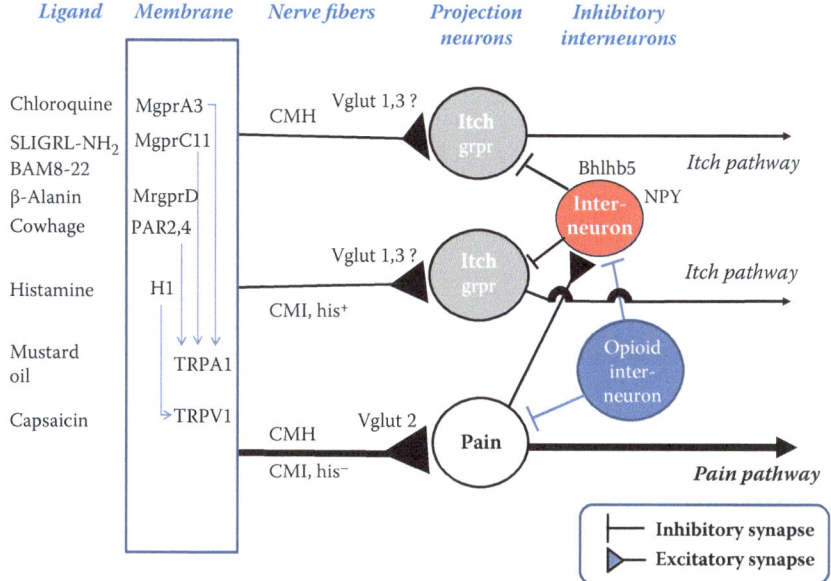

FIGURE 1.1

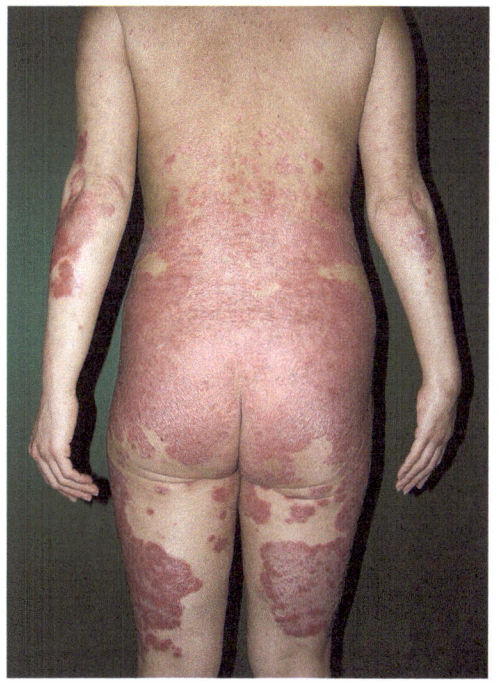

FIGURE 4.1

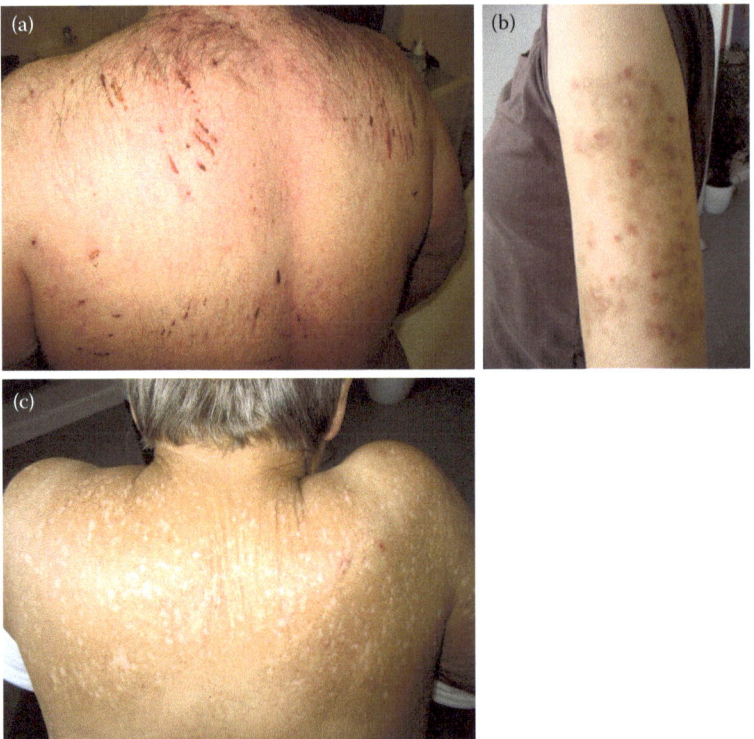

FIGURE 5.1

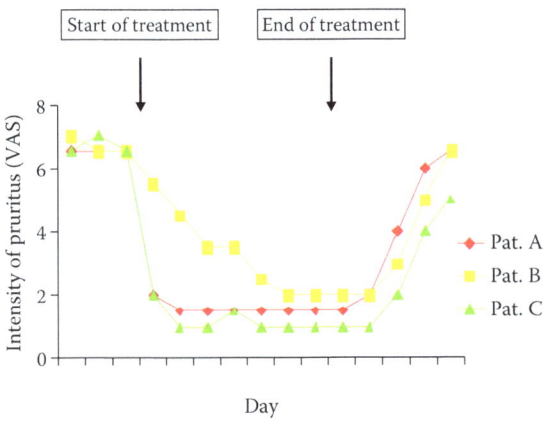

FIGURE 5.4

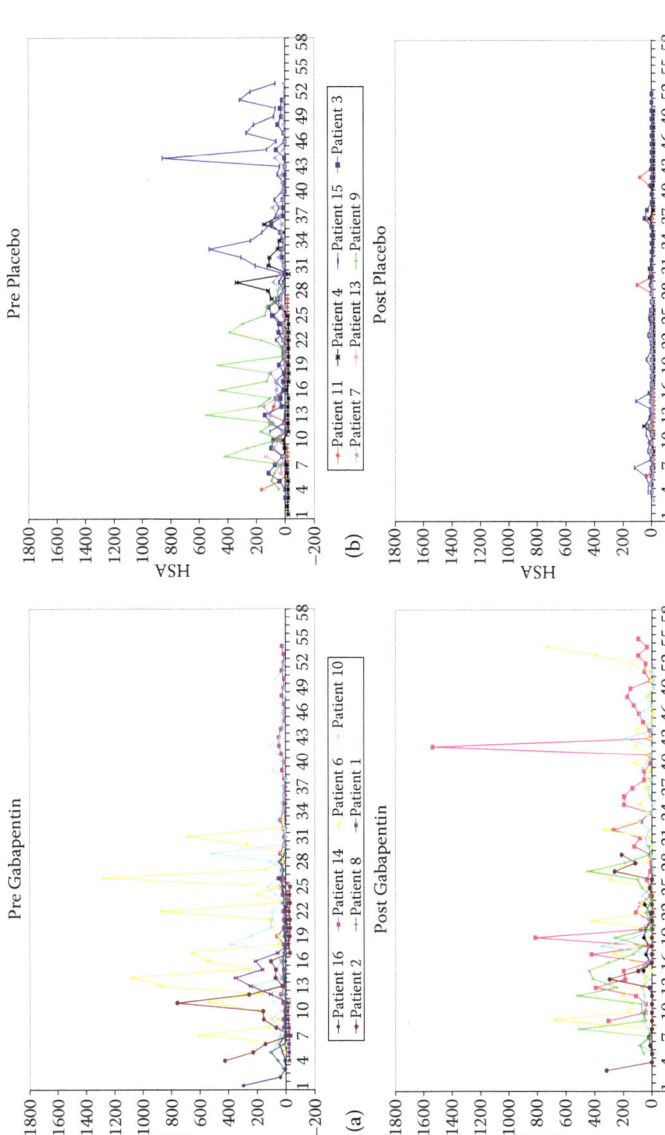

FIGURES 6.2

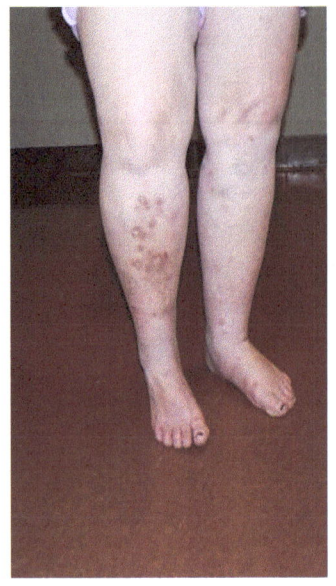

FIGURE 7.1

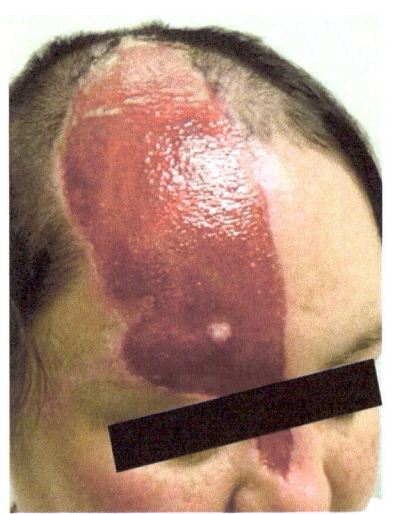

FIGURE 7.2

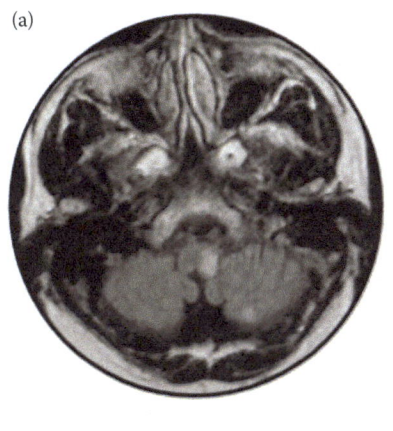

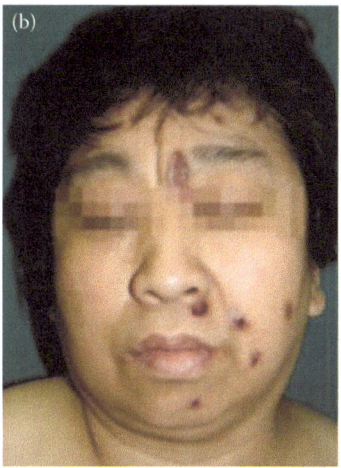

FIGURE 7.3

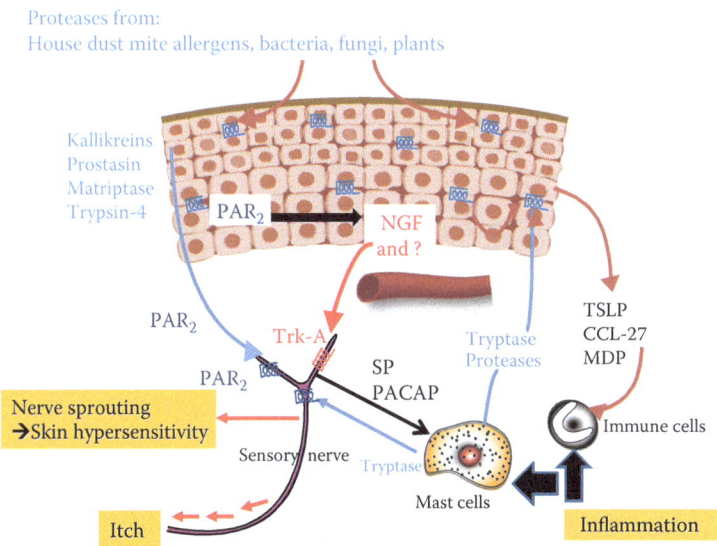

FIGURE 11.1

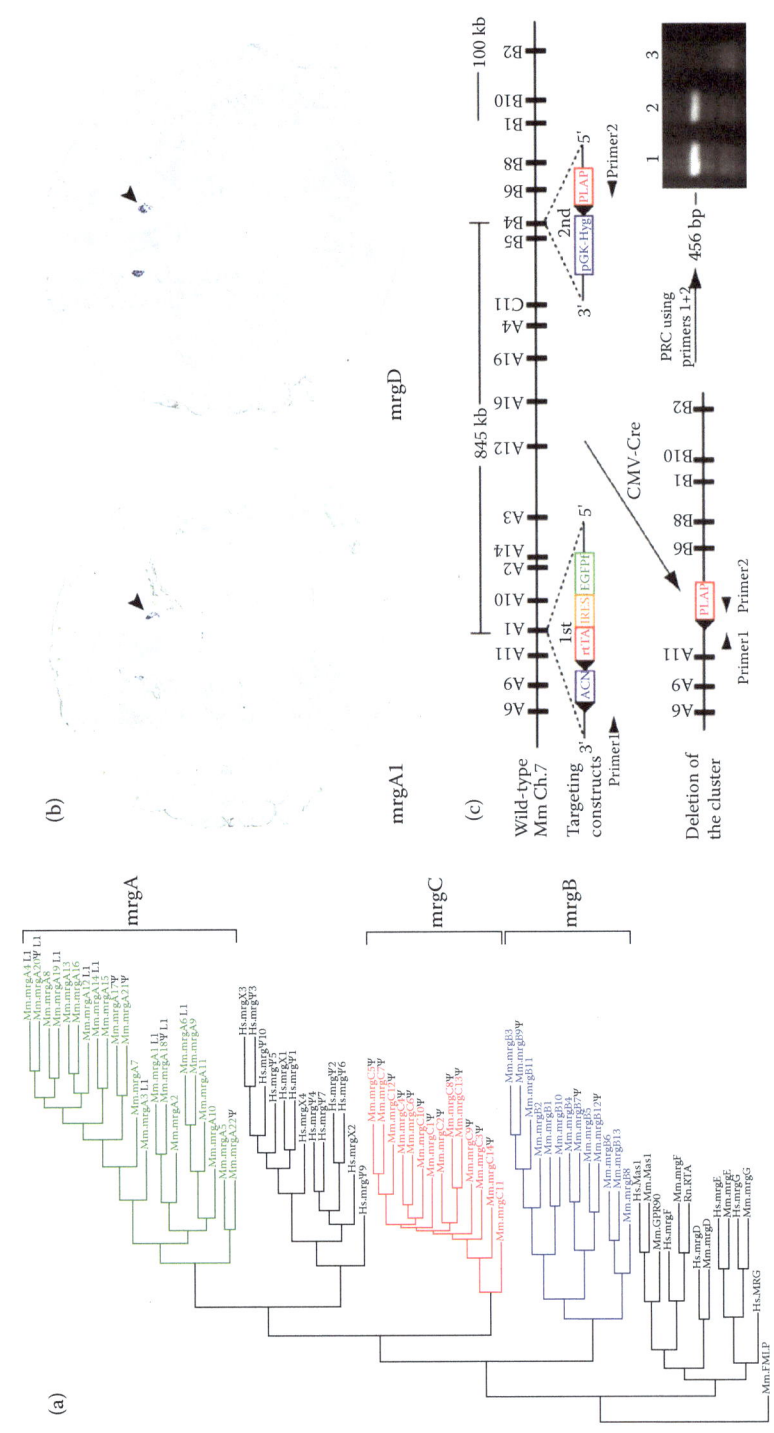

FIGURE 12.1

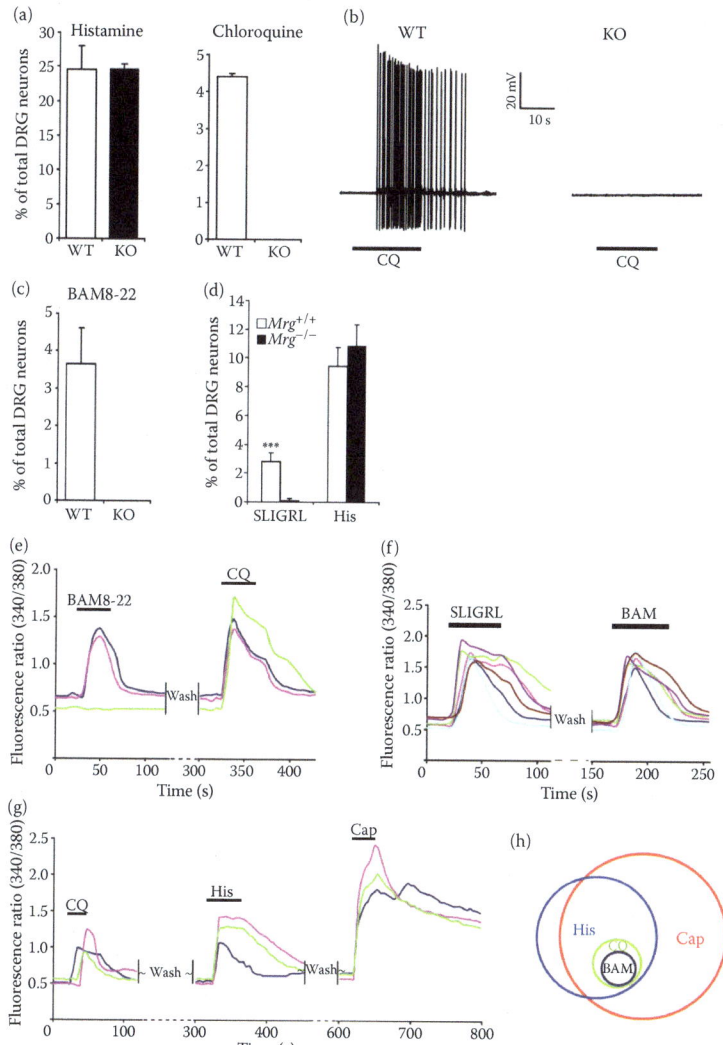

FIGURE 12.3

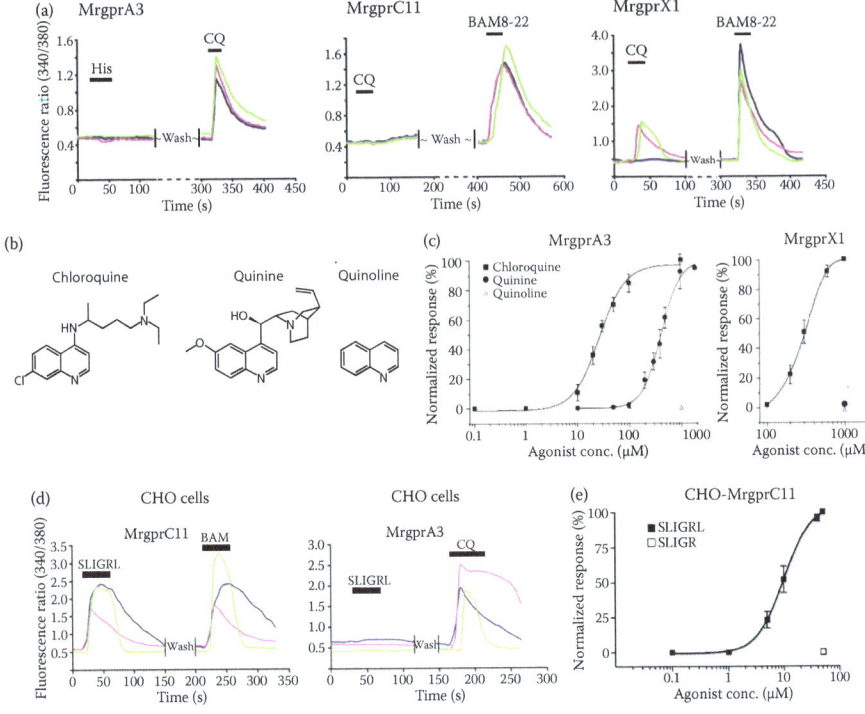

FIGURE 12.4

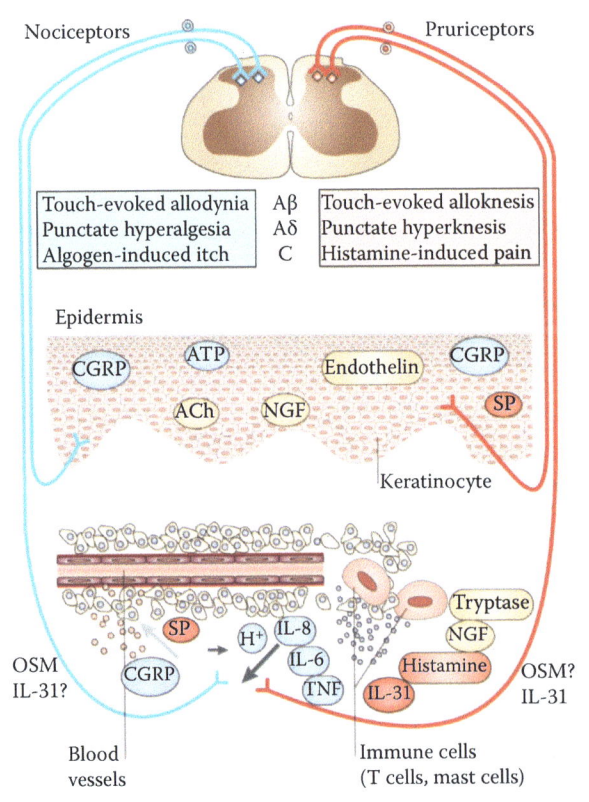

FIGURE 13.1

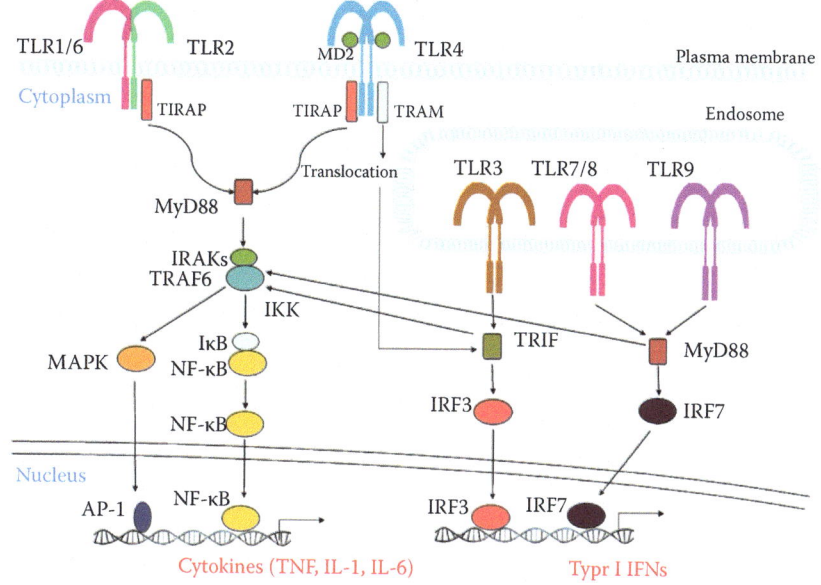

FIGURE 14.1

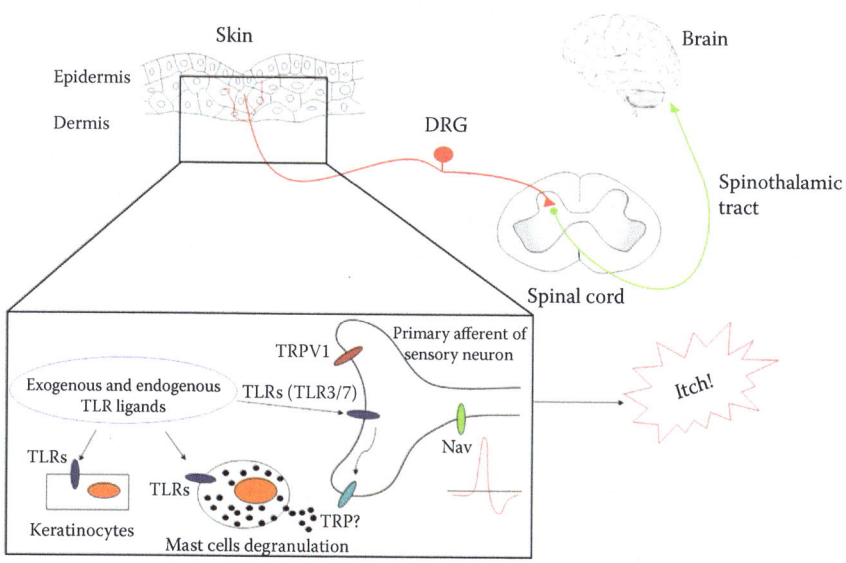

FIGURE 14.2

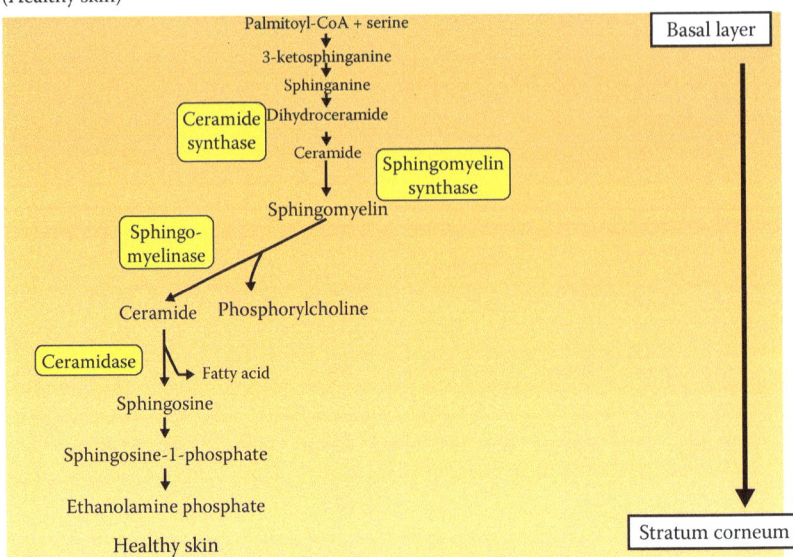

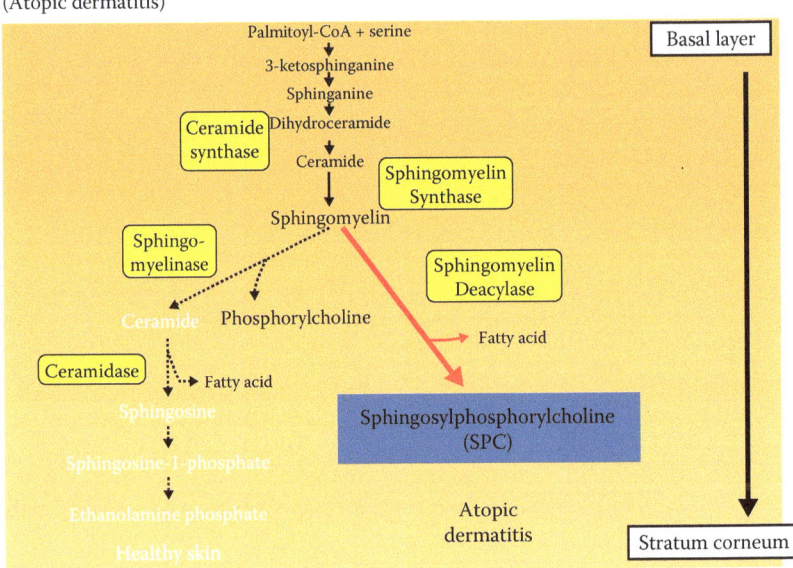

FIGURE 15.1

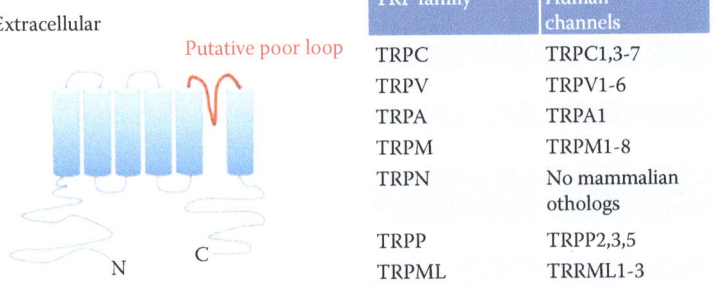

FIGURE 16.1

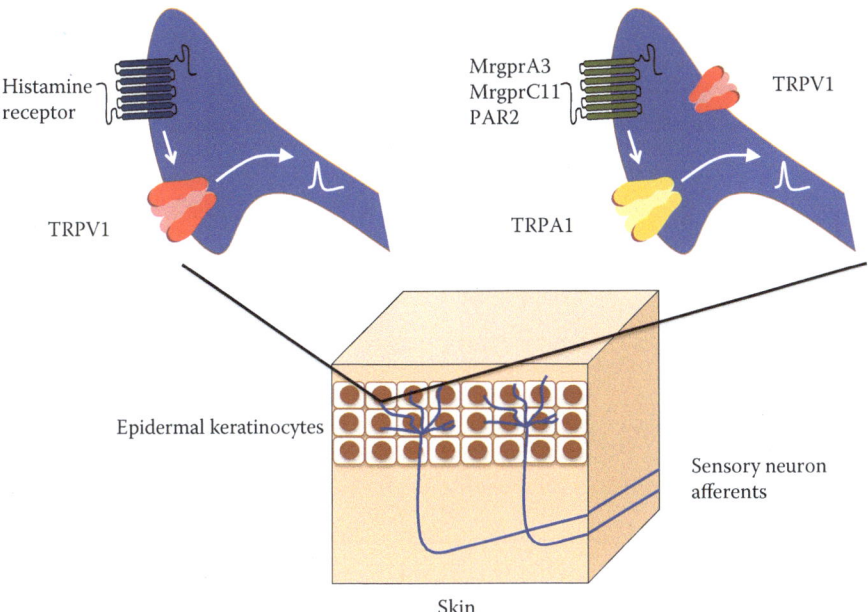

FIGURE 16.2

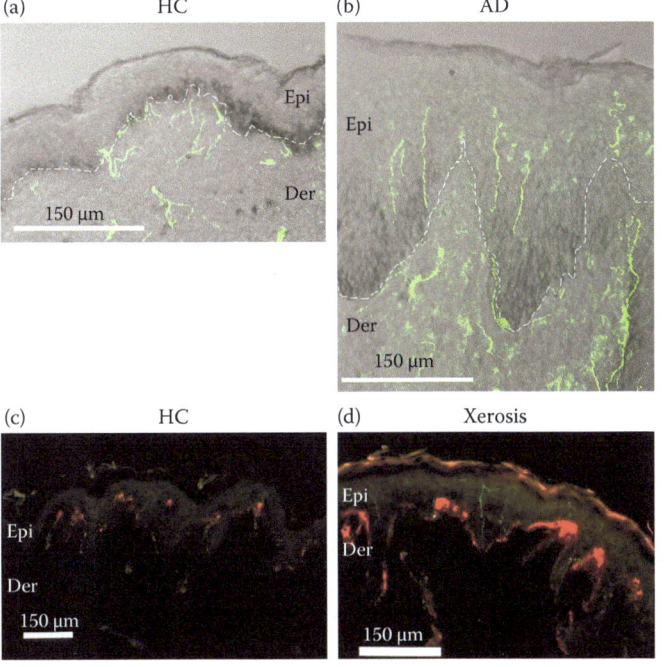

FIGURE 17.1

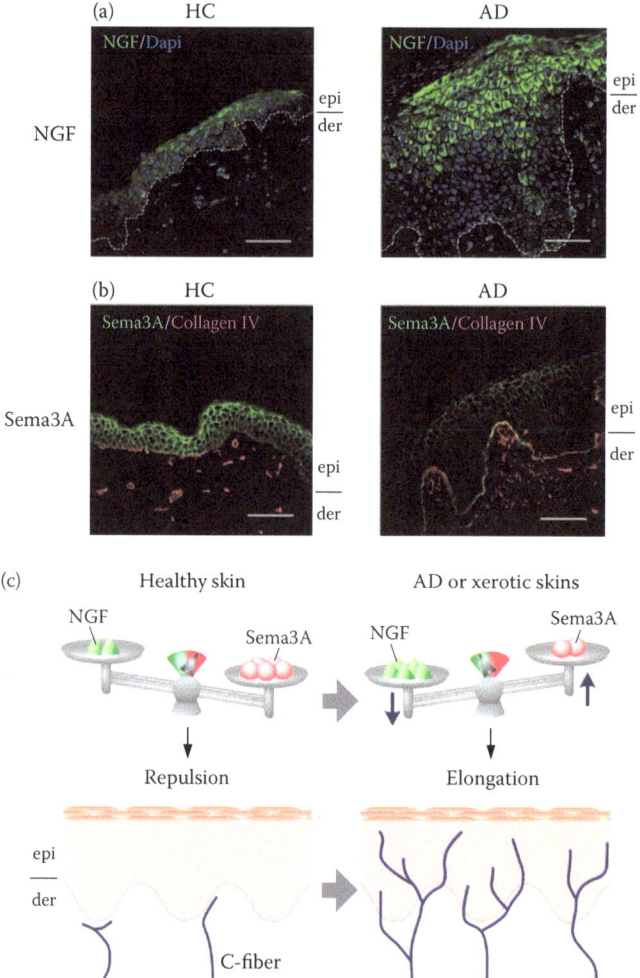

FIGURE 17.2

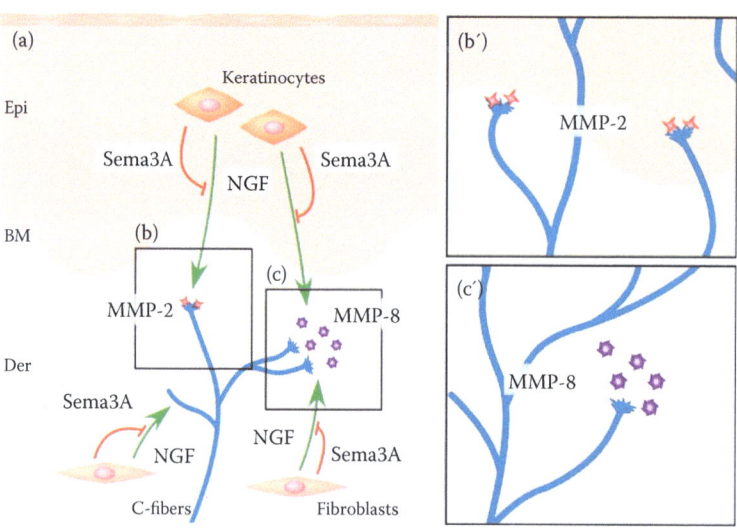

FIGURE 17.3

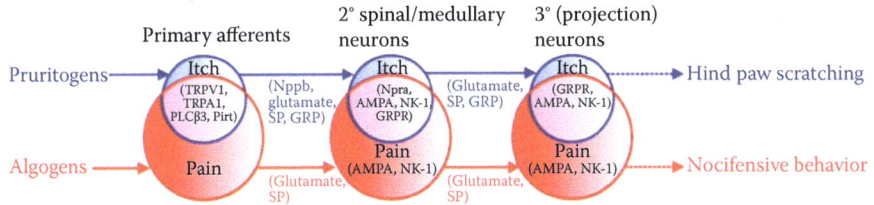

FIGURE 19.2

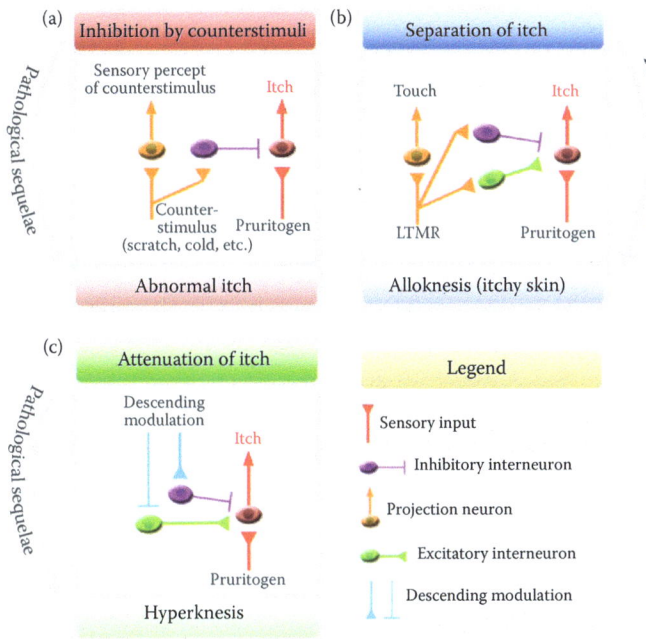

FIGURE 20.1

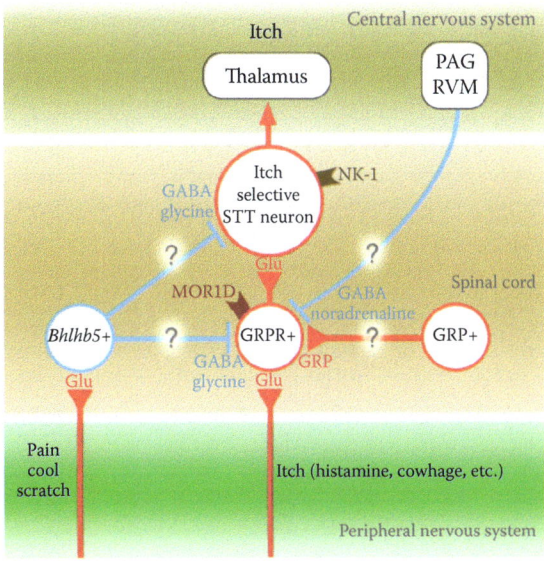

FIGURE 20.2

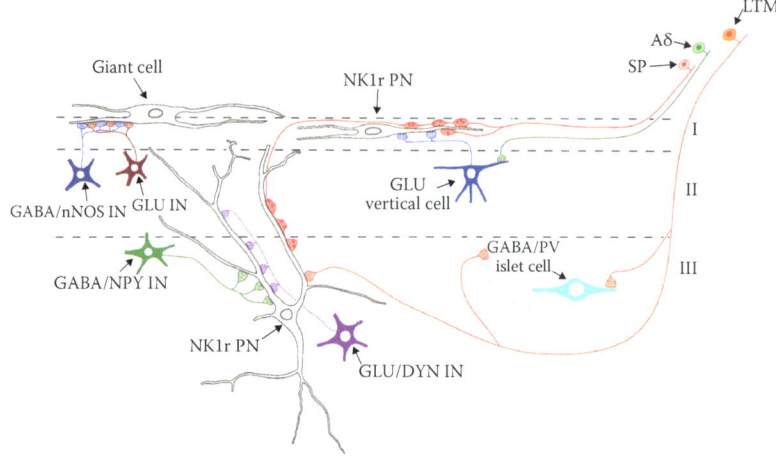

FIGURE 20.4

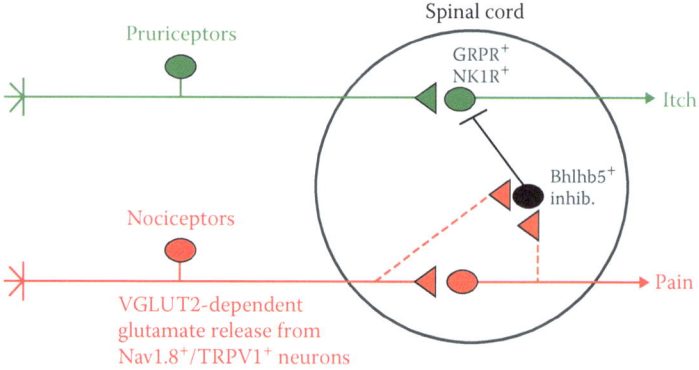

FIGURE 21.1

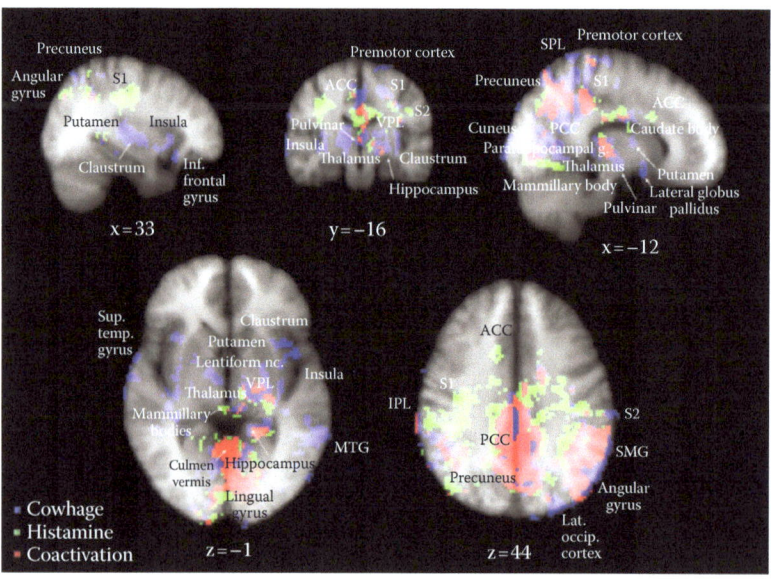

FIGURE 23.1

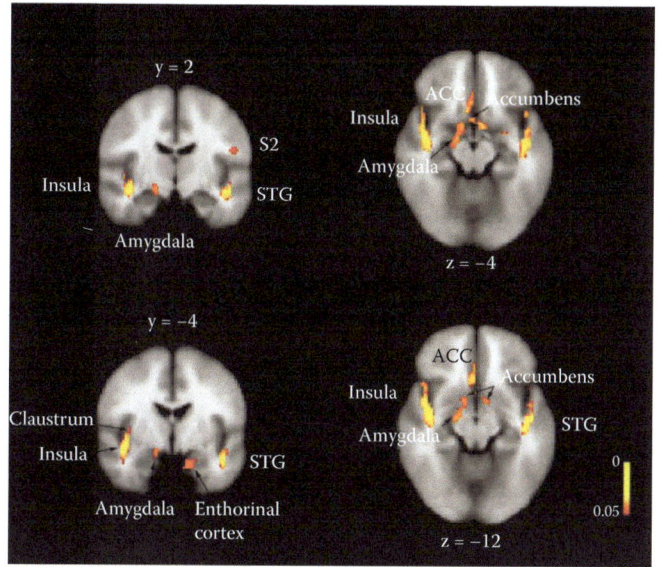

FIGURE 23.2

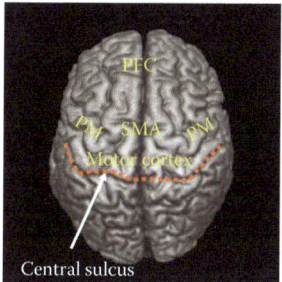

FIGURE 23.3

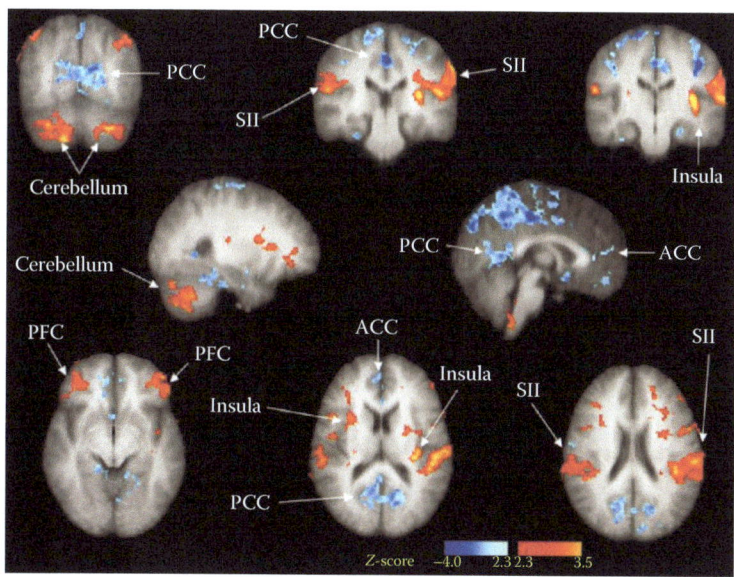

FIGURE 23.4

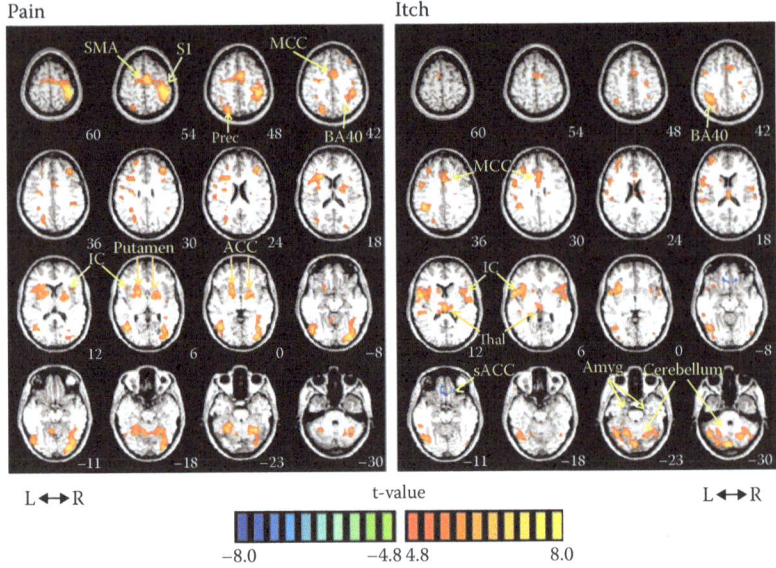

FIGURE 24.1

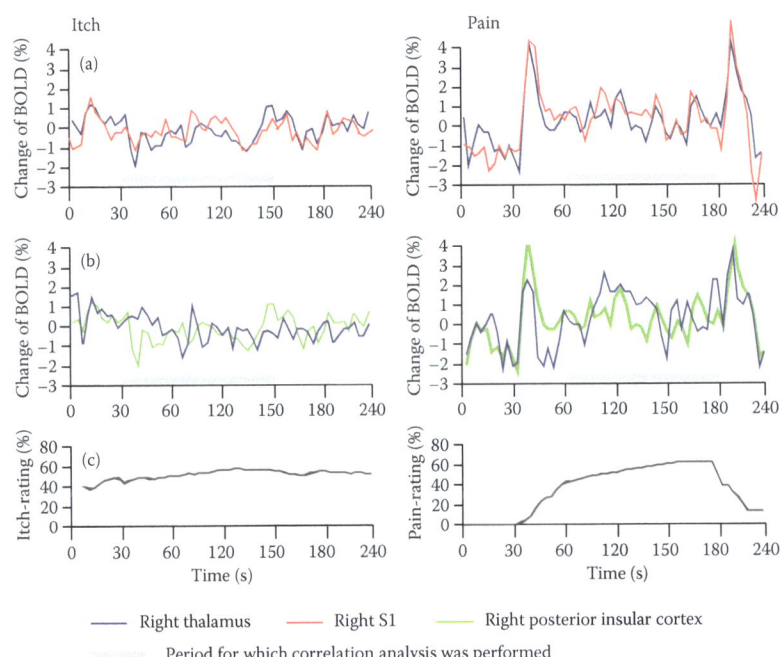

Right thalamus ——— Right S1 ——— Right posterior insular cortex ———

Period for which correlation analysis was performed

FIGURE 24.2

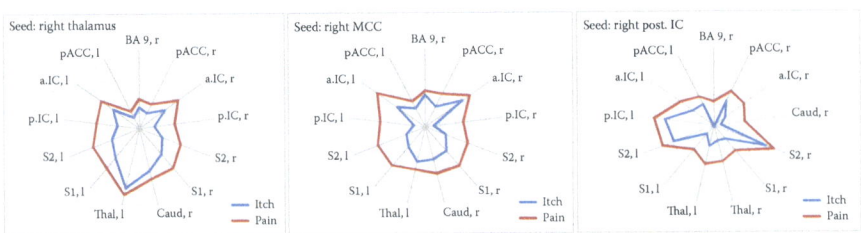

FIGURE 24.3

13 Role of Interleukin-31 and Oncostatin M in Itch and Neuroimmune Communication

Ferda Cevikbas, Cordula Kempkes,
Timo Buhl, Christian Mess,
Joerg Buddenkotte, and Martin Steinhoff

CONTENTS

13.1 INTRODUCTION

13.1.1 SKIN AS A NEUROIMMUNE ORGAN

Generally, cytokines comprise a large family of secreted proteins regulating a variety of cellular functions during inflammation and immune responses. Cytokines activate immune as well as resident skin cells including keratinocytes, endothelial cells, Langerhans cells, mast cells, or fibroblasts. Subsequently, those cells release mediators that potentially communicate with sensory nerves regulating neurogenic inflammation, pain, or itch (Chung et al. 2003, 2004; Dallos et al. 2006; E et al. 2006; Grewe et al. 2000; Huang et al. 2008; Ibrahim et al. 2005; Kakurai et al. 2001; Raychaudhuri et al. 2008; Tanaka et al. 2007). In addition, immune cells (e.g., mast cells) can directly communicate with sensory nerves during inflammation and pruritus (Roosterman et al. 2006) via histamine, tryptase, prostanoids, or RL-peptides (Steinhoff et al. 2000). For example, mediators like histamine or tryptase released by mast cells during inflammation and allergic reactions can directly "talk" to sensory nerves (Steinhoff et al. 2000, 2003) and are thus targets for therapy.

In recent years, new evidence indicates that cytokines or chemokines can also directly communicate with sensory nerves via activation of high-affinity cytokine or chemokine receptors that are involved in pain or inflammation including IL-1, IL-8, IL-10, MCP, CCL2, CCL4, CCL5, MIP1α, for example (Liou et al. 2012, 2013; Liu et al. 2013; Saika et al. 2012; Zhang et al. 2012). Although of importance, information about how immune cells—especially T cells—directly communicate with nerves to regulate neurogenic inflammation, pain, and pruritus is very limited (Bonness and Bieber 2007; Hanifin 2009; Ishiuji et al. 2008; Ong and Leung 2006; Sehra et al. 2008; Yosipovitch et al. 2003; Yosipovitch and Papoiu 2008). Recent studies showed that IL-31, a newly discovered cytokine (Bilsborough et al. 2006; Dillon et al. 2004), and its related cytokine OSM, two members of the IL-6 family of cytokines (Bando et al. 2006; Bilsborough et al. 2006; Boniface et al. 2007; Drogemuller et al. 2008; Grimstad et al. 2009; Morikawa et al. 2004; Neis et al. 2006; Sonkoly et al. 2006; Tamura et al. 2003), are good candidates to understand the role of T-cell-induced neuronal communication. Interestingly, the expression of IL-6 receptor on neurons was described in rats by immunohistochemistry many years ago, although a functional role for IL-6 and its members in sensory nerves is still lacking (Grothe et al. 2000).

13.1.2 IL-6 FAMILY OF CYTOKINES

The IL-6 family of cytokines is comprised of IL-6, IL-21, IL-31, OSM, CNTF, and neuropoietin. OSM (Zhang et al. 2008), as well as "neurotrophic cytokines" such as BDNF (CNTF) or neuropoietin (Bando et al. 2006) have been shown to be important for neuronal regulation. The gp130/IL-6 cytokine family is involved in various important physiological and pathophysiological processes like nerve growth, inflammation, and immune defense (Bando et al. 2006). A role of IL-31 is thus far best understood for its role in atopic diseases and itch.

13.1.2.1 IL-31

The gene encoding human IL-31 is located on chromosome 12q24.31 and the mouse ortholog is situated in a synthetic region of chromosome 5. The IL-31 cDNA is composed of an open reading frame encoding a 164 amino acid (aa) precursor and a predicted 141 aa mature polypeptide containing the four α-helix structure (Steinhoff et al. 2010). In mice, IL-31 mRNA is predominantly released by CD4[+] TH2 helper cells, anti-CD3/CD28-stimulated TH1 and TH2 cells, albeit in CD4[+] TH2 cells with significant higher levels (Dillon et al. 2004). In humans, IL-31 mRNA and protein is predominantly expressed and released by CLA[+](skin-homing[+])-CD45RO[+] memory T cells, and increased expression was found in atopic dermatitis patients (Bilsborough et al. 2006; Dillon et al. 2004). Activated Th2 cells are the main source of IL-31 in humans and mice (Dillon et al. 2004; Sonkoly et al. 2006). Our own and recent studies showed that besides other tissues, IL-31 RNA and protein is also strongly expressed in murine skin and not in DRG neurons (Dillon et al. 2004; Sonkoly et al. 2006). In humans, IL-31 has been detected mainly in different immune cells in the skin. Intriguingly, IFNγ is able to induce the production of IL-31 in human microvascular endothelial cells (Feld et al. 2010). In human mast cells, antimicrobial peptides cathelicidin and β-defensin were reported to affect the IL-31 production, suggesting that IL-31 might derive from different dermal skin types triggered by other cytokines or peptides.

IL-31 seems to be an important cytokine, transmitting itch to the central nervous system. In a murine model, overexpression of IL-31 leads to an atopic dermatitis-like phenotype, including high itch rate, skin lesion formation, increased transepidermal water loss and dry, scaly skin (Dillon et al. 2004). This is of translational value, as patients with atopic dermatitis present high expression of IL-31 in the skin (Nobbe et al. 2012).

13.1.2.2 OSM

OSM has been identified to be significantly involved in several critical physiological and pathophysiological processes, such as inflammation, autoimmunity, tissue remodeling, and cancer (Silver and Hunter 2010). In different murine models, OSM was capable of stimulating collagen production and proliferation of lung and dermal fibroblasts, and OSM overexpression resulted in eosinophilia, lung fibrosis, and lymphocytic infiltration (Bamber et al. 1998; Duncan et al. 1995; Langdon et al. 2000). In patients with rheumatoid arthritis (one of the textbook-classic diseases for Th1-dominated immune dysregulation), OSM is expressed in the synovial fluid and stimulates synovial fibroblasts to proliferate and produce IL-6, MCP-1, and the chemokine CCL13, which contributes to attraction of inflammatory cells (Hintzen et al. 2009; Hui et al. 1997; Ihn and Tamaki 2000). Other studies support relevant involvement of OSM in hematopoiesis (Miyajima et al. 2000; Mukouyama et al. 1998) and hepatocyte differentiation (Kamiya et al. 1999; Miyajima et al. 2000). Interestingly, OSM was also found to have tumor-protective capabilities because OSM can block proliferation in several cancer cell lines, and disease progress in some cancer patients was connected with loss of responsiveness to OSM (Lacreusette et al. 2007; Ouyang et al. 2006). OSM mRNA has been detected in various human tissues and cells, especially in T cells, neutrophils and eosinophils, DRGs, and spinal cord neurons, as well as microglia cells within the central nervous system (Repovic and Benveniste 2002; Tamura et al. 2002). Recent studies also showed that OSM secreted by skin infiltrating T cells is involved in skin inflammation and induce heat hypersensitivity in TRPV1+ neurons (Boniface et al. 2007; Langeslag et al. 2011).

13.1.2.3 Ciliary Neurotrophic Factor

Ciliary neurotrophic factor (CNTF) is produced by astrocytes, Schwann cells, and T cells (Jones et al. 2010; Sleeman et al. 2000; Stockli et al. 1991). It binds to the CNTF receptor α (CNTFRα), triggering a heterodimerization of gp130 and leukemia inhibitory factor receptor (LIFR) (Matthews and Febbraio 2008) with downstream signaling through the JAK/STAT pathway (Davis et al. 1991). The expression of the α receptor is mainly found in cells of the central and peripheral nervous system, and in cells of peripheral tissues such as muscle cells and adipocytes (Davis et al. 1991; Matthews and Febbraio 2008; Sleeman et al. 2000). This broad spectrum of receptor expression reflects the function of CNTF as a modulator of neuronal differentiation (Freda et al. 1990; Vlotides et al. 2004), inhibitor of neurodegeneration (Sendtner et al. 1992), myotrophic effector (Guillet et al. 1999; Helgren et al. 1994), mediator of weight loss (Ettinger et al. 2003; Henderson et al. 1994; Kokoeva et al. 2005), and insulin sensitivity (Matthews and Febbraio 2008; Sleeman et al. 2003). CNTF over-expression is shown to induce increased sensory innervation of the skin (LeMaster et al. 1999).

13.1.2.4 Neuropoietin

The latest member of the IL-6 related cytokine family, the CNTFRα ligand neuro-poietin (NP) was identified by a computational analysis screen effort in 2004 (Derouet et al. 2004). The cloning paper describes high NP expression in the neuro-epithelium during developmental stages of the mouse embryo when CNTF is absent while postnatal production of the protein is not detectable. This observation impli-cates an important role for NP during nervous system development. Subsequent stud-ies link NP to the regulation of murine neuronal differentiation (Ohno et al. 2006), adipogenesis (Derouet et al. 2004), and bone formation (McGregor et al. 2010). In humans, neuropoietin so far appears to be without relevance because, although an NP ortholog is found, all sequenced individuals ($n = 93$) reveal an 8-nt deletion, sug-gesting NP to be a pseudogene in the human genome.

All together, the IL-6 cytokine family including IL-31 and OSM may be impor-tant neuroimmune mediators in the skin regulating inflammation, itch, and probably pain. A role of this cytokine family in cell growth processes has also been verified (Betz et al. 1998; Yoshida et al. 1996). Regarding inflammatory skin diseases and cutaneous tumors, IL-31 has been demonstrated to be involved in atopic dermatitis (AD), pruritus, and cutaneous T-cell lymphoma (Yosipovitch and Bernhard 2013), while a role of OSM has been described in AD, psoriasis, keratoacanthoma, squa-mous cell carcinoma, and Kaposi sarcoma (Ameglio et al. 1997; Bonifati et al. 1998; Miles et al. 1992; Nair et al. 1992; Tran et al. 2000). So far, a direct role of IL-6 family cytokines in itch transmission has only been shown for IL-31 (Dillon et al. 2004; Nobbe et al. 2012).

13.1.3   Receptors for IL-31 and OSM

The activities of cytokines are mediated through ligand-induced dimerization of a tyrosine kinase receptor complex.

13.1.3.1   IL-31RA

IL-31 signals through a heterodimeric receptor composed of IL-31RA (gp-130-like receptor) and the OSMRβ subunit. Several subforms of the IL-31RA exist on chromosome 5q11.2, only 24 kb downstream of gp130 (Zhang et al. 2008). Human IL-31RA mRNA, although undetectable in fresh peripheral blood monocytes, is upregulated substantially in monocytes cultured with interferon-γ (IFN-γ) (Dillon et al. 2004). IFN-γ and lipopolysaccharide together induce the expression of both receptor chains of IL-31R in human monocytes (Dillon et al. 2004). In mice, CD4+ T cells from IL-31RA–/– mice proliferate stronger and secrete more Th2 cytokines when stimulated under neutral or Th2 conditions, suggesting the presence of IL-31R on mouse CD4+ T cells (Perrigoue et al. 2007).

IL-31RA is expressed by DRG neurons, mature dendritic cells, and probably keratinocytes, albeit in the latter two at significantly lower levels (Dillon et al. 2004; Neis et al. 2006; Sonkoly et al. 2006). This is of interest because neuronal expression of IL31RA could indeed mean that IL31 act directly on neuronal expressed IL31RA to induce itch in IL31 overexpressing mice and in atopic dermatitis patients (Figure 13.1).

Very recently, the expression of IL-31RA on DRG neurons has been demonstrated on the RNA (Bando et al. 2006; Sonkoly et al. 2006) and protein level (Bando et al. 2006). On the basis of their morphological findings in murine DRG, some authors hypothesize that IL-31 and OSM may have redundant functions in the development of DRGs, while the expression pattern of their functional receptor complexes is different in the developmental period (Bando et al. 2006). A functional role for a neuronally expressed IL-31RA has not been determined until recently.

13.1.3.2   Functional Role of Neuronally Expressed IL-31RA in Itch

Own preliminary data, however, indicate the expression of a functional IL-31RA on DRG neurons (Cevikbas and Steinhoff 2012). First, we used immunohistochemistry, FACS, and qRTPCR to determine IL-31 expression levels in murine atopic-like dermatitis and in human atopic dermatitis (Cevikbas and Steinhoff 2012). We found that among all immune and resident skin cells examined, IL-31 was produced only by T_H2 and mature dendritic cells and was increased in atopic dermatitis. Likewise, the concentration of IL-31 protein was significantly increased in the SAB-induced as well as ovalbumin-induced atopic dermatitis mouse models (Cevikbas et al. 2013). Moreover, using immunohistochemistry, immunofluorecence, qRTPCR, *in vivo* pharmacology, western blotting, single cell calcium, and electrophysiology, we examined the distribution, functionality, and cellular basis of IL-31RA. In addition to previous studies we found that not only cutaneous (Dillon et al. 2004) but also intrathecal injections of IL-31 evoked intense itch. Interestingly, not only murine but also human DRG neurons express IL-31RA, largely in those that also express TRPV1, indicating a role of IL-31RA in the peptidergic subpopulation of primary afferent

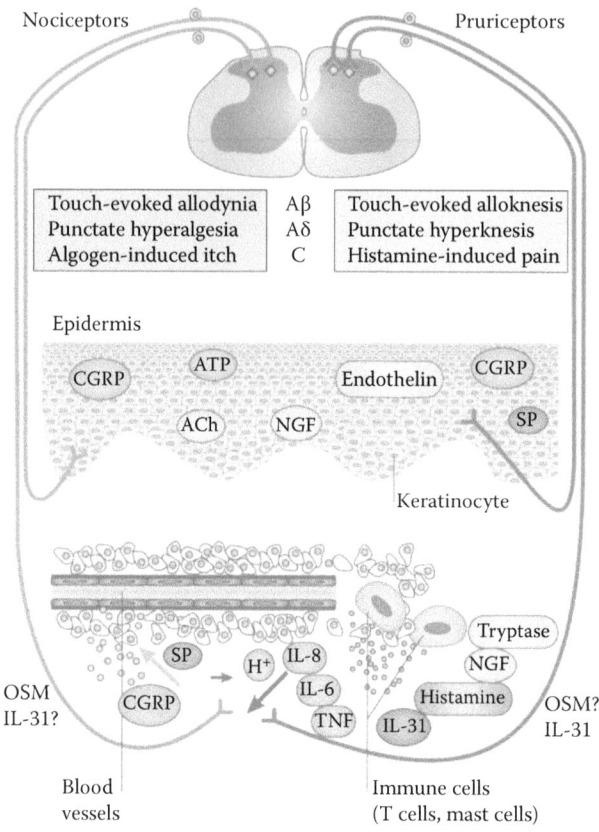

FIGURE 13.1 **(See color insert.)** During T_H2-dominated inflammation (e.g., atopic dermatitis), and probably in certain T_H2-involved tumors (CTCL), as well as probably prurigo, IL-31 is released from activated skin-homing T_H2 cells. IL-31 binds to the IL-31RA receptor (IL31RA/OSMRβ receptor complex) on sensory nerve endings in the skin. As an amplification mechanism, IL-31RA communicates with ion channels such as TRPV1 that result in enhanced Ca signaling, STAT3 activation, and probably other signal transduction pathways in sensory neurons. The transducer of IL-31-mediated itch from central primary afferents to the dorsal horn of spinal cord is currently unknown and probably involves opioids. The role of oncostatin M (OSM) in itch is poorly understood. However, a mutation in familial pruritic amyloidosis, a severely itchy skin disease, has been described for OSMRβ, the receptor for OSM. A receptor cross talk of IL-31RA and TRPV1 may lead to the amplification of neuronal sensitization, enhanced release of neuropeptides (SP, CGRP), and other mediators involved in neurogenic inflammation and itch. (Modified from Cevikbas, F. et al., Submitted to JACI.)

nerve fibers. *In vivo*, we found that IL-31-induced pruritus was significantly reduced in TRPV1- and TRPA1-deficient mice, suggesting for the first time that IL-31RA-mediated itch is linked to TRP channels, namely, TRPV1 and TRPA1 (Cevikbas et al. 2013). Next, we used cultured primary sensory neurons to define the intracellular signaling cascades implicated in IL-31-induced activation of primary afferents and itch. Of note, IL-31 triggered Ca^{2+} release and ERK1/2 phosphorylation, the first linked also to TRPV1 activation, suggesting that IL-31RA-mediated calcium mobilization is associated with intracellular signaling that activates TRPV1 in an intracellular fashion (Cevikbas et al. 2013). In addition, we found that IL-31RA-mediated itch induction is also associated with ERK1/2 phosphorylation because ERK1/2 was significantly activated 5 min after IL-31 application (Cevikbas et al. 2013). Of note, IL-31-induced ERK1/2 activation and scratching behavior was blocked *in vivo* using a specific inhibitor, suggesting that ERK1/2 is involved in IL-31-induced itch transmission *in vitro* and *in vivo*. In conclusion, novel evidence strongly establishes IL-31RA as the first functional neuronally expressed cytokine receptor expressed by a small subpopulation of IL-31RA[+]/TRPV1[+]/TRPA1[+] nerves. Thus, IL-31RA can be seen as the long-discussed neuroimmune link between subtype-2 T helper (T_H2) cells and sensory nerves for the generation of T-cell-mediated itch. These findings have several important clinical implications because it will improve our understanding of itch in atopic dermatitis and other T_H2-mediated diseases and provides a basis for future targeted drug therapy of itch.

Intriguingly, IL-31 was recently found to be increased in the malignant T-cell population of cutaneous T-cell lymphoma patients (Singer et al. 2013). Previous reports have been suggesting that IL-31 is also increased in CTCL (Miyagaki et al. 2013; Ohmatsu et al. 2012). However, the correlation to the incidence of itch in CTCL patients was not evaluated. In the recently published study, the authors have screened for levels of IL-31 mRNA in different cell types of patient samples. Retrospectively, the high IL-31mRNA levels in the patients reflected nicely the experienced pruritus at time of the sample collection. With cell sorting, the predominant cells with high level of IL-31 protein were defined as CD3[+]CD4[+]CD26[-] T lymphocytes that are categorized as malignant cells in CTCL.

Besides the accumulative evidence of IL-31 levels being increased in various pruritic diseases, recent studies also uncovered that genetics might be a potentially relevant aspect in IL-31-related diseases (Hong et al. 2012; Lan et al. 2011). A G-allele variant of IL-31 was linked with higher risks of atopic dermatitis development, suggesting that besides receptor-related missense mutations also the IL-31 gene, variants have impact on the onset of atopic eczema. Indeed, conclusively, a possible treatment against various pruritic diseases might be addressed by neutralizing IL-31 in the patients' body system.

13.1.3.3   OSMRβ

As all members of the IL-6 family, OSM shares the common signaling receptor subunit gp130 (Kishimoto 1994; Silver and Hunter 2010). Murine OSM stimulates a selective heterodimer composed of the β-subunit of the oncostatin M receptor (OSMRβ) and gp130 (Tanaka et al. 1999). In humans, OSM can also activate the OSMRβ/gp130 heterodimer, but additionally, signaling through the leukemia inhibitory factor receptor (LIFR)/gp130 heterodimer may be triggered as well (Ichihara et al. 1997).

While IL-31RA belongs to the IL-6 receptor family, OSMR is a member of the type-1 cytokine receptor family (Homey et al. 2006). IL-31 and OSM are similar regarding their binding capacity to IL-31RA and gp130, respectively. The latter receptor subunits only convert to high-affinity receptors after forming heterodimer complexes with OSMRβ (Diveu et al. 2004; Gearing et al. 1991, 1992; Mosley et al. 1996). OSM and IL-31 do not cross-activate the other receptor but may have additive or synergistic effects by using interfering signal transduction pathways. This remains to be elucidated.

OSMR mRNA is ubiquitously expressed, with the highest levels in trachea, thymus, and skin (Tamura et al. 2003; Boniface et al. 2007; Morikawa 2005). OSMR mRNA expression can be induced in monocytes treated with lipopolysaccharide. OSMRβ is detectable on keratinocytes and on a subset of nociceptive neurons in dorsal root ganglia (DRG) or trigeminal neurons projecting to the skin. Furthermore, OSMRβ-expressing neurons colocalize with both TRPV1 and P2X3, respectively (Tamura et al. 2003). OSM-deficient mice demonstrate reduced responses to noxious stimuli (Morikawa 2005; Morikawa et al. 2004) and are thus involved in nociception. A recent study showed a direct involvement of OSM in induced heat sensitivity. Thereby OSMR-induced activation of gp130 leads to sensitization TRPV1 in sensory neurons (Langeslag et al. 2011). In OSMRβ-expressing DRG both STAT-3 and cyclic AMP-responsive element binding protein (CREbP) are upregulated (Tamura et al. 2005). Our knowledge about the function of OSMRβ at the neuronal level, however, is very poor as of yet. So far, only a role of OSMRβ has been demonstrated for brain development, especially homeostasis of neural precursor cells, and in nerve degeneration and repair process in response to peripheral nerve injury (Beatus et al. 2011; Ito et al. 2000).

13.1.4   ROLE OF IL-31, OSM AND THEIR RECEPTORS IN THE SKIN

The distribution of IL-31 to OSM and of IL-31 RA to OSMRβ in the skin is almost identical in mice (Bando et al. 2006; Morikawa et al. 2004; Tamura et al. 2003). However, the precise distribution of IL-31RA and OSMRβ in the various subtypes of sensory nerves, their codistribution with other important neuronal mediators or receptors, and their functional role in health and disease is far from being understood. Our own preliminary data strongly favor an important role of IL-31 for the regulation of sensory neurons via IL-31RA activation and signaling (Cevikbas et al. 2011, 2013). The precise role for OSM in murine and human itch has to be explored (Bando et al. 2006; Bilsborough et al. 2006; Weiss et al. 2006).

In the skin, IL-31 exerts functions on cultured keratinocytes, activated macrophages, and dendritic cells (Bilsborough et al. 2006; Dillon et al. 2004; Neis et al. 2006; Novak et al. 2003; Sonkoly et al. 2006). Notably, IL-31RA and OSMRβ have also been described to be expressed by DRG neurons, suggesting a link between T cells and nerves in skin diseases (Bando et al. 2006; Morikawa et al. 2004; Sonkoly et al. 2006). In addition, we found IL-31RA mRNA and protein expression also on human DRG neurons (Cevikbas et al. 2013).

13.1.5   ROLE OF IL-31 AND OSM IN ITCH

13.1.5.1   IL-31

In atopic dermatitis, IL-31 is markedly released by T_H2 cells (Bilsborough et al. 2006; Novak et al. 2003) and is upregulated also in prurigo nodularis, a severely itchy dermatosis (Cevikbas et al. 2013). In AD, T_H2 cells infiltrate the inflamed skin and other organs (Brunn et al. 2008; Zhang et al. 2009) where they can be found closely associated with sensory nerve fibers (Cetin et al. 2009; Huang et al. 2003; Pavlovic et al. 2008; Tanaka et al. 2007). Mice overexpressing IL-31 develop a pruritic atopic-like dermatitis (Dillon et al. 2004). Injection of IL-31 into NC/Nga mice results in sustained pruritus, a characteristic of C-fiber activation (Grimstad et al. 2009). IL-31 gene-deficient mice show less pruritus as compared to wild-type controls (Dillon et al. 2004). Genetic studies on polymorphisms in the human IL-31 gene with respect to the genetic susceptibility to eczema suggest that a haplotype of the IL-31 gene is associated with the intrinsic form of pruritic atopic eczema confirming studies in IL-31 overexpressing and IL-31 RA knockout mice (Schulz et al. 2007; Cevikbas et al. 2011, 2013).

13.1.5.2 OSM

OSM released from T cells is likely to activate keratinocytes through OSMRβ in skin inflammation such as psoriasis and atopic dermatitis (Bando et al. 2006; Boniface et al. 2007; Morikawa 2005). Of importance, in addition to the enhanced expression of OSM and OSMRβ in keratinocytes of atopic dermatitis lesions, OSMRβ mutations have been reported in a familial pruritic skin disease (Arita et al. 2008). Here, a missense mutation in the OSMRβ gene has been shown to be associated with localized cutaneous amyloidosis (FPLCA), a disease with severe chronic pruritus. These data suggest the involvement of the OSM/OSMRβ pathway in pruritus and indicate a role of OSMRβ also in human itch. Thus, it will be important to further characterize the impact of functional IL-31RA and OSMR cytokine receptors on peripheral or central nerves in pruritic skin diseases. IL-31 and OSM may directly or indirectly interact with neurons to modulate itch. Whether both IL-6 family cytokines act synergistically or under different conditions in human disease is currently unknown.

13.1.5.3 Mechanisms of IL-31-Induced Itch

There is substantial evidence that IL-31 and OSM regulate nerves and KC in atopic dermatitis (Arita et al. 2008; Bilsborough et al. 2006; Boniface et al. 2007; Dillon et al. 2004; Homey et al. 2006; Neis et al. 2006; Sonkoly et al. 2006). IL-31-overexpressing mice develop spontaneously an atopic-like dermatitis with severe pruritus (Dillon et al. 2004). In a mouse model of AD (Jin et al. 2009), IL-31RA expression was enhanced (Takaoka et al. 2005), and a monoclonal antibody against IL-31 was capable of ameliorating the scratching behavior, indicating a role of IL-31 in pruritus (Grimstad et al. 2009). However, the important question to which extent IL-31RA and OSMR activation on neurons and KC contribute to these effects in skin inflammation including AD and itch is yet to be explored. IL-31 is significantly upregulated in *"pruritic"* AD skin as compared to *"low- or nonpruritic"* psoriatic skin (Sonkoly et al. 2006). Highest IL-31 levels can be detected in prurigo nodularis, representing one of the most pruritic diseases. *In vivo*, exposure to staphylococcal superantigen rapidly induces IL-31 expression in atopic dermatitis patients (Sonkoly et al. 2006). *In vitro*, IL-31 is strongly induced by staphylococcal superantigen stimulation. Our own data show that IL-31 induces scratching behavior in mice, confirming previous reports about an important role of IL-31 in the pathophysiology of pruritus (Bilsborough et al. 2006; Dillon et al. 2004; Grimstad et al. 2009; Ishii et al. 2009; Sonkoly et al. 2006; Cevikbas et al. 2011, 2013).

Our studies revealed a novel mechanism of immune-neuron interaction, suggesting a direct interaction between a cytokine receptor on neuronal cells and neuronal receptors. Together, IL-31 and OSM represent direct links between T cells and sensory nerves in the skin. Accordingly, the role of OSM in AD and pruritus is also poorly understood. So far, it is only known that T cells release OSM to activate cytokine release from KC through OSMRβ (Boniface et al. 2007). In summary, the underlying cellular mechanism and the significance of IL-31- and OSM-induced activation of neurons in human disease, however, are still very poorly understood. Moreover, the impact of IL-31RA and OSMR activation on keratinocyte regulation, pruritus and pain have not been investigated as of yet. However, the fact that IL-31 activates a functional IL-31RA on DRG neurons and induces a chronic pruritic skin disease, namely, atopic-like dermatitis, and that antibodies that block IL-31 are beneficial for the treatment of those skin lesions, and that human DRG express IL-31RA strongly indicate a potential therapeutic role of blocking the IL-31/IL-31RA pathway in human T_H2-mediated skin diseases, such as atopic dermatitis or cutaneous T-cell lymphoma.

13.1.5.4   Potential Role of IL-31, OSM, and Their Receptors in Neurogenic Inflammation

During skin inflammation, sensory nerve endings release neuropeptides that regulate vasodilatation (erythema), plasma extravasation (edema), recruitment of inflammatory cells, keratinocyte function (Hanifin 2009; Jin et al. 2009), pruritus (Ikoma et al. 2006; Paus et al. 2006; Steinhoff et al. 2006), or pain sensations (Cavanaugh et al. 2009; Wang et al. 2009). They also modulate immune cells like Langerhans cells, T cells, macrophages, or mast cells (Hanifin 2009). Neuromediators also regulate skin cells such as KC or endothelial cells during inflammation and AD (Haegerstrand et al. 1989; Hosoi et al. 1993; Scholzen et al. 2004). Whether IL-31 and/or OSM regulate neurogenic inflammation in murine or human skin diseases is currently unknown. However, our own data show that IL-31 induces release of calcitonin gene-related peptide (CGRP) from DRG neurons and interacts with TRPV1, a receptor crucially involved in neurogenic inflammation (Caterina et al. 1997), strongly indicating a role of IL-31 in neurogenic inflammation via neuropeptide release and TRPV1 activation. The precise mechanism of IL-31-mediated neurogenic inflammation and the involvement of OSM in this process still need to be explored.

13.1.5.5   Potential Role of IL-31 and OSM in Pain

It has been demonstrated that receptors for kinins, prostanoids, chemokines, and cytokines are involved in the regulation of pain (Ikoma et al. 2006; Zhang and Oppenheim 2005). Recently, an *in vivo* pain model clearly demonstrated the involvement of IL-6 in the development of heat hyperalgesia by inducing the release of CGRP from rat skin (Opree and Kress 2000). IL-6 shows a sustained increased expression in a model of thermal pain associated with mechanical hyperalgesia suggesting that IL-6 is an important mediator of burn-injury pain (Summer et al. 2008). The blocking of the receptor subunit gp130 by RNA silencing attenuated both burn and intradermal IL-6-induced hyperalgesia (Summer et al. 2008). Our own preliminary data suggest also the involvement of IL-31 in pain transmission (see preliminary results). OSMRβ is expressed in small-sized DRG neurons that are nonpeptidergic but positive for TRPV1 and P2X3, receptors for capsaicin, and ATP, respectively (Tamura et al. 2003). While TRPV1 is activated by protons, heat, and capsaicin, P2X3 is activated by ATP during inflammation or pain. This suggests that OSM produced in the inflamed tissue or during pain may interact with TRPV1 and/ or P2X3 in nociceptive neurons, thereby contributing to neurogenic inflammation, sustained pain, or hyperalgesia during inflammation. Whether the IL-6 cytokine family exerts additive or synergistic effects to regulate acute or chronic pain transmission at the peripheral or central level is currently unknown.

13.1.5.6 IL-31/OSM-Related Signal Transduction Pathways in Sensory Neurons

Cytokine receptors of the IL-6 family activate intracellular calcium, STAT signaling, MAP kinases, and others. IL-31 and OSM do not cross-react the receptor of each other (Zhang et al. 2008). However, in order to understand the significance of the codistribution of IL-31RA and OSMRβ in KC, DRGs, and spinal cord neurons, it will be important to fully explore the shared versus nonshared intracellular signaling pathways induced by IL-31, OSM, or both simultaneously. IL-31 induces the MAPK, PI3K pathway (Diveu et al. 2004) and also stimulates the JAK/STAT pathway, predominantly STAT3 and STAT5 (Dillon et al. 2004; Diveu 2003; Dreuw et al. 2004; Ghilardi et al. 2002; Liu et al. 2001). Of note, the Ca2+, MAPK as well as STAT pathways are involved in important neuronal signaling events in DRG and/or spinal cord neurons (Cevikbas et al. 2011, 2013). Although IL-31 and OSM exert important effects on sensory neurons, little is known about IL-31 and OSM-mediated intracellular signaling in the nervous system, their common effects, as well as their specific effects in itch transmission.

13.1.6 KERATINOCYTES AS A "FOREFRONT" OF NEURONAL SIGNALING

Keratinocytes release various mediators, e.g., neurotrophins (NGF, NT-4), neuropeptides (SP, ET-1), ATP, proteases, or prostanoids, which subsequently modulate neuronal function like pain or pruritus (reviewed in Ikoma et al. 2006; Paus et al. 2006; Steinhoff et al. 2006). In atopic dermatitis patients, keratinocytes produce and secreted increased levels of the chemokines CCL17, CCL27, and the cytokine thymic stromal lymphopoietin, which are known to be involved in initiating and maintaining inflammatory reactions in the skin by, e.g., inducing maturation of T-cell populations (Lee and Yu 2011). Keratinocytes also express IL-31RA and OSMRβ (Heise et al. 2009). Whether IL-31/OSM modulate neurogenic inflammation, pruritus, and pain via direct activation of IL-31RA (OSMR) receptors on DRG and spinal cord neurons and/or indirectly, via keratinocytes, is currently poorly understood. However, a recent study showed that keratinocyte activation by a TLR2 ligand or IFNγ-induced upregulation of both IL31RA and OSMR (Heise et al. 2009; Kasraie et al. 2011). Activation of IL31RA interferes with keratinocyte differentiation by inducing cell cycle arrest as shown in an organotypic skin model (Cornelissen et al. 2012). Further, the authors showed that IL31RA induced cell cycle arrest of keratinocytes, in turn, inducing striking defects in differentiation associated with reduced epidermal thickness, disturbed constitution, and altered alignment of the basal layer and poor development of the stratum granulosum in the skin model (Cornelissen et al. 2012) This is of interest as in both atopic dermatitis patients and in mice models of atopic-like dermatitis, the epidermis is characterized by hyperkeratosis, leading to a thicker epidermis. However, further studies will be necessary to understand the role of IL31RA activation in keratinocytes for pruritic skin diseases. Moreover, upon stimulation, keratinocytes release cytokine of the IL-1 or IL-6 family that may be key mediators for keratinocyte-sensory nerve interactions controlling neurogenic inflammation or itch. Therefore, in the future, it will be of additional importance to analyze the impact of IL-31 and OSM activation on keratinocytes with respect to their communication with sensory nerves during disease state.

13.2 CONCLUSIONS AND FUTURE DIRECTIONS

Recent results strongly suggest that cytokines of the interleukin-6 family, in particular, IL-31 and OSM, as key players in neuroimmune communication including pain, itch, and probably inflammation. Via IL-31, T_H2 cells "talk" to sensory nerves to induce inflammation and pruritus in atopic dermatitis. Thus, IL-31 and its receptor may be effective targets to treat atopic dermatitis and probably other T_H2-mediated pruritic skin diseases. OSM has been shown to be involved in pain conditions, and a missense mutation of the OSMRβ receptor is involved in severe familial itch of FPLCA patients. Understanding the molecular and cellular basis of IL-31RA and OSMRb-mediated activation of sensory nerves, their signaling and regulation will lead to a novel understanding of cytokine-nerve cross talks in many diseases and to the development of new, innovative drugs for humans suffering from atopic dermatitis and recalcitrant chronic itch, and probably for certain chronic pain conditions. With regard to future directions, our understanding of IL-31 and OSM is underscoring the importance of both cytokines in different diseases (Baumann et al. 2012; Hartmann et al. 2012). IL-31 has been proven to be one very important and reliable biomarker for pruritic skin diseases. Studies measuring the serum levels of IL-31 in humans are indeed increasing the idea that targeting IL-31 via neutralization might be a very promising therapeutic improvement in the treatment of itch.

Novel insights into the molecular mechanisms of IL-31-mediated inflammation and itch of an exclusively immune-cell-derived cytokine acting directly on neurons are of critical importance for our understanding of the cross talk between the immune and nervous systems in acute or chronic inflammation, itch, and pain. Various other cytokines (released from immune cells or nonimmune skin cells) might play a similar role in skin diseases and itch.

ACKNOWLEDGMENTS

This work was supported by research grants from the NIH/NIAMS (R01 5R01AR059402-02), DFG (STE 1014/2-1), ZymoGenetics/BMS (to MS), and the DFG (to F.C., C.K.).

REFERENCES

Ameglio, F., Bonifati, C., Fazio, M. et al. 1997. Interleukin-11 production is increased in organ cultures of lesional skin of patients with active plaque-type psoriasis as compared with nonlesional and normal skin. Similarity to interleukin-1 beta, interleukin-6 and interleukin-8. *Arch Dermatol Res* 289:399–403.

Arita, K., South, A.P., Hans-Filho, G. et al. 2008. Oncostatin M receptor-beta mutations underlie familial primary localized cutaneous amyloidosis. *Am J Hum Genet* 82:73–80.

Bamber, B., Reife, R.A., Haugen, H.S. et al. 1998. Oncostatin M stimulates excessive extracellular matrix accumulation in a transgenic mouse model of connective tissue disease. *J Mol Med* 76:61–69.

Bando, T., Morikawa, Y., Komori, T. et al. 2006. Complete overlap of interleukin-31 receptor A and oncostatin M receptor beta in the adult dorsal root ganglia with distinct developmental expression patterns. *Neuroscience* 142:1263–1271.

Baumann, R., Rabaszowski, M., Stenin, I. et al. 2012. The release of IL-31 and IL-13 after nasal allergen challenge and their relation to nasal symptoms. *Clin Transl Allergy* 2:13.

Beatus, P., Jhaveri, D.J., Walker, T.L. et al. 2011. Oncostatin M regulates neural precursor activity in the adult brain. *Dev Neurobiol* 71:619–633.

Betz, U.A., Bloch, W., van den Broek, M. et al. 1998. Postnatally induced inactivation of gp130 in mice results in neurological, cardiac, hematopoietic, immunological, hepatic, and pulmonary defects. *J Exp Med* 188:1955–1965.

Bilsborough, J., Leung, D.Y., Maurer, M. et al. 2006. IL-31 is associated with cutaneous lymphocyte antigen-positive skin homing T cells in patients with atopic dermatitis. *J Allergy Clin Immunol* 117:418–425.

Boniface, K., Diveu, C., Morel, F. et al. 2007. Oncostatin M secreted by skin infiltrating T lymphocytes is a potent keratinocyte activator involved in skin inflammation. *J Immunol* 178:4615–4622.

Bonifati, C., Mussi, A., D'Auria, L. et al. 1998. Spontaneous release of leukemia inhibitory factor and oncostatin-M is increased in supernatants of short-term organ cultures from lesional psoriatic skin. *Arch Dermatol Res* 290:9–13.

Bonness, S., and Bieber, T. 2007. Molecular basis of atopic dermatitis. *Curr Opin Allergy Clin Immunol* 7:382–386.

Brunn, A., Utermohlen, O., Carstov, M. et al. 2008. CD4 T cells mediate axonal damage and spinal cord motor neuron apoptosis in murine p0106-125-induced experimental autoimmune neuritis. *Am J Pathol* 173:93–105.

Caterina, M.J., Schumacher, M.A., Tominaga, M. et al. 1997. The capsaicin receptor: A heat-activated ion channel in the pain pathway. *Nature* 389:816–824.

Cavanaugh, D.J., Lee, H., Lo, L. et al. 2009. Distinct subsets of unmyelinated primary sensory fibers mediate behavioral responses to noxious thermal and mechanical stimuli. *Proc Natl Acad Sci USA* 106(22):9075–9080.

Cetin, E.D., Savk, E., Uslu, M. et al. 2009. Investigation of the inflammatory mechanisms in alopecia areata. *Am J Dermatopathol* 31:53–60.

Cevikbas, F., and Steinhoff, M. 2012. IL-33: A novel danger signal system in atopic dermatitis. *J Invest Dermatol* 132:1326–1329.

Cevikbas, F., Steinhoff, M., and Ikoma, A. 2011. Role of spinal neurotransmitter receptors in itch: New insights into therapies and drug development. *CNS Neurosci Ther* 17:742–749.

Cevikbas, F., Wang, X., Akiyama, T. et al. 2013. A sensory neuron-expressed interleukin-31 receptor mediates T helper cell-dependent pruritus: Involvement of TRPV1 and TRPA1. Submitted to JACI.

Chung, M.K., Lee, H., and Caterina, M.J. 2003. Warm temperatures activate TRPV4 in mouse 308 keratinocytes. *J Biol Chem* 278:32037–32046.

Chung, M.K., Lee, H., Mizuno, A. et al. 2004. TRPV3 and TRPV4 mediate warmth-evoked currents in primary mouse keratinocytes. *J Biol Chem* 279:21569–21575.

Cornelissen, C., Marquardt, Y., Czaja, K. et al. 2012. IL-31 regulates differentiation and filaggrin expression in human organotypic skin models. *J Allergy Clin Immunol* 129:426–433, 33 1-8.

Dallos, A., Kiss, M., Polyanka, H. et al. 2006. Effects of the neuropeptides substance P, calcitonin gene-related peptide, vasoactive intestinal polypeptide and galanin on the production of nerve growth factor and inflammatory cytokines in cultured human keratinocytes. *Neuropeptides* 40:251–263.

Davis, S., Aldrich, T.H., Valenzuela, D.M. et al. 1991. The receptor for ciliary neurotrophic factor. *Science* 253:59–63.

Derouet, D., Rousseau, F., Alfonsi, F. et al. 2004. Neuropoietin, a new IL-6-related cytokine signaling through the ciliary neurotrophic factor receptor. *Proc Natl Acad Sci USA* 101:4827–4832.

Dillon, S.R., Sprecher, C., Hammond, A. et al. 2004. Interleukin 31, a cytokine produced by activated T cells, induces dermatitis in mice. *Nat Immunol* 5:752–760.

Diveu, C., Lak-Hal, A.H., Froger, J. et al. 2004. Predominant expression of the long isoform of GP130-like (GPL) receptor is required for interleukin-31 signaling. *Eur Cytokine Netw* 15:291–302.

Diveu, C., Lelievre, E., Perret, D. et al. 2003. GPL, a novel cytokine receptor related to GP130 and leukemia inhibitory factor receptor. *J Biol Chem* 278:49850–49859.

Dreuw, A., Radtke, S., Pflanz, S. et al. 2004. Characterization of the signaling capacities of the novel gp130-like cytokine receptor. *J Biol Chem* 279:36112–36120.

Drogemuller, K., Helmuth, U., Brunn, A. et al. 2008. Astrocyte gp130 expression is critical for the control of Toxoplasma encephalitis. *J Immunol* 181:2683–2693.

Duncan, M.R., Hasan, A., and Berman, B. 1995. Oncostatin M stimulates collagen and glycosaminoglycan production by cultured normal dermal fibroblasts: Insensitivity of sclerodermal and keloidal fibroblasts. *J Invest Dermatol* 104:128–133.

E Y., Golden, S.C., Shalita, A.R. et al. 2006. Neuropeptide (calcitonin gene-related peptide) induction of nitric oxide in human keratinocytes in vitro. *J Invest Dermatol* 126:1994–2001.

Ettinger, M.P., Littlejohn, T.W., Schwartz, S.L. et al. 2003. Recombinant variant of ciliary neurotrophic factor for weight loss in obese adults: A randomized, dose-ranging study. *JAMA* 289:1826–1832.

Feld, M., Shpacovitch, V.M., Fastrich, M. et al. 2010. Interferon-gamma induces upregulation and activation of the interleukin-31 receptor in human dermal microvascular endothelial cells. *Exp Dermatol* 19:921–923.

Freda, M.C., Andersen, H.F., Damus, K. et al. 1990. Lifestyle modification as an intervention for inner city women at high risk for preterm birth. *J Adv Nurs* 15:364–372.

Gearing, D.P., Comeau, M.R., Friend, D.J. et al. 1992. The IL-6 signal transducer, gp130: An oncostatin M receptor and affinity converter for the LIF receptor. *Science* 255:1434–1437.

Gearing, D.P., Thut, C.J., VandeBos, T. et al. 1991. Leukemia inhibitory factor receptor is structurally related to the IL-6 signal transducer, gp130. *EMBO J* 10:2839–2848.

Ghilardi, N., Li, J., Hongo, J.A. et al. 2002. A novel type I cytokine receptor is expressed on monocytes, signals proliferation, and activates STAT-3 and STAT-5. *J Biol Chem* 277:16831–16836.

Grewe, M., Vogelsang, K., Ruzicka, T. et al. 2000. Neurotrophin-4 production by human epidermal keratinocytes: Increased expression in atopic dermatitis. *J Invest Dermatol* 114:1108–1112.

Grimstad, O., Sawanobori, Y., Vestergaard, C. et al. 2009. Anti-interleukin-31-antibodies ameliorate scratching behaviour in NC/Nga mice: A model of atopic dermatitis. *Exp Dermatol* 18:35–43.

Grothe, C., Heese, K., Meisinger, C. et al. 2000. Expression of interleukin-6 and its receptor in the sciatic nerve and cultured Schwann cells: Relation to 18-kD fibroblast growth factor-2. *Brain Res* 885:172–181.

Guillet, C., Auguste, P., Mayo, W. et al. 1999. Ciliary neurotrophic factor is a regulator of muscular strength in aging. *J Neurosci* 19:1257–1262.

Haegerstrand, A., Jonzon, B., Dalsgaard, C.J. et al. 1989. Vasoactive intestinal polypeptide stimulates cell proliferation and adenylate cyclase activity of cultured human keratinocytes. *Proc Natl Acad Sci USA* 86:5993–5996.

Hanifin, J.M. 2009. Evolving concepts of pathogenesis in atopic dermatitis and other eczemas. *J Invest Dermatol* 129:320–322.

Hartmann, K., Wagner, N., Rabenhorst, A. et al. 2012. Serum IL-31 levels are increased in a subset of patients with mastocytosis and correlate with disease severity in adult patients. *J Allergy Clin Immunol* 132(1):232–235.

Heise, R., Neis, M.M., Marquardt, Y. et al. 2009. IL-31 receptor alpha expression in epidermal keratinocytes is modulated by cell differentiation and interferon gamma. *J Invest Dermatol* 129:240–243.

Helgren, M.E., Squinto, S.P., Davis, H.L. et al. 1994. Trophic effect of ciliary neurotrophic factor on denervated skeletal muscle. *Cell* 76:493–504.

Henderson, J.T., Seniuk, N.A., Richardson, P.M. et al. 1994. Systemic administration of ciliary neurotrophic factor induces cachexia in rodents. *J Clin Invest* 93:2632–2638.

Hintzen, C., Quaiser, S., Pap, T. et al. 2009. Induction of CCL13 expression in synovial fibroblasts highlights a significant role of oncostatin M in rheumatoid arthritis. *Arthritis Rheum* 60:1932–1943.

Homey, B., Steinhoff, M., Ruzicka, T. et al. 2006. Cytokines and chemokines orchestrate atopic skin inflammation. *J Allergy Clin Immunol* 118:178–189.

Hong, C.H., Yu, H.S., Ko, Y.C. et al. 2012. Functional regulation of interleukin-31 production by its genetic polymorphism in patients with extrinsic atopic dermatitis. *Acta Derm Venereol* 92:430–432.

Hosoi, J., Murphy, G.F., Egan, C.L. et al. 1993. Regulation of Langerhans cell function by nerves containing calcitonin gene-related peptide. *Nature* 363:159–163.

Huang, C.H., Kuo, I.C., Xu, H. et al. 2003. Mite allergen induces allergic dermatitis with concomitant neurogenic inflammation in mouse. *J Invest Dermatol* 121:289–293.

Huang, S.M., Lee, H., Chung, M.K. et al. 2008. Overexpressed transient receptor potential vanilloid 3 ion channels in skin keratinocytes modulate pain sensitivity via prostaglandin E2. *J Neurosci* 28:13727–13737.

Hui, W., Bell, M., Carroll, G. 1997. Detection of oncostatin M in synovial fluid from patients with rheumatoid arthritis. *Ann Rheum Dis* 56:184–187.

Ibrahim, M.M., Porreca, F., Lai, J. et al. 2005. CB2 cannabinoid receptor activation produces antinociception by stimulating peripheral release of endogenous opioids. *Proc Natl Acad Sci USA* 102:3093–3098.

Ichihara, M., Hara, T., Kim, H. et al. 1997. Oncostatin M and leukemia inhibitory factor do not use the same functional receptor in mice. *Blood* 90:165–173.

Ihn, H., and Tamaki, K. 2000. Oncostatin M stimulates the growth of dermal fibroblasts via a mitogen-activated protein kinase-dependent pathway. *J Immunol* 165:2149–2155.

Ikoma, A., Steinhoff, M., Stander, S. et al. 2006. The neurobiology of itch. *Nat Rev Neurosci* 7:535–547.

Ishii, T., Wang, J., Zhang, W. et al. 2009. Pivotal role of mast cells in pruritogenesis in patients with myeloproliferative disorders. *Blood* 113(23):5942–5950.

Ishiuji, Y., Coghill, R.C., Patel, T.S. et al. 2008. Repetitive scratching and noxious heat do not inhibit histamine-induced itch in atopic dermatitis. *Br J Dermatol* 158:78–83.

Ito, Y., Yamamoto, M., Li, M. et al. 2000. Temporal expression of mRNAs for neuropoietic cytokines, interleukin-11 (IL-11), oncostatin M (OSM), cardiotrophin-1 (CT-1) and their receptors (IL-11Ralpha and OSMRbeta) in peripheral nerve injury. *Neurochem Res* 25:1113–1118.

Jin, H., He, R., Oyoshi, M. et al. 2009. Animal models of atopic dermatitis. *J Invest Dermatol* 129:31–40.

Jones, J.L., Anderson, J.M., Phuah, C.L. et al. 2010. Improvement in disability after alemtuzumab treatment of multiple sclerosis is associated with neuroprotective autoimmunity. *Brain* 133:2232–2247.

Kakurai, M., Fujita, N., Murata, S. et al. 2001. Vasoactive intestinal peptide regulates its receptor expression and functions of human keratinocytes via type I vasoactive intestinal peptide receptors. *J Invest Dermatol* 116:743–749.

Kamiya, A., Kinoshita, T., Ito, Y. et al. 1999. Fetal liver development requires a paracrine action of oncostatin M through the gp130 signal transducer. *EMBO J* 18:2127–2136.

Kasraie, S., Niebuhr, M., Baumert, K. et al. 2011. Functional effects of interleukin 31 in human primary keratinocytes. *Allergy* 66:845–852.

Kishimoto, T. 1994. Signal transduction through homo- or heterodimers of gp130. *Stem Cells* 12(Suppl 1):37–44; discussion 44–45.

Kokoeva, M.V., Yin, H., and Flier, J.S. 2005. Neurogenesis in the hypothalamus of adult mice: Potential role in energy balance. *Science* 310:679–683.

Lacreusette, A., Nguyen, J.M., Pandolfino, M.C. et al. 2007. Loss of oncostatin M receptor beta in metastatic melanoma cells. *Oncogene* 26:881–892.

Lan, C.C., Tu, H.P., Wu, C.S. et al. 2011. Distinct SPINK5 and IL-31 polymorphisms are associated with atopic eczema and non-atopic hand dermatitis in Taiwanese nursing population. *Exp Dermatol* 20:975–979.

Langdon, C., Kerr, C., Hassen, M. et al. 2000. Murine oncostatin M stimulates mouse synovial fibroblasts in vitro and induces inflammation and destruction in mouse joints in vivo. *Am J Pathol* 157:1187–1196.

Langeslag, M., Constantin, C.E., Andratsch, M. et al. 2011. Oncostatin M induces heat hypersensitivity by gp130-dependent sensitization of TRPV1 in sensory neurons. *Mol Pain* 7:102.

Lee, C.H., and Yu, H.S. 2011. Biomarkers for itch and disease severity in atopic dermatitis. *Curr Probl Dermatol* 41:136–148.

LeMaster, A.M., Krimm, R.F., Davis, B.M. et al. 1999. Overexpression of brain-derived neurotrophic factor enhances sensory innervation and selectively increases neuron number. *J Neurosci* 19:5919–5931.

Liou, J.T., Mao, C.C., Ching-Wah Sum, D. et al. 2013. Peritoneal administration of Met-RANTES attenuates inflammatory and nociceptive responses in a murine neuropathic pain model. *J Pain* 14:24–35.

Liou, J.T., Yuan, H.B., Mao, C.C. et al. 2012. Absence of C-C motif chemokine ligand 5 in mice leads to decreased local macrophage recruitment and behavioral hypersensitivity in a murine neuropathic pain model. *Pain* 153:1283–1291.

Liu, B., Gross, M., ten Hoeve, J. et al. 2001. A transcriptional corepressor of Stat1 with an essential LXXLL signature motif. *Proc Natl Acad Sci USA* 98:3203–3207.

Liu, T., Jiang, C.Y., Fujita, T. et al. 2013. Enhancement by interleukin-1beta of AMPA and NMDA receptor-mediated currents in adult rat spinal superficial dorsal horn neurons. *Mol Pain* 9:16.

Matthews, V.B., and Febbraio, M.A. 2008. CNTF: A target therapeutic for obesity-related metabolic disease? *J Mol Med* 86:353–361.

McGregor, N.E., Poulton, I.J., Walker, E.C. et al. 2010. Ciliary neurotrophic factor inhibits bone formation and plays a sex-specific role in bone growth and remodeling. *Calcif Tissue Int* 86:261–270.

Miles, S.A., Martinez-Maza, O., Rezai, A. et al. 1992. Oncostatin M as a potent mitogen for AIDS-Kaposi's sarcoma-derived cells. *Science* 255:1432–1434.

Miyagaki, T., Sugaya, M., Suga, H. et al. 2013. Increased CCL18 expression in patients with cutaneous T-cell lymphoma: Association with disease severity and prognosis. *J Eur Acad Dermatol Venereol* 27:e60–e67.

Miyajima, A., Kinoshita, T., Tanaka, M. et al. 2000. Role of oncostatin M in hematopoiesis and liver development. *Cytokine Growth Factor Rev* 11:177–183.

Morikawa, Y. 2005. Oncostatin M in the development of the nervous system. *Anat Sci Int* 80:53–59.

Morikawa, Y., Tamura, S., Minehata, K. et al. 2004. Essential function of oncostatin m in nociceptive neurons of dorsal root ganglia. *J Neurosci* 24:1941–1947.

Mosley, B., De Imus, C., Friend, D. et al. 1996. Dual oncostatin M (OSM) receptors. Cloning and characterization of an alternative signaling subunit conferring OSM-specific receptor activation. *J Biol Chem* 271:32635–32643.

Mukouyama, Y., Hara, T., Xu, M. et al. 1998. In vitro expansion of murine multipotential hematopoietic progenitors from the embryonic aorta-gonad-mesonephros region. *Immunity* 8:105–114.

Nair, B.C., DeVico, A.L., Nakamura, S. et al. 1992. Identification of a major growth factor for AIDS-Kaposi's sarcoma cells as oncostatin M. *Science* 255:1430–1432.

Neis, M.M., Peters, B., Dreuw, A. et al. 2006. Enhanced expression levels of IL-31 correlate with IL-4 and IL-13 in atopic and allergic contact dermatitis. *J Allergy Clin Immunol* 118:930–937.

Nobbe, S., Dziunycz, P., Muhleisen, B. et al. 2012. IL-31 expression by inflammatory cells is preferentially elevated in atopic dermatitis. *Acta Derm Venereol* 92:24–28.

Novak, N., Bieber, T., and Leung, D.Y. 2003. Immune mechanisms leading to atopic dermatitis. *J Allergy Clin Immunol* 112:S128–S139.

Ohmatsu, H., Sugaya, M., Suga, H. et al. 2012. Serum IL-31 levels are increased in patients with cutaneous T-cell lymphoma. *Acta Derm Venereol* 92:282–283.

Ohno, M., Kohyama, J., Namihira, M. et al. 2006. Neuropoietin induces neuroepithelial cells to differentiate into astrocytes via activation of STAT3. *Cytokine* 36:17–22.

Ong, P.Y., and Leung, D.Y. 2006. Immune dysregulation in atopic dermatitis. *Curr Allergy Asthma Rep* 6:384–389.

Opree, A., and Kress, M. 2000. Involvement of the proinflammatory cytokines tumor necrosis factor-alpha, IL-1 beta, and IL-6 but not IL-8 in the development of heat hyperalgesia: Effects on heat-evoked calcitonin gene-related peptide release from rat skin. *J Neurosci* 20:6289–6293.

Ouyang, L., Shen, L.Y., Li, T. et al. 2006. Inhibition effect of Oncostatin M on metastatic human lung cancer cells 95-D in vitro and on murine melanoma cells B16BL6 in vivo. *Biomed Res* 27:197–202.

Paus, R., Schmelz, M., Biro, T. et al. 2006. Frontiers in pruritus research: Scratching the brain for more effective itch therapy. *J Clin Invest* 116:1174–1186.

Pavlovic, S., Daniltchenko, M., Tobin, D.J. et al. 2008. Further exploring the brain-skin connection: Stress worsens dermatitis via substance P-dependent neurogenic inflammation in mice. *J Invest Dermatol* 128:434–446.

Perrigoue, J.G., Li, J., Zaph, C. et al. 2007. IL-31-IL-31R interactions negatively regulate type 2 inflammation in the lung. *J Exp Med* 204:481–487.

Raychaudhuri, S.P., Jiang, W.Y., and Raychaudhuri, S.K. 2008. Revisiting the Koebner phenomenon: Role of NGF and its receptor system in the pathogenesis of psoriasis. *Am J Pathol* 172:961–971.

Repovic, P., and Benveniste, E.N. 2002. Prostaglandin E2 is a novel inducer of oncostatin-M expression in macrophages and microglia. *J Neurosci* 22:5334–5343.

Roosterman, D., Goerge, T., Schneider, S.W. et al. 2006. Neuronal control of skin function: The skin as a neuroimmunoendocrine organ. *Physiol Rev* 86:1309–1379.

Saika, F., Kiguchi, N., Kobayashi, Y. et al. 2012. CC-chemokine ligand 4/macrophage inflammatory protein-1beta participates in the induction of neuropathic pain after peripheral nerve injury. *Eur J Pain* 16:1271–1280.

Scholzen, T.E., Steinhoff, M., Sindrilaru, A. et al. 2004. Cutaneous allergic contact dermatitis responses are diminished in mice deficient in neurokinin 1 receptors and augmented by neurokinin 2 receptor blockage. *FASEB J* 18:1007–1009.

Schulz, F., Marenholz, I., Folster-Holst, R. et al. 2007. A common haplotype of the IL-31 gene influencing gene expression is associated with nonatopic eczema. *J Allergy Clin Immunol* 120:1097–1102.

Sehra, S., Tuana, F.M., Holbreich, M. et al. 2008. Scratching the surface: Towards understanding the pathogenesis of atopic dermatitis. *Crit Rev Immunol* 28:15–43.

Sendtner, M., Schmalbruch, H., Stockli, K.A. et al. 1992. Ciliary neurotrophic factor prevents degeneration of motor neurons in mouse mutant progressive motor neuronopathy. *Nature* 358:502–504.

Silver, J.S., and Hunter, C.A. 2010. gp130 at the nexus of inflammation, autoimmunity, and cancer. *J Leukoc Biol* 88:1145–1156.

Singer, E.M., Shin, D.B., Nattkemper, L.A. et al. 2013. Interleukin-31 is produced by the malignant T-cell population in cutaneous T-cell lymphoma and correlates with CTCL pruritus. *J Invest Dermatol* doi:10.1038/jid.2013.227. [Epub ahead of print].

Sleeman, M.W., Anderson, K.D., Lambert, P.D. et al. 2000. The ciliary neurotrophic factor and its receptor, CNTFR alpha. *Pharm Acta Helv* 74:265–272.

Sleeman, M.W., Garcia, K., Liu, R. et al. 2003. Ciliary neurotrophic factor improves diabetic parameters and hepatic steatosis and increases basal metabolic rate in db/db mice. *Proc Natl Acad Sci USA* 100:14297–14302.

Sonkoly, E., Muller, A., Lauerma, A.I. et al. 2006. IL-31: A new link between T cells and pruritus in atopic skin inflammation. *J Allergy Clin Immunol* 117:411–417.

Steinhoff, M., Bienenstock, J., Schmelz, M. et al. 2006. Neurophysiological, neuroimmunological, and neuroendocrine basis of pruritus. *J Invest Dermatol* 126:1705–1718.

Steinhoff, M., Groves, R., LeBoit, P. et al. 2010. Inflammation. In *Rook's Textbook of Dermatology*, 8th ed., Burns, T., Breathnach, S., Cox, N., and Griffiths, C. (Eds.). Blackwell: Malden, MA, pp. 10.1–10.69.

Steinhoff, M., Neisius, U., Ikoma, A. et al. 2003. Proteinase-activated receptor-2 mediates itch: A novel pathway for pruritus in human skin. *J Neurosci* 23:6176–6180.

Steinhoff, M., Vergnolle, N., Young, S.H. et al. 2000. Agonists of proteinase-activated receptor 2 induce inflammation by a neurogenic mechanism. *Nat Med* 6:151–158.

Stockli, K.A., Lillien, L.E., Naher-Noe, M. et al. 1991. Regional distribution, developmental changes, and cellular localization of CNTF-mRNA and protein in the rat brain. *J Cell Biol* 115:447–459.

Summer, G.J., Romero-Sandoval, E.A., Bogen, O. et al. 2008. Proinflammatory cytokines mediating burn-injury pain. *Pain* 135:98–107.

Takaoka, A., Arai, I., Sugimoto, M. et al. 2005. Expression of IL-31 gene transcripts in NC/Nga mice with atopic dermatitis. *Eur J Pharmacol* 516:180–181.

Tamura, S., Morikawa, Y., Miyajima, A. et al. 2002. Expression of oncostatin M in hematopoietic organs. *Dev Dyn* 225:327–331.

Tamura, S., Morikawa, Y., Miyajima, A. et al. 2003. Expression of oncostatin M receptor beta in a specific subset of nociceptive sensory neurons. *Eur J Neurosci* 17:2287–2298.

Tamura, S., Morikawa, Y., and Senba, E. 2003. Localization of oncostatin M receptor beta in adult and developing CNS. *Neuroscience* 119:991–997.

Tamura, S., Morikawa, Y., and Senba, E. 2005. Up-regulated phosphorylation of signal transducer and activator of transcription 3 and cyclic AMP-responsive element binding protein by peripheral inflammation in primary afferent neurons possibly through oncostatin M receptor. *Neuroscience* 133:797–806.

Tanaka, A., Muto, S., Jung, K. et al. 2007. Topical application with a new NF-kappaB inhibitor improves atopic dermatitis in NC/NgaTnd mice. *J Invest Dermatol* 127:855–863.

Tanaka, M., Hara, T., Copeland, N.G. et al. 1999. Reconstitution of the functional mouse oncostatin M (OSM) receptor: Molecular cloning of the mouse OSM receptor beta subunit. *Blood* 93:804–815.

Tran, T.A., Ross, J.S., Sheehan, C.E. et al. 2000. Comparison of oncostatin M expression in keratoacanthoma and squamous cell carcinoma. *Mod Pathol* 13:427–432.

Vlotides, G., Zitzmann, K., Stalla, G.K. et al. 2004. Novel neurotrophin-1/B cell-stimulating factor-3 (NNT-1/BSF-3)/cardiotrophin-like cytokine (CLC)—A novel gp130 cytokine with pleiotropic functions. *Cytokine Growth Factor Rev* 15:325–336.

Wang, X., Ratnam, J., Zou, B. et al. 2009. TrkB signaling is required for both the induction and maintenance of tissue and nerve injury-induced persistent pain. *J Neurosci* 29:5508–5515.

Weiss, T.W., Samson, A.L., Niego, B. et al. 2006. Oncostatin M is a neuroprotective cytokine that inhibits excitotoxic injury in vitro and in vivo. *FASEB J* 20:2369–2371.

Yoshida, K., Taga, T., Saito, M. et al. 1996. Targeted disruption of gp130, a common signal transducer for the interleukin 6 family of cytokines, leads to myocardial and hematological disorders. *Proc Natl Acad Sci USA* 93:407–411.

Yosipovitch, G., and Bernhard, J.D. 2013. Clinical practice. Chronic pruritus. *N Engl J Med* 368:1625–1634.

Yosipovitch, G., Greaves, M.W., and Schmelz, M. 2003. Itch. *Lancet* 361:690–694.

Yosipovitch, G., and Papoiu, A.D. 2008. What causes itch in atopic dermatitis? *Curr Allergy Asthma Rep* 8:306–311.

Zhang, N., and Oppenheim, J.J. 2005. Crosstalk between chemokines and neuronal receptors bridges immune and nervous systems. *J Leukoc Biol* 78:1210–1214.

Zhang, Q., Putheti, P., Zhou, Q. et al. 2008. Structures and biological functions of IL-31 and IL-31 receptors. *Cytokine Growth Factor Rev* 19:347–356.

Zhang, Z.J., Dong, Y.L., Lu, Y. et al. 2012. Chemokine CCL2 and its receptor CCR2 in the medullary dorsal horn are involved in trigeminal neuropathic pain. *J Neuroinflammation* 9:136.

Zhang, Z.Y., Zhang, Z., Fauser, U. et al. 2009. Expression of interleukin-16 in sciatic nerves, spinal roots and spinal cords of experimental autoimmune neuritis rats. *Brain Pathol* 19:205–213.

14 Toll-Like Receptors and Itch

Tong Liu and Ru-Rong Ji

CONTENTS

14.1 INTRODUCTION

Toll-like receptors (TLRs) are best known for their roles in controlling innate immunity (Akira et al. 2006). TLRs are characterized as pattern-recognition receptors (PRRs) to initiate innate immune responses via recognition of pathogen-associated molecular patterns (PAMPs) (Akira et al. 2006). TLRs can also sense endogenous molecules that are released after cellular stress or tissue injury, known as danger-associated molecular patterns (DAMPs). Activation of TLRs in immune cells leads to the synthesis of various proinflammatory cytokines and chemokines via transcriptional regulation. TLRs-mediated innate immune responses are also a prerequisite for the generation of adaptive immune responses (Mills 2011). Thus, TLRs represent the first line of host defense against pathogens and play a key role in both innate and adaptive immunity (Akira et al. 2006).

TLRs are also found to be expressed by various cell types in the central nervous system (CNS) and peripheral nervous system (PNS), such as microglia, astrocytes, oligodendrocytes, Schwann cells, and neurons (Okun et al. 2011; Buchanan et al. 2010; Lehnardt 2010). Activation of TLR signaling in the CNS also results in the production of inflammatory cytokines, enzymes, and other inflammatory mediators, which contributes to the pathogenesis of CNS microbial infection (Suh et al. 2009) as well as noninfective disorders, such as stroke (Caso et al. 2007), Alzheimer's disease (AD) (Tahara et al. 2006), multiple sclerosis (MS) (Prinz et al. 2006), and chronic pain (Guo and Schluesener 2007; Nicotra et al. 2011; Liu et al. 2012b).

TLRs are emerging as important players in acute and chronic itch (Liu et al. 2012b). Our recent study demonstrated that TLRs, including TLR3 and TLR7, are expressed by a subset of primary sensory neurons, which coexpress itch signaling components such as transient receptor potential vanilloid subtype 1 (TRPV1) (Liu et al. 2010, 2012a), and play an important role in itch sensation. TLRs are considered as cellular sensors for detecting exogenous and endogenous ligands/agonists in primary sensory neurons to initiate itch sensation associated with skin infection and tissue injury.

14.2   TLRs, THE LIGANDS, AND INTRACELLULAR SIGNALING

Toll is the first-identified TLR family member in *Drosophila* and plays key roles in antifungal innate immune response and dorsoventral patterning during early develop- · ment (Anderson et al. 1985; Lemaitre et al. 1996). Subsequently, many homologues for Toll were identified across diverse species (Medzhitov and Janeway 2000). To date, 10 (TLR1-10) and 12 (TLR1-9; TLR11-13) functional TLRs have been identified in humans and mice, respectively (Kawai and Akira 2010) (see Table 14.1).

TABLE 14.1
TLRs as Cellular Sensors of Microbial PAMPs and Endogenous DAMPs

TLR	Subcellular Localization	Exogenous Ligands	Endogenous Ligands	Adaptors
TLR1/TLR2	Cell surface	Triacyl lipopeptides (Pam3CSK4)	Unknown	Myd88 TIRAP
TLR2/TLR6	Cell surface	Diacyl lipopeptides PGN LTA Zymosan	HSP-60, 70, 90 HMGB1	Myd88 TIRAP
TLR3	Intracellular Cell surface	dsRNA Poly I:C	mRNA Stathmin	TRIF
TLR4	Cell surface	LPS Lipid A derivatives	HSP-22, -60, -70 HMGB1, fibronection Defensin 2, oxLDL Tenascin C	Myd88 TIRAP TRIF TRAM
TLR5	Cell surface	Flagellin	Unknown	Myd88
TLR7	Intracellular Cell surface	ssRNA Imidazoquinoline Loxoribine Bropirimine	Self RNA MicroRNA	Myd88
TLR8	Intracellular	ssRNA Imidazoquinoline	Self RNA MicroRNA	Myd88
TLR9	Intracellular	Unmethylated CpG-DNA CpG-ODNs	Self DNA HMGB1	Myd88
TLR11	Cell surface	Uropathogenic bacteria Profilin-like molecules	Unknown	Myd88

Note: CpG OND, CpG-containing oligodeoxynucleotides; ds RNA, double-stranded RNA; LPS, lipopolysaccharide; LTA, lipoteichoic acid; n.d., oxLDL, oxidized low-density lipoprotein; PGN, peptidoglycan; polyI:C, polyinosinic-polycytidylic acid; ss RNA, single-stranded RNA.

TLRs are structurally conserved type I transmembrane proteins and comprise an ectodomain with leucine-rich repeats to recognize PAMPs, a transmembrane domain, and a cytosolic Toll-IL-1 receptor (TIR) domain that recruits downstream adaptors (Kawai and Akira 2010). Most TLRs form homodimer between themselves, but some TLRs can also form noncovalent dimmers, such as TLR1/2 and TLR2/6 heterodimers (Oosting et al. 2011; Triantafilou et al. 2007). According to their subcellular localization and respective PAMPs, TLRs can be divided into two groups: (1) cell surface TLRs (TLR1, 2, 4, 5, 6, and 10) that recognize microbial membrane components, such as lipids, lipoproteins, and protein and (2) intracellular TLRs (TLR3, 7/8, and 9) that recognize viral or endogenous nucleic acids. However, some TLRs such as TLR3 and TLR7 could be localized both on the cell surface and in intracellular compartments (Akira et al. 2006).

Different TLRs sense distinct PAMPs derived from viruses, bacteria, mycobacteria, fungi, and parasites. TLR1/2 and TLR6 detect lipoproteins (Alexopoulou et al. 2002; Yamamoto et al. 2002). TLR4 recognizes lipopolysaccharide (LPS) (Shimazu et al. 1999; Poltorak et al. 1998). TLR5 detects flagellin (Hayashi et al. 2001), and TLR11 sense profilin-like protein (Yarovinsky et al. 2005) (Table 14.1). TLR3 and TLR7/8 sense double-stranded (ds) and single-stranded (ss) RNAs, respectively (Diebold et al. 2004; Town et al. 2006; Alexopoulou et al. 2001; Heil et al. 2004). TLR9 senses CpG DNA (Hemmi et al. 2000; Krieg 2002) (Table 14.1). In many noninfectious pathological conditions, TLRs could also recognize endogenous DAMPs that are released during cell necrosis and tissue damage to induce sterile inflammatory responses (Okamura et al. 2001; Imai et al. 2008; Jiang et al. 2005; Midwood et al. 2009; West et al. 2010; Tian et al. 2007; Biragyn et al. 2002; Vabulas et al. 2002; Kariko et al. 2004). Thus, TLRs can recognize both microbial PAMPs and endogenous DAMPs.

TLR signaling is pleiotropic but also tightly regulated. All TLRs, with the exception of TLR3, signal through the adaptor protein MyD88 (Figure 14.1). TLR1/2, 4, and 6 recruit additional adaptor protein TIRAP (Akira and Takeda 2004). In MyD88-dependent pathway, MyD88 recruits the IL-1R associated kinases (IRAKs), which interact with tumor necrosis factor receptor associated factor 6 (TRAF6), leading to the phosphorylation of the inhibitor of NF-κB (IκB) and its degradation and subsequent release and nucleus translocation of NF-κB for gene transcription. The MyD88-dependent pathway activates the mitogen-activated protein kinase (MAPK) signaling pathways, such as p38, and c-Jun N-terminal kinase (JNK), leading to the activation of AP-1 and interferon regulatory factors (IRFs). Activation of TLR signaling produces a wide array of proinflammatory mediators, such as cytokines, chemokines, and reactive oxygen/nitrogen intermediates including nitric oxide (Takeda and Akira 2004) (Figure 14.1).

TLR3 signals through TRIF-dependent pathway. TLR4 could also signal through TRIF-dependent pathways by recruiting additional adaptor TRAM. In the TRIF-dependent pathway, TRIF interacts with TRAF3 to activate IRF3 and IRF7 and initiate the production of Type I interferons (e.g., IFN-α/β), which are hallmarks of the host innate immune response (Yamamoto et al. 2003) (Figure 14.1). TRIF also binds to receptor-interacting protein 1 and TRAF6, leading to the activation of the NF-κB and/or MAPK pathways for the late-phase induction of proinflammatory genes.

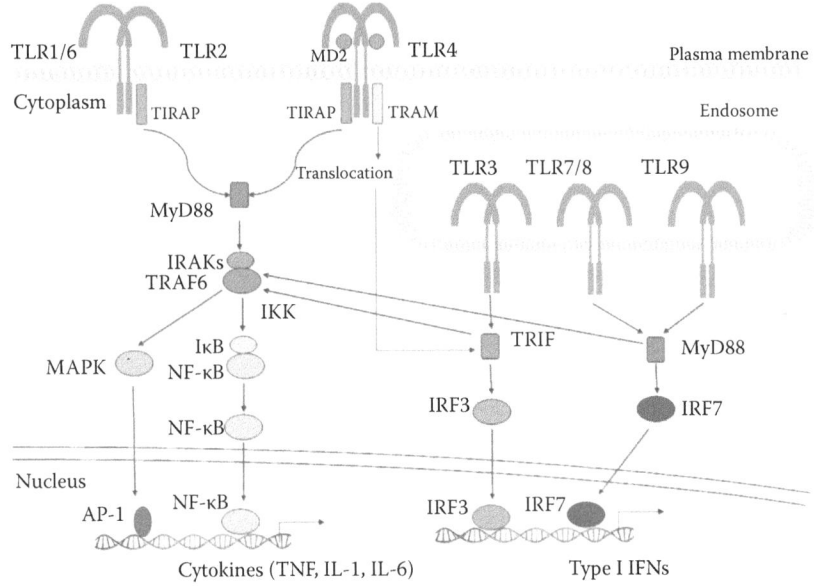

FIGURE 14.1 **(See color insert.)** TLR signaling pathways. TLR1/6, TLR2, TLR4, and TLR5 are expressed on the cell surface for sensing extracellular lipids, lipoproteins, and proteins derived from cell membrane components, whereas TLR3, TLR7/8, and TLR9 are mainly localized in the endosomes for recognizing nucleic acids. All TLRs, except TLR3, recruit the adaptor MyD88. TLR1/6, TLR2, and TLR4 recruit the additional adaptors TIRAP, which links the TIR domain with MyD88. IRAKs and TRAF6 subsequently form a complex. As a result, IκB is phosphorylated and degraded, resulting in NF-κB translocates to nucleus, where it initiates the gene expression (e.g., proinflammatory cytokines). Following LPS stimulation, TLR4 together with TRAM is translocated to the endosome. TLR3 and TLR4 activation initiates the TRIF-dependent pathway and activates NF-κB and IRF3, leading to the induction of proinflammatory cytokines and type I interferon (IFN). Stimulation of TLR7 and TLR9 can also recruit MyD88 and activate the NF-κB and IRF7 pathway for the induction of proinflammatory cytokines and type I IFN.

14.3   TLRs IN CHRONIC PAIN

Chronic pain is generally considered a result of neural plasticity including peripheral sensitization and central sensitization (Ji et al. 2003; Basbaum et al. 2009). Recent studies also suggest that neuroglia interaction is critical for the genesis of neural plasticity and chronic pain (Ren and Dubner 2010; Gao and Ji 2010a; Scholz and Woolf 2007; Tsuda et al. 2005). After tissue or nerve injury, glial cells (e.g., microglia and astrocytes) are activated in the spinal cord and brain stem and contribute to inflammatory and neuropathic pain sensitization (Suter et al. 2007; Ji and Suter 2007; Gao and Ji 2010a, 2010b; Watkins et al. 2009; Romero-Sandoval et al. 2008; Ren and Dubner 2010; Jiang et al. 2009).

TLRs are also expressed in glial cells. Microglia expresses almost all known members of the TLR family identified to date, while astrocytes express a relatively limited TLR repertoire, in part because of their neuroectodermal origin (Farina et al.

2007). Nerve injury and tissue inflammation result in activation of TLRs in microglia and astrocytes, leading to the induction of various proinflammatory mediators including cytokines (e.g., TNF-α), chemokines (e.g., CCL2), and enzymes (e.g., COX-2), as well as other inflammatory mediators (e.g., prostaglandins) (Lehnardt 2010; van Noort and Bsibsi 2009; Basbaum et al. 2009; Nicotra et al. 2011). These mediators are known to generate central sensitization and pain hypersensitivity. In preclinical mouse models, it has been shown that spinal TLR2 and TLR4 contribute to microglia and astrocyte activation and chronic pain hypersensitivity (Kim et al. 2007; Mei et al. 2011; Tanga et al. 2005). Blockade of TLR signaling in the spinal cord by TLR antagonists, specific TLR-targeted antisense oligonucleotides (ODN), or specific siRNA attenuated arthritic pain, neuropathic pain, and bone cancer pain (Christianson et al. 2011; Saito et al. 2010; Qi et al. 2011; Bettoni et al. 2008; Kigerl et al. 2007). Thus, TLRs are promising targets for the control of chronic pain.

14.4 TLRs IN ACUTE AND CHRONIC ITCH

Itch (pruritus) is defined as an unpleasant sensation that elicits desire or reflex to scratch. Although acute itch may sever as a warning and self-protective mechanism (Ikoma et al. 2006), chronic itch is a common clinical problem associated with skin diseases (Bieber 2008; Reich and Szepietowski 2007), liver and kidney diseases (Kremer et al. 2010; Cassano et al. 2010), and metabolism disorders (Yamaoka et al. 2010). It is established that the specific subtype of free nerve endings of primary sensory neurons, whose cell bodies located in dorsal root ganglia (DRG) and trigeminal ganglia (TG), are responsible for detecting itch signals (Ikoma et al. 2006; Paus et al. 2006). The central branches of primary sensory neurons transmit the itch signaling to the superficial dorsal horn of the spinal cord, where the itch signaling can be processed and modulated. Subsequent activation of specific brain regions results in the perception of itch. As the best characterized itch mediator, histamine is released from mast cells and binds H1/H4 receptors on nerve terminals to elicit itch (Shim and Oh 2008) via activation of PLCbeta3 and transient receptor potential vanilloid subtype 1 (TRPV1) (Imamachi et al. 2009; Han et al. 2006). Chloroquine (an antimalaria drug and MrgprA3 agonist), BAM8-22 (an endogenous MrgprC11 agonist), and oxidative stress produce histamine-independent itch via TRPA1 activation (Liu et al. 2009; Wilson et al. 2011; Liu and Ji 2012). TRPV1-expressing nociceptors are required for both histamine-dependent and histamine-independent itch (Imamachi et al. 2009; Mishra et al. 2011).

Recently, we found that TLR7 is functionally expressed by TRPV1+ nociceptors and mediates itch sensation. *In situ* hybridization and immunohistochemistry revealed that TLR7 mRNA and protein were mainly expressed in small-sized DRG neurons and highly colocalized with TRPV1 and gastrin-releasing peptide (GRP), a neuropeptide that is known to elicit itch via GRP receptor expressed by spinal cord superficial dorsal horn neurons (Sun et al. 2009; Sun and Chen 2007). In particular, single-cell real-time PCR (RT-PCR) analysis in small-sized DRG neurons confirmed that all TLR7-positive neurons also expressed GRP and TRPV1 (Liu et al. 2010).

TLR7 is originally identified to recognize imidazoquinoline derivatives, such as imiquimod and resiquimod (R848), and guanine analogs, such as loxoribine (Hemmi et al. 2002). Intradermal injection of imiquimod, R848, and loxoribine induced scratching in wild-type mice, which was reduced in $Tlr7^{-/-}$ mice. Of note, imiqui-mod also elicited TLR7-independent itch (Liu et al. 2010; Kim et al. 2011), in part because of its off-target effects on adenosine receptors, IP3Rs, and potassium chan-nels (Schon et al. 2006; Kaufman et al. 2011; Kim et al. 2011). Imiquimod-induced itch requires TRPV1-expressing C-fibers, but not TRPV1 *per se* (Liu et al. 2010; Kim et al. 2011). Furthermore, $Tlr7^{-/-}$ mice showed a significant reduction in scratch-ing behaviors in response to nonhistaminergic pruritogens, including chloroquine, endothlin-1 (ET-1), and SLIGRL-NH2, an agonist of protease-activated receptor 2 (PAR2). In contrast, scratching evoked by histaminergic pruritogens (e.g., compound 48/80, and histamine) is unaltered in $Tlr7^{-/-}$ mice. Of interest, $Tlr7^{-/-}$ mice exhib-ited normal thermal, mechanical, and inflammatory pain (Liu et al. 2010). Together, these findings suggest that TLR7 may serve as an itch receptor to detect exogenous or endogenous ligands.

Very recently, we also found that TLR3 is functionally expressed by a subset of primary sensory neurons in DRG, as revealed by multiple approaches, including *in situ* hybridization, immunocytochemistry, single-cell RT-PCR, and electrophysiol-ogy (Liu et al. 2012a). Our results showed that TLR3 is highly colocalized with itch signaling components, such as TRPV1 and GRP. Of note, TLR3 and TLR7 are partially colocalized (Liu et al. 2012a). The TLR3 agonist Poly I:C (PIC) is sufficient to induce inward currents and action potentials in DRG neurons and elicit scratching behavior in WT mice but not $Tlr3^{-/-}$ mice. Further, extracted total RNAs induced inward currents in DRG neurons via TLR3 (Liu et al. 2012a). Thus, TLR3 expressed by primary sensory neurons may also serve as itch receptors, like TLR7, to detect foreign pathogens and endogenous ligands (e.g., dsRNAs), leading to rapid defensive responses such as scratching.

Compared to a partial reduction in histamine-independent itch in $Tlr7^{-/-}$ mice, both histamine-dependent and independent itch are reduced in $Tlr3^{-/-}$ mice. $Tlr3^{-/-}$ mice do not have gross anatomical defects, neuronal loss in DRGs and spinal cords, deficits in the skin and spinal cord nerve innervations, impairment in the ascend-ing itch pathway, and changes in skin morphology. Hence, it is unlikely that the itch phenotypes we observed are results of developmental defects in $Tlr3^{-/-}$ mice. Consistently, intrathecal knockdown of TLR3 expression in DRGs also reduces itch in normal wild-type mice (Liu et al. 2012a).

We also demonstrated a critical role of TLR3 in spinal synaptic transmission and central sensitization underlying central itch processing (Liu et al. 2012a). TLR3 defi-ciency resulted in a decrease in spontaneous excitatory postsynaptic currents (sEP-SCs) in spinal lamina II neurons and a failure in the induction of spinal long-term potentiating (LTP) induced by titanic stimulation of sciatic nerve. In contrast, $Tlr7^{-/-}$ mice displayed normal spinal cord synaptic transmission and LTP induction. TLR3 agonist PIC is also sufficient to increased sEPSC frequency via spinal cord TLR3 activation. Given an important role of central sensitization in itch hypersensitivity (Ross et al. 2010), impairment in spinal synaptic plasticity and central sensitization should also contribute to profound itch deficits in $Tlr3^{-/-}$ mice.

TLR signaling in neurons, such as primary sensory neurons, could be distinct from that identified in immune cells (Acosta and Davies 2008). It is clear that activation of TLR3 and TLR7 by their respective ligands can excite DRG neurons via a nontranscriptional mechanism, because this excitation can occur within minutes. We postulate that TLR3 and TLR7 express on the cell surface and are coupled to some unidentified ion channels for inducing inward currents and actions potentials (Figure 14.2). The detailed downstream signaling needs to be determined. Of note, it was also demonstrated that activation TLRs, including TLR3, TLR7, and TLR9 in DRG neurons by their respective ligands may indirectly influence the excitability of DRG neurons by inducing the expression of proinflammatory mediators such as PGE2, CGRP, and IL-1β (Qi et al. 2011). Thus, TLRs expressed by DRG neurons, may regulate neuronal excitability by both transcriptional and nontranscriptional mechanisms.

Increasing evidence showed that TLR4 is also expressed in primary sensory neurons in DRG and TG. Early immunostaining data indicated TLR4 is expressed in TRPV1-expressing trigeminal neurons (Wadachi and Hargreaves 2006). Further studies confirmed that TLR4 agonist LPS binds to trigeminal neurons to elicit

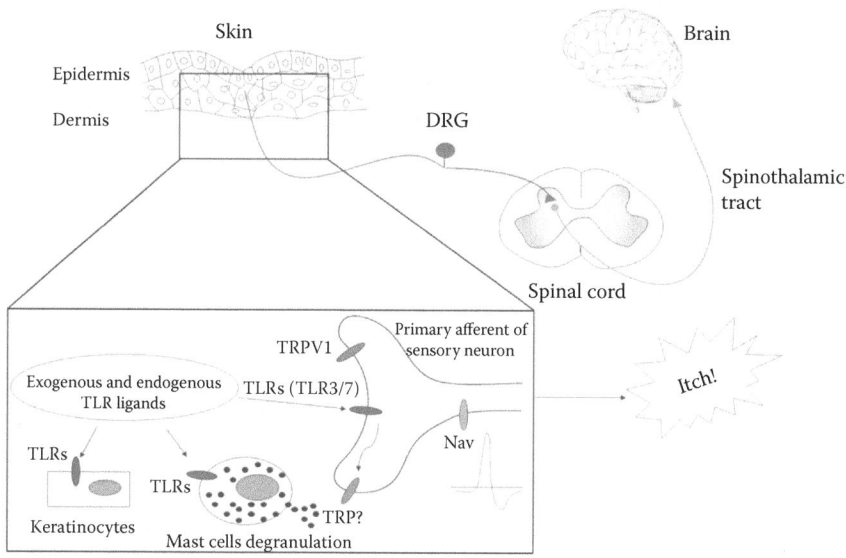

FIGURE 14.2 **(See color insert.)** Schematic of TLR signaling in skin for itch sensation. (1) Skin free nerve endings of primary sensory neurons are responsible for detecting exogenous and endogenous pruritogens, and the itch signal is conducted from periphery to spinal cord through the axons of primary neurons and finally to the brain. (2) TLRs, in particular, TLR3 and TLR7, are expressed in a subset of small-sized primary sensory neurons that coexpress itch signaling components, such as TRPV1, GRP, and MrgprA3. Activation of TLR3 or TLR7 by exogenous ligands (PIC, imiquimod, and loxoribine) or endogenous ligands (dsRNAs and ssRNAs) induces inward currents and action potentials in capsaicin-responsive small-sized DRG neurons, leading to itch sensation. TLRs (e.g., TLR3, TLR4, and TLR7) are also expressed by keratinocytes and mast cells. Activation of these TLRs in skin cells may indirectly contribute to itch through the induction of itch mediators (e.g., proinflammatory cytokines, chemokines, and NGF) in skin cells.

intracellular calcium release and inward currents and increase TRPV1 activity (Diogenes et al. 2011). In addition, TLR4 is also colocalized with calciton gene-related peptide (CGRP) in sensory neurons, and LPS was able to enhance TRPV1-dependent release of CGRP (Ferraz et al. 2011). However, the role of TLR4 in itch requires further investigation.

TLRs in skin may also play an active role in acute and chronic itch. Skin is a critical interface for host defense against microbial invaders. As the key innate immunity receptors for host defense, the expression of TLRs has been shown in different cell types in the epidermis and dermis, including keratinocytes and Langerhans cells in the epidermis and macrophages, dendrite cells, T and B cells, mast cells in the dermis, as well as endothelial cells of the skin microvasculature (Ermertcan et al. 2011; Lai and Gallo 2008; Miller 2008). It appears that each cell type has distinct expression profiles of TLRs and contributes differentially to skin immune responses (Ermertcan et al. 2011).

Keratinocytes and mast cells of skin have been implicated in chronic itch (Ikoma et al. 2006). Epidermis keratinocytes express TLRs 1–6, 9, and 10, which are upregulated in pruritic skin diseases, such as psoriasis and atopic dermatitis (Baker et al. 2003; Ermertcan et al. 2011). Mast cells, predominantly in the dermis, express TLRs 1–7 and 9 and play a key role in IgE-mediated allergic inflammation. The activation of TLRs will result in distinct induction of cytokines and chemokines, recruiting immune cells from the circulation and modulating adaptive immune response. It is likely that cytokines and chemokines induced by TLR activation also contributes to peripheral sensitization and central sensitization in chronic itch conditions (Figure 14.2). Notably, the upregulation of nerve growth factor (NGF) in dry skins of wild-type mice is suppressed in TLR3 knockout mice (Liu et al. 2012a). Given the fact that keratinocytes and mast cells are major sources of NGF (Ikoma et al. 2006), TLR3 in skin cells may regulate NGF expression under dry skin conditions to elicit chronic itch (Figure 14.2).

14.5   FUTURE PERSPECTIVES AND CLINICAL SIGNIFICANCE

TLRs are emerging as important players in regulating acute and chronic itch. However, many questions remain. What are the endogenous ligands for TLRs that are released under chronic itch conditions such as pruritic dermatitis? Is the intracellular signaling of neuronal TLRs distinct from that of immune and glial TLRs? What are the ionic mechanisms underlying the excitatory effects of TLR ligands in sensory neurons? In addition to primary sensory neurons, what is the precise role of TLRs in skin cells for chronic itch? In the past decades, great efforts have been made to develop new TLR agonists and antagonists for immune disorders and anticancer therapies. These compounds may also offer new opportunities for the management of itch-related problems in humans and pets.

ACKNOWLEDGMENTS

The work was supported in part by US National Institutes of Health grants R01-DE17794, R01-NS54362 and R01-NS67686 to RRJ.

CONFLICT OF INTEREST

The authors state no conflict of interest.

REFERENCES

Acosta C, Davies A (2008) Bacterial lipopolysaccharide regulates nociceptin expression in sensory neurons. *J Neurosci Res* 86:1077–1086.

Akira S, Takeda K (2004) Toll-like receptor signalling. *Nat Rev Immunol* 4:499–511.

Akira S, Uematsu S, Takeuchi O (2006) Pathogen recognition and innate immunity. *Cell* 124:783–801.

Alexopoulou L, Holt AC, Medzhitov R et al. (2001) Recognition of double-stranded RNA and activation of NF-kappaB by Toll-like receptor 3. *Nature* 413:732–738.

Alexopoulou L, Thomas V, Schnare M et al. (2002) Hyporesponsiveness to vaccination with Borrelia burgdorferi OspA in humans and in. *Nat Med* 8:878–884.

Anderson KV, Jurgens G, Nusslein-Volhard C (1985) Establishment of dorsal-ventral polarity in the Drosophila embryo: Genetic studies on the role of the Toll gene product. *Cell* 42:779–789.

Baker BS, Ovigne JM, Powles AV et al. (2003) Normal keratinocytes express Toll-like receptors (TLRs) 1, 2 and 5: Modulation of TLR expression in chronic plaque psoriasis. *Br J Dermatol* 148:670–679.

Basbaum AI, Bautista DM, Scherrer G et al. (2009) Cellular and molecular mechanisms of pain. *Cell* 139:267–284.

Bettoni I, Comelli F, Rossini C et al. (2008) Glial TLR4 receptor as new target to treat neuropathic pain: Efficacy of a new receptor antagonist in a model of peripheral nerve injury in mice. *Glia* 56:1312–1319.

Bieber T (2008) Atopic dermatitis. *N Engl J Med* 358:1483–1494.

Biragyn A, Ruffini PA, Leifer CA et al. (2002) Toll-like receptor 4-dependent activation of dendritic cells by beta-defensin 2. *Science* 298:1025–1029.

Buchanan MM, Hutchinson M, Watkins LR et al. (2010) Toll-like receptor 4 in CNS pathologies. *J Neurochem* 114:13–27.

Caso JR, Pradillo JM, Hurtado O et al. (2007) Toll-like receptor 4 is involved in brain damage and inflammation after experimental stroke. *Circulation* 115:1599–1608.

Cassano N, Tessari G, Vena GA et al. (2010) Chronic pruritus in the absence of specific skin disease: An update on pathophysiology, diagnosis, and therapy. *Am J Clin Dermatol* 11:399–411.

Christianson CA, Dumlao DS, Stokes JA et al. (2011) Spinal TLR4 mediates the transition to a persistent mechanical hypersensitivity after the resolution of inflammation in serum-transferred arthritis. *Pain* 152:2881–2891.

Diebold SS, Kaisho T, Hemmi H et al. (2004) Innate antiviral responses by means of TLR7-mediated recognition of single-stranded RNA. *Science* 303:1529–1531.

Diogenes A, Ferraz CC, Akopian AN et al. (2011) LPS sensitizes TRPV1 via activation of TLR4 in trigeminal sensory neurons. *J Dent Res* 90:759–764.

Ermertcan AT, Ozturk F, Gunduz K (2011) Toll-like receptors and skin. *J Eur Acad Dermatol Venereol* 25:997–1006.

Farina C, Aloisi F, Meinl E (2007) Astrocytes are active players in cerebral innate immunity. *Trends Immunol* 28:138–145.

Ferraz CC, Henry MA, Hargreaves KM et al. (2011) Lipopolysaccharide from Porphyromonas gingivalis sensitizes capsaicin-sensitive nociceptors. *J Endod* 37:45–48.

Gao YJ, Ji RR (2010a) Chemokines, neuronal-glial interactions, and central processing of neuropathic pain. *Pharmacol Ther* 126:56–68.

Gao YJ, Ji RR (2010b) Targeting astrocyte signaling for chronic pain. *Neurotherapeutics* 7:482–493.

Guo LH, Schluesener HJ (2007) The innate immunity of the central nervous system in chronic pain: The role of Toll-like receptors. *Cell Mol Life Sci* 64:1128–1136.

Han SK, Mancino V, Simon MI (2006) Phospholipase Cbeta 3 mediates the scratching response activated by the histamine H1 receptor on C-fiber nociceptive neurons. *Neuron* 52:691–703.

Hayashi F, Smith KD, Ozinsky A et al. (2001) The innate immune response to bacterial flagellin is mediated by Toll-like receptor 5. *Nature* 410:1099–1103.

Heil F, Hemmi H, Hochrein H et al. (2004) Species-specific recognition of single-stranded RNA via Toll-like receptor 7 and 8. *Science* 303:1526–1529.

Hemmi H, Kaisho T, Takeuchi O et al. (2002) Small anti-viral compounds activate immune cells via the TLR7 MyD88-dependent signaling pathway. *Nat Immunol* 3:196–200.

Hemmi H, Takeuchi O, Kawai T et al. (2000) A Toll-like receptor recognizes bacterial DNA. *Nature* 408:740–745.

Ikoma A, Steinhoff M, Stander S et al. (2006) The neurobiology of itch. *Nat Rev Neurosci* 7:535–547.

Imai Y, Kuba K, Neely GG et al. (2008) Identification of oxidative stress and Toll-like receptor 4 signaling as a key pathway of acute lung injury. *Cell* 133:235–249.

Imamachi N, Park GH, Lee H et al. (2009) TRPV1-expressing primary afferents generate behavioral responses to pruritogens via multiple mechanisms. *Proc Natl Acad Sci USA* 106:11330–11335.

Ji RR, Kohno T, Moore KA et al. (2003) Central sensitization and LTP: Do pain and memory share similar mechanisms? *Trends Neurosci* 26:696–705.

Ji RR, Suter MR (2007) p38 MAPK, microglial signaling, and neuropathic pain. *Mol Pain* 3:33.

Jiang D, Liang J, Fan J et al. (2005) Regulation of lung injury and repair by Toll-like receptors and hyaluronan. *Nat Med* 11:1173–1179.

Jiang F, Liu T, Cheng M et al. (2009) Spinal astrocyte and microglial activation contributes to rat pain-related behaviors induced by the venom of scorpion Buthus martensi Karch. *Eur J Pharmacol* 623:52–64.

Kariko K, Ni H, Capodici J et al. (2004) mRNA is an endogenous ligand for Toll-like receptor 3. *J Biol Chem* 279:12542–12550.

Kaufman EH, Fryer AD, Jacoby DB (2011) Toll-like receptor 7 agonists are potent and rapid bronchodilators in guinea pigs. *J Allergy Clin Immunol* 127:462–469.

Kawai T, Akira S (2010) The role of pattern-recognition receptors in innate immunity: Update on Toll-like receptors. *Nat Immunol* 11:373–384.

Kigerl KA, Lai W, Rivest S et al. (2007) Toll-like receptor (TLR)-2 and TLR-4 regulate inflammation, gliosis, and myelin sparing after spinal cord injury. *J Neurochem* 102:37–50.

Kim D, Kim MA, Cho IH et al. (2007) A critical role of Toll-like receptor 2 in nerve injury-induced spinal cord glial cell activation and pain hypersensitivity. *J Biol Chem* 282:14975–14983.

Kim SJ, Park GH, Kim D et al. (2011) Analysis of cellular and behavioral responses to imiquimod reveals a unique itch pathway in transient receptor potential vanilloid 1 (TRPV1)-expressing neurons. *Proc Natl Acad Sci USA* 108:3371–3376.

Kremer AE, Martens JJ, Kulik W et al. (2010) Lysophosphatidic acid is a potential mediator of cholestatic pruritus. *Gastroenterology* 139:1008–1018.

Krieg AM (2002) CpG motifs in bacterial DNA and their immune effects. *Annu Rev Immunol* 20:709–760.

Lai Y, Gallo RL (2008) Toll-like receptors in skin infections and inflammatory diseases. *Infect Disord Drug Targets* 8:144–155.

Lehnardt S (2010) Innate immunity and neuroinflammation in the CNS: The role of microglia in Toll-like receptor-mediated neuronal injury. *Glia* 58:253–263.

Lemaitre B, Nicolas E, Michaut L et al. (1996) The dorsoventral regulatory gene cassette spatzle/Toll/cactus controls the potent antifungal response in Drosophila adults. *Cell* 86:973–983.

Liu Q, Tang Z, Surdenikova L et al. (2009) Sensory neuron-specific GPCR Mrgprs are itch receptors mediating chloroquine-induced pruritus. *Cell* 139:1353–1365.

Liu T, Berta T, Xu ZZ et al. (2012a) TLR3 deficiency impairs spinal cord synaptic transmission, central sensitization, and pruritus in mice. *J Clin Invest* 122:2195–2207.

Liu T, Gao YJ, Ji RR (2012b) Emerging role of Toll-like receptors in the control of pain and itch. *Neurosci Bull* 28:131–144.

Liu T, Ji RR (2012) Oxidative stress induces itch via activation of transient receptor potential subtype ankyrin 1 in mice. *Neurosci Bull* 28:145–154.

Liu T, Xu ZZ, Park CK et al. (2010) Toll-like receptor 7 mediates pruritus. *Nat Neurosci* 13:1460–1462.

Medzhitov R, Janeway C, Jr. (2000) Innate immune recognition: Mechanisms and pathways. *Immunol Rev* 173:89–97.

Mei XP, Zhou Y, Wang W et al. (2011) Ketamine depresses Toll-like receptor 3 signaling in spinal microglia in a rat model of neuropathic pain. *Neurosignals* 19:44–53.

Midwood K, Sacre S, Piccinini AM et al. (2009) Tenascin-C is an endogenous activator of Toll-like receptor 4 that is essential for maintaining inflammation in arthritic joint disease. *Nat Med* 15:774–780.

Miller LS (2008) Toll-like receptors in skin. *Adv Dermatol* 24:71–87.

Mills KH (2011) TLR-dependent T cell activation in autoimmunity. *Nat Rev Immunol* 11:807–822.

Mishra SK, Tisel SM, Orestes P et al. (2011) TRPV1-lineage neurons are required for thermal sensation. *EMBO J* 30:582–593.

Nicotra L, Loram LC, Watkins LR et al. (2011) Toll-like receptors in chronic pain. *Exp Neurol* 234:316–329.

Okamura Y, Watari M, Jerud ES et al. (2001) The extra domain A of fibronectin activates Toll-like receptor 4. *J Biol Chem* 276:10229–10233.

Okun E, Griffioen KJ, Mattson MP (2011) Toll-like receptor signaling in neural plasticity and disease. *Trends Neurosci* 34(5):269–281.

Oosting M, Ter HH, Sturm P et al. (2011) TLR1/TLR2 heterodimers play an important role in the recognition of Borrelia spirochetes. *PLoS One* 6:e25998.

Paus R, Schmelz M, Biro T et al. (2006) Frontiers in pruritus research: Scratching the brain for more effective itch therapy. *J Clin Invest* 116:1174–1186.

Poltorak A, He X, Smirnova I et al. (1998) Defective LPS signaling in C3H/HeJ and C57BL/10ScCr mice: Mutations in Tlr4 gene. *Science* 282:2085–2088.

Prinz M, Garbe F, Schmidt H et al. (2006) Innate immunity mediated by TLR9 modulates pathogenicity in an animal model of multiple sclerosis. *J Clin Invest* 116:456–464.

Qi J, Buzas K, Fan H et al. (2011) Painful pathways induced by TLR stimulation of dorsal root ganglion neurons. *J Immunol* 186:6417–6426.

Reich A, Szepietowski JC (2007) Mediators of pruritus in psoriasis. *Mediators Inflamm* 2007:64727.

Ren K, Dubner R (2010) Interactions between the immune and nervous systems in pain. *Nat Med* 16:1267–1276.

Romero-Sandoval EA, Horvath RJ, Deleo JA (2008) Neuroimmune interactions and pain: Focus on glial-modulating targets. *Curr Opin Investig Drugs* 9:726–734.

Ross SE, Mardinly AR, McCord AE et al. (2010) Loss of inhibitory interneurons in the dorsal spinal cord and elevated itch in Bhlhb5 mutant mice. *Neuron* 65:886–898.

Saito O, Svensson CI, Buczynski MW et al. (2010) Spinal glial TLR4-mediated nociception and production of prostaglandin E(2) and TNF. *Br J Pharmacol* 160:1754–1764.

Scholz J, Woolf CJ (2007) The neuropathic pain triad: Neurons, immune cells and glia. *Nat Neurosci* 10:1361–1368.

Schon MP, Schon M, Klotz KN (2006) The small antitumoral immune response modifier imiquimod interacts with adenosine receptor signaling in a TLR7- and TLR8-independent fashion. *J Invest Dermatol* 126:1338–1347.

Shim WS, Oh U (2008) Histamine-induced itch and its relationship with pain. *Mol Pain* 4:29.

Shimazu R, Akashi S, Ogata H et al. (1999) MD-2, a molecule that confers lipopolysaccharide responsiveness on Toll-like receptor 4. *J Exp Med* 189:1777–1782.

Suh HS, Brosnan CF, Lee SC (2009) Toll-like receptors in CNS viral infections. *Curr Top Microbiol Immunol* 336:63–81.

Sun YG, Chen ZF (2007) A gastrin-releasing peptide receptor mediates the itch sensation in the spinal cord. *Nature* 448:700–703.

Sun YG, Zhao ZQ, Meng XL et al. (2009) Cellular basis of itch sensation. *Science* 325:1531–1534.

Suter MR, Wen YR, Decosterd I et al. (2007) Do glial cells control pain? *Neuron Glia Biol* 3:255–268.

Tahara K, Kim HD, Jin JJ et al. (2006) Role of Toll-like receptor signalling in Abeta uptake and clearance. *Brain* 129:3006–3019.

Takeda K, Akira S (2004) TLR signaling pathways. *Semin Immunol* 16:3–9.

Tanga FY, Nutile-McMenemy N, Deleo JA (2005) The CNS role of Toll-like receptor 4 in innate neuroimmunity and painful neuropathy. *Proc Natl Acad Sci USA* 102:5856–5861.

Tian J, Avalos AM, Mao SY et al. (2007) Toll-like receptor 9-dependent activation by DNA-containing immune complexes is mediated by HMGB1 and RAGE. *Nat Immunol* 8:487–496.

Town T, Jeng D, Alexopoulou L et al. (2006) Microglia recognize double-stranded RNA via TLR3. *J Immunol* 176:3804–3812.

Triantafilou M, Uddin A, Maher S et al. (2007) Anthrax toxin evades Toll-like receptor recognition, whereas its cell wall components trigger activation via TLR2/6 heterodimers. *Cell Microbiol* 9:2880–2892.

Tsuda M, Inoue K, Salter MW (2005) Neuropathic pain and spinal microglia: A big problem from molecules in "small" glia. *Trends Neurosci* 28:101–107.

Vabulas RM, Wagner H, Schild H (2002) Heat shock proteins as ligands of Toll-like receptors. *Curr Top Microbiol Immunol* 270:169–184.

van Noort JM, Bsibsi M (2009) Toll-like receptors in the CNS: Implications for neurodegeneration and repair. *Prog Brain Res* 175:139–148.

Wadachi R, Hargreaves KM (2006) Trigeminal nociceptors express TLR-4 and CD14: A mechanism for pain due to infection. *J Dent Res* 85:49–53.

Watkins LR, Hutchinson MR, Rice KC et al. (2009) The "toll" of opioid-induced glial activation: Improving the clinical efficacy of opioids by targeting glia. *Trends Pharmacol Sci* 30:581–591.

West XZ, Malinin NL, Merkulova AA et al. (2010) Oxidative stress induces angiogenesis by activating TLR2 with novel endogenous ligands. *Nature* 467:972–976.

Wilson SR, Gerhold KA, Bifolck-Fisher A et al. (2011) TRPA1 is required for histamine-independent, Mas-related G protein-coupled receptor-mediated itch. *Nat Neurosci* 14:595–602.

Yamamoto M, Sato S, Hemmi H et al. (2003) Role of adaptor TRIF in the MyD88-independent Toll-like receptor signaling pathway. *Science* 301:640–643.

Yamamoto M, Sato S, Mori K et al. (2002) Cutting edge: A novel Toll/IL-1 receptor domain-containing adapter that preferentially activates the IFN-beta promoter in the Toll-like receptor signaling. *J Immunol* 169:6668–6672.

Yamaoka H, Sasaki H, Yamasaki H et al. (2010) Truncal pruritus of unknown origin may be a symptom of diabetic polyneuropathy. *Diabetes Care* 33:150–155.

Yarovinsky F, Zhang D, Andersen JF et al. (2005) TLR11 activation of dendritic cells by a protozoan profilin-like protein. *Science* 308:1626–1629.

15 Lipid Mediators and Itch

Tsugunobu Andoh and Yasushi Kuraishi

CONTENTS

15.1 INTRODUCTION

Lipid mediators have a variety of biological and pathophysiological functions in the skin. Especially well known are the metabolites of arachidonic acid that, when liberated from membrane phospholipids by phospholiplase A_2, are involved in inflammation and pain. In the 1970s, prostaglandin (PG) E_2, one such arachidonic acid metabolite, was reported to elicit itch, probably via histamine release, and to enhance experimentally evoked itch in humans (Hägermark and Strandberg 1977; Fjellner and Hägermark 1979). For a long time after that, only a few reports describing the roles of lipid mediators in pruritus have been published. One reason for such little progress in understanding the lipid itch mediators could be the lack of a reliable method for the behavioral evaluation of itch in animal experiments (Woodward et al. 1985). In 1995, pruritogenic substances, but not algogenic, were shown to elicit hind-paw scratching in mice, raising the possibility that scratching can be used as an index of itch response in rodents (Kuraishi et al. 1995). Animal experiments have revealed the involvement and roles of several lipid mediators for itch. In this chapter, the roles of lipid mediators in human and animal itch are explained.

15.2 ARACHIDONIC ACID METABOLITES

15.2.1 PROSTANOIDS

PGs such as PGE_2, PGI_2, $PGF_{2\alpha}$, and PGD_2 are synthesized from PGH_2, a metabolite of arachidonic acid produced by a reaction catalyzed by cyclooxygenase-1 and -2 (Williams and DuBois 1996). In human subjects, an intradermal injection of

PGE_2 (2.8–71 µM) elicits mild itching, whereas that of PGE_2 (~14 µM) enhances histamine- and serotonin-induced itching (Hägermark and Strandberg 1977; Fjellner and Hägermark 1979). Pruritus in polyerythemia vera is alleviated by aspirin, and it is thought to be mediated by the enhancement of serotonin-induced itching by PGE_2 (Fjellner and Hägermark 1979). In animals, on the other hand, an intradermal injection of PGE_2 (60 nM to 6 mM) does not induce scratching in mice (Andoh and Kuraishi 1998), but a topical application of PGE_2 (~280 µM) inhibits spontaneous scratching in NC mice with chronic dermatitis (Arai et al. 2004). In contrast to an application to the skin, an application of PGE_2 (1.4 and 14 mM) to the ocular surface induces hind-paw scratching in guinea pigs (Woodward et al. 1995, 1996). There are four PGE_2 receptor subtypes, EP_1 to EP_4 (Sugimoto and Narumiya 2007). Although a limited role of the EP_1 receptor subtype has been shown in allergy-induced ocular scratching, the roles of EP receptor subtypes in pruritus remains unclear.

As noted above, an application of PGI_2 (1.4 and 14 mM) to the ocular surface elicits hind-paw scratching in guinea pigs (Woodward et al. 1996); in contrast, a topical application of PGI_2 (~0.28 mM) to the lesional skin suppresses spontaneous scratching in NC mice with chronic dermatitis (Arai et al. 2004).

When travoprost, a synthetic prostaglandin $F_{2\alpha}$ analogue, is instilled into the eyes of patients with open-angle glaucoma or ocular hypertension, the most frequent related adverse event observed is pruritus (Goldberg et al. 2001). However, an application of $PGF_{2\alpha}$ (1.4 and 14 mM) to the ocular surface does not elicit ocular scratching in guinea pigs (Woodward et al. 1995, 1996).

The application of PGD_2 (14–140 mM) or BW245C, a potent PGD_2 agonist, to the ocular surface elicits itch in human subjects (Nakajima et al. 1991). Similarly, an application of PGD_2 (~280 mM) to the ocular surface of guinea pigs induces hind-paw scratching (Hashimoto et al. 2003). In contrast, topical applications to the skin of PGD_2 (2.8–280 µM) and TS-022, a DP_1 receptor agonist suppress both spontaneous scratching in mice with chronic dermatitis (Arai et al. 2004, 2007) and scratching induced by compound 48/80 injection or allergy (Hashimoto et al. 2005).

Thromboxane A_2 (TXA_2) is synthesized from PGH_2 by thromboxane synthase (Needleman et al. 1976), and it is altered spontaneously to inactive TXB_2. Thromboxane synthase is present in epidermal keratinocytes (Andoh et al. 2007), inflammatory cells such as eosinophils (Mita et al. 1999), mast cells (Mita et al. 1999), macrophages (Tone et al. 1994), and platelets (Yokoyama et al. 1991). The serum concentration of TXB_2 is increased in humans with some pruritic diseases (Marks et al. 1984; Mysliwiec et al. 1985; Grekas et al. 1989; Veale et al. 1994). However, it is unknown whether TXA_2 administered to human skin is pruritogenic. An intradermal injection of 9,11-dideoxy-9a,11a-methanoepoxy prostaglandin $F_{2\alpha}$ (U-46619), a stable analogue of TXA_2, elicits hind-paw scratching through its action on the TP prostanoid receptor (Andoh et al. 2007). In the skin, TP receptors are expressed in epidermal keratinocytes and primary afferents (Andoh et al. 2007). TXA_2 may produce itch signaling through its action on primary afferents and enhance itching through its action on epidermal keratinocytes, which produce many itch mediators, including TXA_2 itself.

15.2.2 LEUKOTRIENES

Leukotrienes (LTs) such as LTB_4 and cysteinyl LTs (LTC_4, LTD_4, and LTE_4), are synthesized from arachidonic acid by arachidonate 5-lipoxygenase activated with 5-lipoxygenase-activating protein (FLAP) through 5-hydroperoxyeicosatetraenoic acid (5-HPETE), and then LTA_4. LTB_4, and LTC_4, especially, are derived from LTA_4 via the actions of LTA_4 hydrolase and LTC_4 synthase, respectively (Murphy and Gijón 2007). Furthermore, LTC_4 is rapidly metabolized to LTD_4 and then to LTE_4 via γ-glutamyl transpeptidase and a membrane-bound dipeptidase, respectively (Murphy and Gijón 2007). LTB_4 acts as both a high-affinity BLT1 and a low-affinity BLT2 receptor (Yokomizo et al. 2001a; Toda et al. 2002). Cysteinyl LTs act both on CysLT1 and CysLT2 (Evans 2002).

Zileuton, a 5-lipoxygenase inhibitor, reduces pruritus in Sjögren-Larsson syndrome (Willemsen et al. 2001) and decreases the tendency to produce it in atopic dermatitis (Woodmansee and Simon 1999). Azelastine, an H_1 histamine receptor antagonist with LTB_4 blocking activity, reduces pruritus in hemodialysis patients (Kanai et al. 1995; Matsui et al. 1994). LTB_4 is elevated in the lesional skin of patients with psoriasis or atopic dermatitis (Brain et al. 1984; Ruzicka et al. 1986) and in the urine of patients with Sjögren-Larsson syndrome (Willemsen et al. 2001). These findings, taken together, raise the possibility that LTB_4 is involved in itching in the above mentioned pruritic diseases. An intradermal injection of LTB_4 (3–30 μM) elicits itch in one of six human subjects (Camp et al. 1983). In mice, intradermal injections of LTB_4 (0.06–20 μM) elicit hind-paw scratching; the dose-response curve is bell shaped with a peak effect at 0.6 μM (Andoh and Kuraishi 1998). LTB_4 is involved in spontaneous hind-paw scratching in NC mice with chronic dermatitis (Andoh et al. 2011), in scratching evoked by an allergen challenge in passive cutaneous anaphylaxis (Tsuji et al. 2010), and in contact dermatitis (Tsukumo et al. 2010). It is also associated with hind-paw scratching evoked by intradermal injections of pruritogens such as substance P (Andoh et al. 2001), nociceptin (Andoh et al. 2004), and sphingosylphosphorylcholine (Andoh et al. 2009). As for the eye, subconjunctival injections of LTB_4 (0.5 and 5 μM) elicit hind-paw scratching in mice. Furthermore, LTB_4 is involved in the scratching evoked by allergen challenge in ragweed pollen allergy (Andoh et al. 2012). Applications of LTC_4 (0.8 mM), LTD_4 (0.10 and 1.0 mM), and LTE_4 (0.11 and 1.1 mM) to the eye do not induce hind-paw scratching in guinea pigs (Woodward et al. 1995).

Intradermal injections of LTC_4 and LTD_4 (0.24–0.76 μM) do not elicit itch in human subjects, although they do produce an erythematous reaction (Camp et al. 1983). Intradermal injections of LTD_4 (0.2–6 μM) do not elicit hind-paw scratching in mice (Andoh et al. 2001); moreover, neither the cysteinyl LT antagonist pranlukast nor the LTD_4 antagonist MK-571 suppress hind-paw scratching induced by mosquito allergy in mice (Kuraishi et al. 2007). These accumulated findings suggest that cysteinyl LTs are not pruritogenic in the skin and eye.

15.2.3 12(S)-HYDROPEROXYEICOSA-5Z,8Z,10E,14Z-TETRAENOIC ACID

12(S)-Hydroperoxyeicosa-5Z,8Z,10E,14Z-tetraenoic acid (12(S)-HPETE) is synthesized from arachidonic acid by arachidonate 12-lipoxygenase. An intradermal injection of 12(S)-HPETE (0.3–10 μM) elicits hind-paw scratching mediated by BLT2, but not BLT1, receptors in mice (Kim et al. 2007, 2008b); 12(S)-HPETE acts not only on BLT2 (Yokomizo et al. 2001b) but also on the transient receptor potential vanilloid 1 (TRPV1) (Hwang et al. 2000). BLT2 receptors are expressed in mast cells (Lundeen et al. 2006) but not in sensory neurons (Andoh and Kuraishi 2005). The fact that TRPV1 antagonist does not inhibit 12(S)-HPETE-induced scratching (Kim et al. 2008a) suggests that, in primary afferents, TRPV1 does not play a key role in intradermal 12(S)-HPETE-induced scratching. However, 12(S)-HPETE-induced scratching is inhibited by $5-HT_1$ and $5-HT_2$ receptor antagonists, but not by an H_1 histamine receptor antagonist (Kim et al. 2008a). The pruritogenic activity of serotonin is greater than that of histamine in mice (Akiyama et al. 2010; Maekawa et al. 2000). These pieces of evidence suggest that 12(S)-HPETE acts on BLT2 receptors in mast cells to release serotonin, which, in turn, induces scratching in mice. The involvement of 12(S)-HPETE in human itching is unknown.

15.3   PLATELET-ACTIVATING FACTOR

Platelet-activating factor (PAF) is synthesized in a two-step process catalyzed by the enzymes phospholipase A_2 and lyso-PAF acetyltransferase (Prescott et al. 1990). An intradermal injection of PAF elicits itch in human subjects (Fjellner and Hägermark 1985). Repeated topical application of the PAF antagonist RO-24-0238 was reported to reduce pruritus in patients with atopic dermatitis during the first 2 weeks, but not after 3 and 4 weeks (Abeck et al. 1997). The plasma level of PAF is increased in pruritic diseases such as psoriasis (Izaki et al. 1996) and cold urticaria (Grandel et al. 1985). In animal experiments, the subcutaneous injection of PAF induces scratching in mice (Ishiguro et al. 2002), and the topical application and intraconjunctival injection of PAF induce hind-paw scratching in guinea pigs (Woodward et al. 1995; Kato et al. 2003). The intravenous injection of the PAF antagonist CV-3988 does not inhibit hind-paw scratching induced by mosquito allergy in mice (Kuraishi et al. 2007).

15.4   LYSOPHOSPHATIDIC ACID

There are several potential metabolic routes to lysophosphatidic acid (LPA, mono-acyl-glycerol-3-phosphate). Although extracellular LPA can be produced from phosphatidic acid by phospholipase A_1 and A_2, the most important pathway is the conversion of lysophosphatidylcholine by lysophospholipase D, also known as autotaxin (ATX) (Pebay et al. 2007). An intradermal injection of LPA (4.6 mM) elicits hind-paw scratching in mice; the scratching is inhibited by ketotifen, an H_1 histamine receptor antagonist, and Y-27632, an inhibitor of Rho-associated protein kinase (Hashimoto et al. 2004). The fact that serum ATX levels have been shown to increase in cholestasis patients with pruritus raises the possibility that LPA produced by ATX is involved in cholestatic pruritus (Kremer et al. 2010, 2012). Thus, it is suggested that LPA is involved in cholestatic pruritus. An increase in serum ATX levels is not observed in patients with other pruritic diseases as uremia, Hodgkin's disease, and atopic dermatitis; therefore, it was suggested that LPA has a role in itching in these pruritic diseases (Kremer et al. 2012).

(Healthy skin)

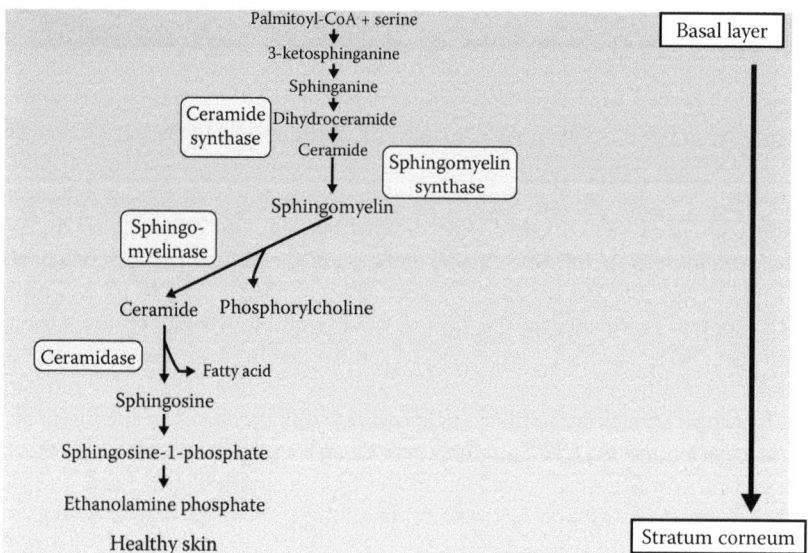

(Atopic dermatitis)

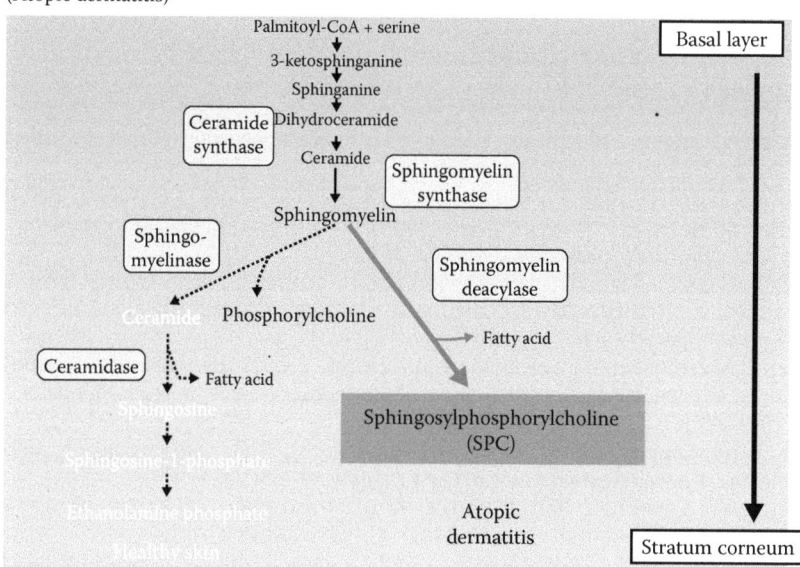

FIGURE 15.1 (See color insert.) Metabolism of sphingomyelin in healthy skin and atopic dermatitis. (Modified schema from Hara, J. et al., *J Invest Dermatol*, 115, 406–413, 2000.)

15.5 SPHINGOLIPIDS

Sphingolipids comprise a complex set of lipids, including sphingomyelin and ceramides, in which fatty acids are linked via amide bonds to sphingoid. Ceramides are produced from sphingomyelin by sphingomyelinase in the stratum corneum, and they play an essential role in structuring and maintaining the water permeability barrier of the skin (Figure 15.1). In atopic dermatitis patients, the activity of sphingomyelin deacylase—an enzyme that converts sphingomyelin into sphingosylphosphorylcholin (SPC) and free fatty acid—is elevated (Murata et al. 1996; Hara et al. 2000), leading to an increase in the SPC content and a decrease in the ceramide content of the stratum corneum of atopic dermatitis (Okamoto et al. 2003) (Figure 15.1). Similarly, the SPC content is elevated in the skin of mice with atopy-like chronic dermatitis (Andoh et al. 2011). However, the activity of sphingomyelin deacylase is not increased in contact dermatitis (Hara et al. 2000). Intradermal injection of SPC (0.6 and 2 mM) elicits hind-paw scratching in mice, an action possibly mediated by the direct action on primary afferents and LTB_4 production in keratinocytes (Andoh et al. 2009). Rho-associated protein kinase involvement in the SPC action has also been previously reported (Kim et al. 2008c). An increase in sphingomyelin deacylase activity may result in skin dryness due to a decrease in ceramides, as well as itching due to SPC production in the skin. Thus, the suppression of sphingomyelin deacylase activity may relieve skin dryness and pruritus in atopic dermatitis.

REFERENCES

Abeck, D., Andersson, T., Grosshans, E. et al. 1997. Topical application of a platelet-activating factor (PAF) antagonist in atopic dermatitis. *Acta Derm Venereol* 77: 449–451.

Akiyama, T., Carstens, M. I., and Carstens, E. 2010. Facial injections of pruritogens and algogens excite partly overlapping populations of primary and second-order trigeminal neurons in mice. *J Neurophysiol* 104: 2442–2450.

Andoh, T., Haza, S., Saito, A., and Kuraishi, Y. 2011. Involvement of leukotriene B_4 in spontaneous itch-related behaviour in NC mice with atopic dermatitis-like skin lesions. *Exp Dermatol* 20: 894–898.

Andoh, T., Katsube, N., Maruyama, M., and Kuraishi, Y. 2001. Involvement of leukotriene B_4 in substance P-induced itch-associated response in mice. *J Invest Dermatol* 117: 1621–1626.

Andoh, T., and Kuraishi, Y. 1998. Intradermal leukotriene B_4, but not prostaglandin E_2, induces itch-associated responses in mice. *Eur J Pharmacol* 353: 93–96.

Andoh, T., and Kuraishi, Y. 2005. Expression of BLT1 leukotriene B_4 receptor on the dorsal root ganglion neurons in mice. *Mol Brain Res* 137: 263–266.

Andoh, T., Nishikawa, Y., Yamaguchi-Miyamoto, T., Nojima, H., Narumiya, S., and Kuraishi, Y. 2007. Thromboxane A_2 induces itch-associated responses through TP receptors in the skin in mice. *J Invest Dermatol* 127: 2042–2047.

Andoh, T., Saito, A., and Kuraishi, Y. 2009. Leukotriene B_4 mediates sphingosylphosphorylcholine-induced itch-associated responses in mouse skin. *J Invest Dermatol* 129: 2854–2860.

Andoh, T., Sakai, K., Urashima, M., Kitazawa, K., Honma, A., and Kuraishi, Y. 2012. Involvement of leukotriene B_4 in itching in a mouse model of ocular allergy. *Exp Eye Res* 98: 97–103.

Andoh, T., Yageta, Y., Takeshima, H., and Kuraishi, Y. 2004. Intradermal nociceptin elicits itch-associated responses through leukotriene B_4 in mice. *J Invest Dermatol* 123: 196–201.

Arai, I., Takano, N., Hashimoto, Y. et al. 2004. Prostanoid DP1 receptor agonist inhibits the pruritic activity in NC/Nga mice with atopic dermatitis. *Eur J Pharmacol* 505: 229–235.

Arai, I., Takaoka, A., Hashimoto, Y. et al. 2007. Effects of TS-022, a newly developed prostanoid DP1 receptor agonist, on experimental pruritus, cutaneous barrier disruptions and atopic dermatitis in mice. *Eur J Pharmacol* 556: 207–214.

Brain, S., Camp, R., Dowd, P., Black, A. K., and Greaves, M. 1984. The release of leukotriene B_4-like material in biologically active amounts from the lesional skin of patients with psoriasis. *J Invest Dermatol* 83: 70–73.

Camp, R. D., Coutts, A. A., Greaves, M. W., Kay, A. B., and Walport, M. J. 1983. Responses of human skin to intradermal injection of leukotrienes C_4, D_4 and B_4. *Br J Pharmacol* 80: 497–502.

Evans, J. F. 2002. Cysteinyl leukotriene receptors. *Prostaglandins Other Lipid Mediat* 68–69: 587–597.

Fjellner, B., and Hägermark, Ö. 1979. Pruritus in polycythemia vera: Treatment with aspirin and possibility of platelet involvement. *Acta Dermatovener* 59: 505–512.

Fjellner, B., and Hägermark, Ö. 1985. Experimental pruritus evoked by platelet activating factor (PAF-acether) in human skin. *Acta Derm Venereol* 65: 409–412.

Goldberg, I., Cunha-Vaz, J., Jakobsen, J. E., Nordmann, J. P., Trost, E., and Sullivan, E. K. 2001. International Travoprost Study Group. Comparison of topical travoprost eye drops given once daily and timolol 0.5% given twice daily in patients with open-angle glaucoma or ocular hypertension. *J Glaucoma* 10: 414–422.

Grandel, K. E., Farr, R. S., Wanderer, A. A., Eisenstadt, T. C., and Wasserman, S. I. 1985. Association of platelet-activating factor with primary acquired cold urticaria. *N Engl J Med* 313: 405–409.

Grekas, D., Alivanis, P., Karamouzis, M., and Pyrpasopoulos, M. 1989. Piracetam as a potent inhibitor of plasma thromboxane B_2 during hemodialysis. *Nephron* 52: 372–373.

Hägermark, Ö., and Strandberg, K. 1977. Pruritogenic activity of prostaglandin E_2. *Acta Dermatovener* 57: 37–43.

Hara, J., Higuchi, K., Okamoto, R., Kawashima, M., and Imokawa, G. 2000. High-expression of sphingomyelin deacylase is an important determinant of ceramide deficiency leading to barrier disruption in atopic dermatitis. *J Invest Dermatol* 115: 406–413.

Hashimoto, T., Igarashi, A., Hoshina, F. et al. 2003. Effects of nonsteroidal anti-inflammatory drugs on experimental allergic conjunctivitis in guinea pigs. *J Ocul Pharmacol Ther* 19: 569–577.

Hashimoto, T., Ohata, H., and Momose, K. 2004. Itch-scratch responses induced by lysophosphatidic acid in mice. *Pharmacology* 72: 51–56.

Hashimoto, Y., Arai, I., Tanaka, M., and Nakaike, S. 2005. Prostaglandin D_2 inhibits IgE-mediated scratching by suppressing histamine release from mast cells. *J Pharmacol Sci* 98: 90–93.

Hwang, S. W., Cho, H., Kwak, J. et al. 2000. Direct activation of capsaicin receptors by products of lipoxygenases: Endogenous capsaicin-like substances. *Proc Natl Acad Sci USA* 97: 6155–6160.

Ishiguro, K., Oku, H., Suitani, A., and Yamamoto, Y. 2002. Effects of conjugated linoleic acid on anaphylaxis and allergic pruritus. *Biol Pharm Bull* 25: 1655–1657.

Izaki, S., Yamamoto, T., Goto, Y. et al. 1996. Platelet-activating factor and arachidonic acid metabolites in psoriatic inflammation. *Br J Dermatol* 134: 1060–1064.

Kanai, H., Nagashima, A., Hirakata, E. et al. 1995. The effect of azelastin hydrochloride on pruritus and leukotriene B_4 in hemodialysis patients. *Life Sci* 57: 207–213.

Kato, M., Kurose, T., Oda, T., and Miyaji, S. 2003. The role of platelet activating factor and the efficacy of apafant ophthalmic solution in experimental allergic conjunctivitis. *J Ocul Pharmacol Ther* 19: 315–324.

Kim, D. K., Kim, H. J., Kim, H. et al. 2008a. Involvement of serotonin receptors 5-HT$_1$ and 5-HT$_2$ in 12(S)-HPETE-induced scratching in mice. *Eur J Pharmacol* 579: 390–394.

Kim, H. J., Kim, D. K., Kim, H. et al. 2008b. Involvement of the BLT2 receptor in the itch-associated scratching induced by 12-(S)-lipoxygenase products in ICR mice. *Br J Pharmacol* 154: 1073–1078.

Kim, D. K., Kim, H. J., Sung, K. S. et al. 2007. 12(S)-HPETE induces itch-associated scratchings in mice. *Eur J Pharmacol* 554: 30–33.

Kim, H. J., Kim, H., Han, E. S. et al. 2008c. Characterizations of sphingosylphosphorylcholine-induced scratching responses in ICR mice using naltrexon, capsaicin, ketotifen and Y-27632. *Eur J Pharmacol* 583: 92–96.

Kremer, A. E., Dijk, R. V., Leckie, P. et al. 2012. Serum autotaxin is increased in pruritus of cholestasis, but not of other origin and responds to therapeutic interventions. *Hepatology* 56: 1391–1400.

Kremer, A. E., Martens, J. J., Kulik, W. et al. 2010. Lysophosphatidic acid is a potential mediator of cholestatic pruritus. *Gastroenterology* 139: 1008–1018.

Kuraishi, Y., Nagasawa, T., Hayashi, K., and Satoh, M. 1995. Scratching behavior induced by pruritogenic but not algesiogenic agents in mice. *Eur J Pharmacol* 275: 229–233.

Kuraishi, Y., Ohtsuka, E., Nakano, T. et al. 2007. Possible involvement of 5-lipoxygenase metabolite in itch-associated response of mosquito allergy in mice. *J Pharmacol Sci* 105: 41–47.

Lundeen, K. A., Sun, B., Karlsson, L., and Fourie, A. M. 2006. Leukotriene B$_4$ receptors BLT1 and BLT2: Expression and function in human and murine mast cells. *J Immunol* 117: 3439–3447.

Maekawa, T., Nojima, H., and Kuraishi, Y. 2000. Itch-associated responses of afferent nerve innervating the murine skin: Different effects of histamine and serotonin in ICR and ddY mice. *Jpn J Pharmacol* 84: 462–466.

Marks, J. G. Jr., Trautlein, J. J., Zwillich, C. W., and Dermers, L. M. 1984. Contact urticarial and airway obstruction from carbon copy paper. *JAMA* 252: 1038–1040.

Matsui, C., Ida, M., Hamada, M., Morohashi, M., and Hasegawa M. 1994. Effects of azelastin on pruritus and plasma histamine levels in hemodialysis patients. *Int J Dermatol* 33: 868–871.

Mita, H., Ishii, T., and Akiyama, K. 1999. Generation of thromboxane A$_2$ from highly purified human sinus mast cells after immunological stimulation. *Prostaglandins Leukot Essent Fatty Acids* 60: 175–180.

Murata, Y., Ogata, J., Higaki, Y. et al. 1996. Abnormal expression of sphingomyelin acylase in atopic dermatitis: An etiologic factor for ceramide deficiency? *J Invest Dermatol* 106: 1242–1249.

Murphy, R. C., and Gijón, M. A. 2007. Biosynthesis and metabolism of leukotrienes. *Biochem J* 405: 379–395.

Mysliwiec, M., Bodzenta, A., Rydzewski, A., and Soszka, J. 1985. Plasma thromboxane in hemodialysis patients. *Folia Haematol Int Mag Klin Morphol Blutforsch* 112: 442–446.

Nakajima, M., Goh, Y., Azuma, I., and Hayaishi, O. 1991. Effects of prostaglandin D$_2$ and its analogue, BW245C, on intraocular pressure in humans. *Graefes Arch Clin Exp Ophthalmol* 229: 411–413.

Needleman, P., Moncada, S., Bunting, S., Vane, J. R., Hamberg, M., and Samuelsson, B. 1976. Identification of an enzyme in platelet microsomes which generates TXA$_2$ from prostaglandin endoperoxides. *Nature* 261: 558–560.

Okamoto, R., Arikawa, J., Ishibashi, M., Kawashima, M., Takagi, Y., and Imokawa, G. 2003. Sphingosylphosphorylcholine is upregulated in the stratum corneum of patients with atopic dermatitis. *J Lipid Res* 44: 93–102.

Pebay, A., Bonder, C. S., and Pitson, S. M. 2007. Stem cell regulation by lysophospholipids. *Prostaglandins Other Lipid Mediat* 84: 83–97.

Prescott, S. M., Zimmerman, G. A., and McIntyre, T. M. 1990. Platelet-activating factor. *J Biol Chem* 265: 17381–17384.

Ruzicka, T., Simmet, T., Peskar, B. A., and Ring, J. 1986. Skin levels of arachidonic acid-derived inflammatory mediators and histamine in atopic dermatitis and psoriasis. *J Invest Dermatol* 86: 105–108.

Sugimoto, Y., and Narumiya, S. 2007. Prostaglandin E receptors. *J Biol Chem* 282: 11613–11617.

Toda, A., Yokomizo, T., and Shimizu, T. 2002. Leukotriene B_4 receptors. *Prostaglandins Other Lipid Mediat* 68–69: 575–585.

Tone, Y., Miyata, A., Hara, S., Yukawa, S., and Tanabe, T. 1994. Abundant expression of thromboxane synthase in rat macrophages. *FEBS Lett* 340: 241–244.

Tsuji, F., Aono, H., Tsuboi, T. et al. 2010. Role of leukotriene B_4 in 5-lipoxygenase metabolite- and allergy-induced itch-associated responses in mice. *Biol Pharm Bull* 33: 1050–1053.

Tsukumo, Y., Harada, D., and Manabe, H. 2010. Pharmacological characterization of itch-associated response induced by repeated application of oxazolone in mice. *J Pharmacol Sci* 113: 255–262.

Veale, D. J., Torley, H. I., Richards, I. M. et al. 1994. A double-blind placebo controlled trial of Efamol Marine on skin and joint symptoms of psoriatic arthritis. *Br J Rheumatol* 33: 954–958.

Willemsen, M. A., Lutt, M. A., Steijlen, P. M. et al. 2001. Clinical and biochemical effects of zileuton in patients with the Sjögren-Larsson syndrome. *Eur J Pediatr* 160: 711–717.

Williams, C. S., and DuBois, R. N. 1996. Prostaglandin endoperoxide synthase: Why two isoforms? *Am J Physiol* 270: G393–G400.

Woodmansee, D. P., and Simon, R. A. 1999. A pilot study examining the role of zileuton in atopic dermatitis. *Ann Allergy Asthma Immunol* 83: 548–552.

Woodward, D. F., Conway, J. L., and Wheeler, L. A. 1985. Cutaneous itching models. In *Models in Dermatology*, vol. 1, H. I. Maibach, and N. J. Lowe (eds.). Basel: Karger, pp. 187–195.

Woodward, D. F., Nieves, A. L., and Friedlaender, M. H. 1996. Characterization of receptor subtypes involved in prostanoid-induced conjunctival pruritus and their role in mediating allergic conjunctival itching. *J Pharmacol Exp Ther* 279: 137–142.

Woodward, D. F., Nieves, A. L., Spada, C. S., Williams, L. S., and Tuckett, R. P. 1995. Characterization of a behavioral model for peripherally evoked itch suggests platelet-activating factor as a potent pruritogen. *J Pharmacol Exp Ther* 272: 758–765.

Yokomizo, T., Izumi, T., and Shimizu, T. 2001a. Leukotriene B_4: Metabolism and signal transduction. *Arch Biochem Biophys* 385: 231–241.

Yokomizo, T., Kato, K., Hagiya, H., Izumi, T., and Shimizu, T. 2001b. Hydroxyeicosanoids bind to and activate the low affinity leukotriene B_4 receptor, BLT2. *J Biol Chem* 276: 12454–12459.

Yokoyama, C., Miyata, A., Ihara, H., Ullrich, V., and Tanabe, T. 1991. Molecular cloning of human platelet thromboxane A synthase. *Biochem Biophys Res Commun* 178: 1479–1484.

16 Role of Transient Receptor Potential Channels in Acute and Chronic Itch

Sarah R. Wilson and Diana M. Bautista

CONTENTS

16.1 INTRODUCTION

Members of the transient receptor potential (TRP) family have emerged as key players in itch transduction in the periphery. TRP family members are tetrameric cation selective channels that are expressed in diverse species, from flies to humans. The founding member of the TRP channel superfamily is *Drosophila* TRP, a transduction channel required for light-evoked excitation of photoreceptors. In phototransduction, activation of the phospholipase C (PLC) pathway leads to the opening of TRP and its homolog TRP-L; flies lacking these channels display no light-evoked transduction currents and are blind. Over 27 members have since been identified in a variety of cell types and tissues (Figure 16.1).[1–5]

TRP channels are divided into seven subgroups based on protein homology rather than function: TRPC, TRPV, TRPM, TRPA, TRPN, TRPP, and TRPML. Generally, TRP channels function as polymodal cellular sensors involved in a wide variety of cellular processes. Many TRPs have been found to participate in sensory transduction pathways, including thermosensation, mechanosensation, taste, perception of pungent compounds, pheromone sensing, and osmolarity regulation. A number of excellent reviews describe the vast roles of TRP channels which will not be discussed.[1–5]

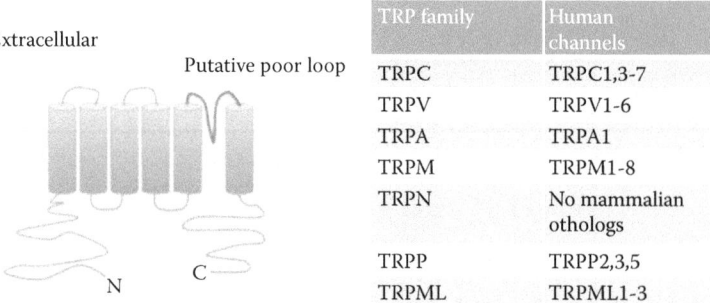

TRP family	Human channels
TRPC	TRPC1,3-7
TRPV	TRPV1-6
TRPA	TRPA1
TRPM	TRPM1-8
TRPN	No mammalian othologs
TRPP	TRPP2,3,5
TRPML	TRPML1-3

FIGURE 16.1 (See color insert.) Transient receptor potential (TRP) ion channel family. TRP channels are tetrameric cation selective channels with six transmembrane spanning helices. Distinct TRP channels have (left) intracellular N- and C-termini that vary dramatically in size and contain a variety of protein interaction and modulatory domains (not shown). The putative pore domain is depicted in red. TRP channels are divided into seven subgroups based on protein homology rather than function: TRPC, TRPV, TRPM, TRPA, TRPN, TRPP, and TRPML (right). Members of six of these subgroups have been identified in a variety of tissue types in humans. This chapter will focus on members of the TRPV, TRPA, and TRPM families and their roles in pruriception.

Here, we discuss the role of four TRP channels that have been proposed to play a role in itch transduction: TRPV1, TRPA1, TRPM8, and TRPV3. Historically, these four channels have been implicated in the transduction of noxious thermal, chemical, and/or mechanical stimuli, and more recent studies have implicated these channels in the transduction of itch.

16.2 TRPV1

TRPV1 was first identified as a receptor for capsaicin, the active ingredient in hot chili peppers that elicits a burning sensation when eaten. TRPV1 is a heat- and ligand-activated, nonselective cation channel.[6] This channel is highly expressed in a subset of temperature-sensitive somatosensory neurons that have cell bodies in the dorsal root and trigeminal sensory ganglia and project afferents to innervate target organs in the periphery, such as the skin.[6,7]

Capsaicin activates TRPV1 by binding to an intracellular region of the channel in between the second and third transmembrane domain.[8] TRPV1 is also activated by heat, with a threshold of activation of approximately $43°C$ and a coefficient of temperature dependence (Q10) of $~40°C$.[6] A number of endogenous ligands also activate TRPV1; protons, anandamide, lipoxygenase products, and N-arachidonoyl dopamine have all been proposed to modulate TRPV1 activity *in vivo*.[9-11] Activation of TRPV1 permits the influx of cations into the peripheral nerve terminal to promote depolarization and action potential propagation to the central nervous system.

The localization of TRPV1 in heat-sensitive neurons, and the ability of TRPV1 to be directly activated by heat, support a model where this channel functions as an *in vivo* thermoreceptor. Indeed, TRPV1-deficient mice display decreased sensitivity to noxious heat in acute behavioral assays.[12,13] TRPV1 also plays a key role in thermal hypersensitivity following injury or inflammation. WT mice display thermal hyperalgesia and allodynia, the increased pain sensitivity to both noxious and previously innocuous stimuli, respectively, after injury or inflammation. Mice lacking TRPV1 fail to develop hyperalgesia and allodynia in these models, demonstrating the importance of TRPV1 in pain hypersensitivity.[12,13]

How does inflammation change TRPV1 activity to promote thermal hypersensitivity? Many of the inflammatory mediators responsible for pain hypersensitivity, including NGF, ATP, chemokines, and prostaglandins, activate G-protein coupled receptors (GPCRs) that signal via PLC.[14] PLC signaling in turn modulates TRPV1 activity, such that the open probability of the channel is increased at body temperature; thus, TRPV1 promotes neuronal excitability in the absence of heat.[8] Indeed, most TRP channels are activated or modulated downstream of GPCRs.[1] As most pruritogens activate GPCRs, and trigger itch via activation of somatosensory afferents, TRP channels, including TRPV1, are attractive candidate itch transducers.[15]

A link between thermal pain and itch has been established for centuries. First, patients with chronic itch conditions have long reported that scalding heat helps to alleviate their pruritus. Second, topical application of the TRPV1 agonist, capsaicin, has been used to treat itch associated with many skin conditions. In 1850, the first formal report of the use of capsaicin to treat itch, and pain, appeared in a publication recommending the use of a hot pepper extract on burning or itching extremities.[16] Today, topical capsaicin formulations are widely used to manage pain and beneficial effects of capsaicin have been reported in chronic, localized pruritic disorders, particularly those of neuropathic origin, such as notalgia paresthetica, brachioradial pruritus, prurigo nodularis, aquagenic pruritus, and pruritus associated with chronic kidney disease.[17–21] Consistent with these treatments in humans, neonatal capsaicin treatment decreases allergy-associated scratching in mice.[22]

The antinociceptive and antipruritic effects of topical capsaicin are thought to manifest through the defunctionalization of TRPV1-expressing primary afferents, mediated by direct desensitization of TRPV1, or voltage gated sodium channels in the short term, and nerve terminal retraction due to excitotoxic terminal damage induced by excessive calcium and inhibition of mitochondrial respiration in the long term.[23] Indeed, immunohistochemical studies using antibodies to nerve terminal proteins, like PGP 9.5, show that capsaicin application induces localized loss of nociceptive nerve fiber terminals in the epidermis and dermis.[24] Therefore, the use of capsaicin to treat pruritus implicates either TRPV1 or TRPV1-containing primary afferents in pruriception.

A number of early studies suggested that TRPV1 may mediate histamine signal transduction in primary sensory neurons. First, many histamine-sensitive fibers are also capsaicin-sensitive. Second, histamine sensitizes primary afferent fibers to heat stimuli. These experiments strongly suggested that the histamine receptor and TRPV1 are coexpressed in sensory afferents.[25–27] Third, histamine-evoked calcium transients in rat dorsal root ganglia neurons are inhibited by the TRPV1 antagonists, capsazepine and SC0030.[28] Trypsin-evoked itch behaviors are attenuated in both capsazepine treated and TRPV1-deficient mice.[29] Fifth, patients with allergic rhinitis display an increased itch response to TRPV1 stimulation from seasonal allergen exposure.[30] Finally, TRPV1 antagonists inhibit pruritus in atopic dermatitis and the commonly prescribed antipruritic, tacrolimus, has been suggested to work in part by inhibiting/desensitizing TRPV1.[31,32]

A definitive role for TRPV1 in histamine-evoked itch behavior was finally established in 2009 when it was shown that TRPV1-deficient mice displayed significantly attenuated itch behavior in response to injection of histamine. This study also found that histamine-evoked scratching behavior is attenuated in mice deficient in PLCß3, the PLC isoform activated downstream of the Gq coupled H1 receptor.[33] As such, a model emerged where activation of histamine H1 G_q-coupled GPCRs signals via PLCß3 to open TRPV1, through an unknown mechanism, in murine primary afferents (Figure 16.2). However, while itch behavior was attenuated in TRPV1 deficient mice, the behavior was not completely ablated and in addition to histamine receptor 1, histamine receptors 3 and 4 are also expressed in primary sensory neurons.[34] Similarly, not all histamine-sensitive sensory neurons are capsaicin sensitive.[28] This suggests that other channels, potentially downstream of other histamine receptors, may be involved in transducing histamine-evoked itch signals.

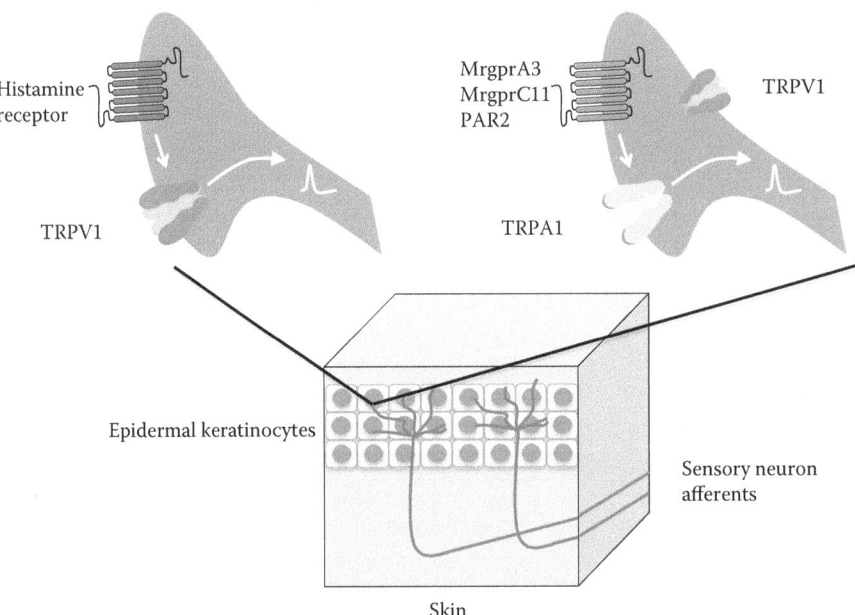

FIGURE 16.2   **(See color insert.)** TRPA1 and TRPV1 play key roles in acute itch stimuli transduction. TRPV1 and TRPA1 are calcium-permeable cation channels that can be activated downstream of numerous pruritogen GPCRs. (left) Fibers expressing TRPV1 are responsible for transmitting histamine-dependent itch; TRPV1 is activated downstream of the histamine receptor via the PLC pathway. (right) Fibers expressing TRPV1 and TRPA1 transmit histamine-independent itch signals; TRPA1 is activated downstream of the Mas-related G protein-coupled receptors, MrgprA3 and MrgprC11, as well as the protease activated receptor 2 (PAR2). While TRPV1 is present in these neurons, it is not required for itch transduction.

16.3 TRPV1-EXPRESSING AFFERENTS AND ACUTE ITCH

While a role for TRPV1 in histamine-evoked itch is well established, many prurito-gens act independently of TRPV1. How do other pruritogens promote excitability in sensory neurons? A hint came from the original TRPV1-directed studies. Two potent pruritogens, serotonin and endothelin-1, were shown to evoke robust scratching in TRPV1-deficient mice. However, itch behavior was significantly attenuated in mice treated with intrathecal capsaicin to ablate TRPV1-positive afferents.[33] Intrathecal capsaicin injections in mice, similar to the topical capsaicin applications in humans discussed above, promotes receptor defunctionalization, as well as excitotoxic neuro-nal ablation; as such, this treatment results in mice lacking TRPV1-positive neurons.[35] Taken together, these data suggest that other channels, expressed in TRPV1-positive afferents, are required for histamine-independent itch. Consistent with this idea, two drugs that evoke antihistamine resistant itch, imiquimod and chloroquine, trigger itch-evoked scratching in TRPV1-deficient mice, but not TRPV1-ablated mice.[33,36,37,38]

16.4 TRPA1 AND HISTAMINE-INDEPENDENT ITCH

The ion channel, TRPA1, is highly expressed in a subset of TRPV1-positive neu-rons and plays a key role in multiple types of histamine-independent itch. TRPA1 is robustly activated by a wide variety of exogenous irritants that cause pain and inflammation. Environmental chemicals that target TRPA1 include allyl isothiocya-nate (AITC), cinnamaldehyde, and allicin, the pungent compounds found in mustard, cinnamon, and garlic extracts, respectively.[39–42] TRPA1 is also a target of endogenous inflammatory agents, such as 15dPGJ2, PGA2, and Δ12-PGJ2 and can be modulated by PLC-coupled receptors mediating inflammation, such as the bradykinin recep-tor.[41–45] Many studies also suggest that TRPA1 can be activated directly by reac-tive oxygen species including hydrogen peroxide and the lipid peroxidation products 4-HNE, 4-ONE, and 4-HHE.[43,46,47] Data from TRPA1-deficient mice have shown that TRPA1 is required both for acute behavioral responses to AITC and for prolonged mechanical and thermal hypersensitivity following AITC exposure.[48] TRPA1 is also required for inflammatory responses to formalin and the α,β-unsaturated aldehyde acrolein, an airway irritant present in tear gas, vehicle exhaust, and smoke.[48,49] These studies show that TRPA1 acts as a general mediator of inflammation that can be activated by a host of endogenous and exogenous irritants.

There is now a growing body of evidence that suggests a significant role for TRPA1 in histamine-independent itch. Members of the novel Mas-related G-protein coupled receptor (Mrgpr) family, MrgprA3 and MrgprC11, are activated by two dif-ferent histamine-independent pruritogens, chloroquine (CQ) and BAM8-22 (BAM), respectively.[50] TRPA1 is required for both CQ- and BAM-evoked calcium signals and action potential firing in somatosensory neurons, as well as CQ- and BAM-evoked scratching in mice (Figure 16.2).[51] TRPA1 has since been shown to be spe-cifically required for scratching in an oxidative stress model of itch.[52] In this model, RTX ablation also reduces scratching, suggesting involvement of C fibers. Genetic or pharmacological blockade of TRPA1 activity, but not TRPV1, decreases oxidative stress-evoked scratching. Taken together, these data demonstrate that TRPA1 is a downstream transduction channel onto which multiple histamine-independent acute itch pathways converge.

16.5   TRPV3 AND KERATINOCYTE-MEDIATED ITCH PATHWAYS

Keratinocytes, the epithelial cells that make up the stratified epidermis of the skin, play a key role in itch by secreting a variety of mediators that target sensory neurons and immune cells.[53] TRPV3 is expressed in keratinocytes and has been proposed to play a role in promoting itch signaling and behaviors.

TRPV3 was originally identified as a heat sensitive ion channel in keratinocytes. Mouse TRPV3 is preferentially activated by innocuous, warm temperature with a threshold of ~33°C.[54] Warmth-evoked currents in cultured keratinocytes display biophysical properties that match those of TRPV3 currents in heterologous cells. Consistent with a role in warm sensing, mice lacking TRPV3 display deficits in responses to innocuous and noxious heat, and keratinocyte-specific TRPV3 knock-in mice display increased avoidance of noxious heat, in the absence of functional TRPV1 channels.[54–57] While it is unclear how TRPV3-dependent signaling in keratinocytes promotes sensory neurons, one model suggests that ATP is the signaling molecule linking keratinocytes and neurons.[58] Like most TRP channels, TRPV3 is a polymodal sensor that can be activated by the natural plant product, camphor, and by nitrates that lead to nitric oxide production in the skin.[55,59–61]

The characterization of rodent strains that spontaneously develop atopic dermatitis-like lesions first implicated TRPV3 in pruritus. Sequencing revealed that these mice had gain of function mutations in TRPV3 (Gly573Ser) that were found to be sufficient to drive AD-like skin alterations.[62,63] Likewise, gain of function mutations in TRPV3 in humans have been linked to Olmsted syndrome, a condition that results in severe chronic itch.[64] Consistent with a role for TRPV3 in chronic itch, TRPV3-deficient mice do not develop chronic itch in a dry skin model of chronic itch and overexpression of TRPV3 in keratinocytes is sufficient to promote secretion of the pruritogen, prostaglandin E2.[57,65] However, how prostaglandin, or other keratinocyte-released mediators, promote pruritus remains unknown.

16.6   TRPM8 IN INHIBITION OF ITCH

Unlike other TRP channels that promote itch signaling, TRPM8, the cold and menthol receptor, is hypothesized to inhibit itch signal transmission. Phenomenologically, cooling is known to soothe and relieve itch sensations. In humans, cooling of the skin inhibits atopic dermatitis-evoked itch in patients.[66] Menthol, icilin, and cooling have also been shown to inhibit histamine and lichenification-evoked itch in humans.[67,68] Likewise, skin cooling attenuates spinal neuron responses to subcutaneous histamine injection.[69]

TRPM8 is activated by a variety of natural plant products that induce cooling sensations, including menthol, menthone, and eucalyptol. It is also activated by cold, with an activation temperature of ~25°C.[70,71] TRPM8 is expressed in a subset of primary afferent sensory neurons.[70–72] Importantly, sensory neurons isolated from

TRPM8-deficient mice have attenuated responses to menthol, icillin, and cold.[70–72] Behavioral studies also reveal severe deficits in cold-evoked responses in these knockout mice, as measured by acetone evaporative cooling, cold plate, and two-choice temperature assays.[73,74] However, TRPM8-deficient mice show normal cold-evoked behavior in response to noxious cold (<10°C), suggesting that other channels may also play a role in cold temperature detection.

The requirement of TRPM8 in menthol and cold sensitivity *in vivo* suggests that TRPM8, or TRPM8-positive neurons, may mediate the attenuation of itch by cold, and cold mimetics. However, future studies are required to test this directly. It will be of particular interest to determine whether these neurons are linked to inhibitory itch interneurons in the spinal cord.[75]

16.7 CONCLUSIONS AND FUTURE PROSPECTS

Research over the last decade shows a critical role of TRP channels in acute itch transduction. As such, TRP channel antagonists may be useful for the selective attenuation of itch. However, most studies have focused on acute rather than chronic itch. Little is known about the molecules that mediate chronic itch in primary sensory neurons and skin. Whether TRP channel signaling contributes to chronic itch is unknown and represents a major question in itch biology.

The TRP channels discussed here have dual roles in itch and in other somatosensory pathways and modalities like pain. A question that emerges from these studies is how any one channel can drive distinct itch or pain behaviors in response to differing stimuli. Multiple models have been proposed to account for this dual role of TRP channels. One is based on population coding, where a TRP agonist would evoke excitation of both itch-specific and pain- or temperature-specific fibers, and computation in the CNS would determine which signal is transmitted. Alternatively, the spatial contrast theory of itch posits that itch is triggered by the activation of a small number of pain fibers in a receptive field and pain is initiated when a larger cohort of cells is activated. Strong support of both itch theories has led to a modified "selectivity" theory of itch that incorporates aspects of both itch models.[76,77] The recent discovery of itch-specific spinal cord neurons suggests that central circuits may generate the specificity observed in itch signaling.[78,79] However, the relationship between itch and pain remains a pressing question in somatosensation.

REFERENCES

1. Clapham DE (2003) TRP channels as cellular sensors. *Nature* 426: 517–524.
2. Julius D (2005) From peppers to peppermints: Natural products as probes of the pain pathway. *Harvey Lect* 101: 89–115.
3. Minke B, Cook B (2002) TRP channel proteins and signal transduction. *Physiol Rev* 82: 429–472.
4. Nilius B, Mahieu F (2006) A road map for TR(I)Ps. *Mol Cell* 22: 297–307.
5. Nishida M, Hara Y, Yoshida T, Inoue R, Mori Y (2006) TRP channels: Molecular diversity and physiological function. *Microcirculation* 13: 535–550.
6. Caterina MJ, Schumacher MA, Tominaga M, Rosen TA, Levine JD et al. (1997) The capsaicin receptor: A heat-activated ion channel in the pain pathway. *Nature* 389: 816–824.

7. Tominaga M, Caterina MJ, Malmberg AB, Rosen TA, Gilbert H et al. (1998) The cloned capsaicin receptor integrates multiple pain-producing stimuli. *Neuron* 21: 531–543.

8. Tominaga M, Tominaga T (2005) Structure and function of TRPV1. *Pflugers Arch* 451: 143–150.

9. Huang SM, Bisogno T, Trevisani M, Al-Hayani A, De Petrocellis L et al. (2002) An endogenous capsaicin-like substance with high potency at recombinant and native vanilloid VR1 receptors. *Proc Natl Acad Sci USA* 99: 8400–8405.

10. Hwang SW, Cho H, Kwak J, Lee SY, Kang CJ et al. (2000) Direct activation of capsaicin receptors by products of lipoxygenases: Endogenous capsaicin-like substances. *Proc Natl Acad Sci USA* 97: 6155–6160.

11. Zygmunt PM, Petersson J, Andersson DA, Chuang H, Sorgard M et al. (1999) Vanilloid receptors on sensory nerves mediate the vasodilator action of anandamide. *Nature* 400: 452–457.

12. Caterina MJ, Leffler A, Malmberg AB, Martin WJ, Trafton J et al. (2000) Impaired nociception and pain sensation in mice lacking the capsaicin receptor. *Science* 288: 306–313.

13. Davis JB, Gray J, Gunthorpe MJ, Hatcher JP, Davey PT et al. (2000) Vanilloid receptor-1 is essential for inflammatory thermal hyperalgesia. *Nature* 405: 183–187.

14. Rohacs T, Thyagarajan B, Lukacs V (2008) Phospholipase C mediated modulation of TRPV1 channels. *Mol Neurobiol* 37: 153–163.

15. Ikoma A, Steinhoff M, Stander S, Yosipovitch G, Schmelz M (2006) The neurobiology of itch. *Nat Rev Neurosci* 7: 535–547.

16. Turnbull A (1850) Tincture of capsaicin as a remedy for chilblains and toothache. *Dublin Free Press* 1: 95–96.

17. Breneman DL, Cardone JS, Blumsack RF, Lather RM, Searle EA et al. (1992) Topical capsaicin for treatment of hemodialysis-related pruritus. *J Am Acad Dermatol* 26: 91–94.

18. Goodless DR, Eaglstein WH (1993) Brachioradial pruritus: Treatment with topical capsaicin. *J Am Acad Dermatol* 29: 783–784.

19. Leibsohn E (1992) Treatment of notalgia paresthetica with capsaicin. *Cutis* 49: 335–336.

20. Lotti T, Teofoli P, Tsampau D (1994) Treatment of aquagenic pruritus with topical capsaicin cream. *J Am Acad Dermatol* 30: 232–235.

21. Stander S, Luger T, Metze D (2001) Treatment of prurigo nodularis with topical capsaicin. *J Am Acad Dermatol* 44: 471–478.

22. Nakano T, Andoh T, Sasaki A, Nojima H, Kuraishi Y (2008) Different roles of capsaicin-sensitive and H1 histamine receptor-expressing sensory neurones in itch of mosquito allergy in mice. *Acta Derm Venereol* 88: 449–454.

23. Anand P, Bley K (2011) Topical capsaicin for pain management: Therapeutic potential and mechanisms of action of the new high-concentration capsaicin 8% patch. *Br J Anaesth* 107: 490–502.

24. Polydefkis M, Hauer P, Sheth S, Sirdofsky M, Griffin JW et al. (2004) The time course of epidermal nerve fibre regeneration: Studies in normal controls and in people with diabetes, with and without neuropathy. *Brain* 127: 1606–1615.

25. Koda H, Minagawa M, Si-Hong L, Mizumura K, Kumazawa T (1996) H1-receptor-mediated excitation and facilitation of the heat response by histamine in canine visceral polymodal receptors studied in vitro. *J Neurophysiol* 76: 1396–1404.

26. Schmelz M, Schmidt R, Weidner C, Hilliges M, Torebjork HE et al. (2003) Chemical response pattern of different classes of C-nociceptors to pruritogens and algogens. *J Neurophysiol* 89: 2441–2448.

27. Mizumura K, Koda H, Kumazawa T (2000) Possible contribution of protein kinase C in the effects of histamine on the visceral nociceptor activities in vitro. *Neurosci Res* 37: 183–190.

28. Kim BM, Lee SH, Shim WS, Oh U (2004) Histamine-induced Ca(2+) influx via the PLA(2)/lipoxygenase/TRPV1 pathway in rat sensory neurons. *Neurosci Lett* 361: 159–162.

29. Costa R, Marotta DM, Manjavachi MN, Fernandes ES, Lima-Garcia JF et al. (2008) Evidence for the role of neurogenic inflammation components in trypsin-elicited scratching behaviour in mice. *Br J Pharmacol* 154: 1094–1103.

30. Alenmyr L, Hogestatt ED, Zygmunt PM, Greiff L (2009) TRPV1-mediated itch in seasonal allergic rhinitis. *Allergy* 64: 807–810.

31. Lim KM, Park YH (2012) Development of PAC-14028, a novel transient receptor potential vanilloid type 1 (TRPV1) channel antagonist as a new drug for refractory skin diseases. *Arch Pharm Res* 35: 393–396.

32. Pereira U, Boulais N, Lebonvallet N, Pennec JP, Dorange G et al. (2010) Mechanisms of the sensory effects of tacrolimus on the skin. *Br J Dermatol* 163: 70–77.

33. Imamachi N, Park GH, Lee H, Anderson DJ, Simon MI et al. (2009) TRPV1-expressing primary afferents generate behavioral responses to pruritogens via multiple mechanisms. *Proc Natl Acad Sci USA* 106: 11330–11335.

34. Rossbach K, Nassenstein C, Gschwandtner M, Schnell D, Sander K et al. (2011) Histamine H1, H3 and H4 receptors are involved in pruritus. *Neuroscience* 190: 89–102.

35. Cavanaugh DJ, Lee H, Lo L, Shields SD, Zylka MJ et al. (2009) Distinct subsets of unmyelinated primary sensory fibers mediate behavioral responses to noxious thermal and mechanical stimuli. *Proc Natl Acad Sci USA* 106: 9075–9080.

36. Kim SJ, Park GH, Kim D, Lee J, Min H et al. (2011) Analysis of cellular and behavioral responses to imiquimod reveals a unique itch pathway in transient receptor potential vanilloid 1 (TRPV1)-expressing neurons. *Proc Natl Acad Sci USA* 108: 3371–3376.

37. Liu T, Xu ZZ, Park CK, Berta T, Ji RR (2010) Toll-like receptor 7 mediates pruritus. *Nat Neurosci* 13: 1460–1462.

38. Lagerstrom MC, Rogoz K, Abrahamsen B, Persson E, Reinius B et al. (2010) VGLUT2-dependent sensory neurons in the TRPV1 population regulate pain and itch. *Neuron* 68: 529–542.

39. Story GM, Peier AM, Reeve AJ, Eid SR, Mosbacher J et al. (2003) ANKTM1, a TRP-like channel expressed in nociceptive neurons, is activated by cold temperatures. *Cell* 112: 819–829.

40. Jordt SE, Bautista DM, Chuang HH, McKemy DD, Zygmunt PM et al. (2004) Mustard oils and cannabinoids excite sensory nerve fibres through the TRP channel ANKTM1. *Nature* 427: 260–265.

41. Bandell M, Story GM, Hwang SW, Viswanath V, Eid SR et al. (2004) Noxious cold ion channel TRPA1 is activated by pungent compounds and bradykinin. *Neuron* 41: 849–857.

42. Bautista DM, Movahed P, Hinman A, Axelsson HE, Sterner O et al. (2005) Pungent products from garlic activate the sensory ion channel TRPA1. *Proc Natl Acad Sci USA* 102: 12248–12252.

43. Andersson DA, Gentry C, Moss S, Bevan S (2008) Transient receptor potential A1 is a sensory receptor for multiple products of oxidative stress. *J Neurosci* 28: 2485–2494.

44. Taylor-Clark TE, Undem BJ, Macglashan DW, Jr., Ghatta S, Carr MJ et al. (2008) Prostaglandin-induced activation of nociceptive neurons via direct interaction with transient receptor potential A1 (TRPA1). *Mol Pharmacol* 73: 274–281.

45. Wang S, Dai Y, Fukuoka T, Yamanaka H, Kobayashi K et al. (2008) Phospholipase C and protein kinase A mediate bradykinin sensitization of TRPA1: A molecular mechanism of inflammatory pain. *Brain* 131: 1241–1251.

46. Macpherson LJ, Xiao B, Kwan KY, Petrus MJ, Dubin AE et al. (2007) An ion channel essential for sensing chemical damage. *J Neurosci* 27: 11412–11415.

47. Trevisani M, Siemens J, Materazzi S, Bautista DM, Nassini R et al. (2007) 4-Hydroxynonenal, an endogenous aldehyde, causes pain and neurogenic inflammation through activation of the irritant receptor TRPA1. *Proc Natl Acad Sci USA* 104: 13519–13524.

48. Bautista DM, Jordt SE, Nikai T, Tsuruda PR, Read AJ et al. (2006) TRPA1 mediates the inflammatory actions of environmental irritants and proalgesic agents. *Cell* 124: 1269–1282.

49. Kwan KY, Allchorne AJ, Vollrath MA, Christensen AP, Zhang DS et al. (2006) TRPA1 contributes to cold, mechanical, and chemical nociception but is not essential for hair-cell transduction. *Neuron* 50: 277–289.

50. Liu Q, Tang Z, Surdenikova L, Kim S, Patel KN et al. (2009) Sensory neuron-specific GPCR Mrgprs are itch receptors mediating chloroquine-induced pruritus. *Cell* 139: 1353–1365.

51. Wilson SR, Gerhold KA, Bifolck-Fisher A, Liu Q, Patel KN et al. (2011) TRPA1 is required for histamine-independent, Mas-related G protein-coupled receptor-mediated itch. *Nat Neurosci* 14: 595–602.

52. Liu T, Ji RR (2012) Oxidative stress induces itch via activation of transient receptor potential subtype ankyrin 1 in mice. *Neurosci Bull* 28: 145–154.

53. Raap U, Stander S, Metz M (2011) Pathophysiology of itch and new treatments. *Curr Opin Allergy Clin Immunol* 11: 420–427.

54. Peier AM, Reeve AJ, Andersson DA, Moqrich A, Earley TJ et al. (2002) A heat-sensitive TRP channel expressed in keratinocytes. *Science* 296: 2046–2049.

55. Moqrich A, Hwang SW, Earley TJ, Petrus MJ, Murray AN et al. (2005) Impaired thermosensation in mice lacking TRPV3, a heat and camphor sensor in the skin. *Science* 307: 1468–1472.

56. Chung MK, Lee H, Mizuno A, Suzuki M, Caterina MJ (2004) TRPV3 and TRPV4 mediate warmth-evoked currents in primary mouse keratinocytes. *J Biol Chem* 279: 21569–21575.

57. Huang SM, Lee H, Chung MK, Park U, Yu YY et al. (2008) Overexpressed transient receptor potential vanilloid 3 ion channels in skin keratinocytes modulate pain sensitivity via prostaglandin E2. *J Neurosci* 28: 13727–13737.

58. Mandadi S, Sokabe T, Shibasaki K, Katanosaka K, Mizuno A et al. (2009) TRPV3 in keratinocytes transmits temperature information to sensory neurons via ATP. *Pflugers Arch* 458: 1093–1102.

59. Chung MK, Lee H, Mizuno A, Suzuki M, Caterina MJ (2004) 2-aminoethoxydiphenyl borate activates and sensitizes the heat-gated ion channel TRPV3. *J Neurosci* 24: 5177–5182.

60. Hu H, Grandl J, Bandell M, Petrus M, Patapoutian A (2009) Two amino acid residues determine 2-APB sensitivity of the ion channels TRPV3 and TRPV4. *Proc Natl Acad Sci USA* 106: 1626–1631.

61. Miyamoto T, Petrus MJ, Dubin AE, Patapoutian A (2011) TRPV3 regulates nitric oxide synthase-independent nitric oxide synthesis in the skin. *Nat Commun* 2: 369.

62. Asakawa M, Yoshioka T, Matsutani T, Hikita I, Suzuki M et al. (2006) Association of a mutation in TRPV3 with defective hair growth in rodents. *J Invest Dermatol* 126: 2664–2672.

63. Yoshioka T, Imura K, Asakawa M, Suzuki M, Oshima I et al. (2009) Impact of the Gly573Ser substitution in TRPV3 on the development of allergic and pruritic dermatitis in mice. *J Invest Dermatol* 129: 714–722.

64. Lin Z, Chen Q, Lee M, Cao X, Zhang J et al. (2012) Exome sequencing reveals mutations in TRPV3 as a cause of Olmsted syndrome. *Am J Hum Genet* 90: 558–564.

65. Yamamoto-Kasai E, Imura K, Yasui K, Shichijou M, Oshima I et al. (2012) TRPV3 as a therapeutic target for itch. *J Invest Dermatol* 132: 2109–2112.

66. Fruhstorfer H, Hermanns M, Latzke L (1986) The effects of thermal stimulation on clinical and experimental itch. *Pain* 24: 259–269.

67. Bromm B, Scharein E, Darsow U, Ring J (1995) Effects of menthol and cold on histamine-induced itch and skin reactions in man. *Neurosci Lett* 187: 157–160.

68. Han JH, Choi HK, Kim SJ (2012) Topical TRPM8 Agonist (icilin) relieved vulva pruritus originating from lichen sclerosus et atrophicus. *Acta Derm Venereol* 92: 561–562.

69. Carstens E, Jinks SL (1998) Skin cooling attenuates rat dorsal horn neuronal responses to intracutaneous histamine. *Neuroreport* 9: 4145–4149.

70. McKemy DD, Neuhausser WM, Julius D (2002) Identification of a cold receptor reveals a general role for TRP channels in thermosensation. *Nature* 416: 52–58.

71. Peier AM, Moqrich A, Hergarden AC, Reeve AJ, Andersson DA et al. (2002) A TRP channel that senses cold stimuli and menthol. *Cell* 108: 705–715.

72. Colburn RW, Lubin ML, Stone DJ, Jr., Wang Y, Lawrence D et al. (2007) Attenuated cold sensitivity in TRPM8 null mice. *Neuron* 54: 379–386.

73. Bautista DM, Siemens J, Glazer JM, Tsuruda PR, Basbaum AI et al. (2007) The menthol receptor TRPM8 is the principal detector of environmental cold. *Nature* 448: 204–208.

74. Dhaka A, Murray AN, Mathur J, Earley TJ, Petrus MJ et al. (2007) TRPM8 is required for cold sensation in mice. *Neuron* 54: 371–378.

75. Ross SE, Mardinly AR, McCord AE, Zurawski J, Cohen S et al. (2010) Loss of inhibitory interneurons in the dorsal spinal cord and elevated itch in Bhlhb5 mutant mice. *Neuron* 65: 886–898.

76. Ma Q (2010) Labeled lines meet and talk: Population coding of somatic sensations. *J Clin Invest* 120: 3773–3778.

77. Ross SE (2011) Pain and itch: Insights into the neural circuits of aversive somatosensation in health and disease. *Curr Opin Neurobiol* 21: 880–887.

78. Sun YG, Chen ZF (2007) A gastrin-releasing peptide receptor mediates the itch sensation in the spinal cord. *Nature* 448: 700–703.

79. Sun YG, Zhao ZQ, Meng XL, Yin J, Liu XY et al. (2009) Cellular basis of itch sensation. *Science* 325: 1531–1534.

17 Sensitization of Itch Signaling

Itch Sensitization—Nerve Growth Factor, Semaphorins

Mitsutoshi Tominaga and Kenji Takamori

CONTENTS

17.1 INTRODUCTION

Itch (or pruritus) has been defined as an unpleasant sensation that provokes the desire to scratch. Itch is also believed to signal danger from various environmental factors or physiological abnormalities. Therefore, it frequently accompanies a variety of inflammatory skin conditions and systemic diseases.

Histamine is the best-known pruritogen in humans and also acts as an experimental itch-causing substance. Clinically, antihistamines, i.e., histamine H_1-receptor blockers, are commonly used to treat all types of itching resulting from renal and liver diseases, as well as from serious skin diseases such as atopic dermatitis. However, antihistamines often lack efficacy in patients with chronic itch that may involve other agonists, including proteases, neuropeptides, cytokines, and opioids, and their cognate receptors, such as thermoreceptors, PARs, Mrgprs, and opioid receptors. Such pruritogenic mediators and modulators released in the periphery may directly activate itch-sensitive fibers, especially C-fibers, by binding to specific receptors on the nerve terminals (Ikoma et al. 2006; Paus et al. 2006; Xiao and Patapoutian 2011). Nerve fibers are also activated by exogenous mechanical, chemical, and biological stimuli, resulting in itch responses (Akiyama et al. 2010; Tominaga and Takamori 2010).

Histological analyses have shown that epidermal nerve densities are increased in patients with atopic dermatitis and xerosis, suggesting that the higher density is partly responsible for itch sensitization in the periphery. Such hyperinnervation is probably caused by an imbalance of nerve elongation factors, such as nerve growth factor (NGF), and nerve repulsion factors, such as semaphorin 3A (Sema3A), produced by keratinocytes (Tominaga and Takamori 2010). These axonal guidance molecules may also act on keratinocytes, immune cells and vascular endothelial cells, and be indirectly involved in the modulation of itching. This chapter presents recent knowledge regarding itch sensitization associated with epidermal nerve density controlled by NGF and Sema3A, especially in atopic dermatitis.

17.2 SKIN DISEASES WITH ITCH INVOLVING EPIDERMAL NERVE FIBERS

Sensory nerve fibers are acceptors of itch sensation as well as pain in the skin. The neuronal mechanisms underlying intractable itch in the periphery have been partly identified. Histological investigations have shown that the density of epidermal nerve fibers is higher in the skin of patients with atopic dermatitis, contact dermatitis, and xerosis than in healthy controls (Figure 17.1) (Ikoma et al. 2006; Tominaga and Takamori 2010), although the nerve density in patients with pruigo nodularis and psoriasis remains debatable (Schuhknecht et al. 2011; Taneda et al. 2011; Kou et al. 2012). Similar findings have been observed in animal models, such as NC/Nga mice, an atopic dermatitis model (Tominaga et al. 2007a, 2009a), and dry skin model mice (Miyamoto et al. 2002; Tominaga et al. 2007b). These findings are indicative of increases in sensory nerve fibers responsive to exogenous trigger factors and to various endogenous pruritogens from immune cells and keratinocytes, suggesting that hyperinnervation is partly responsible for itch sensitization.

Several recent studies have found that epidermal innervation density is reduced in various pruritic skin conditions, such as lichen amyloidosis (Maddison et al. 2008), prurigo nodularis (Schuhknecht et al. 2011), nummular dermatitis (Maddison et al. 2011) and keloids (Tey et al. 2012). Chronic stimulation of itch-transmitting nerve fibers may result in a self-regulated hypoplasia that modulates the intensity and persistence of sensory input. This hypothesis is supported by the results of one study, which showed diminished skin innervation in the skin of patients with neuropathic itch (Wallengren et al. 2002).

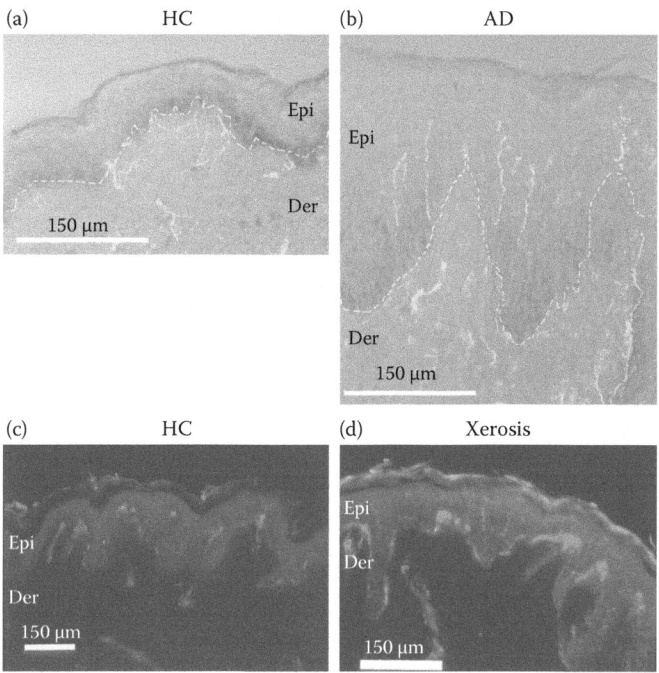

FIGURE 17.1 **(See color insert.)** Distribution of epidermal nerve fibers in healthy and pruritic skin samples. Immunohistochemical staining of the skin of healthy controls (HC), patients with atopic dermatitis (AD), and xerotic patients with antibody to protein gene product 9.5 (anti-PGP9.5). (a, b) Images of nerve fibers overlapped with differential interference microscopic images. PGP9.5-immunoreactive fibers (green) were occasionally present in the epidermis of HC (a, c). Epidermal nerve fibers were observed at higher densities in (b) AD and (d) xerotic patients. (c, d) Basement membrane (red) was stained with antibody to type IV collagen. Scale bars are 150 μm; Epi, epidermis; Der, dermis.

17.3 ROLES OF NGF AND Sema3A IN EPIDERMAL NERVE FIBERS

17.3.1 NGF

NGF is a neurotrophin that affects neurite outgrowth and neuronal survival (Lewin and Mendell 1993). Keratinocyte-derived NGF is a major mediator of cutaneous innervation density, in that local NGF concentrations are higher in the lesional skin of patients with prurigo nodularis, atopic dermatitis, psoriasis, contact dermatitis, and xerosis than in normal skin (Figure 17.2a) (Ikoma et al. 2006). In adult rat primary sensory neurons, NGF has been shown to upregulate neuropeptides, especially substance P (SP) and calcitonin gene-related peptide (CGRP) (Verge et al. 1995), both of which are involved in the hypersensitivity of itch sensation and neurogenic inflammation (Steinhoff et al. 2003). A more recent study demonstrated that human atopic keratinocytes produced elevated levels of NGF and mediated an increased

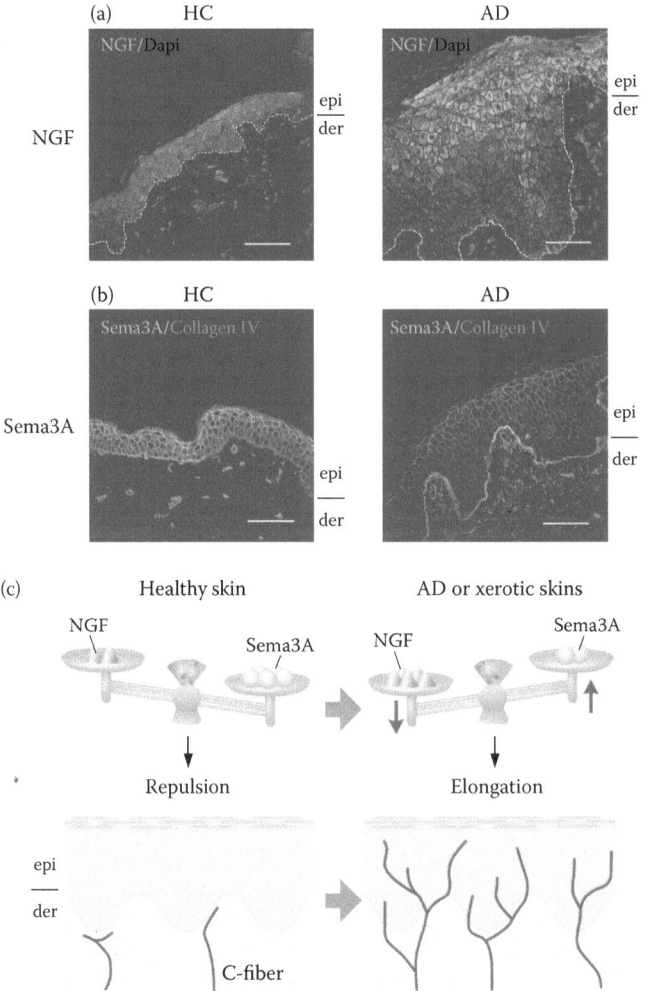

FIGURE 17.2 **(See color insert.)** Epidermal NGF and Sema3A levels in pruritic skin samples. (a) Skin biopsies from HC and patients with AD were stained with anti-NGF antibody. Epidermal NGF levels (green) were higher in AD patients than in HC. Nuclei were counterstained with DAPI (blue). The white dotted line in each panel indicates the border between the epidermis and dermis (basement membrane). (b) Double labeling of Sema3A (green) and type IV collagen (red) in the skin of HCs and AD patients. Epidermal Sema3A levels were lower in AD patients than in HCs. Scale bars are 75 μm; epi, epidermis; der, dermis. (c) A possible regulatory model of sensory nerve fiber penetration into the epidermis by a balance of nerve elongation factors such as NGF and nerve repulsion factors such as Sema3A. Epidermal NGF levels were lower and epidermal Sema3A levels were higher in healthy skin than in atopic and dry skin, suggesting the suppression of penetration and/or elongation of nerve fibers into the normal epidermis. In contrast, epidermal NGF levels were higher and epidermal Sema3A levels were lower in atopic and dry skin than in healthy skin. This mechanism may be involved in the induction or acceleration of penetration and/or elongation of nerve fibers into normal epidermis.

outgrowth of CGRP-immunoreactive sensory fibers, whereas human atopic fibroblasts did not mediate this outgrowth (Roggenkamp et al. 2012). This result indicates that keratinocytes are key factors in hyperinnervation in individuals with atopic dermatitis. Intradermal injection of NGF was shown to sensitize nociceptors for cowhage- but not histamine-induced itch in human skin (Rukwied et al. 2012). Thus, increased NGF in the skin may also sensitize primary afferents, thereby contributing to chronic itch such as atopic dermatitis.

Tumor necrosis factor (TNF)-α has also been found to enhance NGF production by human keratinocytes (Takaoka et al. 2009). TNF-α is a pivotal proinflammatory cytokine in the innate immune response and a key molecule for skin inflammation. Mast cells have been identified as important potential sources of TNF-α (Steinhoff et al. 2003). Plasma TNF-α concentration is higher in individuals with atopic dermatitis than without atopic dermatitis (Sumimoto et al. 1992), and both TNF-α and its receptors are upregulated in the dermal blood vessels of patients with psoriasis (Kristensen et al. 1993). In addition, skin barrier disruption induces the upregulation of TNF-α in the epidermis of acetone-treated mice, an acute dry skin model (Wood et al. 1992). These findings suggest that TNF-α is an important upregulator of NGF in the epidermis, a finding supported by results showing that TNF produced by mast cells promotes the elongation of cutaneous nerve fibers in a mouse model of contact dermatitis (Kakurai et al. 2006).

17.3.2 SEMAPHORIN 3A (SEMA3A)

Class 3 semaphorins, a family of secreted proteins, have been implicated in a variety of biological functions and were originally identified as axonal guidance cues during neural development. Sema3A is the first member of this protein family shown to cause growth cone collapse in neurons, i.e., to function as a nerve repulsion factor, through its interaction with a neuropilin-1 (Nrp-1)/plexin-A receptor complex (Fujisawa 2004; Sharma et al. 2012). We previously reported that Sema3A transcripts are also expressed in cultured normal human epidermal keratinocytes (Tominaga et al. 2008). The proteins are mainly distributed in the suprabasal layer of normal human skin, consistent with findings showing that Sema3A is expressed in differentiated keratinocyte cultures (Fukamachi et al. 2011). Moreover, epidermal Sema3A levels were lower in patients with atopic dermatitis than in healthy controls (Figure 17.2b), concomitant with an increase in epidermal nerve density (Tominaga et al. 2008). Increased epidermal nerve density in an acute dry skin model was recently shown to be associated with decreased levels of Sema3A expression (Kamo et al. 2011a, 2011b). Although the mechanism regulating the Sema3A gene has not yet been determined, these indicate good correlations between epidermal innervation and Sema3A levels. Interestingly, Sema3A has been found to inhibit NGF-induced sprouting of sensory afferents in adult rat spinal cord (Tang et al. 2004), whereas elevated levels of NGF reduced the Sema3A-induced collapse of sensory growth cones (Dontchev and Letourneau 2002). These findings suggest that decreasing the expression of Sema3A can accelerate epidermal nerve growth in patients with atopic dermatitis and xerosis. Thus, epidermal innervation may be regulated by a fine balance between NGF and Sema3A (Figure 17.2c).

17.3.3 Mechanisms by Which Nerve Fibers Penetrate Basement Membranes

NGF stimulation of nerve fiber elongation in pruritic skin may result in the accumulation at the growth cone of integrins, which interact with a variety of extracellular matrix (ECM) components (Grabham and Goldberg 1997; Gardiner 2011). During this process, matrix metalloproteinases (MMPs) would be required for the growth cone to abrogate the three-dimensional ECM barriers. Using an *in vitro* model of basement membrane, consisting of Boyden chambers and Matrigel, and constituting a unique culture system of rat dorsal root ganglion (DRG) neurons, we recently showed that MMP-2 localized on the growth cone was involved in the penetration mechanism (Figure 17.3) (Tominaga et al. 2009b). In Boyden chamber, cultures using type I collagen gels and cultured DRG neurons, MMP-8 secreted by the nerve fibers was shown to be involved in nerve growth within the gels (Figure 17.3) (Tominaga et al. 2011). The expression of MMP-2 and MMP-8 was upregulated by NGF and downregulated by Sema3A. Interestingly, MMP2 expression was induced by its enzymatic substrates, including type IV, collagen, laminin, and fibronectin, as well as by MG. Moreover, MMP-8 expression was upregulated by its substrates types I and III collagens. The expression of both MMP genes was not altered by

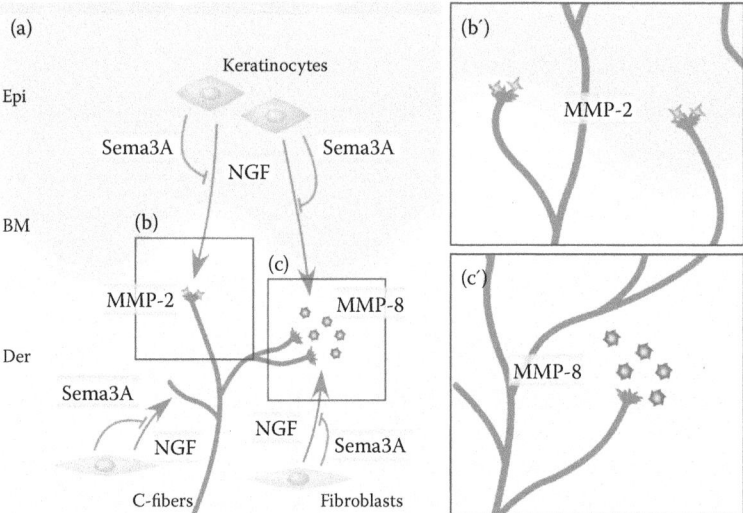

FIGURE 17.3 **(See color insert.)** Possible models of nerve fiber growth into basement membrane or within the dermis. (a) NGF, which is produced by cutaneous cells such as epidermal keratinocytes, immune cells, and fibroblasts, promotes the production of MMP-2 and MMP-8 in sensory nerve fibers. NGF-induced expression of these proteins was likely modulated by the extracellular matrix (ECM) substrates of each enzyme. (b) Activated MMP-2 is localized on the growth cone, while (c) MMP-8 is secreted by nerve fibers. Sema3A produced by keratinocytes and fibroblasts may have the opposite effects on these NGF-dependent events. Activated MMP-2 on the growth cone may contribute to the penetration of nerve fibers into basement membrane (b'). Secreted MMP-8 may contribute to the outgrowth of nerve fibers within the dermal ECM (c'). Epi, epidermis; Der, dermis; BM, basement membrane.

nonsubstrate molecules. The selection and upregulation of MMPs corresponding to the ECM components surrounding the growing nerve fibers are required for efficient nerve fiber penetration, suggesting that the coordinated activation of neurotrophin and ECM-integrin signaling is necessary for efficient and long-distance axon extension (Grabham and Goldberg 1997; Gardiner et al. 2011). Because class 3 semaphorin signaling inhibits integrin-mediated adhesion signaling, Sema3A stimulation of the growing nerve fibers may provide a reverse signaling pathway for these events (Zhou et al. 2008). Thus, although the integrin-mediated regulatory system remains unclear in our *in vitro* studies, this mechanism might be applicable to pruritic skin diseases involving skin hyperinnervation.

17.4   NGF AND Sema3A AS ANTIPRURITIC TARGETS

17.4.1   ANTI-NGF ANTIBODY AND ITS RECEPTOR INHIBITORS

The effects of NGF are mediated by its binding to two classes of transmembrane receptors, a high-affinity receptor (tropomyosin-related kinase A, TrkA) and a low-affinity receptor (p75) (Lewin and Mendell 1993). To date, two anti-NGF approaches have been used to treat pruritus of atopic dermatitis in NC/Nga mice. One study showed that the intraperitoneal administration of anti-NGF neutralizing antibody to atopic NC/Nga mice significantly suppresed both epidermal nerve growth and scratching behavior, but did not ameliorate scratching that had already developed (Takano et al. 2005). Similarly, application of the TrkA inhibitors AG879 and K252a to the backs of the necks of atopic NC/Nga mice five times per week significantly improved established dermatitis and scratching behavior, and decreased the numbers of nerve fibers in the epidermis, suggesting that NGF plays important roles in the pathogenesis of atopic dermatitis-like skin lesions (Takano et al. 2007). Thus, NGF and its receptors may be therapeutic targets in patients with pruritic skin diseases.

17.4.2   SEMA3A REPLACEMENT THERAPY

Decreasing the expression of Sema3A accelerates epidermal nerve growth in patients with some pruritic skin diseases, such as atopic dermatitis and dry skin (Tominaga et al. 2008; Tominaga and Takamori 2010). Therefore, replacement by exogenous Sema3A may have antipruritic effects. Recently, recombinant Sema3A replacement approaches (intradermal injection or ointment application) were found to significantly inhibit scratching behavior and to improve dermatitis score in atopic NC/Nga mice compared with controls (Yamaguchi et al. 2008; Negi et al. 2012). The therapeutic efficacy of exogenous Sema3A on the atopic dermatitis-like symptoms was greater than that of current agents, such as betamethasone and tacrolimus (Negi et al. 2012). Moreover, histological examination showed decreases in (1) the numbers of epidermal PGP9.5- or SP-immunoreactive fibers; (2) the numbers of inflammatory infiltrates, including mast cells, eosinophils and CD4[+] T cells; (3) the production of IL-4; (4) the density of dermal blood vessels; and (5) epidermal thickness in Sema3A-treated lesional skin (Yamaguchi et al. 2008; Negi et al. 2012). These observations suggest that exogenous Sema3A not only affects sensory nerve fibers but also immune cells or other cells, such as endothelial cells and keratinocytes, that express Nrp-1, a coreceptor for Sema3A (Romeo et al. 2002). Therefore, replacement of Sema3A may be useful in therapeutic strategies for patients with pruritic skin conditions such as atopic dermatitis.

17.4.3 OTHERS

Several existing therapies may normalize abnormal levels of NGF and Sema3A in pruritic skin associated with a reduction in epidermal nerve density. Oral administration of olopatadine hydrochloride, a histamine H_1-receptor antagonist, significantly suppressed scratching behavior, improved dermatitis score, and inhibited neurite outgrowth in the lesional skin of mice with atopic dermatitis. Notably, olopatadine treatment increased Sema3A expression in the epidermis (Murota et al. 2010; Ohsawa and Hirasawa 2012). Although it is unclear whether these effects are caused by specific blocking of histamine H_1-receptor signaling, olopatadine may in part improve imbalances of NGF and Sema3A in the epidermis.

Our recent study using dry skin mice showed that application of heparinoid cream (emollient) resulted in greater improvements in epidermal nerve density and epidermal NGF levels than application of petrolatum, although heparinoid cream had no effect on epidermal Sema3A levels (Kamo et al. 2011a). In addition, various types of ultraviolet (UV)-based therapy, including psoralen-ultraviolet A (PUVA) and narrow-band UVB, have been widely used to treat patients with atopic dermatitis (Krutmann 2000). These UV-based therapies were shown to reduce the number of cutaneous nerve fibers, especially in the epidermis, in patients with atopic dermatitis and psoriasis, and to inhibit pruritus (Wallengren and Sundler 2004; Tominaga et al. 2009c). Similar effects of UV-based therapy on epidermal nerve fibers were observed in dry skin mice (Kamo et al. 2011b). The abnormal expression of Sema3A and NGF in the epidermis was normalized by PUVA or narrowband UVB therapy (Tominaga et al. 2009c; Kamo et al. 2011b).

17.5 CONCLUSION

Nerve density in the epidermis may be involved in itch sensitization in pruritic skin diseases. Epidermal innervation is thought to be regulated by a fine balance between nerve elongation factors and nerve repulsion factors through the regulation of expression of MMPs. Treatment with anti-NGF agents, Sema3A replacement and other treatments such as UV-based therapies may normalize epidermal nerve fiber density. These findings may expand the knowledge of potential therapeutic strategies for ameliorating pruritus associated with epidermal nerve density, including patients with atopic dermatitis and xerosis.

ACKNOWLEDGMENTS

This work was supported by a Health Labor Sciences Research Grant for Research on Allergic Disease and Immunology from the Japanese Ministry of Health, Labor and Welfare, by a KAKENHI (20591354 and 2079081) and a "High-Tech Research

Center" Project for Private Universities. Matching fund subsidy from MEXT and by a JSPS Research Fellow.

REFERENCES

Akiyama, T., Carstens, M.I., and Carstens, E. 2010. Enhanced scratching evoked by PAR-2 agonist and 5-HT but not histamine in a mouse model of chronic dry skin itch. *Pain* 151:378–83.

Dontchev, V.D., and Letourneau, P.C. 2002. Nerve growth factor and semaphorin 3A signaling pathways interact in regulating sensory neuronal growth cone motility. *J Neurosci* 22:6659–69.

Fujisawa, H. 2004. Discovery of semaphorin receptors, neuropilin and plexin, and their functions in neural development. *J Neurobiol* 59:24–33.

Fukamachi, S., Bito, T., Shiraishi, N. et al. 2011. Modulation of semaphorin 3A expression by calcium concentration and histamine in human keratinocytes and fibroblasts. *J Dermatol Sci* 61:118–23.

Gardiner, N.J. 2011. Integrins and the extracellular matrix: Key mediators of development and regeneration of the sensory nervous system. *Dev Neurobiol* 71:1054–72.

Grabham, P.W., and Goldberg, D.J. 1997. Nerve growth factor stimulates the accumulation of beta1 integrin at the tips of filopodia in the growth cones of sympathetic neurons. *J Neurosci* 17:5455–65.

Ikoma, A., Steinhoff, M., Ständer, S., Yosipovitch, G., and Schmelz, M. 2006. The neurobiology of itch. *Nat Rev Neurosci* 7:535–47.

Kakurai, M., Monteforte, R., Suto, H., Tsai, M., Nakae, S., and Galli, S.J. 2006. Mast cell-derived tumor necrosis factor can promote nerve fiber elongation in the skin during contact hypersensitivity in mice. *Am J Pathol* 169:1713–21.

Kamo, A., Tominaga, M., Negi, O., Tengara, S., Ogawa, H., and Takamori, K. 2011a. Topical application of emollients prevents dry skin-inducible intraepidermal nerve growth in acetone-treated mice. *J Dermatol Sci* 62:64–6.

Kamo, A., Tominaga, M., Tengara, S., Ogawa, H., and Takamori, K. 2011b. Inhibitory effects of UV-based therapy on dry skin-inducible nerve growth in acetone-treated mice. *J Dermatol Sci* 62:91–7.

Kou, K., Nakamura, F., Aihara, M. et al. 2012. Decreased expression of semaphorin-3A, a neurite-collapsing factor, is associated with itch in psoriatic skin. *Acta Derm Venereol* 92:521–8, doi:10.2340/00015555-1350.

Kristensen, M., Chu, C.Q., Eedy, J., Feldmann, M., Brennan, F.M., and Breathnach, S.M. 1993. Localization of tumour necrosis factor-alpha (TNF-alpha) and its receptors in normal and psoriatic skin: Epidermal cells express the 55-kD but not the 75-kD TNF receptor. *Clin Exp Immunol* 94:354–62.

Krutmann, J. 2000. Phototherapy for atopic dermatitis. *Clin Exp Dermatol* 25:552–8.

Lewin, G.R., and Mendell, L.M. 1993. Nerve growth factor and nociception. *Trends Neurosci* 16:353–9.

Maddison, B., Namazi, M.R., Samuel, L.S. et al. 2008. Unexpected diminished innervation of epidermis and dermoepidermal junction in lichen amyloidosus. *Br J Dermatol* 159:403–6.

Maddison, B., Parsons, A., Sangueza, O., Sheehan, D.J., and Yosipovitch, G. 2011. Retrospective study of intraepidermal nerve fiber distribution in biopsies of patients with nummular eczema. *Am J Dermatopathol* 33:621–3.

Miyamoto, T., Nojima, H., Shinkado, T., Nakahashi, T., and Kuraishi, Y. 2002. Itch-associated response induced by experimental dry skin in mice. *Jpn J Pharmacol* 88:285–92.

Murota, H., El-latif, M.A., Tamura, T., Amano, T., and Katayama, I. 2010. Olopatadine hydro-chloride improves dermatitis score and inhibits scratch behavior in NC/Nga mice. *Int Arch Allergy Immunol* 153:121–32.

Negi, O., Tominaga, M., Tengara, S. et al. 2012. Topically applied semaphorin 3A ointment inhibits scratching behavior and improves skin inflammation in NC/Nga mice with atopic dermatitis. *J Dermatol Sci* 66:37–43.

Ohsawa, Y., and Hirasawa, N. 2012. The antagonism of histamine H1 and H4 receptors ame-liorates chronic allergic dermatitis via anti-pruritic and anti-inflammatory effects in NC/Nga mice. *Allergy* 67:1014–22.

Paus, R., Schmelz, M., Bíró, T., and Steinhoff, M. 2006. Frontiers in pruritus research: Scratching the brain for more effective itch therapy. *J Clin Invest* 116:1174–86.

Roggenkamp, D., Falkner, S., Stäb, F., Petersen, M., Schmelz, M., and Neufang, G. 2012. Atopic keratinocytes induce increased neurite outgrowth in a coculture model of porcine dorsal root ganglia neurons and human skin cells. *J Invest Dermatol* 132:1892–900.

Romeo, P.H., Lemarchandel, V., and Tordjman, R. 2002. Neuropilin-1 in the immune system. *Adv Exp Med Biol* 515:49–54.

Rukwied, R.R., Main, M., Weinkauf, B., and Schmelz, M. 2012. NGF sensitizes nociceptors for cowhage- but not histamine-induced itch in human skin. *J Invest Dermatol* 133:268–70, doi:10.1038/jid.2012.242.

Schuhknecht, B., Marziniak, M., Wissel, A. et al. 2011. Reduced intraepidermal nerve fibre density in lesional and nonlesional prurigo nodularis skin as a potential sign of subclini-cal cutaneous neuropathy. *Br J Dermatol* 165:85–91.

Sharma, A., Verhaagen, J., and Harvey, A.R. 2012. Receptor complexes for each of the Class 3 Semaphorins. *Front Cell Neurosci* 6:28.

Steinhoff, M., Ständer, S., Seeliger, S., Ansel, J.C., Schmelz, M., and Luger, T. 2003. Modern aspects of cutaneous neurogenic inflammation. *Arch Dermatol* 139:1479–88.

Sumimoto, S., Kawai, M., Kasajima, Y., and Hamamoto, T. 1992. Increased plasma tumour necrosis factor-alpha concentration in atopic dermatitis. *Arch Dis Child* 67:277–9.

Takano, N., Sakurai, T., and Kurachi, M. 2005. Effects of anti-nerve growth factor antibody on symptoms in the NC/Nga mouse, an atopic dermatitis model. *J Pharmacol Sci* 99:277–86.

Takano, N., Sakurai, T., Ohashi, Y., and Kurachi, M. 2007. Effects of high-affinity nerve growth factor receptor inhibitors on symptoms in the NC/Nga mouse atopic dermatitis model. *Br J Dermatol* 156:241–6.

Takaoka, K., Shirai, Y., and Saito, N. 2009. Inflammatory cytokine tumor necrosis factor-alpha enhances nerve growth factor production in human keratinocytes, HaCaT cells. *J Pharmacol Sci* 111:381–91.

Taneda, K., Tominaga, M., Negi, O. et al. 2011. Evaluation of epidermal nerve density and opioid receptor levels in psoriatic itch. *Br J Dermatol* 165:277–84.

Tang, X.Q., Tanelian, D.L., and Smith, G.M. 2004. Semaphorin3A inhibits nerve growth factor-induced sprouting of nociceptive afferents in adult rat spinal cord. *J Neurosci* 24:819–27.

Tey, H.L., Maddison, B., Wang, H. et al. 2012. Cutaneous innervation and itch in keloids. *Acta Derm Venereol* 92:529–31, doi:10.2340/00015555-1336.

Tominaga, M., Kamo, A., Tengara, S., Ogawa, H., and Takamori, K. 2009b. In vitro model for penetration of sensory nerve fibres on a Matrigel basement membrane: Implications for possible application to intractable pruritus. *Br J Dermatol* 161:1028–37.

Tominaga, M., Ogawa, H., and Takamori, K. 2008. Decreased production of semaphorin 3A in the lesional skin of atopic dermatitis. *Br J Dermatol* 158:842–4.

Tominaga, M., Ogawa, H., and Takamori, K. 2009a. Histological characterization of cutane-ous nerve fibers containing gastrin-releasing peptide in NC/Nga mice: An atopic derma-titis model. *J Invest Dermatol* 129:2901–5.

Tominaga, M., Ozawa, S., Ogawa, H., and Takamori, K. 2007a. A hypothetical mechanism of intraepidermal neurite formation in NC/Nga mice with atopic dermatitis. *J Dermatol Sci* 46:199–210.

Tominaga, M., Ozawa, S., Tengara, S., Ogawa, H., and Takamori, K. 2007b. Intraepidermal nerve fibers increase in dry skin of acetone-treated mice. *J Dermatol Sci* 48:103–11.

Tominaga, M., and Takamori, K. 2010. Recent advances in pathophysiological mechanisms of itch. *Expert Rev Dermatol* 5:197–212.

Tominaga, M., Tengara, S., Kamo, A., Ogawa, H., and Takamori, K. 2009c. Psoralen-ultraviolet A therapy alters epidermal Sema3A and NGF levels and modulates epidermal innervation in atopic dermatitis. *J Dermatol Sci* 55:40–6.

Tominaga, M., Tengara, S., Kamo, A., Ogawa, H., and Takamori, K. 2011. Matrix metalloproteinase-8 is involved in dermal nerve growth: Implications for possible application to pruritus from in vitro models. *J Invest Dermatol* 131:2105–12.

Verge, V.M., Richardson, P.M., Wiesenfeld-Hallin, Z., and Hokfelt, T. 1995. Differential influence of nerve growth factor on neuropeptide expression in vivo: A novel role in peptide suppression in adult sensory neurons. *J Neurosci* 15:2081–96.

Wallengren, J., and Sundler, F. 2004. Phototherapy reduces the number of epidermal and CGRP-positive dermal nerve fibres. *Acta Derm Venereol* 84:111–5.

Wallengren, J., Tegner, E., and Sundler, F. 2002. Cutaneous sensory nerve fibers are decreased in number after peripheral and central nerve damage. *J Am Acad Dermatol* 46:215–7.

Wood, L.C., Jackson, S.M., Elias, P.M., Grünfeld, C., and Feingold, K.R. 1992. Cutaneous barrier perturbation stimulates cytokine production in the epidermis of mice. *J Clin Invest* 90:482–7.

Xiao, B., and Patapoutian, A. 2011. Scratching the surface: A role of pain-sensing TRPA1 in itch. *Nat Neurosci* 14:540–2.

Yamaguchi, J., Nakamura, F., Aihara, M. et al. 2008. Semaphorin3A alleviates skin lesions and scratching behavior in NC/Nga mice, an atopic dermatitis model. *J Invest Dermatol* 128:2842–9.

Zhou, Y., Gunput, R.A., and Pasterkamp, R.J. 2008. Semaphorin signaling: Progress made and promises ahead. *Trends Biochem Sci* 33:161–70.

18 Peripheral Opioids

Paul L. Bigliardi and Mei Bigliardi-Qi

CONTENTS

18.1 INTRODUCTION

The sensation of itch is difficult to define but is generally accepted as an unpleasant cutaneous sensation, leading to the desire to scratch. It has clear survival value as it has been conserved across many mammalian species through different evolutionary pathways. There are many different manifestations of itch or other related sensations, such as tingling, crawling, or irritation. Some of these more diffuse sensations are initiated in the central nervous system (CNS), but most originate from the periphery, in particular, the skin. Cutaneous itch has many different causes and triggers, and it is crucial to understand that the interactions between the peripheral nonmyelinated, sensory C-fibers and different skin cells is the initial step for the initiation of the itch sensation in skin. Numerous skin cells are involved in this nerve–skin interaction, ranging from keratinocytes, melanocytes, Merkel cells, Langerhans cells to various dermal cells such as mastocytes, endothelial cells, fibroblasts, and cells in skin appendages. In previous publications we proposed the existence of a keratinocyte-nerve "unit," consisting of very fine and superficial nerve fibers in the epidermis connecting to keratinocytes that may be specialized in function. These keratinocytes could act as sensors and send signals either to other keratinocytes or to the epidermal C-fibers. The interaction between different cell systems in the epidermis might be crucial for the initiation of various peripheral sensations such as pain, itching, burning, tickling, and tingling. All these sensations have very distinct functions, but most of these sensation qualities do not require an immediate reflex mechanical withdrawal reaction such as the stimulation of dermal myelinated A-delta fibers with conduction of deep, well-defined injuries. Sensations such as itching, not well localized pain burning, tickling or tingling are danger signals that do not require an immediate withdrawal action but rather notify of a danger that needs to be removed by swiping or scratching. This clearly implies that the skin

305

has very sophisticated mechanisms to sense different levels of danger and to react in different ways.

The brain provides a conscious realization that there is a sensation of itch, after which we will react by rubbing or scratching to remove the noxious stimulus. The itch signal from the periphery will be modulated during its journey through the peripheral nerves, dorsal root ganglia and spinal cord to the higher centers in the brain. Inflammation of the skin as well as constant stimulation of the peripheral nerve system will modify signals emanating from the skin and in transmission of these signals to the CNS; indeed, the threshold and the irritability of the nerve fibers will change under the influence of various cytokines, growth factors, neuropeptides, and neurotransmitters. The different perceptions of itch and pain seem to have very distinct pathways in the CNS. The removal of pain is not a prerequisite for induction of itch (Liu et al. 2011). It is very difficult to separate the peripheral events in itch and pain from the processes in the CNS, but it is clear that these various sensations can be modulated at every level of transduction (Sun and Chen 2007). The events in the periphery appear to be equally important as mechanisms occurring in the CNS, and this is especially true for the involvement of the opioid receptor system in modulation of sensory function both in the periphery and CNS. While the opioid system, and in particular its influence on pain, has been well studied in the CNS, very little is known about the role of opioid receptors in skin, and in addition we are at the infancy of understanding the nerve systems in the skin.

The goal of this chapter is to describe the role of the opioid receptor system in the induction and regulation of the peripheral components of pain and itching mechanisms, with strong focus on itch. Also discussed are possibilities of treating very cumbersome sensations (itch, pain, and tingling) that have been peripherally induced by topical applications of opioids, with the intention of limiting the side effects generated by opioidergic activity within the CNS.

18.2   OPIOID RECEPTORS IN SKIN

Opioid receptors are G-protein coupled receptors that mediate the effects not only by endogenous opioid peptides but also by exogenous opiate alkaloids such as morphine. The opiate receptor system consists of three major receptor types with corresponding endogenous ligands; the μ-(mu opioid receptor [MOR]/Oprm1) with endorphins, κ-(kappa opioid receptor [KOR]/Oprk1) with dynorphins, and δ-(delta opioid receptor [DOR]/Oprd1) with enkephalins. These endogenous ligands are not, however, exclusively specific for the corresponding receptor, for example, endorphins cross-react between KOR and DOR and the enkephalins between DOR and MOR. Some of the ligands can act as agonists on one receptor type and show antagonist activity on others.

In general terms the opioid receptor system has been strongly related to the CNS, particularly in the context of pain. More recently, however, there is growing evidence that the opioid receptor system can also produce potent and specific analgesic effects outside the CNS. In addition, emerging research has shown that opioid receptors are not only involved in the sensation of pain and itch; these receptors also influence inflammation, proliferation, differentiation, and apoptosis of various cells in skin.

Opioid receptors are clearly present on peripheral nerve fibers located in the dermis and epidermis, around hair follicles (Bigliardi et al. 2009), bone, joint tissue, and in dental pulp (Sehgal et al. 2011). Coggeshall et al. (1997) reported MOR expression in 29% to 38% of cutaneous unmyelinated sensory axons, and the existence of MOR on peripheral nerve fibers in skin was later confirmed by several other groups (Stander et al. 2002; Bigliardi-Qi et al. 2004).

Opioid peptides are involved in pathophysiologic responses to stress and inflammation, and they can act on the peripheral nerve system by decreasing the excitability of sensory nerves and/or inhibiting release of proinflammatory cytokines and neuropeptides (Sehgal et al. 2011). Opioid receptors are not only present in neuronal cells and tissue but also in immune cells (macrophages and neutrophils), gastrointestinal system, and various skin cells, such as melanocytes, hair follicle epithelium, fibroblasts, and keratinocytes (Bigliardi et al. 2009).

Over the years, several different opioid ligands have been found in human skin, such as beta-endorphin (Wintzen et al. 2001; Kauser et al. 2004), enkephalins (Zagon et al. 1996; Schulze et al. 1997; Nissen and Kragballe 1997; Tachibana and Nawa 2005), and dynorphins (Tominaga et al. 2007). Bigliardi-Qi et al. (1999) reported for the first time the presence of MOR in human skin cells. During the following years the existence of MOR in nonneuronal cells in human skin was confirmed by the same and other groups (Bigliardi et al. 2002; Stander et al. 2002; Bigliardi and Bigliardi-Qi 2004; Bigliardi-Qi et al. 2005; Kauser et al. 2003), with DOR (Bigliardi-Qi et al. 2006; Bigliardi et al. 2009) and KOR (Tominaga et al. 2007; Bigliardi et al. 2009) being found subsequently. Bigliardi et al. (2009) have shown by real-time polymerase chain reaction that MOR expression in ectoderm-derived skin cells (keratinocytes and melanocytes) is higher than DOR expression. By contrast, mesoderm-derived fibroblasts express higher levels of DOR than of MOR. The distinct expression of opioid receptors in various skin cells suggests differing roles for these receptor systems in skin physiology and pathology. Functionally, DOR, and to a lesser extent MOR, are involved in skin differentiation and homeostasis and profoundly modulate wound healing (Bigliardi et al. 2003, 2009; Bigliardi-Qi et al. 2006). Wound healing is often accompanied by strong itch sensation, which is a major problem in burn wounds, and this pruritus is histamine independent (Cheng et al. 2011). This could suggest a leading role for the opioid system in itch.

18.3   OPIOID RECEPTORS IN SKIN DISEASES

Several publications describe the regulation of opioid receptor expression in various skin disorders associated with pruritus. Bigliardi et al. (2007) measured the expression of MOR in epidermis of patients with pruritic, chronic atopic dermatitis with a specific antibody (Diasorin) (Bigliardi-Qi et al. 2005; Bigliardi et al. 2007). A significant downregulation and internalization of the MOR expression by these antibodies was observed; in addition, *in situ* hybridization revealed a distinct distribution pattern of the mRNA for MOR in patients with chronic atopic dermatitis. In atopic dermatitis the mRNA for MOR is concentrated in the subcorneal layer of the epidermis, and in normal skin it is more highly expressed in the suprabasal layer.

Other groups have made different observations on the distribution of MOR in skin of patients with atopic dermatitis. There was no apparent difference in MOR (Stander et al. 2002; Tominaga et al. 2007) and KOR (Tominaga et al. 2007) expression measured by immunohistochemistry in patients with atopic dermatitis compared to healthy volunteers. KOR expression was, however, downregulated in skin of atopic dermatitis, and PUVA treatment resulted in a downregulation of MOR and a restoration of KOR (Tominaga et al. 2007).

Similar discrepancies were found with MOR in psoriasis and prurigo nodularis; Bigliardi et al. (2000) observed a significant downregulation of MOR immunoreactivity in lesional skin of psoriasis (Bigliardi-Qi et al. 2000), prurigo nodularis (Bigliardi and Bigliardi-Qi 2004) and chronic nonhealing wounds (Bigliardi et al. 2003). The nonlesional skin of patients with plaque psoriasis and also in acute wounds shows similar MOR expressions as normal skin. However, alternative studies showed MOR immunoreactivity in skin of patients with psoriasis (Stander et al. 2002; Taneda et al. 2011) and Prurigo nodularis (Stander et al. 2002) as not differing from normal skin.

There are several possibilities to explain the different results on expression of MOR in atopic dermatitis. First, Bigliardi et al. (2005) selected patients with chronic atopic dermatitis, and, second, they used different polyclonal antibodies against MOR (Diasorin) than Staender (Biotrend), and Tominaga/Taneda (Santa Cruz). In addition, the permeabilization of the cells could result in different staining patterns. Bigliardi et al. (2005) describe in their earlier publication an internalization of the MOR in atopic dermatitis. With certain staining protocols the internalized receptor, which has no functional activity, will be stained as well. Finally, all the analyses of the MOR expression in the papers of Bigliardi et al. (2004), and Bigliardi-Qi (2005) were done by confocal microscopy with less background and three-dimensional reconstruction. Nevertheless, regardless of the reasons for the different staining pattern for opioid receptors in the skin, further studies are required to elucidate the exact role the opioid receptor system in pruritic skin diseases.

18.4 PERIPHERAL μ-OPIOID RECEPTORS (MOR) AND ITCH

The MOR system has been used for centuries to treat chronic and acute pain, and it is still one of the most effective antinociceptive treatment strategies. The side effects can be numerous and debilitating such as constipation, nausea, sedation, and pruritus. The prevalence of pruritus depends on the opioid used and the method of administration and may be influenced by genetic factors. Pruritus occurs in 2% to 10% of the patients treated systemically with opioids (Reich and Szepietowski 2010). The highest prevalence for opioid-induced pruritus is associated with intrathecal administration of morphine with neuroaxial itching mostly around the nose and upper parts of the face. Systemically applied MOR antagonists are successfully used in the treatment of various dermatologic and systemic diseases, such as chronic urticaria, prurigo nodularis, mycosis fungoides, and postburn itch (Phan et al. 2010), with several reports existing related to atopic dermatitis (Brune et al. 2004; Malekzad et al. 2009) and cholestatic pruritus. There are even some controlled trials with the opioid antagonists naltrexone and naloxone in cholestatic pruritus (Bergasa et al. 1995; Bergasa 2004). MOR antagonists are not, however, currently recommended in the treatment of uremic pruritus (Phan et al. 2010). The question remains connected to the use of systemic applications of naloxone or naltrexone. Are the antipruritic effects of naloxone and naltrexone purely restricted to the CNS or does the opioid receptor system in the skin also modulate itch sensation?

The first question was whether the MOR system is really active in the peripheral nervous system or if the effects are purely related to the CNS. The following discussion should shed some light on this topic. Akiyama et al. (2010) published a mouse model that could distinguish between topical nociceptive or pruritogenic stimulation. The opioid antagonist naloxone did not affect the pain component (wiping activity) after capsaicin treatment, but it significantly reduced the already low level of capsaicin-evoked scratching behavior. These results from the mouse cheek model are consistent with human observations that capsaicin induces a burning pain, eventually also leading to itch, but can induce an initial itch response in some individuals. Histamine- and touch-evoked scratching was inhibited by the MOR antagonist naltrexone (Akiyama et al. 2012). In this study the pruritogens were applied locally on cheek skin of mice and the MOR antagonists were applied subcutaneously suggesting a peripheral action, although additional effects through CNS could not be excluded.

There are, however, several clear indications that opioid receptors, especially MOR, are involved in the peripheral induction and modulation of itch sensation. Methylnaltrexone is a peripherally restricted, relatively potent MOR antagonist with moderately good selectivity for MOR versus KOR and no effects at DOR (Goodman et al. 2007). Methylnaltrexone does not cross the blood–brain barrier and has been FDA approved for the treatment of opioid-induced constipation without affecting the analgesic effects (Holzer 2012). Interestingly, clinical and laboratory studies performed during the development of methylnaltrexone have indicated that this drug also influences nausea, vomiting, and pruritus (Phan et al. 2010). Orally administered methylnaltrexone has been shown to decrease morphine-induced itching in one small, double-blind randomized, placebo-controlled study (Yuan et al. 1998), strongly challenging the theory that opioid related itching is purely related to CNS. Further studies are required to elucidate the role of peripherally active MOR antagonists on opioid induced side effects and pruritus mechanisms. New compounds such as Alvimopan (Goodman et al. 2007) are in the pipeline and will lead to further studies.

Yamamoto and Sugimoto (2010) have shown that intradermal injection of MOR agonists DAMGO (very specific) and loperamide elicits dose-dependent scratching behavior in mice, while the intradermal injection of specific DOR agonist (DPDPE) and KOR agonist (U-50488H) does not. This scratching behavior was inhibited by peripherally restricted naloxone-methionine, and therefore the authors concluded that MOR plays a primary role in peripheral pruritus. This conclusion is directly supported by other studies performed by the group of Bigliardi et al. (2007). Topically applied naltrexone has been shown to be effective in the treatment of pruritus in patients with atopic dermatitis (Bigliardi et al. 2007). The double-blind, placebo-controlled, crossover trial on 40 patients demonstrated clearly that cream containing naltrexone had an overall 29.4% better effect compared with placebo. The formulation containing naltrexone required a median of 46 minutes to reduce the itch symptoms to 50% (placebo 74 min). After careful analysis of the patient's data, another important observation was made. Atopic dermatitis patients with chronic pruritus (>6 weeks) had a significant 45% alleviation of the pruritus by the topical naltrexone formulation using visual analogue scale (VAS) measurements compared with placebo ($n = 24$). Reduction of pruritus by topically applied naltrexone in atopic dermatitis patients with acute pruritus (<6 weeks), however, was not significant (7%, $n = 15$). This indicates clearly that the influence of various modulators of itch change over time in patients with the same pruritic dermatologic disorder. Clinical experience demonstrates that antihistamines might work in acute flare-ups of pruritus in atopic dermatitis, but they are often ineffective in chronic pruritus in patients with atopic dermatitis. The same clinical study (Bigliardi et al. 2007) has confirmed previous results (Bigliardi-Qi et al. 2005) that the significant downregulation of MOR expression is especially pronounced in chronic pruritic skin disorders with long-lasting history of pruritus. Treatment of the chronic pruritic skin disorders for 2 weeks with topically applied naltrexone resulted in increase of epidermal MOR staining (Bigliardi et al. 2007). The regulation of the epidermal opioid receptor correlated clearly with the clinical assessment and indicates a role of the MOR in epidermis in modulation of the peripheral components of itch sensation.

There are animal models also suggesting that MOR is involved in the modulation of itch sensation in chronic, but not acute skin disorders. Miyamoto et al. (2002) described a new dry skin pruritus mouse model, which correlates to a chronic atopic dermatitis with marked increase of trans epidermal water loss (TEWL), epithelial hypertrophy, and spontaneous scratching. Subcutaneous administered opioid antagonists (naloxone and naltrexone) significantly suppressed spontaneous scratching in mice with dry skin dermatitis. There was no change of scratching behavior in mast cell-deficient mice, suggesting that the role of mast cells and histamine is not relevant in this dry skin itch model (Miyamoto et al. 2002). Experiments using MOR knockout mice (Bigliardi-Qi et al. 2007) have shown that these mice reveal a phenotype of significantly thinner epidermis with higher density of epidermal nerve fibers stained by PGP 9.5 compared to wild-type mice. This suggests that MOR has effects on skin homeostasis and also cutaneous innervation. In addition, Bigliardi-Qi et al. (2007) decided to induce the dry skin dermatitis model described by Miyamoto on these mice, on the basis of the observations from the clinical study that topically applied naltrexone affected primarily pruritus in chronic atopic dermatitis. The histological analysis of the skin revealed that the epidermal hypertrophy, induced by the dry skin dermatitis, was significantly less developed in MOR knockout mice than in wild-type mice. Behavioral experiments also demonstrated that MOR knockout mice scratch significantly less after induction of dry skin dermatitis than wild-type mice, and this was not due to T-cell induced inflammation and independent of mast cell counts in the dermis. In conclusion, there are various indications that the MOR system in skin plays an important role in skin homeostasis, epidermal nerve fiber regulation, and most importantly, in the initiation and modulation of itch sensation in skin.

Naltrexone and naloxone have antagonistic effects on all opioid receptors and therefore a discussion is necessary on whether the peripheral effects of methylnaltrexone or naltrexone alone on pruritus are purely by blockage of MOR or if KOR is also involved in peripheral components of itch sensation. New data with dogs using new and highly selective MOR antagonists (3-Azabicyclo [3.1.0] hexane compounds) that produced rapid and dramatic reduction in pruritic behavior in dogs with flea allergy dermatitis add new evidence that pruritus in skin is mostly linked to MOR (Lunn et al. 2012) rather than the other opioid receptors.

18.5   PERIPHERAL κ-OPIOID RECEPTORS (KOR) AND ITCH

There are few reports linking KOR and its ligand dynorphin to pruritic skin diseases, such as atopic dermatitis. KOR immunostaining was downregulated in the epidermis of atopic dermatitis and in lesional skin of patients with pruritic psoriasis. PUVA therapy did not change KOR levels, but it increased the expression of KOR ligand dynorphin A (Tominaga et al. 2007). However, a recent study has shown no significant correlation between atopic dermatitis and pruritus in 211 patients with atopic dermatitis by analysis of pro-dynorphin promoter polymorphism (Greisenegger et al. 2009). These results show that the role of KOR in pruritic skin diseases is still under debate and further studies on patients with chronic pruritus have to follow.

In general, the exact role of KOR in the pathogenesis of pruritus is still unknown, and there are some data from animal and human studies that can shed some light into this topic. The density of free nerve endings in the epidermis is higher in KOR knockout mice compared to wild-type mice. In addition, scratching behavior of KOR knockout mice with dry skin dermatitis is reduced compared to wild-type mice. Using the dry skin model with chronic itch the MOR and KOR knockout mice behaved in a similar manner. Interestingly, the same KOR deficient animals showed no obvious alteration in the perception of thermal or mechanical pain, to the contrary of MOR knockout mice (Gaveriaux-Ruff and Kieffer 2002). Disruption of the KOR gene in mice enhances sensitivity to chemical visceral pain, impairs pharmacological actions of the selective KOR agonist U-50488 and attenuates morphine withdrawal (Simonin et al. 1998). These observations are in accordance to reports indicating a role of KOR in visceral pain.

However, in other more CNS-related itch models and in controlled human studies the KOR agonists have an antipruritic activity, in contrast with pruritogenic activity of MOR agonists. In addition, application of peripherally restricted KOR agonist ICI 204,448 inhibited chloroquin-induced pruritus in mice (Inan and Cowan 2004). This suggests that KOR also has action against pruritus in the periphery and that the opioid receptor system generally has very distinct effects on the different types of pruritus.

KOR agonists are used in the treatment of uremic pruritus, prurigo nodularis, paraneoplastic, and cholestatic pruritus (Phan et al. 2012). Nalfurafine hydrochloride (TRK 820 or Remitch®) is a KOR agonist with partial MOR and DOR agonism and has been shown to be effective in placebo-controlled studies against uremic pruritus. This drug has approval in the Japanese market for treatment of pruritus during hemodialysis. Another ligand with agonistic activity on KOR and antagonistic activity on MOR is Butophanol. This drug has been shown to be highly effective in the treatment of intractable pruritus (Dawn and Yosipovitch 2006; Lim et al. 2008), but has also demonstrated induction of pruritus (Bernstein and Grinzi 1981). With all of these observations and treatments it is literally impossible to separate peripheral KOR effects from activity within the CNS. Some authors postulate that the itch sensation in the periphery and CNS are the results of an imbalance between MOR and KOR activity; indeed, it is widely accepted that KOR signaling suppresses itch, while MOR signaling can stimulate itch. If KOR signaling was truly itch suppressive, however, one would expect enhanced scratching behavior in KOR-KO mice rather than a reduction. KOR- and MOR-KO mice do not differ in their scratching behavior after induction of dry skin dermatitis (Bigliardi-Qi et al. 2007). An obvious explanation of this finding is not clear, but pharmacological effects of putative agonists or antagonists in *in vitro* binding assays do not reflect the functional *in vivo* situation, or that *in vivo* there are compensatory mechanisms masking the actual functional effects of knockout of the respective opioidergic receptors. Alternatively, the different *in vivo* models used may not be directly comparable; moreover, KOR agonists have been clinically tested on acute forms of itching, while KOR- and MOR-KO mice studies have used a chronic form of itching (i.e., a dry skin itch model) (Bigliardi et al. 2009).

18.6   PERIPHERAL δ-OPIOID RECEPTORS (DOR) AND ITCH

As with KOR, the role of DOR in pruritus is poorly understood. In DOR-deficient animals, analgesia induced by the two agonists DPDPE and deltorphin was either abolished, reduced, or maintained, depending on the nociceptive assay and route of administration (Zhu et al. 1999). The observation that activity of DOR agonists remains prevalent in DOR-deficient animals, and that the activity of the same compounds in MOR-deficient mice is decreased, strongly suggests cross-reactivity of agonists between DOR and MOR *in vivo*. Results from DOR agonists therefore are complex and should be placed into context in combination with MOR effects. There are, however, a few reports linking endogenous DOR ligands to certain types of pruritus, for example. Hemodialysis patients with pruritus revealed significant higher opioid levels in plasma compared to hemodialysis patients without pruritus and normal controls. The increased opioid level was mainly restricted to increased plasma enkephalins (DOR) and not β-endorphin (MOR) or dynorphin (KOR) (Odou et al. 2001). Thornton and Losowsky (1988) found increased plasma levels of the DOR agonists methenkephalin and leu-enkephalin in patients with primary biliary cirrhosis. Other actions of DOR in skin are completely unknown except for profound effects on wound healing, skin homeostasis and skin differentiation (Bigliardi-Qi et al. 2006).

18.7   CONCLUSIONS

One of the major questions concerning opioid receptors and itch sensation will always be the relative involvement of the opioid receptor systems in the periphery versus the CNS. The most likely explanation is that it is a mixture between both settings, and this leads to the obvious challenge of dissecting out the different effects. Normal itch sensation is obviously initiated in the periphery, in particular the skin as a sensory organ, and is then transmitted by slow-conducting, nonmyelinated, sensory C-fibers through the DRG into the spinal cord and eventually to the brain.

In the brain we will perceive the real or conscious sensation of itch. The signal underlying itch is open to alteration, suppression, and modulation in all the different components of this pathway. The exact interactions of the opioid receptor system and the involvement of other receptor systems can only be elucidated by developing new *in vitro* coculture models of skin and nerve (Pereira et al. 2010) and comparing them to new animal models as described by Akiyama and Carstens (Akiyama et al. 2012), while keeping a close watch over emerging clinical observations and trial data.

Most of the current papers describe the KOR system as itch suppressing and the MOR system as itch stimulating; however, the itch intensity of KOR- and MOR-knockout mice in response to dry skin dermatitis did not differ (Bigliardi-Qi et al. 2007). Several explanations could account for this discrepancy. The *in vitro* pharmacological effects of putative opioid agonists or antagonists may not reflect the *in vivo* situation, especially at the level of keratinocytes and peripheral nerve fibers. Alternatively, the different *in vivo* models might not be directly comparable. This is especially true for different species and different types or modalities of pruritus such as chronic versus acute pruritus or that induced by various pruritic stimuli. Topically applied naltrexone works mostly on chronic itch and not on acute itch in patients with atopic dermatitis (Bigliardi et al. 2007). It seems that acute itching is mainly related to stimulations by histamine, prostaglandins, leucotrienes, proteinases, CGRP, SP, or KOR agonists, while MOR seems to be more effective in chronic itch sensations. Future studies are needed to prove if this is due to changes of receptors and/or endogenous ligands in skin, in particular, in or on keratinocytes or peripheral cutaneous C-fibers, and what role the keratinocyte-nerve unit plays in the peripheral induction of itch sensations.

Another explanation for diverse reactions to opioid ligands in pain and itch treatments will be genetic variants of opioid receptors in different species and even individuals. There are several new indications for the importance of genetics in responses to opioids. Studies of twins have shown significant heritability for respiratory depression, nausea, drug disliking, or even pruritus (Angst et al. 2012). In addition, the pruritus related to application of opioid ligands seems to be linked to single nucleotide polymorphisms, such as for SNP A118 (Wei et al. 2008; Sia et al. 2008; Tsai et al. 2010). Most of the authors have concluded that these SNP variations are responsible for changed susceptibility to central pruritus. This may not be the full story, and further studies are needed to examine the role of genetic variations such as SNPs in the threshold for induction of peripheral itch sensations, especially in pruritic skin diseases such as atopic dermatitis.

Opioids are not only involved in the sensations of pain or itch. There are various indications that peripheral opioids are also involved in wound healing and anti-inflammatory responses (Rachinger-Adam et al. 2011). These effects are closely linked to pain and itch sensations, and more and more evidence is evolving toward the idea of the peripheral opioids being crucial in the regulation of inflammation, wound healing, and unpleasant itch or pain sensations. Targeting these opioids by topical or peripheral treatments will lead to novel strategies without the well-known and dreadful side effects on the CNS (Bigliardi et al. 2009; Reich and Szepietowski 2012).

REFERENCES

Akiyama, T., Carstens, M. I. and Carstens, E. 2010. Differential itch- and pain-related behavioral responses and micro-opoid modulation in mice. *Acta Derm Venereol,* 90, 575–81.

Akiyama, T., Carstens, M. I., Ikoma, A., Cevikbas, F., Steinhoff, M. and Carstens, E. 2012. Mouse model of touch-evoked itch (alloknesis). *J Invest Dermatol,* 132, 1886–91.

Angst, M. S., Lazzeroni, L. C., Phillips, N. G., Drover, D. R., Tingle, M., Ray, A., Swan, G. E. and Clark, J. D. 2012. Aversive and reinforcing opioid effects: A pharmacogenomic twin study. *Anesthesiology,* 117, 22–37.

Bergasa, N. V. 2004. Treatment of the pruritus of cholestasis. *Curr Treat Options Gastroenterol,* 7, 501–8.

Bergasa, N. V., Alling, D. W., Talbot, T. L., Swain, M. G., Yurdaydin, C., Turner, M. L., Schmitt, J. M., Walker, E. C. and Jones, E. A. 1995. Effects of naloxone infusions in patients with the pruritus of cholestasis. A double-blind, randomized, controlled trial. *Ann Intern Med,* 123, 161–7.

Bernstein, J. E. and Grinzi, R. A. 1981. Butorphanol-induced pruritus antagonized by naloxone. *J Am Acad Dermatol,* 5, 227–8.

Bigliardi, P. L. and Bigliardi-Qi, M. 2004. Peripheral opiate receptor system in human epidermis and itch. In Yosipovitch, G., Greaves, M. W., Fleischer, A. B., Jr. and McGlone, F. (eds.) *Itch: Basic Mechanisms and Therapy.* New York: Marcel Dekker.

Bigliardi, P. L., Buchner, S., Rufli, T. and Bigliardi-Qi, M. 2002. Specific stimulation of migration of human keratinocytes by mu-opiate receptor agonists. *J Recept Signal Transduct Res,* 22, 191–9.

Bigliardi, P. L., Stammer, H., Jost, G., Rufli, T., Buchner, S. and Bigliardi-Qi, M. 2007. Treatment of pruritus with topically applied opiate receptor antagonist. *J Am Acad Dermatol,* 56, 979–88.

Bigliardi, P. L., Sumanovski, L. T., Buchner, S., Rufli, T. and Bigliardi-Qi, M. 2003. Different expression of mu-opiate receptor in chronic and acute wounds and the effect of beta-endorphin on transforming growth factor beta type II receptor and cytokeratin 16 expression. *J Invest Dermatol,* 120, 145–52.

Bigliardi, P. L., Tobin, D. J., Gaveriaux-Ruff, C. and Bigliardi-Qi, M. 2009. Opioids and the skin—Where do we stand? *Exp Dermatol,* 18, 424–30.

Bigliardi-Qi, M., Bigliardi, P. L., Buchner, S. and Rufli, T. 1999. Characterization of mu-opiate receptor in human epidermis and keratinocytes. *Ann N Y Acad Sci,* 885, 368–71.

Bigliardi-Qi, M., Bigliardi, P. L., Eberle, A. N., Buchner, S. and Rufli, T. 2000. Beta-endorphin stimulates cytokeratin 16 expression and downregulates mu-opiate receptor expression in human epidermis. *J Invest Dermatol,* 114, 527–32.

Bigliardi-Qi, M., Gaveriaux-Ruff, C., Pfaltz, K., Bady, P., Baumann, T., Rufli, T., Kieffer, B. L. and Bigliardi, P. L. 2007. Deletion of mu- and kappa-opioid receptors in mice changes epidermal hypertrophy, density of peripheral nerve endings, and itch behavior. *J Invest Dermatol,* 127, 1479–88.

Bigliardi-Qi, M., Gaveriaux-Ruff, C., Zhou, H., Hell, C., Bady, P., Rufli, T., Kieffer, B. and Bigliardi, P. 2006. Deletion of delta-opioid receptor in mice alters skin differentiation and delays wound healing. *Differentiation,* 74, 174–85.

Bigliardi-Qi, M., Lipp, B., Sumanovski, L. T., Buechner, S. A. and Bigliardi, P. L. 2005. Changes of epidermal mu-opiate receptor expression and nerve endings in chronic atopic dermatitis. *Dermatology,* 210, 91–9.

Bigliardi-Qi, M., Sumanovski, L. T., Buchner, S., Rufli, T. and Bigliardi, P. L. 2004. Mu-opiate receptor and Beta-endorphin expression in nerve endings and keratinocytes in human skin. *Dermatology,* 209, 183–9.

Brune, A., Metze, D., Luger, T. A. and Stander, S. 2004. [Antipruritic therapy with the oral opioid receptor antagonist naltrexone. Open, non-placebo controlled administration in 133 patients]. *Hautarzt,* 55, 1130–6.

Cheng, B., Liu, H. W. and Fu, X. B. 2011. Update on pruritic mechanisms of hypertrophic scars in postburn patients: The potential role of opioids and their receptors. *J Burn Care Res,* 32, e118–25.

Coggeshall, R. E., Zhou, S. and Carlton, S. M. 1997. Opioid receptors on peripheral sensory axons. *Brain Res,* 764, 126–32.

Dawn, A. G. and Yosipovitch, G. 2006. Butorphanol for treatment of intractable pruritus. *J Am Acad Dermatol,* 54, 527–31.

Gaveriaux-Ruff, C. and Kieffer, B. L. 2002. Opioid receptor genes inactivated in mice: The highlights. *Neuropeptides,* 36, 62–71.

Goodman, A. J., Le Bourdonnec, B. and Dolle, R. E. 2007. Mu opioid receptor antagonists: Recent developments. *ChemMedChem,* 2, 1552–70.

Greisenegger, E. K., Zimprich, A., Zimprich, F., Stingl, G. and Kopp, T. 2009. Analysis of the prodynorphin promoter polymorphism in atopic dermatitis and disease-related pruritus. *Clin Exp Dermatol,* 34, 728–30.

Holzer, P. 2012. Non-analgesic effects of opioids: Management of opioid-induced constipation by peripheral opioid receptor antagonists: Prevention or withdrawal? *Curr Pharm Des,* 18, 6010–20.

Inan, S. and Cowan, A. 2004. Kappa opioid agonists suppress chloroquine-induced scratching in mice. *Eur J Pharmacol,* 502, 233–7.

Kauser, S., Schallreuter, K. U., Thody, A. J., Gummer, C. and Tobin, D. J. 2003. Regulation of human epidermal melanocyte biology by beta-endorphin. *J Invest Dermatol,* 120, 1073–80.

Kauser, S., Thody, A. J., Schallreuter, K. U., Gummer, C. L. and Tobin, D. J. 2004. Beta-endorphin as a regulator of human hair follicle melanocyte biology. *J Invest Dermatol,* 123, 184–95.

Lim, G. J., Ishiuji, Y., Dawn, A., Harrison, B., Kim Do, W., Atala, A. and Yosipovitch, G. 2008. In vitro and in vivo characterization of a novel liposomal butorphanol formulation for treatment of pruritus. *Acta Derm Venereol,* 88, 327–30.

Liu, X. Y., Liu, Z. C., Sun, Y. G., Ross, M., Kim, S., Tsai, F. F., Li, Q. F., Jeffry, J., Kim, J. Y., Loh, H. H. and Chen, Z. F. 2011. Unidirectional cross-activation of GRPR by MOR1D uncouples itch and analgesia induced by opioids. *Cell,* 147, 447–58.

Lunn, G., Roberts, L. R., Content, S., Critcher, D. J., Douglas, S., Fenwick, A. E., Gethin, D. M. et al. 2012. SAR and biological evaluation of 3-azabicyclo[3.1.0]hexane derivatives as mu opioid ligands. *Bioorg Med Chem Lett,* 22, 2200–3.

Malekzad, F., Arbabi, M., Mohtasham, N., Toosi, P., Jaberian, M., Mohajer, M., Mohammadi, M. R., Roodsari, M. R. and Nasiri, S. 2009. Efficacy of oral naltrexone on pruritus in atopic eczema: A double-blind, placebo-controlled study. *J Eur Acad Dermatol Venereol,* 23, 948–50.

Miyamoto, T., Nojima, H., Shinkado, T., Nakahashi, T. and Kuraishi, Y. 2002. Itch-associated response induced by experimental dry skin in mice. *Jpn J Pharmacol,* 88, 285–92.

Nissen, J. B. and Kragballe, K. 1997. Enkephalins modulate differentiation of normal human keratinocytes in vitro. *Exp Dermatol,* 6, 222–9.

Odou, P., Azar, R., Luyckx, M., Brunet, C. and Dine, T. 2001. A hypothesis for endogenous opioid peptides in uraemic pruritus: Role of enkephalin. *Nephrol Dial Transplant,* 16, 1953–4.

Pereira, U., Boulais, N., Lebonvallet, N., Lefeuvre, L., Gougerot, A. and Misery, L. 2010. Development of an in vitro coculture of primary sensitive pig neurons and keratinocytes for the study of cutaneous neurogenic inflammation. *Exp Dermatol,* 19, 931–5.

Phan, N. Q., Lotts, T., Antal, A., Bernhard, J. D. and Stander, S. 2012. Systemic kappa opioid receptor agonists in the treatment of chronic pruritus: A literature review. *Acta Derm Venereol*, 92, 555–60.

Phan, N. Q., Siepmann, D., Gralow, I. and Stander, S. 2010. Adjuvant topical therapy with a cannabinoid receptor agonist in facial postherpetic neuralgia. *J Dtsch Dermatol Ges,* 8, 88–91.

Rachinger-Adam, B., Conzen, P. and Azad, S. C. 2011. Pharmacology of peripheral opioid receptors. *Curr Opin Anaesthesiol,* 24, 408–13.

Reich, A. and Szepietowski, J. C. 2010. Opioid-induced pruritus: An update. *Clin Exp Dermatol,* 35, 2–6.

Reich, A. and Szepietowski, J. C. 2012. Non-analgesic effects of opioids: Peripheral opioid receptors as promising targets for future anti-pruritic therapies. *Curr Pharm Des,* 18, 6021–4.

Schulze, E., Witt, M., Fink, T., Hofer, A. and Funk, R. H. 1997. Immunohistochemical detection of human skin nerve fibers. *Acta Histochem,* 99, 301–9.

Sehgal, N., Smith, H. S. and Manchikanti, L. 2011. Peripherally acting opioids and clinical implications for pain control. *Pain Physician,* 14, 249–58.

Sia, A. T., Lim, Y., Lim, E. C., Goh, R. W., Law, H. Y., Landau, R., Teo, Y. Y. and Tan, E. C. 2008. A118G single nucleotide polymorphism of human mu-opioid receptor gene influences pain perception and patient-controlled intravenous morphine consumption after intrathecal morphine for postcesarean analgesia. *Anesthesiology,* 109, 520–6.

Simonin, F., Valverde, O., Smadja, C., Slowe, S., Kitchen, I., Dierich, A., Le Meur, M., Roques, B. P., Maldonado, R. and Kieffer, B. L. 1998. Disruption of the kappa-opioid receptor gene in mice enhances sensitivity to chemical visceral pain, impairs pharmacological actions of the selective kappa-agonist U-50,488H and attenuates morphine withdrawal. *Embo J,* 17, 886–97.

Stander, S., Gunzer, M., Metze, D., Luger, T. and Steinhoff, M. 2002. Localization of micro-opioid receptor 1A on sensory nerve fibers in human skin. *Regul Pept,* 110, 75–83.

Sun, Y. G. and Chen, Z. F. 2007. A gastrin-releasing peptide receptor mediates the itch sensation in the spinal cord. *Nature,* 448(7154), 700–3.

Tachibana, T. and Nawa, T. 2005. Immunohistochemical reactions of receptors to met-enkephalin, VIP, substance P, and CGRP located on Merkel cells in the rat sinus hair follicle. *Arch Histol Cytol,* 68, 383–91.

Taneda, K., Tominaga, M., Negi, O., Tengara, S., Kamo, A., Ogawa, H. and Takamori, K. 2011. Evaluation of epidermal nerve density and opioid receptor levels in psoriatic itch. *Br J Dermatol,* 165, 277–84.

Thornton, J. R. and Losowsky, M. S. 1988. Opioid peptides and primary biliary cirrhosis. *BMJ,* 297, 1501–4.

Tominaga, M., Ogawa, H. and Takamori, K. 2007. Possible roles of epidermal opioid systems in pruritus of atopic dermatitis. *J Invest Dermatol,* 127, 2228–35.

Tsai, F. F., Fan, S. Z., Yang, Y. M., Chien, K. L., Su, Y. N. and Chen, L. K. 2010. Human opioid mu-receptor A118G polymorphism may protect against central pruritus by epidural morphine for post-cesarean analgesia. *Acta Anaesthesiol Scand,* 54, 1265–9.

Wei, L. X., Floreani, A., Variola, A., El Younis, C. and Bergasa, N. V. 2008. A study of the mu opioid receptor gene polymorphism A118G in patients with primary biliary cirrhosis with and without pruritus. *Acta Derm Venereol,* 88, 323–6.

Wintzen, M., De Winter, S., Out-Luiting, J. J., Van Duinen, S. G. and Vermeer, B. J. 2001. Presence of immunoreactive beta-endorphin in human skin. *Exp Dermatol,* 10, 305–11.

Yamamoto, A. and Sugimoto, Y. 2010. Involvement of peripheral mu opioid receptors in scratching behavior in mice. *Eur J Pharmacol,* 649, 336–41.

Yuan, C. S., Foss, J. F., O'Connor, M., Osinski, J., Roizen, M. F. and Moss, J. 1998. Efficacy of orally administered methylnaltrexone in decreasing subjective effects after intravenous morphine. *Drug Alcohol Depend,* 52, 161–5.

Zagon, I. S., Wu, Y. and McLaughlin, P. J. 1996. The opioid growth factor [Met5]-enkephalin, and the zeta opioid receptor are present in human and mouse skin and tonically act to inhibit DNA synthesis in the epidermis. *J Invest Dermatol,* 106, 490–7.

Zhu, Y., King, M. A., Schuller, A. G., Nitsche, J. F., Reidl, M., Elde, R. P., Unterwald, E., Pasternak, G. W. and Pintar, J. E. 1999. Retention of supraspinal delta-like analgesia and loss of morphine tolerance in delta opioid receptor knockout mice. *Neuron,* 24(1), 243–52.

19 Spinal Coding of Itch and Pain

Tasuku Akiyama and E. Carstens

CONTENTS

19.1 INTRODUCTION

Itch is defined as "an unpleasant cutaneous sensation which provokes the desire to scratch" (Rothman 1941, p. 357) and differs from pain sensation, defined as an "unpleasant sensory and emotional experience associated with actual or potential tissue damage" (IASP Taxonomy). Nevertheless, the neuronal mechanisms that distinguish itch from pain are not fully understood. There is a long-standing debate between labeled-line versus and population-coding theories of itch (Ma 2010). The concept of labeled-line coding holds that itch-specific primary afferents or "pruriceptors" transmit information to central ascending sensory neurons that are dedicated to signaling itch sensation. In contrast, population-coding mechanisms are based on the assumption that itch is signaled by a subpopulation of neurons that may receive both pruriceptive and nociceptive input. The thermosensitive transient receptor potential (TRP) channels TRPV1 and TRPA1, which bind capsaicin and allyl isothiocyanate (mustard oil), respectively, are important for pain transmission. Interestingly, knockout mice lacking TRPV1 or TRPA1 exhibit less histamine- or chloroquine-evoked scratching behavior, respectively (Imamachi et al. 2009; Shim et al. 2007; Wilson et al. 2011), suggesting that pruriceptors express TRPV1 or TRPA1 that are presumably downstream of the immediate itch transduction process. Moreover, these findings imply that both pruriceptors and nociceptors express these algogen-sensitive TRP channels. Consistent with this, *in vivo* electrophysiology studies revealed that most pruritogen-responsive neurons in the spinal cord dorsal horn also respond to algogens in mice (Akiyama et al. 2009a, 2012b), rats (Jinks and Carstens 2002), and monkeys (Davidson et al. 2007, 2012; Simone et al. 2004). It was hypothesized that itch is signaled by the subpopulation of pruritogen- and algogen-sensitive neurons, while pain is signaled by the subpopulation of neurons responsive to algogens but not pruritogens (Akiyama et al. 2010c; Davidson et al. 2012). This chapter will discuss spinal coding of itch and pain as well as modulation of itch by noxious counterstimuli.

19.2   RELATIONSHIP TO PAIN

Itch and pain are similar in that they signal the organism of potentially dangerous stimuli and are associated with protective motor responses. While only pain but not itch occurs in deep tissues (e.g., muscle, joints, or inner organs), both itch and pain occur at the surface of the body including skin, ocular surface, and other oral and anogenital mucosal surfaces. The density of sensory spots uniquely sensitive to itch (itch points) in the skin is equal to that of pain spots (around $1/mm^2$) (Shelley and Arthur 1957; von Frey 1922). Itch and pain appear to share common ascending sensory pathways such as the spinothalamic tract (STT) (Davidson et al. 2007; Simone et al. 2004). Cordotomy, clinical sectioning of the STT, abolishes both itch and pain but not touch (Bickford 1937). Interestingly, genetically pain-insensitive patients are also insensitive to itch (Kunkle and Chapman 1943). Recent brain imaging studies revealed that many common brain regions are active under conditions of acute experimentally evoked itch or pain (Carstens 2009; Herde et al. 2007). However, itch and pain differ significantly. Importantly, itch-inducing stimuli typically elicit scratching or biting to remove the stimulus, whereas algogenic stimuli typically elicit withdrawal of the stimulated body area away from the stimulus and/or other integrated escape or aggressive motor responses. Pain is attenuated by μ-opioids which can elicit or exacerbate itch (Stander and Schmelz 2006). Conversely, μ-opioid antagonists suppress itch but not pain (Heyer et al. 1997).

19.3   PERIPHERAL PATHWAYS

Itch appears to be determined by the location of the sensory receptor rather than what activates it. Itch is induced by a variety of stimuli, including mechanical, chemical, thermal, and electrical stimulation of the skin. Both algogens (preferentially eliciting pain) and pruritogens (preferentially eliciting itch) cause either pain or itch, depending on the concentrations of chemicals, and the delivery methods. For instance, when histamine is delivered to the epidermis by iontophoresis, by inactivated histamine-laden cowhage spicule, or by intradermal injection, it elicits itch, but when delivered to the deep dermis by intradermal injection, histamine usually elicits pain (Hosogi et al. 2006; Keele 1970; Sikand et al. 2011). A high concentration of histamine elicits pain rather than itch (Keele 1970). Conversely, intradermal injection of capsaicin normally elicits burning pain (Simone et al. 1989). However, when capsaicin is topically applied by soaked filter paper or delivered by cowhage spicules, it can produce itch (Green 1990; Sikand et al. 2011; Wang et al. 2010). It has been consistently demonstrated that removal of the epidermis eliminates the ability to perceive itch (Shelley and Arthur 1957). Insertion of a cowhage spicule at the dermo-epidermal junction produced maximal itch, while deep insertion in the dermis was without effect (Shelley and Arthur 1957). Thus, itch is presumably produced by the activation of subpopulations of sensory fibers located in upper skin layers.

Itch is mediated by unmyelinated C-fiber afferents as well as thinly myelinated Aδ-fiber afferents. Mechano-insensitive C-fibers preferentially respond to histamine but not cowhage (Namer et al. 2008; Schmelz et al. 1997). In contrast, mechanosensitive, polymodal C-fibers readily respond to cowhage with lesser or no responses to histamine (Johanek et al. 2008; Namer et al. 2008). Thus, cowhage and histamine apparently activate separate populations of C-fibers. While pruritogen-responsive polymodal C-fibers can respond to noxious mechanical stimuli and thus are not pruritogen-specific, histamine-responsive mechano-insensitive C-fibers may be pruritogen specific, although most of them additionally respond to algogens such as capsaicin or bradykinin (Schmelz et al. 2003b). Mechanosensitive A-fibers responded more vigorously to cowhage than to histamine, but some exclusively responded to histamine (Ringkamp et al. 2011). C-fibers and Aδ-fibers may convey two distinct qualities of itch, a slow component (burning) and a fast component (pricking), respectively (Graham et al. 1951). Local anesthesia by procaine abolished the slow component without affecting the fast component. Either local pressure-evoked ischemia or anesthesia by cold produced an area where slow, but not fast, itch can be elicited. However, other studies are inconsistent with these observations. The local anesthetic, chloroprocaine, enhanced itch following an intradermal injection of histamine (Atanassoff et al. 1999). Conduction block of myelinated fibers by nerve compression reduced pricking as well as burning sensations (Ringkamp et al. 2011).

C-fibers have been divided into peptidergic and nonpeptidergic subsets mainly on the basis of neurochemical criteria. The peptidergic neurons are mostly marked with neuropeptides including substance P (SP) and calcitonin gene-related peptide (CGRP), while nonpeptidergic neurons are commonly labeled with the purinergic P2X3 receptor and the plant lectin isolectin B4 (IB4) (Dussor et al. 2008; Hunt and Mantyh 2001). Peptidergic and nonpeptidergic neurons exhibit anatomically distinct distribution patterns in the dorsal horn of the spinal cord as well as in skin. A large subset of nonpeptidergic neurons expresses Mas-related G-protein-coupled receptor D (MrgprD), and these MrgprD-expressing neurons mostly terminate in inner lamina II (IIi), while peptidergic neurons mainly terminate in lamina I and outer lamina II (IIo) (Zylka et al. 2005). Within the skin, MrgprD-expressing neurons innervate the most superficial layer of epidermis, the stratum granulosum, while peptidergic neurons innervate the underlying stratum spinosum (Zylka et al. 2005). The sensory function of these two subsets of neurons has been examined and is apparently different. Ablation of MrgprD-expressing neurons selectively reduces mechanical nociception without affecting thermal nociception, even though they respond to noxious thermal and mechanical stimuli and innervate diverse types of spinal neurons (Cavanaugh et al. 2009; Rau et al. 2009; Wang et al. 2013). Conversely, a large subset of peptidergic neurons expresses TRPV1 and removal of TRPV1-expressing neurons results in a reduction in thermal nociception without affecting mechanical nociception (Cavanaugh et al. 2009; Rau et al. 2009). Itch is likely transmitted by both peptidergic and nonpeptidergic neurons. TRPV1 knockout (TRPV1KO) mice show deficits in histamine-evoked scratching (Han et al. 2006; Imamachi et al. 2009). In contrast, TRPA1 is expressed by a subset of TRPV1-expressing neurons and TRPA1KO mice exhibit a reduction in scratch responses to the MrgprA3 agonist, chloroquine and the MrgprC11 agonist, bovine adrenal medulla peptide (BAM) BAM8-22, in a histamine-independent manner (Liu et al. 2009; Wilson et al. 2011). Thus, histaminergic and nonhistaminergic itch pathway utilize the distinct channels, namely, TRPV1 and TRPA1, respectively. Roles for neuropeptides in the spinal transmission of itch signaling are described in the following section. The involvement of nonpeptidergic neurons in itch was reported by Dong's group (Liu et al. 2012). The MrgprD agonist, β-alanine, elicited itch in humans and scratching in mice; the latter was abolished in MrgprDKO mice. All MrgprD-expressing neurons exhibited responses to noxious mechanical stimuli. Within these neurons, about 40% exhibited responses to β-alanine as well as noxious heat stimuli, while other MrgprD-expressing neurons responded to neither β-alanine nor heat stimuli. Overall, both peptidergic and nonpeptidergic neurons that respond to pruritogens are also sensitive to noxious stimuli. Most primary afferent pruriceptors do not specifically respond only to pruritogens but additionally respond to algogens. Consistent with this, pruriceptors apparently share some of the same signal transduction molecules with nociceptors. Phospholipase C (PLC) plays a key role in intracellular signaling by G-protein-coupled receptors (GPCRs). PLCβ3 contributes to certain types of itch as well as inflammatory and neuropathic, but not thermal, pain (Han et al. 2006; Imamachi et al. 2009; Joseph et al. 2007; Shi et al. 2008; Xie et al. 1999). Phosphoinositide interacting regulator of TRP (Pirt) binds to phosphatidylinositol (4,5)-bisphosphate (PIP_2), TRPV1, and other ion channels and plays a role in thermal nociception through regulation of TRPV1. Pirt KO mice exhibited significant reductions in scratch responses to a variety of pruritogens (Patel et al. 2011).

19.4   SPINAL NEUROTRANSMITTERS

The neurotransmitters involved in spinal or trigeminal transmission of itch have recently come under investigation, with particular emphasis on gastrin releasing peptide (GRP), SP, and glutamate. Neurotoxic ablation of neurokinin-1 (NK-1) receptor-expressing neurons in the superficial dorsal horn of rats attenuated 5-HT-evoked scratching (Carstens et al. 2010) (Figure 19.1), and selective NK-1 antagonists reduced scratching elicited by chloroquine but not histamine, in mice, implying a role for SP in nonhistaminergic itch (Akiyama et al. 2013). Moreover, SP was expressed in pruritogen-sensitive dorsal root ganglion (DRG) neurons (Akiyama et al. 2013), consistent with a role as neuropeptide transmitter released from the intraspinal terminals of pruriceptors. Neurotoxic ablation of GRP receptor (GRPR)-expressing neurons in mice attenuated scratching elicited by a variety of pruritogens without affecting pain-related behaviors (Sun et al. 2009). The GRPR is extensively colocalized with the μ-opioid receptor (MOR) isoform MOR1D in the superficial spinal cord, and antagonism of the GRPR abolished morphine-induced scratching (Liu et al. 2011), implicating the GRPR in opioid-induced itch. Selective GRPR antagonists and knockout of the GRPR partially reduced scratching evoked by chloroquine but

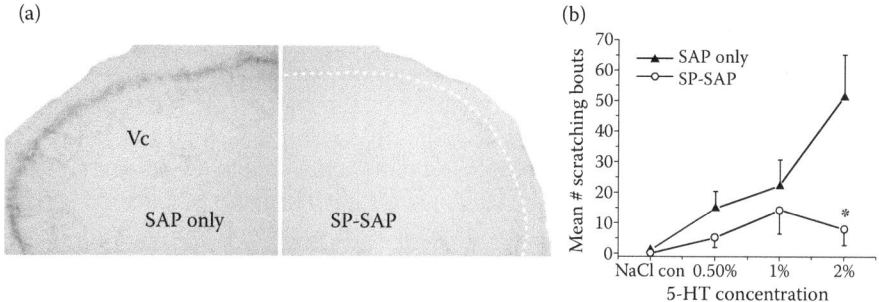

FIGURE 19.1 Ablation of superficial NK-1 receptor-expressing neurons attenuates 5-HT-evoked scratching in rats. (a) Left section through caudal medulla at level of trigeminal subnucleus caudains (Vc) from a rat pretreated with saporin (SAP only) shows extensive NK-1 immunostaining in superficial medullary dorsal horn. Right section: absence of NK-1 immunoreactivity in rat treated with SP-SAP. (b) Graph plots mean number of scratch bouts elicited by different concentrations of intradermally injected 5-HT in rats treated with SP-SAP (O) or SAP only (▲). (From Carstens, E.E. et al., *Neuroreport*, 21(4), 303–308, 2010. With permission.)

not histamine, implying a partial role for GRP in nonhistaminergic itch (Akiyama et al. 2013; Sun and Chen 2007). GRP was also expressed by a fraction of pruritogen-responsive DRG cells, consistent with it being released from pruriceptors (Akiyama et al. 2013; Simone et al. 2004). Glutamate acting at the AMPA/kainate receptor is also a likely candidate for spinal itch transmission (Koga et al. 2011). AMPA/kainate receptor antagonists abolished neuronal activity evoked by electrical stimulation of dorsal roots in histamine-responsive spinal neurons. Thus SP, GRP, and glutamate are potential neurotransmitters released from primary afferent pruriceptors (Figure 19.2) and represent good targets for developing novel treatments for itch.

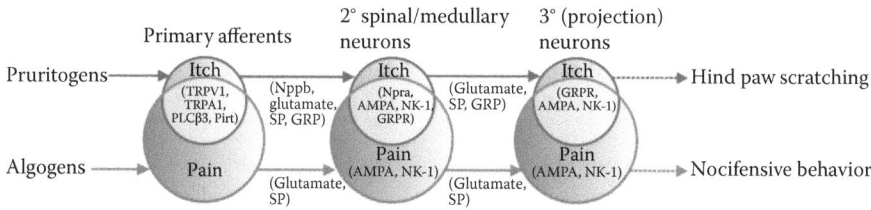

FIGURE 19.2 **(See color insert.)** Schematic diagram of population coding of itch and pain. A noxious stimulus activates nociceptive primary sensory afferents that in turn activate second- and third-order nociceptive spinal or trigeminal neurons (red circles). A pruritic stimulus activates a subset of nociceptive primary afferents, followed by the activation of second- and third-order spinal or trigeminal neurons (blue circles). The CNS decodes the activation of these subpopulations as pain or itch, and initiates nocifensive (e.g., forelimb wiping) or hind paw scratching behaviors, respectively. Itch transmission requires receptors, ion channels, signal transduction molecules, and spinal neurotransmitters, many of which are also utilized during nociceptive transmission. Nppb and GRP and their respective receptors Npra and GRPR may possibly be utilized only during itch transmission.

It was very recently reported that natriuretic polypeptide B (Nppb) is importantly involved in spinal itch but not pain transmission (Mishra and Hoon 2013). Nppb is expressed in a subset of small diameter sensory neurons that coexpress TRPV1, PLCβ3, and MrgprA3 (Mishra and Hoon 2013). The receptor for Nppb, natriuretic peptide receptor A (Npra) is expressed primarily in lamina I. Nppb knockout mice exhibited lack of scratching responses to histamine, chloroquine, endothelin-1, 5-HT, the protease-activated receptor-2 (PAR-2) and MrgprC11 agonist SLIGRL-NH2, and compound 48/80 but exhibited normal nocifensive behaviors. Moreover, intrathecal injection of Nppb elicited scratching. Mice lacking the Npra receptor exhibited impaired scratching to histamine as well as Nppb. The authors also confirmed the necessity of GRPR-expressing neurons for itch. Collectively, these data suggest that Nppb plays a major role in spinal itch transmission. The authors proposed that Nppb is released from primary afferent pruriceptors to excite second-order Npra-expressing spinal neurons that, in turn, ultimately excite downstream GRPR-expressing spinal neurons that are required for the transmission of itch signals to higher centers (Figure 19.2). It was recently reported that most GRP is expressed by spinal neurons (Fleming et al. 2012) rather than (or in addition to) primary afferents as noted earlier (Akiyama et al. 2013). Genetic ablation of testicular orphan nuclear receptor 4 (TR4) resulted in the loss of spinal neurons expressing SP, the vesicular glutamate transporter-2 (VGLUT2), and GRPR, consistent with the possibility that they function as excitatory interneurons (Maxwell et al. 2007; Todd et al. 2003; Wang et al. 2013). Figure 19.2 summarizes the above findings by indicating roles for Nppb, SP, GRP, and glutamate as neurotransmitters released from primary afferent pruriceptors, as well as possible roles for glutamate, SP, and GRP as neurotransmitters released by spinal interneurons. Further work is needed to determine the precise neurocircuitry and sites of action of these neurotransmitters in spinal itch signaling.

Neurokinin B encoded by the tachykinin 2 gene is a member of the tachykinin peptides along with SP and is a possible candidate as a neuropeptide transmitter in spinal itch transmission. However, a recent study revealed that tachykinin 2 null mice exhibited normal scratch responses to compound 48/80, chloroquine, SLIGRL-NH,2 and methyl-serotonin as well as normal responses to noxious heat and mechanical stimuli, implying that neurokinin B is not involved in the spinal transmission of itch or pain signaling (Mar et al. 2012). Neuromedin B is a member of the mammalian bombesin family of peptides along with GRP and is another candidate as a neuropeptide transmitter in spinal itch transmission. Intrathecal or intracerebroventricular administration of neuromedin B elicited marked scratching (Cridland and Henry 1992; Sun and Ko 2011; Van Wimersma Greidanus and Maigret 1991), while intrathecal administration of neuromedin B produced a transient decrease followed by a delayed increase to above baseline in tail flick latency (Cridland and Henry 1992). Neurotoxic ablation of neuromedin B receptor-expressing neurons in the superficial dorsal horn did not affect histamine H1 receptor agonist-evoked scratching but reduced noxious heat-evoked behavioral responses (Mishra et al. 2012), implying a role for neuromedin B in thermal pain. It would be interesting to test whether neuromedin B is involved in the spinal transmission of nonhistaminergic itch. CGRP, a member of the calcitonin family of peptides, is a widely recognized marker for peptidergic neurons and may contribute to the spinal transmission of pain as well as itch. CGRPα was expressed by 61% to 73% of pruritogen-responsive DRG neurons and 27% to 83% of algogen-responsive DRG neurons (McCoy et al. 2012).

Overall, pathways for itch and pain signaling both appear to use glutamate and SP, while recent evidence indicates that itch may be selectively signaled by GRP and Nppb. Moreover, most pruritogen-responsive spinal neurons are activated by nociceptive primary afferents.

19.5 PRURITOGEN-RESPONSIVE SPINAL AND MEDULLARY DORSAL HORN NEURONS

The dorsal horn is the major site relaying information from primary sensory afferents. Superficial dorsal horn neurons (laminae I-II) receive direct input from most nociceptive Aδ- and C-fibers, while deep dorsal horn neurons (laminae III-V) receive direct input from Aβ-fibers (Todd 2002). Recent molecular studies have further categorized the central projections of nociceptive C-fibers and low-threshold mechanoreceptors (LTMRs) in the spinal cord (Basbaum et al. 2009; Li et al. 2011). Spinal cord neurons in lamina I-IIo, IIm, IIi, III, or III-V receive projections from peptidergic C-fibers, nonpeptidergic C-fibers, C-fiber-LTMRs, Aδ-LTMRs, or Aβ-LTMRs, respectively. Dorsal horn neurons can be classified into four general categories according to their responses to mechanical stimuli: mechano-insensitive (MI) neurons that respond to neither noxious nor innocuous mechanical stimuli, low-threshold (LT) neurons that do not respond to noxious mechanical stimuli, wide dynamic range (WDR) neurons that respond at higher firing rate to noxious than to innocuous mechanical stimuli, and nociceptive specific (NS) neurons that respond to noxious but not innocuous mechanical stimuli. WDR and NS neurons are located in both superficial and deep dorsal horn (Chudler et al. 1991; Dado et al. 1994; Price et al. 1978). LT neurons are located mainly in the deep dorsal horn (laminae III-IV) (Chudler et al. 1991; Price et al. 1978). Pruritogen-responsive neurons are mostly either NS or WDR, but a few of them are MI (Table 19.1) (Akiyama et al. 2009a, b, 2010c, 2012b; Andrew and Craig 2001; Davidson et al. 2012; Jinks and Carstens 2002). Andrew and Craig recorded 190 lamina I STT neurons in cats and identified 18 that were mechanically and thermally insensitive. Interestingly, 10 of the latter responded to iontophoretically applied histamine and two of four histamine-responsive STT neurons tested did not respond to the algogen, mustard oil (capsaicin was not tested in this study). Jinks and Carstens (2002) identified 21 5-HT-responsive neurons in the superficial dorsal horn of rats, of which 15 were classified as WDR, five as NS and one as MI. The MI unit responded to noxious chemical and heat stimuli. Davidson et al. (2012) recorded from 111 STT neurons in adult macaques; 32 responded to either histamine or cowhage and two responded to both. All of the histamine- and cowhage-responsive STT neurons were either WDR or NS. Thus, approximately 30% of nociceptive STT neurons are pruritogen responsive as well. In studies conducted in our laboratory, we have collectively recorded from 17 histamine-, seven 5-HT-, 58 SLIGRL-NH2-, and 10 chloroquine-responsive lumber spinal neurons (Akiyama 2009a, b, 2012b). Of these, five histamine- and three SLIGRL-NH2-responsive spinal neurons were MI.

TABLE 19.1

Pruritogenic Activation of Different Classes of Spinal and Trigeminal Neurons

Animals	Pruritogens	WDR	NS	MI	MI				References
					Capsaicin	AITC	Heat	Cold	
Cat	Histamine	N.T.	N.T.	10		2/4			Andrew and Craig 2001
Rat (lumbar)	5-HT	15	5	1	1/1	1/1	1/1		Jinks and Carstens 2002
Rat (Vc)	5-HT	12	5	2	1/2	2/2	1/2	1/2	Klein et al. 2001
Monkey	Histamine	14	8	0					Davidson et al. 2012
	Cowhage	10	4	0					Davidson et al. 2012
Mice	Histamine	9	3	5[a]	2/4	0/3	0/2	0/1	Akiyama et al. 2009a
(Lumbar)	5-HT	6	1	0					Akiyama et al. 2009b
	SLIGRL-NH2	40	15	3	1/1	1/1			Akiyama et al. 2009a, 2009b
	Chloroquine	4	6	0					Akiyama et al. 2012b
Mice (Vc)	Histamine	8	6	4[b]	3/4	1/1			Akiyama et al. 2010c
	SLIGRL-NH2	8	6	0					Akiyama et al. 2010c

Note: N.T., not tested.

[a] Two MI responded to capsaicin. Others did not receive noxious thermal stimuli.

[b] All MI responded to either capsaicin or mustard oil.

We also recorded from 58 neurons in the superficial dorsal horn of trigeminal subnucleus caudalis (Vc) with afferent input from the cheek (Akiyama et al. 2010c). Intradermal cheek injection of pruritogens or algogens differentially elicits hind paw scratching or forelimb wipes, presumably reflecting distinct itch and pain sensations, respectively (Akiyama et al. 2010a) (see Figure 19.2). Out of 32 pruritogen-responsive Vc neurons, four were MI and responded to either capsaicin or mustard oil (Figure 19.3). In this study, a subpopulation of nociceptive neurons was isolated using an algogen (mustard oil) search stimulus and subsequently tested with several pruritogens. Only a minority of these nociceptive neurons (13%–41%) additionally responded to the pruritogens histamine, SLIGRL-NH2, or 5-HT (Figure 19.3c). Overall, the vast majority of pruritogen-responsive spinal neurons, including WDR, NS, and MI (but not LT) subtypes, additionally responds to noxious mechanical, thermal, and/or chemical stimuli and thus appears to be a subset of nociceptive spinal neurons.

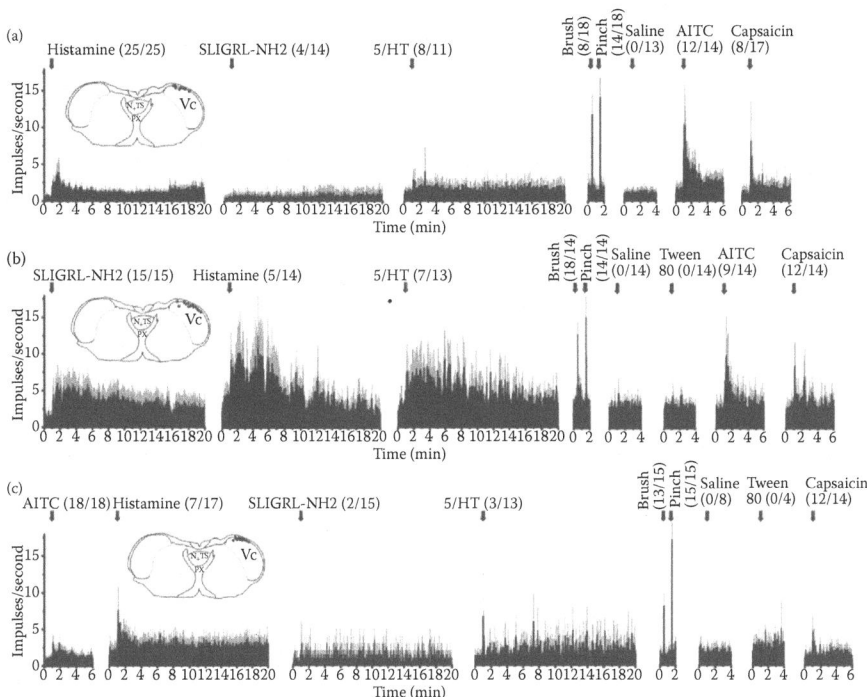

FIGURE 19.3 Averaged responses of Vc units isolated by chemical search strategy. (a) Histamine. Averaged PSTHs (bin width: 1 s) of 25 units isolated by histamine search strategy. Gray error bars: SE. PSTHs to right show averaged responses to indicated stimuli. Numbers in parentheses: # responsive units/all units tested. Inset: recording sites in trigeminal subnucleus caudalis (Vc). (b) PAR-2 agonist SLIGRH-NH2 (format as in a). (c) AITC (format as in a and b). (Reprinted with permission from Akiyama, T. et al., *Journal of Neurophysiology*, 104(5), 2442–2450, 2010. Copyright 2010 by the American Physiological Society.)

19.6 MODULATION OF ITCH BY NOXIOUS COUNTERSTIMULI

Itch can be inhibited by various types of noxious counterstimuli of a thermal, mechanical, chemical, or electrical nature. Recent studies support a spinal site of action by which scratching inhibits pruritogen-responsive neurons. Monkey STT neurons are modulated in a state-dependent manner by cutaneous scratching; scratching inhibited responses elicited by the pruritogen histamine but did not inhibit responses of the same neurons to the algogen capsaicin (Davidson et al. 2009). Consistent with this, pruritogen-responsive mouse spinal neurons are modulated in a state-dependent and also site-dependent manner by cutaneous scratching (Akiyama et al. 2012a) (Figure 19.4).

In a human study, localization of the site of itch evoked by histamine iontophoresis was accurate to within approximately 7 mm and matched the accuracy of localization of a painful stimulus (Koltzenburg et al. 1993). Thus, scratch movements may

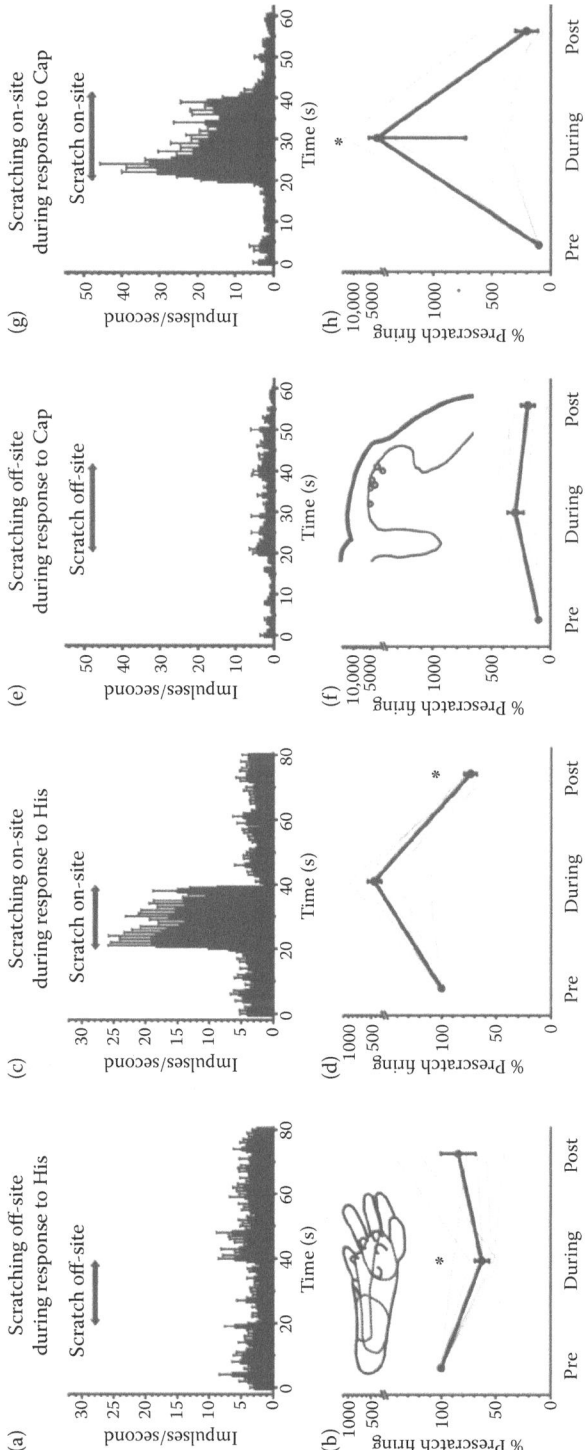

FIGURE 19.4 State- and site-dependent inhibition of histamine-responsive superficial dorsal horn neurons. (a) Averaged PSTH (bins, 1 s) of response to intradermal histamine (His), before, during and after scratching off-site. (b) Individual (thin lines) and mean (thick line, ± SEM) normalized responses to histamine. Asterisk indicates significantly different from the prescratch baseline ($P < 0.05$). Figurine shows receptive fields. (c) On-site scratching during response to histamine. (d) Graph as in B. Asterisk indicates significantly different from the prescratch baseline ($P < 0.05$). (e) Lack of effect of off-site scratching during response to capsaicin (Cap). (f) Graph as in B. Inset: recording sites. (g) Facilitatory effect of on-site scratching during response to capsaicin. (h) Graph as in D. Asterisk indicates significantly different from prescratching ($P < 0.05$). (From Akiyama, T. et al., *European Journal of Neuroscience*, 36(3), 2311–2316, 2012. With permission.)

be accurately directed to a site of itch to remove the causal stimulus, such as an insect or plant spicule. However, itch can also be inhibited by scratching a wider area surrounding the source of itch. Our recent study showed that pruritogen-evoked activity of spinal neurons was inhibited by scratching within an area 5–17 mm distal from the injection site of the pruritogen (Akiyama et al. 2012a) (Figure 19.4). Older studies reported that itch elicited by cowhage is suppressed by a pinprick applied within a large surrounding area as far as 24 cm away (Chapmann et al. 1960; Graham et al. 1951). Acupuncture stimulation more effectively suppressed itch when applied within rather than outside the dermatome in which histamine was applied (Lundeberg et al. 1987). Consistent with these observations, scratch marks are often aligned longitudinally along dermatomal lines, particularly in the distal extremities (Savin et al. 1991), and the length of scratch marks was correlated with the two-point discrimination threshold (Cornbleet 1953).

The inhibition of itch-signaling spinal neurons is thought to be mediated at the spinal level by interneurons that release the inhibitory neurotransmitters glycine and/or GABA and possibly also dynorphin. Spinal application of glycine and GABA-A and GABA-B receptor antagonists attenuated or abolished scratch-evoked inhibition of spontaneous activity in dorsal horn neurons with input from dry skin (Akiyama et al. 2011b). Conceivably, spinal inhibitory interneurons are tonically active, based on recent studies showing that decreased activity in, or deletion of, inhibitory spinal interneurons is associated with enhanced itch. Loss of a population of inhibitory interneurons in the superficial dorsal horn of knockout mice lacking the transcription factor Bhlhb5 (Ross et al. 2010), as well as knockout of the glutamate transporter VGLUT2 in certain types of nociceptors (Lagerstrom et al. 2010; Liu et al. 2010), both resulted in excessive scratching behavior. These findings suggest that a reduction in nociceptive input decreases spinal inhibition to disinhibit itch transmission.

Both segmental and supraspinal circuits appear to be involved in the scratch-evoked inhibition, at least under chronic itch conditions. Cold-block or complete transection of the upper cervical spinal cord reduced scratch-evoked inhibition of spontaneous activity in dorsal horn neurons with input from dry skin by 30% and 50%, respectively. This implies that scratch-evoked inhibition is mediated partially via activation of supraspinal neurons that, in turn, engage descending pathways to result in spinal release of glycine and GABA. The supraspinal circuit is unknown but may involve the same neurons in the rostral ventromedial medulla that give rise to descending projections to the spinal cord to modulate nociceptive transmission (Ossipov et al. 2010). Descending noradrenergic, but not serotonergic, pathways were reported to exert tonic inhibition of itch signaling in the spinal cord (Gotoh et al. 2011). In a human brain imaging study, the midbrain periaqueductal gray (PAG) was activated during the inhibition of histamine-evoked itch by a noxious cold stimulus (Mochizuki et al. 2003), suggesting that the PAG, a well-known center for descending modulation of pain, may also be involved in modulating itch.

19.7   SENSITIZATION OF ITCH-SIGNALING PATHWAYS

Peripheral and central sensitization play important roles in the establishment of chronic pain, and it has been suggested that sensitization of itch-signaling pathways may contribute to various types of chronic itch. Many chronic pain states are associated with ongoing spontaneous pain, hyperalgesia, and allodynia (touch-evoked pain). These conditions can also be experimentally reproduced in human skin by intradermal injection of capsaicin, which similarly enhanced the responses of monkey STT neurons to noxious thermal and innocuous mechanical stimuli, as well as to electrical nerve stimulation, suggesting enhanced excitability of the STT neurons (Simone et al. 1991).

Similar to chronic pain, chronic itch can be associated with spontaneous itch, hyperknesis (enhanced itch to a normally itchy stimulus), and alloknesis (itch elicited by an innocuous touch stimulus). There are few studies addressing whether primary afferent pruriceptors are sensitized under chronic itch conditions. Pruritogen-sensitive C-fibers recorded in atopic dermatitis patients exhibited high levels of spontaneous firing (Schmelz et al. 2003a). To our knowledge, few if any animal studies have addressed whether peripheral or central itch-signaling neurons are sensitized under chronic itch conditions (with the exception of the genetic models noted in the previous section). Using the dry skin model of chronic itch, we reported that mice exhibited significantly greater scratching following intradermal injections of 5-HT and SLIGRL (an agonist of PAR-2 and MrgprC11) delivered in the dry skin treatment area compared to control (water-)treated mice (Akiyama et al. 2010b). Interestingly, histamine-evoked scratching was not significantly enhanced in dry skin-treated mice. DRG cells taken from the dry skin-treated mice exhibited significantly greater responses to 5-HT and SLIGRL but not histamine, consistent with the behavioral results.

To investigate if chronic itch sensitizes spinal neurons, we recorded from superficial dorsal horn neurons receiving afferent input from a dry skin-treated hind paw (Akiyama et al. 2011a). These neurons exhibited significantly enhanced responses to SLIGRL but not histamine, compared to units recorded in control animals (Figure 19.5). However, in this study, neurons with receptive fields on the dry skin-treated hind paw did not exhibit enhanced responses to innocuous mechanical stimulation, suggesting that the enhanced response to SLIGRL may be attributed to peripheral sensitization of pruriceptors projecting to the recorded dorsal horn neurons.

Alloknesis is a common and often distressing symptom of many chronic itch patients. Alloknesis is postulated to be mediated by tactile afferents contacting hypersensitive itch-signaling spinal neurons that send an amplified itch signal, but there is currently no evidence for this. To begin to investigate alloknesis, we recently developed an animal model (Akiyama et al. 2012a). C57BL6 mice do not normally respond to innocuous mechanical stimulation of the rostral back. However, following intradermal injection of histamine and certain other pruritogens, low-threshold mechanical stimuli delivered to skin around the injection site reliably elicited discrete hind paw scratch bouts directed to the stimulus. The time course of touch-evoked scratching had a slower onset and longer duration compared to the "spontaneous" scratching following the pruritogen injection that usually ceased within 30 min. Touch-evoked scratching was observed following histamine, 5-HT, a PAR-4 agonist, and BAM8-22 but not SLIGRL or chloroquine. We also observed touch-evoked scratching in dry skin-treated animals, suggesting that this model of itch is associated with alloknesis.

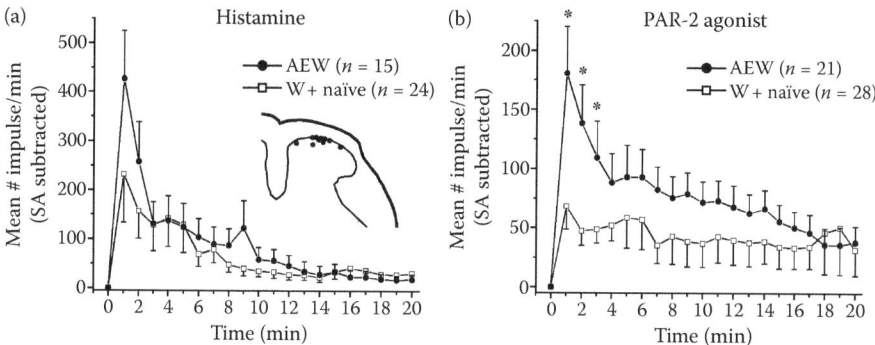

FIGURE 19.5 Enhancement of responses to PAR-2 agonist but not histamine in AEW-treated skin. (a) Averaged histamine-evoked responses (error bars: SEM) of superficial dorsal horn neurons recorded in dry skin (AEW) treated animals or controls. Spontaneous activity (SA) during 1 min prior to histamine injection was subtracted. Average firing following id histamine was significantly greater than prestimulus baseline out to 6 min postinjection for AEW and control (naïve or water-treated) groups ($p < 0.05$, paired t-test). There was no significant difference between the AEW and control (naïve or water-treated) groups at any time point. Inset shows recording sites for units isolated using the histamine search stimulus. (b) Graph as in A for units responding to a PAR-2 agonist. Asterisk indicates significant difference between AEW and control groups during the first through third minutes postinjection ($p < 0.05$, unpaired t-test). (Reprinted with permission from Akiyama, T. et al., *Journal of Neurophysiology*, 105(6), 2811–2817, 2011. Copyright 2011 by the American Physiological Society.)

As more animal models of chronic itch conditions become available, it will be important to determine if the affected subjects exhibit ongoing itch, hyperknesis, and alloknesis, and if peripheral and/or central itch-signaling neurons exhibit sensitization. If so, identification of the molecular players involved in the development of sensitization may represent important future targets for developing treatments for chronic itch.

19.8 SPECIFICITY VERSUS POPULATION CODING THEORIES OF ITCH AND PAIN

Several hypotheses have been proposed to explain how itch is differentiated from pain. Labeled-line or specificity theory postulates that each sensory modality is processed along a direct-linkage system from skin to brain. The existence of histamine-sensitive MI neurons in the cat spinal cord supports this hypothesis (Andrew and Craig 2001). A recent striking observation also supports the concept that certain peripheral pruritogen-sensitive sensory afferents may be dedicated to specifically signal itch. In TRPV1 knockout mice in which TRPV1 was reexpressed selectively in MrgprA3-expressing sensory neurons, capsaicin, which normally evokes pain instead elicited itch-related scratching behavior (Han et al. 2012). And as noted earlier in this chapter, GRP and Nppb may be specifically involved in signaling itch but not pain.

However, as discussed above, many of the available molecular-genetic, behavioral, and electrophysiological studies suggest that pruriceptors are a subset of nociceptors that respond to noxious mechanical, chemical, and/or thermal stimuli. This raises the question as to how noxious stimuli elicit pain without simultaneously eliciting itch. One possibility is that noxious stimuli activate inhibitory interneurons that suppress itch transmission. Although noxious stimuli can inhibit responses of spinal neurons to pruritogens, this inhibition is state dependent (Akiyama et al. 2011b, 2012b; Davidson et al. 2009). Most pruritogen-sensitive spinal neurons are also excited by capsaicin, so that scratching becomes an excitatory rather than inhibitory stimulus to further excite the neuron (Figure 19.4). If such neurons are dedicated to only signal itch, then it remains difficult to explain how a noxious stimulus only elicits pain. Another possibility is that noxious stimuli do elicit both itch and pain simultaneously, whereby the sensation of itch is masked or occluded by the larger pain signal in order to discriminate between the two sensory qualities. An additional concept is population coding, expressed in Figure 19.2. Itch is postulated to be signaled by the activation of a subset of spinal neurons that responds to both pruritogens and noxious stimuli. Noxious stimulation activates a larger population of nociceptive spinal neurons, including those responsive to pruritogens, to signal pain. The CNS decodes activity in the former and latter neuronal populations as itch and pain, respectively. According to this idea, MrgprA3-expressing sensory neurons would be expected to project to a subset of GRPR-expressing spinal neurons that respond to both pruritogens and noxious stimuli. MrgprA3-expressing sensory neurons do respond to noxious mechanical and chemical stimuli, and their central spinal terminals make synaptic contact with GRPR-expressing spinal neurons (Han et al. 2012). Because ablation of neither MrgprA3-expressing neurons nor GRPR-expressing spinal neurons has any effect on nociception (Han et al. 2012; Sun et al. 2009), the CNS may decode activity in a larger population of nociceptive spinal neurons as pain regardless of activity in a smaller subset of spinal neurons responsive to both pruritogens and noxious stimuli.

Additional long-postulated theories for discriminating between itch and pain have recently been revisited, including frequency (intensity) coding and firing pattern (Davidson et al. 2012). According to frequency or intensity theory, the CNS decodes weak (low frequency) and strong (high frequency) activation of neurons as itch and pain, respectively. Consistent with this theory, histamine elicited less activation of thalamus and anterior cingulated cortex than a noxious cold stimulus (Mochizuki et al. 2007). The CNS may additionally utilize the neuronal firing pattern, such as bursting, to accomplish quality discrimination. STT neurons consistently exhibited weaker responses to pruritogens with longer interburst intervals compared to responses elicited by algogens (Davidson et al. 2012). Although available data from recording and stimulation of primary afferent fibers argue against these theories (Handwerker et al. 1991; Tuckett 1982), the properties of neural activity coming from primary afferents might be altered by processing in spinal circuits to result in itch- and pain-selective patterns of activity.

19.9 CONCLUSION

Although itch and pain are distinct somatosensory qualities, they share many common elements including ion channels, signaling molecules, neurotransmitters, and even sensory neurons. About one third of nociceptive spinal neurons additionally respond to pruritogens. Although it is still a debatable question as to how itch is differentiated from pain, a small subset of pruriceptive and nociceptive spinal neurons may prove to play a crucial role in itch transmission. Chronic itch and pain are major clinical problems. A better understanding the neuronal processing of and interactions between itch and pain will provide novel targets for the development of more effective treatments for these often debilitating conditions.

ACKNOWLEDGMENTS

The work was supported by grants from the National Institutes of Health DE013685 and AR057194.

REFERENCES

Akiyama T, Carstens MI, Carstens E. 2009a. Excitation of mouse superficial dorsal horn neurons by histamine and/or PAR-2 agonist: Potential role in itch. *Journal of Neurophysiology* 102(4):2176–2183.

Akiyama T, Carstens MI, Carstens E. 2010a. Differential itch- and pain-related behavioral responses and micro-opoid modulation in mice. *Acta Dermato-Venereologica* 90(6):575–581.

Akiyama T, Carstens MI, Carstens E. 2010b. Enhanced scratching evoked by PAR-2 agonist and 5-HT but not histamine in a mouse model of chronic dry skin itch. *Pain* 151(2):378–383.

Akiyama T, Carstens MI, Carstens E. 2010c. Facial injections of pruritogens and algogens excite partly overlapping populations of primary and second-order trigeminal neurons in mice. *Journal of Neurophysiology* 104(5):2442–2450.

Akiyama T, Carstens MI, Carstens E. 2011a. Enhanced responses of lumbar superficial dorsal horn neurons to intradermal PAR-2 agonist but not histamine in a mouse hind paw dry skin itch model. *Journal of Neurophysiology* 105(6):2811–2817.

Akiyama T, Carstens MI, Carstens E. 2011b. Transmitters and pathways mediating inhibition of spinal itch-signaling neurons by scratching and other counterstimuli. *PloS One* 6(7):e22665.

Akiyama T, Carstens MI, Ikoma A, Cevikbas F, Steinhoff M, Carstens E. 2012a. Mouse model of touch-evoked itch (Alloknesis). *Journal of Investigative Dermatology* 132(7):1886–1891.

Akiyama T, Merrill AW, Carstens MI, Carstens E. 2009b. Activation of superficial dorsal horn neurons in the mouse by a PAR-2 agonist and 5-HT: Potential role in itch. *Journal of Neuroscience* 29(20):6691–6699.

Akiyama T, Tominaga M, Carstens MI, Carstens EE. 2012b. Site-dependent and state-dependent inhibition of pruritogen-responsive spinal neurons by scratching. *European Journal of Neuroscience* 36(3):2311–2316.

Akiyama T, Tominaga M, Davoodi A, Nagamine M, Blansit K, Horwitz A, Carstens MI, Carstens E. 2013. Roles for substance P and gastrin releasing peptide as neurotransmitters released by primary afferent pruriceptors. *Journal of Neurophysiology* 109(3):742–748.

Andrew D, Craig A. 2001. Spinothalamic lamina I neurons selectively sensitive to histamine: A central neural pathway for itch. *Nature Neuroscience* 4(1):72–77.

Atanassoff P, Brull S, Zhang J, Greenquist K, Silverman D, Lamotte R. 1999. Enhancement of experimental pruritus and mechanically evoked dysesthesiae with local anesthesia. *Somatosensory and Motor Research* 16(4):291–298.

Basbaum A, Bautista D, Scherrer G, Julius D. 2009. Cellular and molecular mechanisms of pain. *Cell* 139(2):267–284.

Bickford RG. 1937. Experiments relating to the itch sensation, itch peripheral mechanism and central pathways. *Clinical Science* 3:377–386.

Carstens E. 2009. Neurobiology of itch and pain: Scratching for answers. In: J Castro-Lopes, ed., *Current Topics in Pain: 12th World Congress on Pain*. Seattle: IASP Press, pp. 73–93.

Carstens EE, Carstens MI, Simons CT, Jinks SL. 2010. Dorsal horn neurons expressing NK-1 receptors mediate scratching in rats. *Neuroreport* 21(4):303–308.

Cavanaugh D, Lee H, Lo L, Shields S, Zylka M, Basbaum A, Anderson D. 2009. Distinct subsets of unmyelinated primary sensory fibers mediate behavioral responses to noxious thermal and mechanical stimuli. *Proceedings of the National Academy of Sciences of the United States of America* 106(22):9075–9080.

Chapmann L, Goodell H, Wolff HG. 1960. Structures and processes involved in the sensation of itch. In: W Montagna, ed., *Advances in Biology of Skin*, Vol. 1. NY: Pergamon, pp. 161–188.

Chudler E, Foote W, Poletti C. 1991. Responses of cat C1 spinal cord dorsal and ventral horn neurons to noxious and non-noxious stimulation of the head and face. *Brain Research* 555(2):181–192.

Cornbleet T. 1953. Scratching patterns. 1. Influence of site. *Journal of Investigative Dermatology* 20(2):105–110.

Cridland R, Henry J. 1992. Bombesin, neuromedin C and neuromedin B given intrathecally facilitate the tail flick reflex in the rat. *Brain Research* 584(1–2):163–168.

Dado R, Katter J, Giesler G. 1994. Spinothalamic and spinohypothalamic tract neurons in the cervical enlargement of rats. II. Responses to innocuous and noxious mechanical and thermal stimuli. *Journal of Neurophysiology* 71(3):981–1002.

Davidson S, Zhang X, Khasabov SG, Moser HR, Honda CN, Simone DA, Giesler GJ. 2012. Pruriceptive spinothalamic tract neurons: Physiological properties and projection targets in the primate. *Journal of Neurophysiology* 108(6):1711–1723.

Davidson S, Zhang X, Khasabov SG, Simone DA, Giesler GJ, Jr. 2009. Relief of itch by scratching: State-dependent inhibition of primate spinothalamic tract neurons. *Nature Neuroscience* 12(5):544–546.

Davidson S, Zhang X, Yoon CH, Khasabov SG, Simone DA, Giesler GJ, Jr. 2007. The itch-producing agents histamine and cowhage activate separate populations of primate spinothalamic tract neurons. *Journal of Neuroscience* 27(37):10007–10014.

Dussor G, Zylka M, Anderson D, McCleskey E. 2008. Cutaneous sensory neurons expressing the Mrgprd receptor sense extracellular ATP and are putative nociceptors. *Journal of Neurophysiology* 99(4):1581–1589.

Fleming MS, Ramos D, Han SB, Zhao J, Son YJ, Luo W. 2012. The majority of dorsal spinal cord gastrin releasing peptide is synthesized locally whereas neuromedin B is highly expressed in pain- and itch-sensing somatosensory neurons. *Molecular Pain* 8:52.

Gotoh Y, Omori Y, Andoh T, Kuraishi Y. 2011. Tonic inhibition of allergic itch signaling by the descending noradrenergic system in mice. *Journal of Pharmacological Sciences* 115(3):417–420.

Graham D, Goodell H, Wolff H. 1951. Neural mechanisms involved in itch, itchy skin, and tickle sensations. *Journal of Clinical Investigation* 30(1):37–49.

Green BG. 1990. Spatial summation of chemical irritation and itch produced by topical application of capsaicin. *Perception and Psychophysics* 48(1):12–18.

Han L, Ma C, Liu Q, Weng HJ, Cui Y, Tang Z, Kim Y et al. 2012. A subpopulation of nociceptors specifically linked to itch. *Nature Neuroscience* 16(2):174–182.

Han S-K, Mancino V, Simon M. 2006. Phospholipase C-beta 3 mediates the scratching response activated by the histamine H1 receptor on C-fiber nociceptive neurons. *Neuron* 52(4):691–703.

Handwerker HO, Forster C, Kirchhoff C. 1991. Discharge patterns of human C-fibers induced by itching and burning stimuli. *Journal of Neurophysiology* 66(1):307–315.

Herde L, Forster C, Strupf M, Handwerker HO. 2007. Itch induced by a novel method leads to limbic deactivations a functional MRI study. *Journal of Neurophysiology* 98(4):2347–2356.

Heyer G, Dotzer M, Diepgen TL, Handwerker HO. 1997. Opiate and H1 antagonist effects on histamine induced pruritus and alloknesis. *Pain* 73(2):239–243.

Hosogi M, Schmelz M, Miyachi Y, Ikoma A. 2006. Bradykinin is a potent pruritogen in atopic dermatitis: A switch from pain to itch. *Pain* 126(1–3):16–23.

Hunt S, Mantyh P. 2001. The molecular dynamics of pain control. *Nature Reviews Neuroscience* 2(2):83–91.

IASP Taxonomy. http://www.iasp-pain.org/Content/NavigationMenu/GeneralResourceLinks/PainDefinitions/default.htm#Pain.

Imamachi N, Park GH, Lee H, Anderson DJ, Simon MI, Basbaum AI, Han SK. 2009. TRPV1-expressing primary afferents generate behavioral responses to pruritogens via multiple mechanisms. *Proceedings of the National Academy of Sciences of the United States of America* 106(27):11330–11335.

Jinks S, Carstens E. 2002. Responses of superficial dorsal horn neurons to intradermal serotonin and other irritants: Comparison with scratching behavior. *Journal of Neurophysiology* 87(3):1280–1289.

Johanek LM, Meyer RA, Friedman RM, Greenquist KW, Shim B, Borzan J, Hartke T, LaMotte RH, Ringkamp M. 2008. A role for polymodal C-fiber afferents in nonhistaminergic itch. *Journal of Neuroscience* 28(30):7659–7669.

Joseph E, Bogen O, Alessandri-Haber N, Levine J. 2007. PLC-beta 3 signals upstream of PKC epsilon in acute and chronic inflammatory hyperalgesia. *Pain* 132(1–2):67–73.

Keele C. 1970. Chemical causes of pain and itch. *Annual Review of Medicine* 21:67–74.

Klein A, Carstens MI, Carstens E. 2011. Facial injections of pruritogens or algogens elicit distinct behavior responses in rats and excite overlapping populations of primary sensory and trigeminal subnucleus caudalis neurons. *Journal of Neurophysiology* 106(3):1078–1088.

Koga K, Chen T, Li XY, Descalzi G, Ling J, Gu J, Zhuo M. 2011. Glutamate acts as a neurotransmitter for gastrin releasing peptide-sensitive and insensitive itch-related synaptic transmission in mammalian spinal cord. *Molecular Pain* 7:47.

Koltzenburg M, Handwerker HO, Torebjork HE. 1993. The ability of humans to localise noxious stimuli. *Neuroscience Letters* 150(2):219–222.

Kunkle EC, Chapman WP. 1943. Insensitivity to pain in man. *Proceedings of the Association for Research in Nervous and Mental Diseases* 23:100–109.

Lagerstrom MC, Rogoz K, Abrahamsen B, Persson E, Reinius B, Nordenankar K, Olund C et al. 2010. VGLUT2-dependent sensory neurons in the TRPV1 population regulate pain and itch. *Neuron* 68(3):529–542.

Li L, Rutlin M, Abraira V, Cassidy C, Kus L, Gong S, Jankowski M et al. 2011. The functional organization of cutaneous low-threshold mechanosensory neurons. *Cell* 147(7):1615–1627.

Liu Q, Sikand P, Ma C, Tang Z, Han L, Li Z, Sun S, LaMotte R, Dong X. 2012. Mechanisms of itch evoked by β-alanine. *Journal of Neuroscience* 32(42):14532–14537.

Liu Q, Tang Z, Surdenikova L, Kim S, Patel KN, Kim A, Ru F et al. 2009. Sensory neuron-specific GPCR Mrgprs are itch receptors mediating chloroquine-induced pruritus. *Cell* 139(7):1353–1365.

Liu XY, Liu ZC, Sun YG, Ross M, Kim S, Tsai FF, Li QF, Jeffry J, Kim JY, Loh HH, Chen ZF. 2011. Unidirectional cross-activation of GRPR by MOR1D uncouples itch and analgesia induced by opioids. *Cell* 147(2):447–458.

Liu Y, Abdel Samad O, Zhang L, Duan B, Tong Q, Lopes C, Ji RR, Lowell BB, Ma Q. 2010. VGLUT2-dependent glutamate release from nociceptors is required to sense pain and suppress itch. *Neuron* 68(3):543–556.

Lundeberg T, Bondesson L, Thomas M. 1987. Effect of acupuncture on experimentally induced itch. *British Journal of Dermatology* 117(6):771–777.

Ma Q. 2010. Labeled lines meet and talk: Population coding of somatic sensations. *Journal of Clinical Investigation* 120(11):3773–3778.

Mar L, Yang F-C, Ma Q. 2012. Genetic marking and characterization of Tac2-expressing neurons in the central and peripheral nervous system. *Molecular Brain* 5:3.

Maxwell DJ, Belle MD, Cheunsuang O, Stewart A, Morris R. 2007. Morphology of inhibitory and excitatory interneurons in superficial laminae of the rat dorsal horn. *Journal of Physiology* 584(Pt 2):521–533.

McCoy ES, Taylor-Blake B, Zylka MJ. 2012. CGRPalpha-expressing sensory neurons respond to stimuli that evoke sensations of pain and itch. *PloS One* 7(5):e36355.

Mishra S, Holzman S, Hoon M. 2012. A nociceptive signaling role for neuromedin B. *Journal of Neuroscience* 32(25):8686–8695.

Mishra SK, Hoon MA. 2013. The cells and circuitry for itch responses in mice. *Science* 340(6135):968–971.

Mochizuki H, Sadato N, Saito D, Toyoda H, Tashiro M, Okamura N, Yanai K. 2007. Neural correlates of perceptual difference between itching and pain: A human fMRI study. *NeuroImage* 36(3):706–717.

Mochizuki H, Tashiro M, Kano M, Sakurada Y, Itoh M, Yanai K. 2003. Imaging of central itch modulation in the human brain using positron emission tomography. *Pain* 105(1–2):339–346.

Namer B, Carr R, Johanek LM, Schmelz M, Handwerker HO, Ringkamp M. 2008. Separate peripheral pathways for pruritus in man. *Journal of Neurophysiology* 100(4):2062–2069.

Ossipov MH, Dussor GO, Porreca F. 2010. Central modulation of pain. *Journal of Clinical Investigation* 120(11):3779–3787.

Patel K, Liu Q, Meeker S, Undem B, Dong X. 2011. Pirt, a TRPV1 modulator, is required for histamine-dependent and -independent itch. *PloS One* 6(5):e20559.

Price D, Hayes R, Ruda M, Dubner R. 1978. Spatial and temporal transformations of input to spinothalamic tract neurons and their relation to somatic sensations. *Journal of Neurophysiology* 41(4):933–947.

Rau K, McIlwrath S, Wang H, Lawson J, Jankowski M, Zylka M, Anderson D, Koerber H. 2009. Mrgprd enhances excitability in specific populations of cutaneous murine polymodal nociceptors. *Journal of Neuroscience* 29(26):8612–8619.

Ringkamp M, Schepers R, Shimada S, Johanek L, Hartke T, Borzan J, Shim B, LaMotte R, Meyer R. 2011. A role for nociceptive, myelinated nerve fibers in itch sensation. *Journal of Neuroscience* 31(42):14841–14849.

Ross SE, Mardinly AR, McCord AE, Zurawski J, Cohen S, Jung C, Hu L et al. 2010. Loss of inhibitory interneurons in the dorsal spinal cord and elevated itch in Bhlhb5 mutant mice. *Neuron* 65(6):886–898.

Rothman S. 1941. Physiology of itching. *Physiological Reviews* 21(2):357–381.

Savin JA, Aoki T, Bielska CA. 1999. Some observations on the direction of scratch marks. *Journal of the American Academy of Dermatology* 41(6):1035–1039.

Schmelz M, Hilliges M, Schmidt R, Orstavik K, Vahlquist C, Weidner C, Handwerker HO, Torebjork HE. 2003a. Active "itch fibers" in chronic pruritus. *Neurology* 61(4):564–566.

Schmelz M, Schmidt R, Bickel A, Handwerker HO, Torebjork HE. 1997. Specific C-receptors for itch in human skin. *Journal of Neuroscience* 17(20):8003–8008.

Schmelz M, Schmidt R, Weidner C, Hilliges M, Torebjork H, Handwerker H. 2003b. Chemical response pattern of different classes of C-nociceptors to pruritogens and algogens. *Journal of Neurophysiology* 89(5):2441–2448.

Shelley W, Arthur R. 1957. The neurohistology and neurophysiology of the itch sensation in man. *AMA Archives of Dermatology* 76(3):296–323.

Shi T-JS, Liu S-XL, Hammarberg H, Watanabe M, Xu Z-QD, Hökfelt T. 2008. Phospholipase C{beta}3 in mouse and human dorsal root ganglia and spinal cord is a possible target for treatment of neuropathic pain. *Proceedings of the National Academy of Sciences of the United States of America* 105(50):20004–20008.

Shim WS, Tak MH, Lee MH, Kim M, Koo JY, Lee CH, Oh U. 2007. TRPV1 mediates histamine-induced itching via the activation of phospholipase A2 and 12-lipoxygenase. *Journal of Neuroscience* 27(9):2331–2337.

Sikand P, Shimada S, Green B, LaMotte R. 2011. Sensory responses to injection and punctate application of capsaicin and histamine to the skin. *Pain* 152(11):2485–2494.

Simone DA, Baumann TK, LaMotte RH. 1989. Dose-dependent pain and mechanical hyperalgesia in humans after intradermal injection of capsaicin. *Pain* 38(1):99–107.

Simone DA, Sorkin LS, Oh U, Chung JM, Owens C, LaMotte RH, Willis WD. 1991. Neurogenic hyperalgesia: Central neural correlates in responses of spinothalamic tract neurons. *Journal of Neurophysiology* 66(1):228–246.

Simone DA, Zhang X, Li J, Zhang JM, Honda CN, LaMotte RH, Giesler GJ, Jr. 2004. Comparison of responses of primate spinothalamic tract neurons to pruritic and algogenic stimuli. *Journal of Neurophysiology* 91(1):213–222.

Stander S, Schmelz M. 2006. Chronic itch and pain—Similarities and differences. *European Journal of Pain* 10(5):473–478.

Su P-Y, Ko M-C. 2011. The role of central gastrin-releasing peptide and neuromedin B receptors in the modulation of scratching behavior in rats. *Journal of Pharmacology and Experimental Therapeutics* 337(3):822–829.

Sun YG, Chen ZF. 2007. A gastrin-releasing peptide receptor mediates the itch sensation in the spinal cord. *Nature* 448(7154):700–703.

Sun YG, Zhao ZQ, Meng XL, Yin J, Liu XY, Chen ZF. 2009. Cellular basis of itch sensation. *Science* 325(5947):1531–1534.

Todd A. 2002. Anatomy of primary afferents and projection neurones in the rat spinal dorsal horn with particular emphasis on substance P and the neurokinin 1 receptor. *Experimental Physiology* 87(2):245–249.

Todd AJ, Hughes DI, Polgar E, Nagy GG, Mackie M, Ottersen OP, Maxwell DJ. 2003. The expression of vesicular glutamate transporters VGLUT1 and VGLUT2 in neurochemically defined axonal populations in the rat spinal cord with emphasis on the dorsal horn. *European Journal of Neuroscience* 17(1):13–27.

Tuckett R. 1982. Itch evoked by electrical stimulation of the skin. *Journal of Investigative Dermatology* 79(6):368–373.

Van Wimersma Greidanus T, Maigret C. 1991. Neuromedin-induced excessive grooming/ scratching behavior is suppressed by naloxone, neurotensin and a dopamine D1 receptor antagonist. *European Journal of Pharmacology* 209(1–2):57–61.

Von Frey M. 1922. Zur Physiologie der Juckempfindung. *Archives Neerlandaises de Physiologie de l'Homme et des Animaux* 7:142–145.

Wang H, Papoiu A, Coghill R, Patel T, Wang N, Yosipovitch G. 2010. Ethnic differences in pain, itch and thermal detection in response to topical capsaicin: African Americans display a notably limited hyperalgesia and neurogenic inflammation. *British Journal of Dermatology* 162(5):1023–1029.

Wang H, Zylka M. 2009. Mrgprd-expressing polymodal nociceptive neurons innervate most known classes of substantia gelatinosa neurons. *Journal of Neuroscience* 29(42):13202–13209.

Wang X, Zhang J, Eberhart D, Urban R, Meda K, Solorzano C, Yamanaka H, Rice D, Basbaum AI. 2013. Excitatory superficial dorsal horn interneurons are functionally heterogeneous and required for the full behavioral expression of pain and itch. *Neuron* 78(2):312–324.

Wilson SR, Gerhold KA, Bifolck-Fisher A, Liu Q, Patel KN, Dong X, Bautista DM. 2011. TRPA1 is required for histamine-independent, Mas-related G protein-coupled receptor-mediated itch. *Nature Neuroscience* 14(5):595–602.

Xie W, Samoriski G, McLaughlin J, Romoser V, Smrcka A, Hinkle P, Bidlack J, Gross R, Jiang H, Wu D. 1999. Genetic alteration of phospholipase C beta3 expression modulates behavioral and cellular responses to mu opioids. *Proceedings of the National Academy of Sciences of the United States of America* 96(18):10385–10390.

Zylka M, Rice F, Anderson D. 2005. Topographically distinct epidermal nociceptive circuits revealed by axonal tracers targeted to Mrgprd. *Neuron* 45(1):17–25.

20 Spinal Microcircuits and the Regulation of Itch

Sarah E. Ross, Junichi Hachisuka,
and Andrew J. Todd

CONTENTS

Itch is a somatosensory percept that is triggered by irritants at the skin's surface. However, the manner in which itch is coded in the nervous system remains almost completely unknown. Recent work has uncovered a key role of spinal interneurons in the modulation of itch. Here we discuss these recent discoveries in the context of our understanding of spinal microcircuitry, highlighting the possible roles of the dorsal horn in the processing of pruritic input.

While it is not known which specific subsets of primary afferents underlie itch, there is good evidence that the main receiving zone for these afferents is within laminae I and II of the spinal cord. For instance, itch sensation is only lost when the conduction of all fibers (including C-fibers) is blocked, implying that itch is mediated in large part by fine diameter fibers, which are known terminate in superficial laminae. In particular, many of these itch-mediating C-fibers are likely to be sensory afferents that express TrpV1 and/or TrpA1, and the primary afferents that express these channels have synaptic connections with lamina I and lamina II neurons (Yang et al. 1998; Nakatsuka et al. 2002; Kosugi et al. 2007; Shim et al. 2007; Imamachi et al. 2009; Uta et al. 2010; Patel et al. 2011; Wilson et al. 2011). As further evidence, neurons in the superficial dorsal horn, show fos induction upon intradermal injection of itch-evoking chemicals such as serotonin and SLIGRL as well as in a dry skin model of pruritus (Nojima et al. 2003, 2004; Akiyama et al. 2009a,b). Last, *in vivo* responses to various pruritogens such as histamine, serotonin, SLIGRL, and chloroquine, as well as the response to mosquito allergy have been recorded from neurons in the superficial dorsal horn (Jinks and Carstens 2000, 2002; Akiyama et al. 2009a,b; Omori et al. 2009; Akiyama et al. 2012c). Together these findings suggest that neurons in the superficial dorsal horn receive the somatosensory input that gives rise to itch sensation.

Laminae I and II of the spinal cord contain numerous functional populations of neurons. However, projection neurons that convey information to the brain represent around one percent of the total number of neurons in this region (Todd 2010). Thus, the vast majority of neurons in the superficial dorsal horn are interneurons with local (and in some cases also long propriospinal) axonal projections, and these are involved in the processing of sensory information. Does the existence of a complex network of spinal interneurons imply that itch is decoded within the spinal cord? To what extent is itch modulated in the spinal cord? While we do not yet know the answers to these key questions, there are a number of important psychophysical experiments that are suggestive of the idea that itch is modulated by spinal microcircuits.

20.1 PUTATIVE SPINAL MICROCIRCUITS INVOLVED IN THE MODULATION OF ITCH

20.1.1 INHIBITION OF ITCH BY COUNTERSTIMULI

The idea that counterstimuli can relieve itch is familiar to anyone who has ever found himself scratching in response to a mosquito bite. In experimental settings, stimuli that have been shown to block histamine-induced itch include scratching, noxious heat and cold, pinprick, menthol, capsaicin, mustard oil, and cutaneous field stimulation (Bickford 1937; Graham et al. 1951; Bromm et al. 1995; Ward et al. 1996; Nilsson et al. 1997; Yosipovitch et al. 2007). What these stimuli have in common is that they robustly activate subsets of C-fibers. This idea has led to the suggestion that certain types of C-fiber input may inhibit itch. Importantly, counterstimuli can inhibit itch when delivered quite a significant distance from the itch itself—10 cm or more away (Bickford 1937). The finding that the inhibition of itch by counterstimuli occurs at such a large distance (larger than the receptive field of an individual primary afferent) suggests that this inhibition occurs through the integration of sensory input, potentially by local circuits at the level of the spinal cord.

Because all C-fibers release the excitatory neurotransmitter glutamate, the simplest explanation for how activity within excitatory C-fibers could result in the inhibition of itch within the spinal cord is via inhibitory spinal interneurons. According to this model, the activation of subsets of C-fibers in response to a counterstimulus such as scratching would not only activate scratch-sensitive (nociceptive) projection neurons but also activate inhibitory spinal interneurons that function to inhibit pruritoception and thereby reduce the perception of itch (Figure 20.1a).

In support of this idea, when Davidson et al. (2009) recorded from spinothalamic projection neurons, they found that scratching the skin resulted in the inhibition of histamine-evoked responses. Furthermore, scratching at a distance from the site of itch suppressed the firing rate of itch-responsive neurons in the superficial dorsal horn, indicating that mechanical inhibition of itch is evoked in an area that is wider

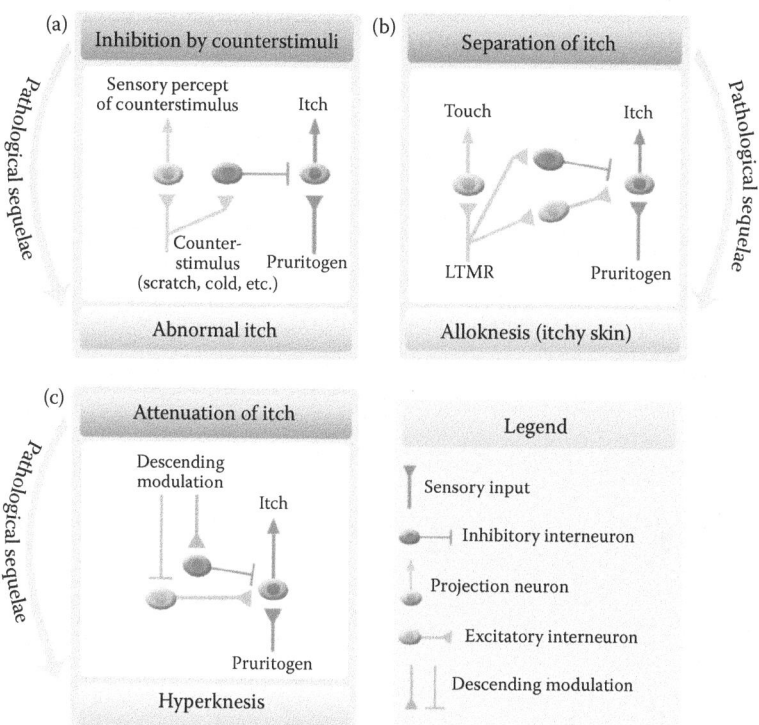

FIGURE 20.1 **(See color insert.)** Putative spinal microcircuits involved in the modulation of itch. (a) Counterstimuli such as scratching and cold may inhibit itch by activating inhibitory interneurons (purple) that function to inhibit itch within the spinal cord. (b) Though itch and touch are normally separate, there may be spinal neural circuits that allow the coupling of these sensations. Under normal circumstances, inhibition (purple) may predominate. However, following strong pruritogenic input, the synaptic connections may be reweighted such that excitation (green) predominates and the activation of low threshold mechanoreceptors (LTMRs) results in alloknesis. (c) There are likely to be multiple mechanisms that are involved in the attenuation of itch. Some of these likely include descending modulation that may act directly on projection neurons (not shown) or via excitatory (green) or inhibitory (purple) interneurons in the spinal cord to inhibit itch and prevent hyperknesis. These conceptual models are based in part on ideas described in a review by Sandhulker (2009) on the role of inhibition within the nociceptive system.

than the site of the itch itself (Akiyama et al. 2012b). Electrophysiological data also suggest that inhibitory input is widely distributed (Kato et al. 2004, 2011). *In vivo* patch clamp recording of neurons in the dorsal horn reveals that mechanical stimulation of the skin evokes IPSCs (Narikawa et al. 2000). Together these studies hinted at possible circuitry underlying the inhibition of itch by counterstimuli such as scratching, but the specific neurons, mediators, and pathways involved remained unclear.

Recently, we identified a subset of spinal inhibitory neurons that may be involved in mediating the inhibition of itch. This work began when we discovered that mice lacking the transcription factor *Bhlhb5* develop self-inflicted skin lesions due to excessive licking and scratching (Ross et al. 2010). Because this behavior was suggestive of abnormal itch, we tested the response of these mice to pruritogens and found that *Bhlhb5* mutant mice showed significantly elevated scratching responses compared to littermate controls.

To understand which neurons are responsible for this elevated itch phenotype, we generated a conditional *Bhlhb5* knockout mouse, allowing us to selectively ablate *Bhlhb5* in specific regions of the nervous system. When we ablated *Bhlhb5* in either the forebrain or in primary sensory neurons, the resulting mice were completely normal with respect to itch. However, we found that loss of *Bhlhb5* within inhibitory neurons of the spinal cord was sufficient for elevated itch (Ross et al. 2010). Finally, to investigate what happens to these spinal interneurons in the absence of *Bhlhb5*, we generated a *Bhlhb5*-cre knockout mouse, thereby allowing us to follow the fate of *Bhlhb5*-expressing cells throughout the life of the animal. Using this approach, we discovered that *Bhlhb5* is selectively required for the survival of specific neurons in the dorsal horn—without *Bhlhb5*, these neurons undergo apoptosis during embryonic development. Together, these findings suggest that *Bhlhb5* is required for the survival of a subset of spinal inhibitory interneurons that function to inhibit itch (Ross et al. 2010). We and others have speculated that these neurons might be involved in mediating the inhibition of itch by counterstimuli (Ross 2011; Ma 2010; Lagerstrom et al. 2010; Liu et al. 2010; Patel and Dong 2010). However, while this remains an attractive hypothesis, there is as of yet no direct evidence for this idea.

In addition to the possible role of a subset of *Bhlhb5*-expressing inhibitory interneurons in the inhibition of itch by counterstimuli, there are likely to be numerous other mechanisms working in parallel. Evidence for this idea comes from Akiyama et al. (2011) who performed *in vivo* recordings from spontaneously active neurons in the superficial dorsal horn in a dry skin model of itch. Dry skin is known to drive spontaneous activity in itch-responsive primary sensory fibers, which would presumably provide tonic drive to itch circuits within the cord. Because neurons in the dorsal horn are normally silent in the absence of a stimulus, the spontaneously active neurons in the dorsal horn of mice with dry skin are likely to be involved in mediating itch. In the mouse model, cutaneous scratching inhibited these spontaneously active neurons, consistent with the idea that these neurons may be involved in integrating input to mediate the inhibition of itch by scratching. Importantly, scratch-evoked inhibition of the spontaneous activity of such neurons was reduced upon blockade of either GABA or glycine receptors, implicating both types of inhibitory neurotransmitter in the inhibition of itch. Furthermore, cervical transection to block descending inhibition also attenuated scratch-evoked inhibition. These findings suggest that multiple mechanisms are involved in the inhibition of itch.

When we scratch an itch, the relief is almost instantaneous, implying that neurotransmitters acting on a timescale of milliseconds are likely involved. However, there is also strong evidence that fast-acting inhibitory neurotransmission is not the only mechanism at play. Lewis et al. (1927) found that, following the electrical stimulation of the skin, histamine was no longer able to elicit itch. This state, which Bickford (1937) termed an antipruritic state, can be achieved when the counterstimulus is delivered either before or after the pruritogen and can last for minutes or even hours (Graham et al. 1951). For instance, brief noxious heating of the skin caused a 50% decrease in perceived itch that lasted more than 30 min (Ward et al. 1996). Analogously, cutaneous field stimulation, which causes a burning and pricking sensation, abolishes itch for up to four hours (Nilsson et al. 1997). These long-lasting effects suggest that, in addition to fast-acting neurotransmitters, there may also be neuromodulatory mechanisms involved in the inhibition of itch by counterstimuli. Though little is known about the neuromodulators of itch, there is good evidence that the mu opioid receptor is involved. In particular, morphine-induced scratching was recently shown to be mediated by a specific isoform of the mu opioid receptor that interacts functionally with gastrin-releasing peptide receptor (GRPR) (Liu et al. 2011).

20.1.2 SEPARATION OF ITCH

In one of the early studies on the psychophysical properties of itch, Bickford (1937) noted that the skin surrounding a site of itching (caused by a gnat bite) was abnormally sensitive to rubbing. Indeed, rubbing of the surrounding skin elicited abnormal itch. This common phenomenon is often referred to as itchy skin, and it has been studied in the context of both histamine- and cowhage-induced itch (Graham et al. 1951; Simone et al. 1991). Because itchy skin is not the norm but rather a state that can be elicited under specific circumstances, this implies that touch and itch are normally subserved by largely separate neural circuits that have the capacity to influence each other (Figure 20.1b).

Important insight into these neural circuits came from Simone et al. (1991), who performed a quantitative assessment of itchy skin and noted commonalities between this phenomenon and allodynia. In light of this similarity, they proposed the term alloknesis to describe itch in response to innocuous mechanical stimulation. One key implication of this idea is that allodynia and alloknesis may be mediated by parallel neural circuits. Indeed, when an aversive stimulus causes pain, the pain is accompanied by allodynia, whereas when an aversive stimulus causes itch, the itch is accompanied by alloknesis. Thus, perhaps the strong activation of pruritoceptors causes alloknesis just as strong activation of nociceptors causes allodynia.

What types of primary afferents mediate alloknesis? Handwerker (1992) proposed that light-touch evoked itch was likely mediated by A-delta fibers. However, the observation by Bickford (1937) that itchy skin is only *completely* lost at the same time that pain is lost implies that some C-fibers must also be capable of eliciting alloknesis. In this regard, C-low threshold mechanoreceptors may be a good candidate.

Evidence for the existence of spinal circuits that mediate alloknesis comes from *in vivo* recordings of spinothalamic neurons. When Andrew and Craig (2001) investigated the response properties of putative itch-mediating projection neurons, they found that those that responded to histamine application were not responsive to mechanical stimuli, at least not initially. However, once the skin was treated with histamine, these neurons became sensitized such that light mechanical stimulation was now sufficient to trigger a response. This finding suggests that alloknesis is mediated by neural circuits in the spinal cord.

Though the neural basis of alloknesis is poorly understood, some of the underlying mechanisms of its nociceptive counterpart have been elucidated, and this insight into the mechanisms of allodynia may be revealing. Noxious and innocuous inputs are mainly conveyed to distinct laminae, but it is clear that polysynaptic connections exist that connect low threshold input in deeper laminae to noxious input in superficial laminae (Takazawa and MacDermott 2010). Although the transmission across this polysynaptic circuit is normally silenced by inhibitory interneurons, sustained noxious inputs are thought to cause allodynia through the release of this inhibition. Recently, one of the underlying mechanisms for this loss of inhibition has been elucidated in detail. Specifically, when noxious C-fiber input into the dorsal horn is abnormally sustained, this strong activity in nociceptors results in the release of endocannabinoids from the postsynaptic target, which act upon nearby inhibitory neurons to inhibit neurotransmitter release (Pernia-Andrade et al. 2009). As a consequence, endocannabinoid-mediated disinhibition allows low threshold input to activate pain circuits in the superficial dorsal horn, resulting in allodynia. Whether endocannabinoids are likewise involved in mediating alloknesis is not known.

Recently, Akiyama et al. (2012a) developed a mouse model of alloknesis. In this model, following the intradermal injection of a pruritogen, nearby mechanical stimulation with fine von Frey filaments elicits a scratching response. Importantly, this effect persists for hours, long after scratching due to the pruritogen itself has passed. This mouse model is an important step forward because it will allow us to investigate the precise neural mechanisms underlying alloknesis.

20.1.3 ATTENUATION OF ITCH

Drawing on the parallels between itch and pain, there are likely to be multiple mechanisms that can modulate itch, ensuring the proper response level to a pruritogen. Some of these mechanisms are likely to be mediated in part by descending pathways that may act to limit itch (Figure 20.1c). This idea is supported by experiments in which the electrical stimulation of the periaqueductal gray (a key region for descending control) was found to suppress the activity of neurons in the dorsal horn to an intradermal injection of histamine (Carstens 1997). Furthermore, in a dry skin model, cervical transection reduced the effect of scratching on activity of spontaneously active neurons in the dorsal horn (Akiyama et al. 2011). These observations suggest that descending input from supraspinal areas such as midbrain periaqueductal gray and the rostral ventromedial medulla may inhibit pruritoception, just as they inhibit nociception (Yoshimura and Furue 2006; Heinricher et al. 2009).

One of the key mediators of this descending inhibition is likely to be noradrenaline acting through α_2-adrenoceptors. There is anecdotal evidence that postherpetic itch can be relieved by intrathecal delivery of the α_2-agonist, clonidine (Elkersh et al. 2003). Likewise in mice, both serotonin and mosquito allergy mediated itch were inhibited by intrathecal clonidine, whereas intracisternal injection had no effect (Gotoh et al. 2011a,b,c). Furthermore, these effects were blocked by yohimbine, an α_2-adrenoceptor antagonist. These studies suggest that spinal neurons are the target of tonic inhibition via a descending noradrenergic system through stimulation of α_2-adrenoceptors. Noradrenaline is known to act both at presynaptic sites and postsynaptic sites (Sonohata et al. 2004), but the specific targets of its antipruritic effect are not known.

20.2 ROLE OF GRPR NEURONS AND ITCH

One of the most exciting recent discoveries in the field of itch is the key role of a specific population of neurons—those that express GRPR. When GRPR-expressing neurons were ablated using a bombesin-saporin conjugate, the scratching responses to a variety of pruritogens were almost completely blocked, whereas the responses to painful stimuli were normal (Sun et al. 2009). This important finding suggests that GRPR neurons are absolutely required for itch; yet little else is known about them.

Within the dorsal horn, GRPR-expressing neurons are located predominantly within lamina I, which contains the majority of projection neurons. However, GRPR-expressing neurons are likely distinct from projection neurons because the expression of neurokinin 1 receptor (NK1r) (which is present on ~80% of anterolateral tract [ALT] neurons) was not affected by bombesin-saporin (Sun et al. 2009). Furthermore, projection neurons tend to have large cells bodies, whereas GRPR-expressing neurons appear to be small in size. Thus, GRPR-expressing neurons are likely to be a population of (presumably excitatory) spinal interneurons. If so, this raises the interesting possibility that itch input is relayed through a class of spinal interneuron before being transmitted through projection cells to the brain (Figure 20.2).

In addition to the essential role of GRPR-expressing neurons, it appears that the receptor itself is required for normal itch. Mice with a constitutive loss of GRPR show a significant reduction in their response to SLIGRL, chloroquine, and compound 48/80 (Sun and Chen 2007). Furthermore, the abnormal itch that is seen upon loss of VGLUT2 in subsets of primary sensory neurons is rescued in compound mutant mice that are also lacking GRPR (Lagerstrom et al. 2010). This genetic evidence is complemented by pharmacological manipulations showing that GRPR agonists cause scratching, whereas treatment with a GRPR antagonist reduces (but does not eliminate) pruritogen-evoked responses (Sun and Chen 2007).

Intriguingly, GRPR appears to play a greater role for histamine-independent itch than histamine-dependent itch. In fact, GRPR mutant mice show completely normal itch behavior in response to histamine (see supplemental data in Sun et al. [2009]). Furthermore, while bombesin-saporin significantly lowered the number of fos-positive cells in dorsal horn after stimulation with histamine and chloroquine, the reduction of fos expression was greater for chloroquine than for histamine (Han et al. 2012). Finally, systemic or intrathecal injection of a GRPR antagonist attenuated itch evoked by chloroquine and SLIGRL, but not that evoked by BAM8-22 or histamine (Akiyama et al. 2012c). These findings imply that GRPR is not necessarily involved in all types of itch. Intriguingly, neuromedin B and bombesin (two peptides that are related to gastrin-releasing peptide [GRP]) also cause itch and, importantly, their effects appear to be mediated by the neuromedin B receptor (NMBR) rather than GRPR (Su and Ko 2011). Thus, it is possible that different bombesin-like peptides are involved in distinct types of itch.

The key role of GRPR signaling in certain types of itch, together with the apparent expression of GRP in subsets of sensory neurons, raised the possibility that GRP released from primary afferents might mediate itch (Sun et al. 2009). However, this interpretation remains a matter of controversy. Part of this controversy is based on experiments using *in situ* hybridization, which show that spinal interneurons (rather

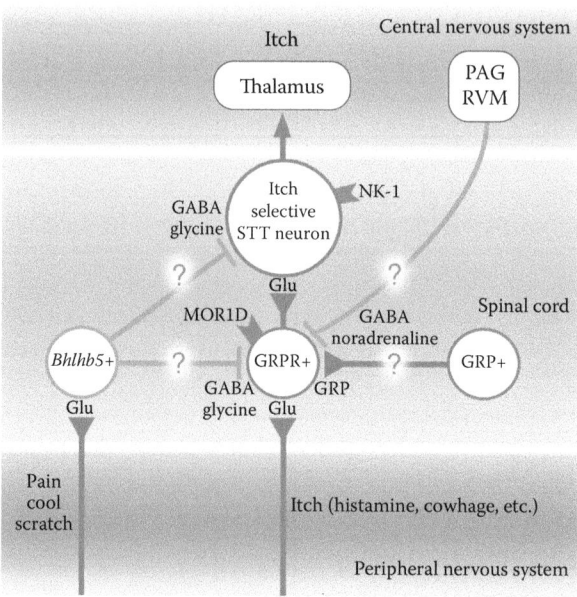

FIGURE 20.2 (**See color insert.**) Speculative model of the spinal microcircuits of itch. Itch appears to be mediated at least in part by an unknown subset of NK-1-expressing, itch selective spinothalamic tract (STT) neurons. GRPR-expressing neurons, which are likely interneurons, also appear to be required for itch and might therefore be involved in the transmission of itch between primary sensory neurons and projection neurons. Though the origin of GRP that activates GRPR remains controversial, one possibility is that spinal interneurons are an important source (shown); alternatively, GRP may come from primary sensory afferents that also release glutamate (not shown). *Bhlhb5* is transiently expressed in a subset of inhibitory interneurons that appear to inhibit itch, though their specific target is unknown. These inhibitory neurons are well positioned to mediate the inhibition of itch by counterstimuli, as illustrated.

than primary sensory neurons) are the major source of GRP mRNA (Fleming et al. 2012). Moreover, the idea that GRP released from sensory neurons mediates itch has recently been challenged by a new study that performed patch clamp recordings from neurons in the superficial dorsal horn that respond to GRP. Using this approach, Koga et al. (2011) showed that activation of GRP-responsive neurons by primary afferents is mediated by glutamate, not GRP. Thus, while there is abundant evidence that GRPR plays a key role in pruritus, more research is required to settle the controversy of how GRP and its receptor fit within the neural circuits of itch.

20.3 BRIEF OVERVIEW OF NEURONAL ORGANIZATION AND CIRCUITRY WITHIN THE DORSAL HORN

While the exact nature of spinal circuits that modulate itch remains a matter of speculation, it is clear that itch input is processed by a highly complex network of neurons in the dorsal horn. The main limitation to our understanding of the circuitry has been the difficulty of defining functional populations among the diverse array of dorsal horn neurons. Without a satisfactory scheme for classifying these cells, it is impossible to define their inputs and outputs, and therefore their functions within these circuits. However, recent work has begun to shed light on some aspects of the neuronal organization and circuitry.

The dorsal horn consists of a continuous column of gray matter on either side of the midline that extends throughout the length of the spinal cord and merges with the spinal trigeminal nucleus in the medulla. The neural circuits in this region are made up from four basic components: (1) primary afferent axons which provide sensory (including pruritic) input; (2) axons descending from the brain that can modulate transmission of this input; (3) projection neurons, with axons that travel to the brain and represent the major output from the region; and (4) interneurons, which form the great majority of dorsal horn neurons and have axons that remain within the spinal cord (Todd 2010).

20.3.1 PROJECTION NEURONS

The main ascending pathway that is responsible for perception of itch is the ALT. As evidence of this idea, severing of the ALT results in the complete loss of itch sensation (as well as pain). This tract consists of neurons that are concentrated in lamina I and scattered throughout the deeper laminae (III-VI) of the dorsal horn, and it is not yet clear which of these projection neurons are involved in itch sensation. However, the finding of lamina I ALT neurons that respond to pruritic stimuli (Andrew and Craig 2001; Davidson et al. 2009) implicates a subset of ALT cells in this lamina.

Axons in the ALT cross the midline and ascend in the ventrolateral white matter to reach several brain regions, including the thalamus, periaqueductal gray matter, lateral parabrachial area, and various nuclei within the medulla. Individual neurons within this pathway typically send their axons to several of these brain targets (Todd 2010). Many of the lamina I ALT neurons also give rise to local axon collaterals within the spinal cord, although the postsynaptic targets of these are not yet known (Szucs et al. 2010).

The majority of ALT neurons in lamina I, together with a distinctive population of large ALT cells in lamina III, express the NK1r, the main receptor for substance P (Todd et al. 2000). Both types are densely innervated by substance P-containing primary afferents (Naim et al. 1997; Todd et al. 2002), which provide approximately half of the excitatory synapses on these cells (Baseer et al. 2012; Polgar et al. 2010). The lamina III cells also receive a few synapses from myelinated low-threshold mechanoreceptive (LTM) primary afferents (Naim et al. 1998). The importance of these NK1r-expressing projection neurons has been demonstrated by selectively ablating them *in vivo* with intrathecal injections of substance P conjugated to saporin, which prevents the development of hyperalgesia (Mantyh et al. 1997) and may also lead to the loss of some forms of itch (Carstens et al. 2010). Although NK1rs are also expressed by many excitatory interneurons in the superficial laminae, this is generally at a much lower level than that seen on the projection neurons (Al Ghamdi et al. 2009), and it is therefore likely that the effects of substance P-saporin result from loss of NK1r-expressing ALT cells in laminae I and III.

Not all lamina I ALT cells express the NK1r, and among those that do not, we have identified a population of giant cells that are characterized by their very high density of synapses from both inhibitory and excitatory interneurons, but little if any primary afferent input (Polgar et al. 2008; Puskar et al. 2001). Whether this intriguing population plays a role in itch is not yet known.

20.3.2 INTERNEURONS

Within the lumbar enlargement, virtually all of the neurons in lamina II and around 95% of those in lamina I have axons that remain in the spinal cord and are therefore classified as interneurons. However, although these cells all give rise to local axonal arbors (Grudt and Perl 2002; Maxwell et al. 2007; Yasaka et al. 2007), many of them also have axons that travel for several segments within the spinal cord (Bice and Beal 1997a,b). Many small neurons in laminae I-II in the cervical enlargement can be retrogradely labeled from the medulla (Lima et al. 1991), and these may correspond to the lumbar interneurons with long propriospinal axons. Interneurons in the superficial dorsal horn are therefore not only involved in local circuits. However, the functional significance of these long intersegmental pathways in itch is still poorly understood.

Interneurons can be divided into two broad classes: inhibitory cells that use GABA and/or glycine as their fast transmitter, and excitatory (glutamatergic) neurons. Immunocytochemical studies suggest that 25% of the neurons in lamina I and 30% of those in lamina II are GABAergic inhibitory interneurons, with some using glycine as a cotransmitter (Polgar et al. 2003; Todd and Sullivan 1990). However, even though glycine and GABA are likely to be coreleased at synapses formed by these cells, the selective distribution of postsynaptic receptors may mean that some of their synapses are purely glycinergic (Keller et al. 2001). Although there are no reliable immunocytochemical markers for the cell bodies of glutamatergic interneurons, their axons express vesicular glutamate transporter 2 (VGLUT2), which can be used to identify these cells in combined electrophysiological/anatomical studies (Maxwell et al. 2007; Yasaka et al. 2010). Because projection neurons make up ~5% of the neurons in lamina I, but are largely absent from lamina II, excitatory interneurons presumably account for around 70% of the neurons in both laminae.

Both excitatory and inhibitory interneuron classes are made up from several functionally distinct populations. For example, it has been suggested that inhibitory interneurons have a number of different antinociceptive roles (Sandkuhler 2009) and also mediate the inhibition of itch by noxious counterstimulation (Ross et al. 2010). These tasks are probably performed by different cell types. In addition, while most inhibitory interneurons mediate postsynaptic inhibition through axodendritic or axosomatic synapses on dorsal horn neurons, some presynaptically inhibit primary afferents via axo-axonic synapses (Hughes et al. 2012). Considerable effort has therefore gone into trying to identify interneuron populations within the superficial dorsal horn (Graham et al. 2007; Todd 2010).

The most widely accepted classification scheme is that developed by Grudt and Perl (2002), who combined an electrophysiological recording of lamina II neurons in spinal cord slices from hamsters with anatomical reconstruction of their dendritic and axonal trees. They identified four main classes: islet cells, with dendrites that were highly elongated in the rostrocaudal axis, central cells (which resembled islet cells but had far shorter dendritic trees), vertical cells, with a dorsally located soma and ventrally directed dendrites, and radial cells, with short radiating dendrites (Figure 20.3). Subsequent studies have identified cells with similar morphology in other species (Heinke et al. 2004; Maxwell et al. 2007; Yasaka et al. 2007, 2010; Wang and Zylka 2009), but a major limitation of this scheme is that in most of these studies many of the recorded cells (typically ~30%) did not fit into any of these classes.

An alternative approach has been to identify cells that show various firing patterns during injection of depolarizing current pulses (Heinke et al. 2004; Maxwell et al. 2007; Wang and Zylka 2009; Yasaka et al. 2007, 2010). These patterns include tonic, initial bursting, single-spike, reluctant, delayed, and gap-firing and depend on the expression of specific ion channels (Graham et al. 2007). Recent studies have related morphological and physiological properties of lamina II interneurons to their neurotransmitter phenotype (Maxwell et al. 2007; Yasaka et al. 2010). This approach has shown that all islet cells are inhibitory, while radial cells and most vertical cells are excitatory, and this is consistent with physiological results obtained from paired recordings from neurons in spinal cord slices (Lu and Perl 2003, 2005). However, both inhibitory and excitatory interneuron classes include some central cells as well as other cells that are morphologically diverse. Interestingly, the reluctant, delayed and gap firing patterns, which are associated with A-type potassium (I_A) currents, are largely restricted to excitatory interneurons (Yasaka et al. 2010).

Another way of classifying interneurons involves using the rich array of neurochemical markers (neuropeptides, receptors, and other proteins) that are expressed in the superficial dorsal horn (Table 20.1). Some of these are restricted to either inhibitory or excitatory interneurons, while others are found among both types (Todd and Koerber 2005). It is important to note that a single marker may not define

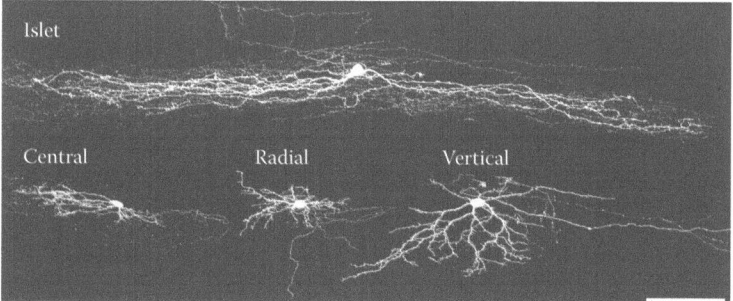

FIGURE 20.3 Morphological types of lamina II interneurons. Confocal images of four lamina II interneurons labeled with neurobiotin following whole-cell patch-clamp recording in a sagittal slice from an adult rat (Yasaka et al. 2010). These correspond to the four main morphological classes identified by Grudt and Perl (2002). The dendritic trees of islet cells are highly elongated along the rostrocaudal axis, with little medio-lateral or dorso-ventral spread. Central cells are similar, but with much smaller dendritic trees. Radial cells have short radiating dendrites. Vertical cells have a dendritic tree that fans ventrally from a dorsally located cell body.

TABLE 20.1

Expression of Various Neuropeptides, Receptors, and Other Proteins by Interneurons in the Superficial Dorsal Horn

	Inhibitory (GABAergic)	Excitatory (Glutamatergic)
Neuropeptides	NPY	Somatostatin
	Galanin	Neurotensin
	Enkephalin	Neurokinin B
	Dynorphin	Substance P
		Enkephalin
		Dynorphin
Neuropeptide receptors	sst_{2A}	NK1
	NK3	NK3
		MOR-1
		NPY-Y1
Calcium binding proteins	Parvalbumin	Calbindin
		Calretinin
Other proteins	nNOS	nNOS
	ChAT	PKCγ

Note: ChAT, choline acetyltransferase; MOR-1, μ-opioid receptor 1; NK1, neuro-kinin 1; NK3 neurokinin 3; nNOS, neuronal form of nitric oxide synthase; NPY, neuropeptide Y; PKC-γ, protein kinase C-γ; sst_{2A}, somatostatin receptor 2A.

a specific population, and that combinations of markers may be more useful. For example, dynorphin is expressed by both excitatory and inhibitory interneurons, and the inhibitory cells also contain galanin (Sardella et al. 2011; Brohl et al. 2008). The presence of galanin can therefore be used to distinguish these two populations. We have recently shown that neuropeptide Y (NPY), galanin, neuronal nitric oxide synthase (nNOS), and parvalbumin are present in nonoverlapping populations of inhibitory interneurons, and that between them these account for around half of the inhibitory interneurons in laminae I-II in the rat (Sardella et al. 2011; Tiong et al. 2011). We also know that there are differences between the postsynaptic targets for these different populations. Specifically, some NPY neurons preferentially target the NK1r-expressing ALT cells in lamina III (Polgar et al. 1999; Polgar et al. 2011), nNOS-containing inhibitory interneurons innervate the giant lamina I projection neurons (Puskar et al. 2001), while axons of the parvalbumin cells form axoaxonic synapses on the central terminals of myelinated LTM afferent (Hughes et al. 2012). The parvalbumin neurons correspond to a subset of islet cells (Antal et al. 1990) and because they are also innervated by low-threshold afferents (Hughes et al. 2012), they are presumably involved in regulation of tactile input. Because most (if not all) ALT neurons in laminae I and III respond to noxious stimulation, NPY and nNOS cells are likely to be involved in limiting the transmission of pain signals through the dorsal horn.

There is also evidence that excitatory interneurons have specific postsynaptic targets. We have recently shown that over half of the input from excitatory interneurons to lamina III NK1r+ ALT cells comes from the neurons that express dynorphin (Baseer et al. 2012), while Lu and Perl (2005) have shown that lamina II vertical cells, which are innervated by small myelinated (Aδ) primary afferents, are presynaptic to NK1r-expressing lamina I projection neurons. These circuits are illustrated in Figure 20.4. This circuit diagram is clearly far from complete, since interneurons are shown as having only a single type of postsynaptic target, while in reality they will presumably innervate a number of different neuronal types. Also, several classes of neuron are not shown in this diagram because we know little about their synaptic connections. However, it reflects the fact that neuronal circuits in the dorsal

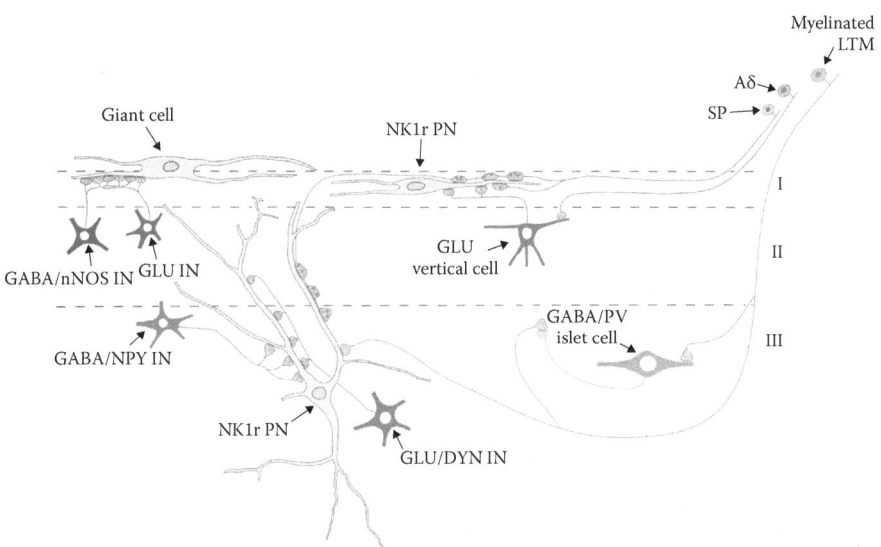

FIGURE 20.4 **(See color insert.)** A diagram showing some of the synaptic connections that have been identified in the rodent superficial dorsal horn. Three projection neurons are shown in grey. One of these is a lamina I giant cell, while the other two are NK1r-expressing projection neurons (PN) with their cell bodies in laminae I and III. Note that the lamina III cell has prominent dorsal dendrites that enter the superficial laminae. Both types of NK1r-expressing PN are densely innervated by substance P-containing primary afferents, while the lamina III cells also receive an input from myelinated low threshold mechanoreceptive (LTM) afferents. The giant cells receive dense synaptic input from GABAergic inhibitory interneurons, many of which contain neuronal nitric oxide synthase (GABA/nNOS), and from glutamatergic excitatory interneurons of unknown origin (GLU IN). The lamina III NK1r PN is densely innervated by NPY-containing GABAergic interneurons and dynorphin (DYN)-containing glutamatergic interneurons. Glutamatergic vertical cells, which receive monosynaptic input from small myelinated (Aδ) afferents are presynaptic to the lamina I NK1r PNs. Parvalbumin-containing GABAergic islet cells (GABA/PV) receive direct input from myelinated LTM afferents and their axons form axoaxonic synapses with the same type of afferent. For further information, see text.

horn do appear to be arranged in a selective (although highly complex) way, and it provides a basis from which a more complete picture will emerge.

Although we are just at the early stages of identifying and classifying the distinct neuronal populations found within the dorsal horn and understanding their function, progress is being made at a rapid pace. The emerging picture is that itch is likely to be modulated by numerous types of spinal interneurons, which have well-organized patterns of input and output, thus forming highly specialized circuits. Delineating these spinal microcircuits is an important first step in understanding how itch is coded and may eventually lead to therapies that provide some much needed relief for chronic itch.

REFERENCES

Akiyama, T., M. I. Carstens, and E. Carstens. 2011. Transmitters and pathways mediating inhibition of spinal itch-signaling neurons by scratching and other counterstimuli. *PloS One* 6 (7):e22665.

Akiyama, T., M. I. Carstens, A. Ikoma, F. Cevikbas, M. Steinhoff, and E. Carstens. 2012a. Mouse model of touch-evoked itch (alloknesis). *Journal of Investigative Dermatology* 132 (7):1886–91.

Akiyama, T., A. W. Merrill, M. I. Carstens, and E. Carstens. 2009a. Activation of superficial dorsal horn neurons in the mouse by a PAR-2 agonist and 5-HT: Potential role in itch. *Journal of Neuroscience* 29 (20):6691–9.

Akiyama, T., A. W. Merrill, K. Zanotto, M. I. Carstens, and E. Carstens. 2009b. Scratching behavior and Fos expression in superficial dorsal horn elicited by protease-activated receptor agonists and other itch mediators in mice. *Journal of Pharmacology and Experimental Therapeutics* 329 (3):945–51.

Akiyama, T., M. Tominaga, M. I. Carstens, and E. E. Carstens. 2012b. Site-dependent and state-dependent inhibition of pruritogen-responsive spinal neurons by scratching. *European Journal of Neuroscience* 36 (3):2311–6.

Akiyama, T., M. Tominaga, A. Davoodi, M. Nagamine, K. Blansit, A. Horwitz, M. I. Carstens, and E. Carstens. 2012c. Roles for substance P and gastrin releasing peptide as neurotransmitters released by primary afferent pruriceptors. *Journal of Neurophysiology* 109 (3):742–8.

Al Ghamdi, K. S., E. Polgar, and A. J. Todd. 2009. Soma size distinguishes projection neurons from neurokinin 1 receptor-expressing interneurons in lamina I of the rat lumbar spinal dorsal horn. *Neuroscience* 164 (4):1794–804.

Andrew, D., and A. D. Craig. 2001. Spinothalamic lamina I neurons selectively sensitive to histamine: A central neural pathway for itch. *Nature Neuroscience* 4 (1):72–7.

Antal, M., T. F. Freund, and E. Polgar. 1990. Calcium-binding proteins, parvalbumin- and calbindin-D 28k-immunoreactive neurons in the rat spinal cord and dorsal root ganglia: A light and electron microscopic study. *Journal of Comparative Neurology* 295 (3):467–84.

Baseer, N., E. Polgar, M. Watanabe, T. Furuta, T. Kaneko, and A. J. Todd. 2012. Projection neurons in lamina III of the rat spinal cord are selectively innervated by local dynorphin-containing excitatory neurons. *Journal of Neuroscience* 32 (34):11854–63.

Bice, T. N., and J. A. Beal. 1997a. Quantitative and neurogenic analysis of neurons with supraspinal projections in the superficial dorsal horn of the rat lumbar spinal cord. *Journal of Comparative Neurology* 388 (4):565–74.

Bice, T. N., and J. A. Beal. 1997b. Quantitative and neurogenic analysis of the total population and subpopulations of neurons defined by axon projection in the superficial dorsal horn of the rat lumbar spinal cord. *Journal of Comparative Neurology* 388 (4):550–64.

Bickford, R. G. 1937. Experiments relating to the itch sensation, it's periphral mechanism, and central pathways. *Clinical Science* 3:337–86.

Brohl, D., M. Strehle, H. Wende, K. Hori, I. Bormuth, K. A. Nave, T. Muller, and C. Birchmeier. 2008. A transcriptional network coordinately determines transmitter and peptidergic fate in the dorsal spinal cord. *Developmental Biology* 322 (2):381–93.

Bromm, B., E. Scharein, U. Darsow, and J. Ring. 1995. Effects of menthol and cold on histamine-induced itch and skin reactions in man. *Neuroscience Letters* 187 (3):157–60.

Carstens, E. 1997. Responses of rat spinal dorsal horn neurons to intracutaneous microinjection of histamine, capsaicin, and other irritants. *Journal of Neurophysiology* 77 (5):2499–514.

Carstens, E. E., M. I. Carstens, C. T. Simons, and S. L. Jinks. 2010. Dorsal horn neurons expressing NK-1 receptors mediate scratching in rats. *Neuroreport* 21 (4):303–8.

Davidson, S., X. Zhang, S. G. Khasabov, D. A. Simone, and G. J. Giesler, Jr. 2009. Relief of itch by scratching: State-dependent inhibition of primate spinothalamic tract neurons. *Nature Neuroscience* 12 (5):544–6.

Elkersh, M. A., T. T. Simopoulos, A. B. Malik, E. H. Cho, and Z. H. Bajwa. 2003. Epidural clonidine relieves intractable neuropathic itch associated with herpes zoster-related pain. *Regional Anesthesia and Pain Medicine* 28 (4):344–6.

Fleming, M. S., D. Ramos, S. B. Han, J. Zhao, Y. J. Son, and W. Luo. 2012. The majority of dorsal spinal cord gastrin releasing peptide is synthesized locally whereas neuromedin B is highly expressed in pain- and itch-sensing somatosensory neurons. *Molecular Pain* 8:52.

Gotoh, Y., T. Andoh, and Y. Kuraishi. 2011a. Clonidine inhibits itch-related response through stimulation of alpha(2)-adrenoceptors in the spinal cord in mice. *European Journal of Pharmacology* 650 (1):215–9.

Gotoh, Y., T. Andoh, and Y. Kuraishi. 2011b. Noradrenergic regulation of itch transmission in the spinal cord mediated by alpha-adrenoceptors. *Neuropharmacology* 61 (4):825–31.

Gotoh, Y., Y. Omori, T. Andoh, and Y. Kuraishi. 2011c. Tonic inhibition of allergic itch signaling by the descending noradrenergic system in mice. *Journal of Pharmacological Sciences* 115 (3):417–20.

Graham, B. A., A. M. Brichta, and R. J. Callister. 2007. Moving from an averaged to specific view of spinal cord pain processing circuits. *Journal of Neurophysiology* 98 (3):1057–63.

Graham, D. T., H. Goodell, and H. G. Wolff. 1951. Neural mechanisms involved in itch, itchy skin, and tickle sensations. *Journal of Clinical Investigation* 30 (1):37–49.

Grudt, T. J., and E. R. Perl. 2002. Correlations between neuronal morphology and electrophysiological features in the rodent superficial dorsal horn. *Journal of Physiology* 540 (Pt 1):189–207.

Han, N., J. Y. Zu, and J. Chai. 2012. Spinal bombesin-recognized neurones mediate more nonhistaminergic than histaminergic sensation of itch in mice. *Clinical and Experimental Dermatology* 37 (3):290–5.

Handwerker, H. O. 1992. Pain and allodynia, itch and alloknesis: An alternative hypothesis. *American Pain Society Journal* 1:115–26.

Heinke, B., R. Ruscheweyh, L. Forsthuber, G. Wunderbaldinger, and J. Sandkuhler. 2004. Physiological, neurochemical and morphological properties of a subgroup of GABAergic spinal lamina II neurones identified by expression of green fluorescent protein in mice. *Journal of Physiology* 560 (Pt 1):249–66.

Heinricher, M. M., I. Tavares, J. L. Leith, and B. M. Lumb. 2009. Descending control of nociception: Specificity, recruitment and plasticity. *Brain Research Reviews* 60 (1):214–25.

Hughes, D. I., S. Sikander, C. M. Kinnon, K. A. Boyle, M. Watanabe, R. J. Callister, and B. A. Graham. 2012. Morphological, neurochemical and electrophysiological features of parvalbumin-expressing cells: A likely source of axo-axonic inputs in the mouse spinal dorsal horn. *Journal of Physiology* 590 (Pt 16):3927–51.

Imamachi, N., G. H. Park, H. Lee, D. J. Anderson, M. I. Simon, A. I. Basbaum, and S. K. Han. 2009. TRPV1-expressing primary afferents generate behavioral responses to pruritogens via multiple mechanisms. *Proceedings of the National Academy of Sciences of the United States of America* 106 (27):11330–5.

Jinks, S. L., and E. Carstens. 2000. Superficial dorsal horn neurons identified by intracutaneous histamine: Chemonociceptive responses and modulation by morphine. *Journal of Neurophysiology* 84 (2):616–27.

Jinks, S. L., and E. Carstens. 2002. Responses of superficial dorsal horn neurons to intradermal serotonin and other irritants: Comparison with scratching behavior. *Journal of Neurophysiology* 87 (3):1280–9.

Kato, G., H. Furue, T. Katafuchi, T. Yasaka, Y. Iwamoto, and M. Yoshimura. 2004. Electrophysiological mapping of the nociceptive inputs to the substantia gelatinosa in rat horizontal spinal cord slices. *Journal of Physiology* 560 (Pt 1):303–15.

Kato, G., M. Kosugi, M. Mizuno, and A. M. Strassman. 2011. Separate inhibitory and excitatory components underlying receptive field organization in superficial medullary dorsal horn neurons. *Journal of Neuroscience* 31 (47):17300–5.

Keller, A. F., J. A. Coull, N. Chery, P. Poisbeau, and Y. De Koninck. 2001. Region-specific developmental specialization of GABA-glycine cosynapses in laminas I-II of the rat spinal dorsal horn. *Journal of Neuroscience* 21 (20):7871–80.

Koga, K., T. Chen, X. Y. Li, G. Descalzi, J. Ling, J. Gu, and M. Zhuo. 2011. Glutamate acts as a neurotransmitter for gastrin releasing peptide-sensitive and insensitive itch-related synaptic transmission in mammalian spinal cord. *Molecular Pain* 7:47.

Kosugi, M., T. Nakatsuka, T. Fujita, Y. Kuroda, and E. Kumamoto. 2007. Activation of TRPA1 channel facilitates excitatory synaptic transmission in substantia gelatinosa neurons of the adult rat spinal cord. *Journal of Neuroscience* 27 (16):4443–51.

Lagerstrom, M. C., K. Rogoz, B. Abrahamsen, E. Persson, B. Reinius, K. Nordenankar, C. Olund et al. 2010. VGLUT2-dependent sensory neurons in the TRPV1 population regulate pain and itch. *Neuron* 68 (3):529–42.

Lewis, T., R. T. Grant, and H. M. Marvin. 1927. Vascular reactions of the skin to injury. X. The intervention of a chemical stimulus illustrated especially by the flare. The response to faradism. *Heart* 14:139–60.

Lima, D., J. A. Mendes-Ribeiro, and A. Coimbra. 1991. The spino-latero-reticular system of the rat: Projections from the superficial dorsal horn and structural characterization of marginal neurons involved. *Neuroscience* 45 (1):137–52.

Liu, X. Y., Z. C. Liu, Y. G. Sun, M. Ross, S. Kim, F. F. Tsai, Q. F. Li et al. 2011. Unidirectional cross-activation of GRPR by MOR1D uncouples itch and analgesia induced by opioids. *Cell* 147 (2):447–58.

Liu, Y., O. Abdel Samad, L. Zhang, B. Duan, Q. Tong, C. Lopes, R. R. Ji, B. B. Lowell, and Q. Ma. 2010. VGLUT2-dependent glutamate release from nociceptors is required to sense pain and suppress itch. *Neuron* 68 (3):543–56.

Lu, Y., and E. R. Perl. 2003. A specific inhibitory pathway between substantia gelatinosa neurons receiving direct C-fiber input. *Journal of Neuroscience* 23 (25):8752–8.

Lu, Y., and E. R. Perl. 2005. Modular organization of excitatory circuits between neurons of the spinal superficial dorsal horn (laminae I and II). *Journal of Neuroscience* 25 (15):3900–7.

Ma, Q. 2010. Labeled lines meet and talk: Population coding of somatic sensations. *Journal of Clinical Investigation* 120 (11):3773–8.

Mantyh, P. W., S. D. Rogers, P. Honore, B. J. Allen, J. R. Ghilardi, J. Li, R. S. Daughters, D. A. Lappi, R. G. Wiley, and D. A. Simone. 1997. Inhibition of hyperalgesia by ablation of lamina I spinal neurons expressing the substance P receptor. *Science* 278 (5336):275–9.

Maxwell, D. J., M. D. Belle, O. Cheunsuang, A. Stewart, and R. Morris. 2007. Morphology of inhibitory and excitatory interneurons in superficial laminae of the rat dorsal horn. *Journal of Physiology* 584 (Pt 2):521–33.

Naim, M. M., S. A. Shehab, and A. J. Todd. 1998. Cells in laminae III and IV of the rat spinal cord which possess the neurokinin-1 receptor receive monosynaptic input from myelinated primary afferents. *European Journal of Neuroscience* 10 (9):3012–9.

Naim, M., R. C. Spike, C. Watt, S. A. Shehab, and A. J. Todd. 1997. Cells in laminae III and IV of the rat spinal cord that possess the neurokinin-1 receptor and have dorsally directed dendrites receive a major synaptic input from tachykinin-containing primary afferents. *Journal of Neuroscience* 17 (14):5536–48.

Nakatsuka, T., H. Furue, M. Yoshimura, and J. G. Gu. 2002. Activation of central terminal vanilloid receptor-1 receptors and alpha beta-methylene-ATP-sensitive P2X receptors reveals a converged synaptic activity onto the deep dorsal horn neurons of the spinal cord. *Journal of Neuroscience* 22 (4):1228–37.

Narikawa, K., H. Furue, E. Kumamoto, and M. Yoshimura. 2000. In vivo patch-clamp analysis of IPSCs evoked in rat substantia gelatinosa neurons by cutaneous mechanical stimulation. *Journal of Neurophysiology* 84 (4):2171–4.

Nilsson, H. J., A. Levinsson, and J. Schouenborg. 1997. Cutaneous field stimulation (CFS): A new powerful method to combat itch. *Pain* 71 (1):49–55.

Nojima, H., J. M. Cuellar, C. T. Simons, M. I. Carstens, and E. Carstens. 2004. Spinal c-fos expression associated with spontaneous biting in a mouse model of dry skin pruritus. *Neuroscience Letters* 361 (1–3):79–82.

Nojima, H., C. T. Simons, J. M. Cuellar, M. I. Carstens, J. A. Moore, and E. Carstens. 2003. Opioid modulation of scratching and spinal c-fos expression evoked by intradermal serotonin. *Journal of Neuroscience* 23 (34):10784–90.

Omori, Y., T. Andoh, H. Shirakawa, H. Ishida, T. Hachiga, and Y. Kuraishi. 2009. Itch-related responses of dorsal horn neurons to cutaneous allergic stimulation in mice. *Neuroreport* 20 (5):478–81.

Patel, K. N., and X. Dong. 2010. An itch to be scratched. *Neuron* 68 (3):334–9.

Patel, K. N., Q. Liu, S. Meeker, B. J. Undem, and X. Dong. 2011. Pirt, a TRPV1 modulator, is required for histamine-dependent and -independent itch. *PloS One* 6 (5):e20559.

Pernia-Andrade, A. J., A. Kato, R. Witschi, R. Nyilas, I. Katona, T. F. Freund, M. Watanabe et al. 2009. Spinal endocannabinoids and CB1 receptors mediate C-fiber-induced heterosynaptic pain sensitization. *Science* 325 (5941):760–4.

Polgar, E., K. S. Al Ghamdi, and A. J. Todd. 2010. Two populations of neurokinin 1 receptor-expressing projection neurons in lamina I of the rat spinal cord that differ in AMPA receptor subunit composition and density of excitatory synaptic input. *Neuroscience* 167 (4):1192–204.

Polgar, E., K. M. Al-Khater, S. Shehab, M. Watanabe, and A. J. Todd. 2008. Large projection neurons in lamina I of the rat spinal cord that lack the neurokinin 1 receptor are densely innervated by VGLUT2-containing axons and possess GluR4-containing AMPA receptors. *Journal of Neuroscience* 28 (49):13150–60.

Polgar, E., D. I. Hughes, J. S. Riddell, D. J. Maxwell, Z. Puskar, and A. J. Todd. 2003. Selective loss of spinal GABAergic or glycinergic neurons is not necessary for development of thermal hyperalgesia in the chronic constriction injury model of neuropathic pain. *Pain* 104 (1–2):229–39.

Polgar, E., T. C. Sardella, M. Watanabe, and A. J. Todd. 2011. Quantitative study of NPY-expressing GABAergic neurons and axons in rat spinal dorsal horn. *Journal of Comparative Neurology* 519 (6):1007–23.

Polgar, E., S. A. Shehab, C. Watt, and A. J. Todd. 1999. GABAergic neurons that contain neuropeptide Y selectively target cells with the neurokinin 1 receptor in laminae III and IV of the rat spinal cord. *Journal of Neuroscience* 19 (7):2637–46.

Puskar, Z., E. Polgar, and A. J. Todd. 2001. A population of large lamina I projection neurons with selective inhibitory input in rat spinal cord. *Neuroscience* 102 (1):167–76.

Ross, S. E. 2011. Pain and itch: Insights into the neural circuits of aversive somatosensation in health and disease. *Current Opinion in Neurobiology* 21 (6):880–7.

Ross, S. E., A. R. Mardinly, A. E. McCord, J. Zurawski, S. Cohen, C. Jung, L. Hu et al. 2010. Loss of inhibitory interneurons in the dorsal spinal cord and elevated itch in *Bhlhb5* mutant mice. *Neuron* 65 (6):886–98.

Sandkuhler, J. 2009. Models and mechanisms of hyperalgesia and allodynia. *Physiological Reviews* 89 (2):707–58.

Sardella, T. C., E. Polgar, F. Garzillo, T. Furuta, T. Kaneko, M. Watanabe, and A. J. Todd. 2011. Dynorphin is expressed primarily by GABAergic neurons that contain galanin in the rat dorsal horn. *Molecular Pain* 7:76.

Shim, W. S., M. H. Tak, M. H. Lee, M. Kim, J. Y. Koo, C. H. Lee, and U. Oh. 2007. TRPV1 mediates histamine-induced itching via the activation of phospholipase A2 and 12-lipoxygenase. *Journal of Neuroscience* 27 (9):2331–7.

Simone, D. A., M. Alreja, and R. H. LaMotte. 1991. Psychophysical studies of the itch sensation and itchy skin ("alloknesis") produced by intracutaneous injection of histamine. *Somatosensory and Motor Research* 8 (3):271–9.

Sonohata, M., H. Furue, T. Katafuchi, T. Yasaka, A. Doi, E. Kumamoto, and M. Yoshimura. 2004. Actions of noradrenaline on substantia gelatinosa neurones in the rat spinal cord revealed by in vivo patch recording. *Journal of Physiology* 555 (Pt 2):515–26.

Su, P. Y., and M. C. Ko. 2011. The role of central gastrin-releasing peptide and neuromedin B receptors in the modulation of scratching behavior in rats. *Journal of Pharmacology and Experimental Therapeutics* 337 (3):822–9.

Sun, Y. G., and Z. F. Chen. 2007. A gastrin-releasing peptide receptor mediates the itch sensation in the spinal cord. *Nature* 448 (7154):700–3.

Sun, Y. G., Z. Q. Zhao, X. L. Meng, J. Yin, X. Y. Liu, and Z. F. Chen. 2009. Cellular basis of itch sensation. *Science* 325 (5947):1531–4.

Szucs, P., L. L. Luz, D. Lima, and B. V. Safronov. 2010. Local axon collaterals of lamina I projection neurons in the spinal cord of young rats. *Journal of Comparative Neurology* 518 (14):2645–65.

Takazawa, T., and A. B. MacDermott. 2010. Synaptic pathways and inhibitory gates in the spinal cord dorsal horn. *Annals of the New York Academy of Sciences* 1198:153–8.

Tiong, S. Y., E. Polgar, J. C. van Kralingen, M. Watanabe, and A. J. Todd. 2011. Galanin-immunoreactivity identifies a distinct population of inhibitory interneurons in laminae I-III of the rat spinal cord. *Molecular Pain* 7:36.

Todd, A. J. 2010. Neuronal circuitry for pain processing in the dorsal horn. *Nature Reviews. Neuroscience* 11 (12):823–36.

Todd, A. J., M. M. McGill, and S. A. Shehab. 2000. Neurokinin 1 receptor expression by neurons in laminae I, III and IV of the rat spinal dorsal horn that project to the brainstem. *European Journal of Neuroscience* 12 (2):689–700.

Todd, A. J., Z. Puskar, R. C. Spike, C. Hughes, C. Watt, and L. Forrest. 2002. Projection neurons in lamina I of rat spinal cord with the neurokinin 1 receptor are selectively innervated by substance p-containing afferents and respond to noxious stimulation. *Journal of Neuroscience* 22 (10):4103–13.

Todd, A. J., and A. C. Sullivan. 1990. Light microscope study of the coexistence of GABA-like and glycine-like immunoreactivities in the spinal cord of the rat. *Journal of Comparative Neurology* 296 (3):496–505.

Todd, A. J., and H. R. Koerber. 2005. *Wall and Melzack's Textbook of Pain.* Edited by S. B. McMahon, and M. Koltzenburgh. Edinburgh: Elsevier.

Uta, D., H. Furue, A. E. Pickering, M. H. Rashid, H. Mizuguchi-Takase, T. Katafuchi, K. Imoto, and M. Yoshimura. 2010. TRPA1-expressing primary afferents synapse with a morphologically identified subclass of substantia gelatinosa neurons in the adult rat spinal cord. *European Journal of Neuroscience* 31 (11):1960–73.

Wang, H., and M. J. Zylka. 2009. Mrgprd-expressing polymodal nociceptive neurons inner-vate most known classes of substantia gelatinosa neurons. *Journal of Neuroscience* 29 (42):13202–9.

Ward, L., E. Wright, and S. B. McMahon. 1996. A comparison of the effects of noxious and innocuous counterstimuli on experimentally induced itch and pain. *Pain* 64 (1):129–38.

Wilson, S. R., K. A. Gerhold, A. Bifolck-Fisher, Q. Liu, K. N. Patel, X. Dong, and D. M. Bautista. 2011. TRPA1 is required for histamine-independent, Mas-related G protein-coupled receptor-mediated itch. *Nature Neuroscience* 14 (5):595–602.

Yang, K., E. Kumamoto, H. Furue, and M. Yoshimura. 1998. Capsaicin facilitates excitatory but not inhibitory synaptic transmission in substantia gelatinosa of the rat spinal cord. *Neuroscience Letters* 255 (3):135–8.

Yasaka, T., G. Kato, H. Furue, M. H. Rashid, M. Sonohata, A. Tamae, Y. Murata, S. Masuko, and M. Yoshimura. 2007. Cell-type-specific excitatory and inhibitory circuits involv-ing primary afferents in the substantia gelatinosa of the rat spinal dorsal horn in vitro. *Journal of Physiology* 581 (Pt 2):603–18.

Yasaka, T., S. Y. Tiong, D. I. Hughes, J. S. Riddell, and A. J. Todd. 2010. Populations of inhibi-tory and excitatory interneurons in lamina II of the adult rat spinal dorsal horn revealed by a combined electrophysiological and anatomical approach. *Pain* 151 (2):475–88.

Yoshimura, M., and H. Furue. 2006. Mechanisms for the anti-nociceptive actions of the descending noradrenergic and serotonergic systems in the spinal cord. *Journal of Pharmacological Sciences* 101 (2):107–17.

Yosipovitch, G., M. I. Duque, K. Fast, A. G. Dawn, and R. C. Coghill. 2007. Scratching and noxious heat stimuli inhibit itch in humans: A psychophysical study. *British Journal of Dermatology* 156 (4):629–34.

21 Itch Modulation by VGLUT2-Dependent Glutamate Release from Somatic Sensory Neurons

Qiufu Ma

CONTENTS

21.1 INTRODUCTION

Somatic sensory neurons in the dorsal root ganglia (DRG) are composed of a variety of sensory modalities, such as pain-related nociceptors, itch-related pruriceptors, thermoreceptors and mechanoreceptors (Basbaum et al. 2009; Delmas et al. 2011). This chapter will focus on the neurotransmitter basis of somatic sensory information processing. Most, if not all, DRG neurons are glutamatergic excitatory neurons and release glutamate onto postsynaptic neurons in the dorsal horn of the spinal cord (Broman et al. 1993; De Biasi and Rustioni 1988; Schneider and Perl 1988; Yoshimura and Jessell 1990). A subset of DRG neurons additionally release neuropeptide transmitters (Hökfelt 1991). Accordingly, neurons involved with pain, itch, and thermoception are divided into two subtypes, "peptidergic" and "nonpeptidergic." Peptidergic neurons release one or two "classic" neuropeptides, the substance P (SP) and the calcitonin gene-related peptide (CGRP), whereas many nonpeptidergic neurons bind the isolection B4 or IB4 (Basbaum et al. 2009). This classic definition is, however, not entirely accurate, due to the existence of other neuropeptides. For example, while the gastrin-releasing peptide (GRP) is expressed in putative CGRP+ pruriceptors, the expression of the GRP-related peptide Neuromendin B (NMB) is associated with both CGRP+ and IB4+ neurons (Fleming et al. 2012; Sun and Chen 2007). Here, we will discuss the roles of distinct transmitters in processing itch- and pain-related sensory information.

21.2 DISCOVERY OF VGLUT1-3 AND THEIR EXPRESSION IN DRG

Glutamate is the principle fast excitatory neurotransmitter used in the vertebrate nervous system. Thus, the discovery of a family of transporter proteins, VGLUT1-3, that package glutamate into the synaptic vesicles and control excitatory neurotransmission represents a major breakthrough in neuroscience (Fremeau et al. 2004). VGLUT1 was initially identified as the brain-specific Na^+-dependent inorganic phosphate transporter (Ni et al. 1994) and later was demonstrated to be located in synaptic vesicles and function as a glutamate transporter (Bellocchio et al. 1998, 2000). Subsequently, multiple groups identified the related VGLUT2 and VGLUT3 proteins as synaptic glutamate transporters as well (Aihara et al. 2000; Bai et al. 2001; Fremeau et al. 2001, 2002; Gras et al. 2002; Hayashi et al. 2001; Herzog et al. 2001; Schäfer et al. 2002; Takamori et al. 2001; Varoqui et al. 2002). In the brain, VGLUT1 and VGLUT2 are expressed in a largely complementary manner and collectively mark most excitatory neurons, whereas VGLUT3 is expressed in more restricted cells, such as the cholinergic and serotonergic neurons, as well as certain glial cells (El Mestikawy et al. 2011; Fremeau et al. 2004).

All three VGLUT proteins are expressed in DRG. *In situ* hybridization shows that about ~40% of DRG neurons expressed VGLUT1 (Liu et al. 2010; Morris et al. 2005; Oliveira et al. 2003). The VGLUT1 protein is expressed in most, if not all, low-threshold myelinated mechanoreceptors and proprioceptors (Brumovsky et al. 2007; Li et al. 2003; Oliveira et al. 2003) but also in a subset of CGRP+ neurons and SP+ terminals (Li et al. 2003; Liu et al. 2010), although other studies did not detect VGLUT1 in CGRP+ neurons (Brumovsky et al. 2007; Morris et al. 2005; Oliveira et al. 2003). *In situ* hybridization shows that most DRG neurons show detectable VGLUT2 mRNA, albeit at variable levels (Landry et al. 2004; Scherrer et al. 2010). As a result of this variability of expression levels, different groups report quite different percentages of DRG neurons or their central terminals expressing the VGLUT2 protein (Brumovsky et al. 2007; Lagerström et al. 2010; Li et al. 2003; Liu et al. 2010; Morris et al. 2005; Todd et al. 2003; Tong et al. 2001; Wu et al. 2004). In most sensitive studies, the VGLUT2 protein was detected in ~92% of DRG neurons and nearly in all of the sensory neurons involved with pain, itch, and thermoception, such as CGRP-expressing (CGRP+) peptidergic and IB4-binding (IB4+) nonpeptidergic neurons (Brumovsky et al. 2007; Liu et al. 2010). A notable exception is that VGLUT2 expression appears to be excluded from proprioceptors innervating muscles and spinal motor neurons, and these proprioceptors instead express VGLUT1 (Wu et al. 2004). VGLUT3 is expressed in 10% to 16% of DRG neurons, and its expression is excluded from myelinated mechanoreceptors, as well as from CGRP[+] and IB4[+] neurons (Seal et al. 2009; Shields et al. 2012). These VGLUT3[+] neurons represent unmyelinated low-threshold mechanoreceptors (C-LTMRs) and form the longitudinal lanceolate endings around hair follicles (Li et al. 2011; Seal et al. 2009). Thus, singular VGLUT1-3 expression appears to be associated with distinct sensory modalities, although a subset of VGLUT2[+] DRG neurons likely coexpress VGLUT1 and/or VGLUT3.

21.3   VGLUT2-DEPENDENT GLUTAMATE RELEASE FROM DRG NEURONS IS REQUIRED TO SENSE ACUTE AND CHRONIC PAIN

To address the role of synaptic glutamate release from DRG neurons, several groups made conditional knockout mice in which the *Vglut2* gene was removed from specific subsets of DRG neurons (Lagerström et al. 2010, 2011; Liu et al. 2010; Rogoz et al. 2012; Scherrer et al. 2010). This was achieved by crossing mice carrying a conditional floxed *Vglut2* allele with a series of *Cre* mouse lines, whose *Cre*-dependent DNA recombination will eliminate VGLUT2 expression and function. Subsequent behavioral analyses have provided considerable insights into how somatic sensory information is processed. Here, we first discuss how *Vglut2* knockout impacts on pain.

The Kullander group used the *Ht-Pa-Cre* mice to remove *Vglut2* from all DRG neurons (Rogoz et al. 2012). These mice show marked deficits in sensing acute pain, including light mechanical pain measured by the von Frey fibers, intense mechanical pain measured by the Randall-Selitto assay, cold pain measured by the –15°C tail withdrawal test, heat pain measured by the Hargreaves assay, and chemical pain evoked by formalin (Rogoz et al. 2012). In addition, these mutant mice showed an abolition or attenuation of mechanical and thermal hypersensitivity induced by inflammatory reagents or nerve lesions (Rogoz et al. 2012). Thus, VGLUT2-dependent glutamate release from DRG neurons is essential for proper processing of pain-related information, which in turn suggests that VGLUT1- and VGLUT3-dependent glutamate release from a subset of DRG neurons is insufficient to mediate acute or chronic pain. Nonetheless, an earlier study showed that mice lacking *Vglut3* exhibited marked deficits in mechanical hypersensitivity induced by inflammation, tissue injury or nerve lesions (Seal et al. 2009). Because of exclusive expression of VGLUT3 in C-LTMRs within DRG, it was proposed that glutamate release from these unmyelinated low threshold mechanoreceptors is involved in the expression of injury-induced mechanical hypersensitivity (Seal et al. 2009). However, this interpretation might be complicated by VGLUT3 expression in other parts of the nervous system, such as the dorsal horn of the developing spinal cord (the Allen Brain Atlas: http://mousespinal.brain-map.org) and the descending serotoninergic neurons (Cheng et al. 2003; Fremeau et al. 2002; Gras et al. 2002; Schäfer et al. 2002). Thus, VGLUT2-dependent glutamate release from DRG neurons is critical for pain-information processing, and a definite role of VGLUT3 in C-LTMRs will wait for results upon a conditional knockout of *Vglut3* in these unmyelinated low threshold mechanoreceptors.

Several groups subsequently used distinct *Cre* lines to remove *Vglut2* in subsets of DRG neurons, and analyses of these mice have provided insight into molecular and cellular identities of DRG neurons that process specific pain modalities (summarized in Figure 21.1a) (Lagerström et al. 2010, 2011; Liu et al. 2010; Scherrer et al. 2010). For example, heat pain is impaired following *Vglut2* removal by using *Cre*

(a)

Phenotypes \ *Cre* lines	Ht-Pa	TH	TRPV1	Nav1.8-K	Nav1.8-W	Peripherin
Mechanical pain (v.F)	↓	–	–	–	–	↓
Mechanical pain (R.S.)	↓	–	–	↓	↓	↓
Heat pain (48°C–50°C)	N.M.	↓	N.M.	–	N.M.	↓
Heat pain (52°C–55°C)	N.M.	↓	N.M.	↓	–	↓
Heat pain (Hargreave)	↓	N.M.	↓	↓	–	↓
Chemical pain (formalin)	↓	↓	–	N.M.	–	N.M.
Neuropathic (heat)	↓	N.M.	N.M.	N.M.	–	↓
Neuropathic (mechanical)	↓	N.M.	N.M.	↓	–	–
Inflammatory (heat)	↓	N.M.	N.M.	↓	↓	↓
Inflammatory (mechanical)	↓	N.M.	N.M.	↓	N.M.	–
Itch sensitization	+	+	+	+	+	N.M.
Skin lesions	+	+	+	+	–	–

(b)

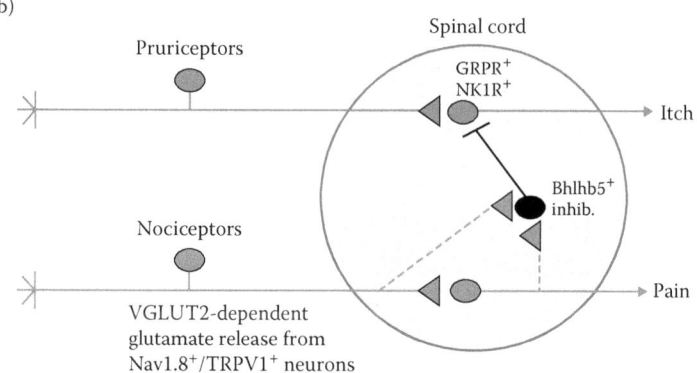

FIGURE 21.1 **(See color insert.)** The role of VGLUT2-dependent glutamate release in processing pain versus itch and the population coding of somatic sensations. (a) A summary of pain and itch phenotypes in *Vglut2* conditional knockout mice. Six *Cre* lines were used to remove *Vglut2* in partially overlapped subsets of DRG neurons, and pain- and itch-related phenotypes are listed. Abbreviations: v.F, the von Frey assay; R.S., the Randall-Selitto assay; ↓, impaired pain behaviors; –, unaffected pain or itch behaviors; N.M, unmeasured pain or itch behaviors; +, spontaneous development of itch sensitization or skin lesions. (b) Population coding of pain versus itch [modified from (Akiyama et al. 2012; Handwerker 2010; Ma 2010, 2012; McMahon and Koltzenburg 1992). Pain and itch are processed along two different labeled lines. VGLUT2-dependent glutamate release from Nav1.8+ or TRPV1+ neurons is required for pain sensation, and these pain-processing neurons can dominantly suppress itch, possibly by activating Bhlhb5-dependent inhibitory neurons in the dorsal spinal cord.

lines driven by the promoters of genes encoding the tyrosine hydroxylase (TH), the transient potential receptor ion channel TRPV1, peripherin, or Nav1.8 (the transgenic *Cre* mice made by the Kuner lab), suggesting that VGLUT2-dependent glutamate release from neurons that coexpress TH, TRPV1, peripherin, and Nav1.8 (irrespective of persistent or transient expression) is required to sense heat pain. The data summarized in Figure 21.1 also suggests that (1) intense mechanical pain measured by the Randall-Selitto assay is mediated by neurons that coexpress Nav1.8 and peripherin, but not TRPV1 or TH; (2) light mechanical pain measured by the von Frey assay is mediated by neurons that express peripherin, but not Nav1.8, TH or TRPV1; and (3) formalin-induced chemical pain is mediated by neurons expressing TH and Nav1.8, but not TRPV1 or peripherin.

21.4   ITCH INHIBITION BY VGLUT2-DEPENDENT GLUTAMATE RELEASE FROM DRG NEURONS

A major surprise in analyzing various *Vglut2* conditional knockouts is the finding that the loss of pain (and other unexamined somatic sensations) is accompanied with spontaneous development of excessive scratching and skin lesions (Lagerström et al. 2010; Liu et al. 2010; Rogoz et al. 2012). This excessive scratching phenotype was observed upon *Vglut2* removal using the following four *Cre* lines: *Ht-Pa-Cre*, *Nav1.8-Cre-K* (transgenic mice made by the Kuner lab), *TH-Cre* and *TRPV1-Cre*. Skin lesions were, however, not observed when a different *Nav1.8-Cre-W* mouse line (the knock-in mice made by the Wood lab) was used, even though itch was sensitized in these mice (Lagerström et al. 2010). The transgenic *Nav1.8-Cre-K* mouse line is able to drive reporter expression in ~81% of lumbar DRG (Agarwal et al. 2004; Liu et al. 2010), whereas the knock-in *Nav1.8-Cre-W* mice appear to drive reporter expression in a smaller subset of DRG neurons (Lagerström et al. 2010; Shields et al. 2012; Stirling et al. 2005); this difference is likely explained by the fact that *Nav1.8-Cre-K* may drive *Cre* expression at higher levels than *Nav1.8-Cre-W* does, such that neurons with low levels of Nav1.8 expression will contain enough *Cre* activity only in *Nav1.8-Cre-K* mice. Excessive scratching normally developed during the second month after birth and preceded the eventual skin lesions (Lagerström et al. 2010; Liu et al. 2010). Intradermal injections of itching compounds demonstrated that both histamine-dependent and histamine-independent itch pathways are sensitized in these knockout mice (Liu et al. 2010). Thus, VGLUT2-dependent glutamate release from DRG neurons is required to sense pain and inhibit itch. As discussed below, these findings raise a number of intriguing questions about how itch information is processed.

21.5   WHAT IS THE NEUROTRANSMITTER BASIS OF ITCH?

Itch sensitization occurs upon a complete loss of VGLUT2 in all DRG neurons (Rogoz et al. 2012). As mentioned above, about 40% to 50% of DRG neurons express VGLUT1 and/or VGLUT3. Thus, pruriceptors might use VGLUT1 or VGLUT3 to release glutamate and to mediate sensitized itch upon VGLUT2 loss. However, this explanation cannot be entirely true. Pruriceptors that are marked by the expression of the G-protein coupled receptor MrgprA3 express VGLUT2, but not VGLUT1 or VGLUT3 (Liu et al. 2009, 2010; Lou and Ma, unpublished data). Nonetheless, upon VGLUT2 loss, itch evoked by chloroquine, which acts through Mrgpra3, is not attenuated (Liu et al. 2009). Moreover, genetic ablation of MrgprA3+ neurons suggests a broad role of these neurons in mediating both histamine-dependent and histamine-independent itch (Han et al. 2013). In other words, upon VGLUT2 removal in all DRG neurons, glutamate release is apparently dispensable for MrgprA3+ pruriceptors to process itch. Two scenarios need to be pointed out.

First, pruriceptors may use transmitters other than glutamate. Indeed, several neuropeptides have been implicated in itch, such as GRP and SP (Smith et al. 2010; Sun and Chen 2007). Mice lacking GRPR exhibited marked deficits in histamine-independent itch (Sun and Chen 2007). Although no itch deficits had been reported in mice lacking SP (Cuellar et al. 2003), possibly due to unknown redundancy, a role of SP in processing itch in the dorsal spinal cord was strongly indicated in naked more rats (Smith et al. 2010). Both the GRP receptor (GRPR) and the SP receptor (NK1R) are expressed in a subset of neurons located in the superficial laminae of the dorsal spinal cord (Sun and Chen 2007; Todd 2010). Ablation of GRPR+ neurons in mice upon intrathecal injection of the bombesin peptide conjugated with the saporin toxin (bombesin-saporin), which binds GRPR, leads to a virtual abolition of scratching responses evoked by a range of itching compounds (Sun et al. 2009). Similarly, ablation of NK1R+ neurons in rats by SP-saporin injections resulted in a loss of serotonin-evoked itch (Carstens et al. 2010). Notably, NK1R+ and GRPR+ neurons appear to intermingle with each other, because SP-saporin failed to ablate GRPR+ neurons and vice versus (Mishra et al. 2012; Sun et al. 2009). Thus GRPR+ and NK1R+ neurons might form a circuit to process itch. However, cautiousness is needed in interpreting these ablation results. (1) NK1R is additionally expressed in DRG neurons (Li and Zhao 1998; Zhang et al. 2007), and central NK1R+ terminals appeared to be concurrently ablated following intracisternal injection of SP-saporin (Carstens et al. 2010). Whether or not GRPR is also expressed in DRG neurons has not yet been carefully examined. Thus, the cellular basis of itch loss following SP-saporin or bombesin-saporin injections needs further investigation. (2) GRP and SP are expressed in both DRG and the dorsal spinal cord (Fleming et al. 2012; Hökfelt 1991); accordingly, the sources of GRP and SP used for itch processing remain to be determined. (3) Mrgpra3+ pruriceptors express NMB, a GRP-related peptide (Fleming et al. 2012). Although ablation of dorsal horn neurons expressing the NMB receptor did not affect histamine-evoked itch, histamine-independent itch has not yet been examined in these mice (Mishra et al. 2012). Thus, it remains unclear about what nonglutamate transmitters are used for by peripheral pruriceptors.

Second, even though MrgprA3+ pruriceptors use nonglutamate transmitters, VGLUT2-dependent glutamate might still play a role in wild-type mice. The spontaneous development of itch sensitization in *Vglut2* conditional knockouts suggests that spinal itch neurons may receive tonic inhibition from circuits activated by glutamate release from other sensory modalities. In wild-type mice, glutamate release from pruriceptors might be required to overcome this tonic inhibition, thereby establishing a competent state for nonglutamate transmitters to activate the itch pathway. Indeed, GPR-sensitive spinal itch neurons receive glutamatergic inputs from unmyelinated

C fibers or thinly myelinated Aδ fibers (Koga et al. 2011). This glutamate release would become dispensable in *Vglut2* knockouts due to preelimination of tonic itch inhibition. To further test this possibility, we need to selectively eliminate *Vglut2* in pruriceptors, such as Mrgpra3[+] neurons (that do not express VGLUT1 or VGLUT3), followed by determining whether scratching responses by chloroquine and/or other itching compounds would be affected or not.

21.6 HOW IS ITCH INHIBITED BY VGLUT2-DEPENDENT GLUTAMATE RELEASE FROM DRG NEURONS?

The first issue is which populations of DRG neurons are involved with itch inhibition. As mentioned above, itch sensitization was observed in mice in which *Vglut2* was removed by using four *Cre* lines: *Ht-Pa-Cre, Nav1.8-Cre-K, TH-Cre,* and *TRPV1-Cre* (Lagerström et al. 2010; Liu et al. 2010; Rogoz et al. 2012). *Ht-Pa-Cre* marks all DRG neurons (Rogoz et al. 2012). *Nav1.8-Cre-K* marks 81% of DRG neurons, including ~92% of TRPV1-persistent neurons (Liu et al. 2010). *TH-Cre* marks ~93% of TRPV1-persistent neurons and other DRG neurons (Lagerström et al. 2010). Because of dynamic TRPV1 expression during development, the TRPV1 lineage neurons include both peptidergic and nonpeptidergic neurons, including TRPV1-persistent heat-sensitive nociceptors, Mrgprd[+] polymodal nociceptors, MrgprA3[+] pruriceptors, TRPM8[+] cold-sensitive thermoceptors, and others (Cavanaugh et al. 2011; Lagerström et al. 2010; Mishra et al. 2011). Because of the heterogeneity of DRG neurons marked by all these *Cre* lines, the identities of DRG neurons involved with itch inhibition remain unclear. Nonetheless, coinjection of capsaicin, an agonist of TRPV1, is able to suppress itch evoked by itching compounds and this inhibition is impaired upon VGLUT2 loss (Liu et al. 2010), suggesting that a portion of TRPV1-persistent DRG neurons is involved with itch inhibition. TRPV1-negative DRG neurons, such as nociceptive mechanoreceptors, might inhibit itch as well, because painful mechanical stimuli, such as scratching, are able to suppress itch, and TRPV1-persistent DRG neurons are dispensable for mechanical pain (Akiyama et al. 2012; Cavanaugh et al. 2009; Davidson and Giesler 2010; Davidson et al. 2009). Thus, multiple groups of DRG neurons could be involved with itch inhibition.

The second issue is how a loss of glutamate releases from DRG neurons leads to spontaneous itch sensitization and excessive scratching? It needs to be pointed out that itch inhibition might occur in two distinct states. First, histamine-sensitive neurons in the spinal cord normally stay silent, at least in cats, possibly by receiving tonic inhibitory inputs from spinal neurons that process other sensory modalities (Andrew and Craig 2001). Second, scratching can suppress the evoked activity of histamine-sensitive spinal neurons, in a location- and state-dependent manner (Akiyama et al. 2012; Davidson et al. 2009). At this moment, it is not known whether VGLUT2-dependent glutamate release from DRG neurons is involved with tonic or state-dependent itch inhibition. Regardless, VGLUT2-dependent glutamate release likely activates spinal inhibitory circuits to suppress itch, based on the following observations. First, scratching-mediated itch inhibition is dependent on the activation of inhibitory neurons (Akiyama et al. 2011). Second, the transcription factor

Bhlhb5 is required for proper development of a group of spinal cord inhibitory neurons, and mice lacking Bhlhb5 develop spontaneous itch sensitization, excessive scratching, and skin lesions (Ross et al. 2010), analogous to the phenotypes seen in *Vglut2* conditional knockouts. Third, capsaicin injection activates multiple groups of spinal inhibitory neurons, including those marked by the expression of the neuropeptide Y (Liu et al. 2010; Zou et al. 2002). Upon VGLUT2 loss, capsaicin failed to activate NPY[+] neurons (Liu et al. 2010), raising the possibility that capsaicin-mediated itch inhibition might act by activating these inhibitory neurons (Figure 21.1b).

21.7 POPULATION CODING OF ITCH VERSUS PAIN OR OTHER SENSORY MODALITIES

The loss of pain and the concurrent gain of itch in *Vglut2* knockout mice provide a strong support for the population coding theory, also called the selectivity hypothesis, in explaining how distinct sensory modalities are encoded in the nervous system (Akiyama et al. 2009; Handwerker 2010; Ma 2010, 2012; McMahon and Koltzenburg 1992; Patel and Dong 2010; Ross 2011; Sikand et al. 2011; Wood et al. 2009), although alternative or more complex models also exist (Davidson and Giesler 2010; McMahon and Koltzenburg 1992; Prescott and Ratté 2012). This population coding theory is composed of the following three features (Figure 21.1b) (Ma 2010, 2012). First, each sensory modality is processed along a specific neural circuit or network, also called a sensory labeled line. The existence of an itch labeled line is supported by microneurographic studies in humans (Handwerker 2010; Schmelz et al. 1997) and electrophysiologial studies in cats (Andrew and Craig 2001) and by the finding that ablation of GRPR[+] spinal neurons affect itch, but not pain (Sun et al. 2009). Second, most sensory neurons are polymodal. In other words, a given sensory receptor could be associated with multiple sensory modalities. For example, TRPV1[+] neurons include histamine-dependent pruriceptors and heat-sensitive nociceptors (Costa et al. 2008; Green and Shaffer 1993; Imamachi et al. 2009; Schmelz et al. 2003; Shim et al. 2007; Shimada and LaMotte 2008; Sikand et al. 2011). Similarly, TRPA1 is associated with histamine-independent pruriceptors and pain-related fibers (Liu et al. 2009, 2011; Wilson et al. 2011). Thus, a pruriceptor can paradoxically respond to stimuli that normally causes pain. In other words, an itch labeled line is defined by a specific neural circuit rather than by neurons responding selectively to itching stimuli. Third, activation of one sensory modality can modulate other sensory modalities. For example, itch can be dominantly suppressed by a range of painful stimuli, including scratching (Akiyama et al. 2012; Atanassoff et al. 1999; Davidson et al. 2009; Graham et al. 1951; Liu et al. 2010; Sikand et al. 2011; Ward et al. 1996). Accordingly, itch will be evoked only if a stimulus selectively activates itch-related fibers, or the concurrent activation of other sensory modalities is insufficient to mask itch. In this sense, it is interesting to note that while intradermal injection of capsaicin evokes pain, delivery of a small amount of capsaicin into the skin epidermis evokes both itch and nociceptive sensations (Sikand et al. 2009, 2011), suggesting that capsaicin-sensitive fibers in the epidermis might be enriched for itch fibers, whereas pain-related fibers might be enriched in the dermis, as suggested by other studies (Ikoma et al. 2011).

The population-coding model might explain a key feature of human chronic itch. In such patients, painful stimuli, such as scratching, not just fail to suppress itch, but paradoxically promote itch (Hosogi et al. 2006; Ikoma et al. 2004; Ishiuji et al. 2008). It was proposed that pain-induced itch inhibition is lost in these patients (Ikoma et al. 2006). Because pruriceptors are polymodal and capable of responding to painful stimuli, a loss of itch inhibition by pain may allow painful stimuli to directly activate the normally masked itch pathways. Interestingly, mice with a knockout of *Vglut2* in DRG neurons share exactly this key feature, with capsaicin injection in the cheeks evoking itch-indicated scratching responses, rather than pain-indicated wiping responses seen in wild-type mice (Liu et al. 2010).

21.8 CONCLUDING REMARKS

Analyses of a series of *Vglut2* conditional knockouts provide important insight into the processing of itch versus other somatic sensory modalities. VGLUT2-dependent glutamate release from DRG neurons is essential for pain sensation and itch inhibition. Removal of this neural component creates a neurogenic chronic itch model that mimics symptoms seen in human patients, such as paradoxical promotion of itch by painful stimuli that normally should suppress itch. Upon a loss of itch inhibition, MrgprA3⁺ pruriceptors are able to use unknown nonglutamate transmitters to evoke itch, but it remains a possibility that glutamate release from pruriceptors could play a modulatory role in the wild-type context, such as overcoming the tonic inhibition of itch by other sensory modalities. The loss of pain and the concurrent gain of itch in *Vglut2* knockouts provide a strong support for the population coding theory of somatic sensations, which emphasizes the existence of modality-specific neural circuits/networks as well as antagonistic interactions among different sensory modalities.

REFERENCES

Agarwal, N., Offermanns, S., and Kuner, R. (2004). Conditional gene deletion in primary nociceptive neurons of trigeminal ganglia and dorsal root ganglia. *Genesis 38*, 122–129.

Aihara, Y., Mashima, H., Onda, H., Hisano, S., Kasuya, H., Hori, T., Yamada, S. et al. (2000). Molecular cloning of a novel brain-type Na(+)-dependent inorganic phosphate cotransporter. *J Neurochem 74*, 2622–2625.

Akiyama, T., Carstens, M.I., and Carstens, E. (2011). Transmitters and pathways mediating inhibition of spinal itch-signaling neurons by scratching and other counterstimuli. *PLoS One 6*, e22665.

Akiyama, T., Merrill, A.W., Carstens, M.I., and Carstens, E. (2009). Activation of superficial dorsal horn neurons in the mouse by a PAR-2 agonist and 5-HT: Potential role in itch. *J Neurosci 29*, 6691–6699.

Akiyama, T., Tominaga, M., Carstens, M.I., and Carstens, E.E. (2012). Site-dependent and state-dependent inhibition of pruritogen-responsive spinal neurons by scratching. *Eur J Neurosci 36*, 2311–2316.

Andrew, D., and Craig, A.D. (2001). Spinothalamic lamina I neurons selectively sensitive to histamine: A central neural pathway for itch. *Nat Neurosci 4*, 72–77.

Atanassoff, P.G., Brull, S.J., Zhang, J., Greenquist, K., Silverman, D.G., and Lamotte, R.H. (1999). Enhancement of experimental pruritus and mechanically evoked dysesthesiae with local anesthesia. *Somatosens Mot Res 16*, 291–298.

Bai, L., Xu, H., Collins, J.F., and Ghishan, F.K. (2001). Molecular and functional analysis of a novel neuronal vesicular glutamate transporter. *J Biol Chem 276*, 36764–36769.

Basbaum, A.I., Bautista, D.M., Scherrer, G., and Julius, D. (2009). Cellular and molecular mechanisms of pain. *Cell 139*, 267–284.

Bellocchio, E.E., Hu, H., Pohorille, A., Chan, J., Pickel, V.M., and Edwards, R.H. (1998). The localization of the brain-specific inorganic phosphate transporter suggests a specific presynaptic role in glutamatergic transmission. *J Neurosci 18*, 8648–8659.

Bellocchio, E.E., Reimer, R.J., Fremeau, R.T.J., and Edwards, R.H. (2000). Uptake of glutamate into synaptic vesicles by an inorganic phosphate transporter. *Science 289*, 957–960.

Broman, J., Anderson, S., and Ottersen, O.P. (1993). Enrichment of glutamate-like immunoreactivity in primary afferent terminals throughout the spinal cord dorsal horn. *Eur J Neurosci 5*, 1050–1061.

Brumovsky, P., Watanabe, M., and Hökfelt, T. (2007). Expression of the vesicular glutamate transporters-1 and -2 in adult mouse dorsal root ganglia and spinal cord and their regulation by nerve injury. *Neuroscience 147*, 469–490.

Carstens, E.E., Carstens, M.I., Simons, C.T., and Jinks, S.L. (2010). Dorsal horn neurons expressing NK-1 receptors mediate scratching in rats. *Neuroreport 21*, 303–308.

Cavanaugh, D.J., Chesler, A.T., Bráz, J.M., Shah, N.M., Julius, D., and Basbaum, A.I. (2011). Restriction of transient receptor potential vanilloid-1 to the peptidergic subset of primary afferent neurons follows its developmental downregulation in nonpeptidergic neurons. *J Neurosci 31*, 10119–10127.

Cavanaugh, D.J., Lee, H., Lo, L., Shields, S.D., Zylka, M.J., Basbaum, A.I., and Anderson, D.J. (2009). Distinct subsets of unmyelinated primary sensory fibers mediate behavioral responses to noxious thermal and mechanical stimuli. *Proc Natl Acad Sci USA 106*, 9075–9080.

Cheng, L., Chen, C.L., Luo, P., Tan, M., Qiu, M., Johnson, R., and Ma, Q. (2003). Lmx1b, Pet-1, and Nkx2.2 coordinately specify serotonergic neurotransmitter phenotype. *J Neurosci 23*, 9961–9967.

Costa, R., Marotta, D.M., Manjavachi, M.N., Fernandes, E.S., Lima-Garcia, J.F., Paszcuk, A.F., Quintão, N.L., Juliano, L., Brain, S.D., and Calixto, J.B. (2008). Evidence for the role of neurogenic inflammation components in trypsin-elicited scratching behaviour in mice. *Br J Pharmacol 154*, 1094–1103.

Cuellar, J.M., Jinks, S.L., Simons, C.T., and Carstens, E. (2003). Deletion of the preprotachykinin A gene in mice does not reduce scratching behavior elicited by intradermal serotonin. *Neurosci Lett 339*, 72–76.

Davidson, S., and Giesler, G.J. (2010). The multiple pathways for itch and their interactions with pain. *Trends Neurosci 33*, 550–8.

Davidson, S., Zhang, X., Khasabov, S.G., Simone, D.A., and Giesler, G.J.J. (2009). Relief of itch by scratching: State-dependent inhibition of primate spinothalamic tract neurons. *Nat Neurosci 12*, 544–546.

De Biasi, S., and Rustioni, A. (1988). Glutamate and substance P coexist in primary afferent terminals in the superficial laminae of spinal cord. *Proc Natl Acad Sci USA 85*, 7820–7824.

Delmas, P., Hao, J., and Rodat-Despoix, L. (2011). Molecular mechanisms of mechanotransduction in mammalian sensory neurons. *Nat Rev Neurosci 12*, 139–153.

El Mestikawy, S., Wallén-Mackenzie, A., Fortin, G.M., Descarries, L., and Trudeau, L.E. (2011). From glutamate co-release to vesicular synergy: Vesicular glutamate transporters. *Nat Rev Neurosci 12*, 204–216.

Fleming, M.S., Ramos, D., Han, S.B., Zhao, J., Son, Y.J., and Luo, W. (2012). The majority of dorsal spinal cord gastrin releasing peptide is synthesized locally whereas neuromedin B is highly expressed in pain- and itch-sensing somatosensory neurons. *Mol Pain 8*, 52.

Fremeau, R.T.J., Burman, J., Qureshi, T., Tran, C.H., Proctor, J., Johnson, J., Zhang, H. et al. (2002). The identification of vesicular glutamate transporter 3 suggests novel modes of signaling by glutamate. *Proc Natl Acad Sci USA 99*, 14488–14493.

Fremeau, R.T.J., Troyer, M.D., Pahner, I., Nygaard, G.O., Tran, C.H., Reimer, R.J., Bellocchio, E.E., Fortin, D., Storm-Mathisen, J., and Edwards, R.H. (2001). The expression of vesicular glutamate transporters defines two classes of excitatory synapse. *Neuron 31*, 247–260.

Fremeau, R.T.J., Voglmaier, S., Seal, R.P., and Edwards, R.H. (2004). VGLUTs define subsets of excitatory neurons and suggest novel roles for glutamate. *Trends Neurosci 27*, 98–103.

Graham, D.T., Goodell, H., and Wolff, H.G. (1951). Neural mechanisms involved in itch, itchy skin, and tickle sensations. *J Clin Invest 30*, 37–49.

Gras, C., Herzog, E., Bellenchi, G.C., Bernard, V., Ravassard, P., Pohl, M., Gasnier, B., Giros, B., and El Mestikawy, S. (2002). A third vesicular glutamate transporter expressed by cholinergic and serotoninergic neurons. *J Neurosci 22*, 5442–5451.

Green, B.G., and Shaffer, G.S. (1993). The sensory response to capsaicin during repeated topical exposures: Differential effects on sensations of itching and pungency. *Pain 53*, 323–334.

Han, L., Ma, C., Liu, Q., Weng, H.J., Cui, Y., Tang, Z., Kim, Y. et al. (2013). A subpopulation of nociceptors specifically linked to itch. *Nat Neurosci 16*(2), 174–182.

Handwerker, H.O. (2010). Microneurography of pruritus. *Neurosci Lett 470*, 193–196.

Hayashi, M., Otsuka, M., Morimoto, R., Hirota, S., Yatsushiro, S., Takeda, J., Yamamoto, A., and Moriyama, Y. (2001). Differentiation-associated Na+-dependent inorganic phosphate cotransporter (DNPI) is a vesicular glutamate transporter in endocrine glutamatergic systems. *J Biol Chem 276*, 43400–43406.

Herzog, E., Bellenchi, G.C., Gras, C., Bernard, V., Ravassard, P., Bedet, C., Gasnier, B., Giros, B., and El Mestikawy, S. (2001). The existence of a second vesicular glutamate transporter specifies subpopulations of glutamatergic neurons. *J Neurosci 21*, RC181.

Hökfelt, T. (1991). Neuropeptides in perspective: The last ten years. *Neuron 7*, 867–879.

Hosogi, M., Schmelz, M., Miyachi, Y., and Ikoma, A. (2006). Bradykinin is a potent pruritogen in atopic dermatitis: A switch from pain to itch. *Pain 126*, 16–23.

Ikoma, A., Cevikbas, F., Kempkes, C., and Steinhoff, M. (2011). Anatomy and neurophysiology of pruritus. *Semin Cutan Med Surg 30*, 64–70.

Ikoma, A., Fartasch, M., Heyer, G., Miyachi, Y., Handwerker, H., and Schmelz, M. (2004). Painful stimuli evoke itch in patients with chronic pruritus: Central sensitization for itch. *Neurology 62*, 212–217.

Ikoma, A., Steinhoff, M., Ständer, S., Yosipovitch, G., and Schmelz, M. (2006). The neurobiology of itch. *Nat Rev Neurosci 7*, 535–547.

Imamachi, N., Park, G.H., Lee, H., Anderson, D.J., Simon, M.I., Basbaum, A.I., and Han, S.K. (2009). TRPV1-expressing primary afferents generate behavioral responses to pruritogens via multiple mechanisms. *Proc Natl Acad Sci USA 106*, 11330–11335.

Ishiuji, Y., Coghill, R.C., Patel, T.S., Dawn, A., Fountain, J., Oshiro, Y., and Yosipovitch, G. (2008). Repetitive scratching and noxious heat do not inhibit histamine-induced itch in atopic dermatitis. *Br J Dermatol 158*, 78–83.

Koga, K., Chen, T., Li, X.Y., Descalzi, G., Ling, J., Gu, J., and Zhuo, M. (2011). Glutamate acts as a neurotransmitter for gastrin releasing peptide-sensitive and insensitive itch-related synaptic transmission in mammalian spinal cord. *Mol Pain 7*, 47.

Lagerström, M.C., Rogoz, K., Abrahamsen, B., Lind, A.L., Olund, C., Smith, C., Mendez, J.A., Wallén-Mackenzie, Å., Wood, J.N., and Kullander, K. (2011). A sensory subpopulation depends on vesicular glutamate transporter 2 for mechanical pain, and together with substance P, inflammatory pain. *Proc Natl Acad Sci USA 108*, 5789–5794.

Lagerström, M.C., Rogoz, K., Abrahamsen, B., Persson, E., Reinius, B., Nordenankar, K., Olund, C. et al. (2010). VGLUT2-dependent sensory neurons in the TRPV1 population regulate pain and itch. *Neuron 68*, 529–542.

Landry, M., Bouali-Benazzouz, R., El Mestikawy, S., Ravassard, P., and Nagy, F. (2004). Expression of vesicular glutamate transporters in rat lumbar spinal cord, with a note on dorsal root ganglia. *J Comp Neurol 468*, 380–394.

Li, H.S., and Zhao, Z.Q. (1998). Small sensory neurons in the rat dorsal root ganglia express functional NK-1 tachykinin receptor. *Eur J Neurosci 10*, 1292–1299.

Li, J.L., Fujiyama, F., Kaneko, T., and Mizuno, N. (2003). Expression of vesicular glutamate transporters, VGluT1 and VGluT2, in axon terminals of nociceptive primary afferent fibers in the superficial layers of the medullary and spinal dorsal horns of the rat. *J Comp Neurol 457*, 236–249.

Li, L., Rutlin, M., Abraira, V.E., Cassidy, C., Kus, L., Gong, S., Jankowski, M.P. et al. (2011). The functional organization of cutaneous low-threshold mechanosensory neurons. *Cell 147*, 1615–1627.

Liu, Q., Tang, Z., Surdenikova, L., Kim, S., Patel, K.N., Kim, A., Ru, F. et al. (2009). Sensory neuron-specific GPCR Mrgprs are itch receptors mediating chloroquine-induced pruritus. *Cell 139*, 1353–1365.

Liu, Q., Weng, H.J., Patel, K.N., Tang, Z., Bai, H., Steinhoff, M., and Dong, X. (2011). The distinct roles of two GPCRs, MrgprC11 and PAR2, in itch and hyperalgesia. *Sci Signal 4*, ra45.

Liu, Y., Abdel Samad, O., Duan, B., Zhang, L., Tong, Q., Ji, R.R., Lowell, B., and Ma, Q. (2010). VGLUT2-dependent glutamate release from peripheral nociceptors is required to sense pain and suppress itch. *Neuron 68*, 543–556.

Ma, Q. (2010). Labeled lines meet and talk: Population coding of somatic sensations. *J Clin Invest 120*, 3773–3778.

Ma, Q. (2012). Population coding of somatic sensations. *Neurosci Bull 28*, 91–99.

McMahon, S.B., and Koltzenburg, M. (1992). Itching for an explanation. *Trends Neurosci 15*, 497–501.

Mishra, S.K., Holzman, S., and Hoon, M.A. (2012). A nociceptive signaling role for neuromedin B. *J Neurosci 32*, 8686–8695.

Mishra, S.K., Tisel, S.M., Orestes, P., Bhangoo, S.K., and Hoon, M.A. (2011). TRPV1-lineage neurons are required for thermal sensation. *EMBO J 30*, 582–593.

Morris, J.L., König, P., Shimizu, T., Jobling, P., and Gibbins, I.L. (2005). Most peptide-containing sensory neurons lack proteins for exocytotic release and vesicular transport of glutamate. *J Comp Neurol 483*, 1–16.

Ni, B., Rosteck, P.R.J., Nadi, N.S., and Paul, S.M. (1994). Cloning and expression of a cDNA encoding a brain-specific Na(+)-dependent inorganic phosphate cotransporter. *Proc Natl Acad Sci USA 91*, 5607–5611.

Oliveira, A.L., Hydling, F., Olsson, E., Shi, T., Edwards, R.H., Fujiyama, F., Kaneko, T., Hökfelt, T., Cullheim, S., and Meister, B. (2003). Cellular localization of three vesicular glutamate transporter mRNAs and proteins in rat spinal cord and dorsal root ganglia. *Synapse 50*, 117–129.

Patel, K.N., and Dong, X. (2010). An itch to be scratched. *Neuron 68*, 334–339.

Prescott, S.A., and Ratté, S. (2012). Pain processing by spinal microcircuits: Afferent combinatorics. *Curr Opin Neurobiol 22*, 631–639.

Rogoz, K., Lagerström, M.C., Dufour, S., and Kullander, K. (2012). VGLUT2-dependent glutamatergic transmission in primary afferents is required for intact nociception in both acute and persistent pain modalities. *Pain 153*, 1525–1536.

Ross, S.E. (2011). Pain and itch: Insights into the neural circuits of aversive somatosensation in health and disease. *Curr Opin Neurobiol 21*, 880–887.

Ross, S.E., Mardinly, A.R., McCord, A.E., Zurawski, J., Cohen, S., Jung, C., Hu, L. et al. (2010). Loss of inhibitory interneurons in the dorsal spinal cord and elevated itch in Bhlhb5 mutant mice. *Neuron 65*, 65.

Schäfer, M.K., Varoqui, H., Defamie, N., Weihe, E., and Erickson, J.D. (2002). Molecular cloning and functional identification of mouse vesicular glutamate transporter 3 and its expression in subsets of novel excitatory neurons. *J Biol Chem 277*, 50734–50748.

Scherrer, G., Low, S.A., Wang, X., Zhang, J., Yamanaka, H., Urban, R., Solorzano, C. et al. (2010). VGLUT2 expression in primary afferent neurons is essential for normal acute pain and injury-induced heat hypersensitivity. *Proc Natl Acad Sci USA 107*, 22296–22301.

Schmelz, M., Schmidt, R., Bickel, A., Handwerker, H.O., and Torebjörk, H.E. (1997). Specific C-receptors for itch in human skin. *J Neurosci 17*, 8003–8008.

Schmelz, M., Schmidt, R., Weidner, C., Hilliges, M., Torebjork, H.E., and Handwerker, H.O. (2003). Chemical response pattern of different classes of C-nociceptors to pruritogens and algogens. *J Neurophysiol 89*, 2441–2448.

Schneider, S.P., and Perl, E.R. (1988). Comparison of primary afferent and glutamate excitation of neurons in the mammalian spinal dorsal horn. *J Neurosci 8*, 2062–2073.

Seal, R.P., Wang, X., Guan, Y., Raja, S.N., Woodbury, C.J., Basbaum, A.I., and Edwards, R.H. (2009). Injury-induced mechanical hypersensitivity requires C-low threshold mechanoreceptors. *Nature 462*, 651–655.

Shields, S.D., Ahn, H.S., Yang, Y., Han, C., Seal, R.P., Wood, J.N., Waxman, S.G., and Dib-Hajj, S.D. (2012). Na(v)1.8 expression is not restricted to nociceptors in mouse peripheral nervous system. *Pain 153*, 2017–2030.

Shim, W.S., Tak, M.H., Lee, M.H., Kim, M., Kim, M., Koo, J.Y., Lee, C.H., Kim, M., and Oh, U. (2007). TRPV1 mediates histamine-induced itching via the activation of phospholipase A2 and 12-lipoxygenase. *J Neurosci 27*, 2331–2337.

Shimada, S.G., and LaMotte, R.H. (2008). Behavioral differentiation between itch and pain in mouse. *Pain 139*, 681–687.

Sikand, P., Shimada, S.G., Green, B.G., and LaMotte, R.H. (2009). Similar itch and nociceptive sensations evoked by punctate cutaneous application of capsaicin, histamine and cowhage. *Pain 144*, 66–75.

Sikand, P., Shimada, S.G., Green, B.G., and LaMotte, R.H. (2011). Sensory responses to injection and punctate application of capsaicin and histamine to the skin. *Pain 152*, 2485–2494.

Smith, E.S., Blass, G.R., Lewin, G.R., and Park, T.J. (2010). Absence of histamine-induced itch in the African naked mole-rat and "rescue" by substance P. *Mol Pain 6*, 2.

Stirling, L.C., Forlani, G., Baker, M.D., Wood, J.N., Matthews, E.A., Dickenson, A.H., and Nassar, M.A. (2005). Nociceptor-specific gene deletion using heterozygous NaV1.8-Cre recombinase mice. *Pain 113*, 27–36.

Sun, Y.G., and Chen, Z.F. (2007). A gastrin-releasing peptide receptor mediates the itch sensation in the spinal cord. *Nature 448*, 700–703.

Sun, Y.G., Zhao, Z.Q., Meng, X.L., Yin, J., Liu, X.Y., and Chen, Z.F. (2009). Cellular basis of itch sensation. *Science 325*, 1531–1534.

Takamori, S., Rhee, J.S., Rosenmund, C., and Jahn, R. (2001). Identification of differentiation-associated brain-specific phosphate transporter as a second vesicular glutamate transporter (VGLUT2). *J Neurosci 21*, RC182.

Todd, A.J. (2010). Neuronal circuitry for pain processing in the dorsal horn. *Nat Rev Neurosci 11*, 823–836.

Todd, A.J., Hughes, D.I., Polgár, E., Nagy, G.G., Mackie, M., Ottersen, O.P., and Maxwell, D.J. (2003). The expression of vesicular glutamate transporters VGLUT1 and VGLUT2 in neurochemically defined axonal populations in the rat spinal cord with emphasis on the dorsal horn. *Eur J Neurosci 17*, 13–27.

Tong, Q., Ma, J., and Kirchgessner, A.L. (2001). Vesicular glutamate transporter 2 in the brain-gut axis. *Neuroreport 12*, 3929–3934.

Varoqui, H., Schäfer, M.K., Zhu, H., Weihe, E., and Erickson, J.D. (2002). Identification of the differentiation-associated Na+/PI transporter as a novel vesicular glutamate transporter expressed in a distinct set of glutamatergic synapses. *J Neurosci 22*, 142–155.

Ward, L., Wright, E., and McMahon, S.B. (1996). A comparison of the effects of noxious and innocuous counterstimuli on experimentally induced itch and pain. *Pain 64*, 129–138.

Wilson, S.R., Gerhold, K.A., Bifolck-Fisher, A., Liu, Q., Patel, K.N., Dong, X., and Bautista, D.M. (2011). TRPA1 is required for histamine-independent, Mas-related G protein-coupled receptor-mediated itch. *Nat Neurosci 14*, 595–602.

Wood, G.J., Akiyama, T., Carstens, E., Oaklander, A.L., and Yosipovitch, G. (2009). An insatiable itch. *J Pain 10*, 792–797.

Wu, S.X., Koshimizu, Y., Feng, Y.P., Okamoto, K., Fujiyama, F., Hioki, H., Li, Y.Q., Kaneko, T., and Mizuno, N. (2004). Vesicular glutamate transporter immunoreactivity in the central and peripheral endings of muscle-spindle afferents. *Brain Res 1011*, 247–251.

Yoshimura, M., and Jessell, T. (1990). Amino acid-mediated EPSPs at primary afferent synapses with substantia gelatinosa neurones in the rat spinal cord. *J Physiol 430*, 315–335.

Zhang, H., Cang, C.L., Kawasaki, Y., Liang, L.L., Zhang, Y.Q., Ji, R.R., and Zhao, Z.Q. (2007). Neurokinin-1 receptor enhances TRPV1 activity in primary sensory neurons via PKCepsilon: A novel pathway for heat hyperalgesia. *J Neurosci 27*, 12067–12077.

Zou, X., Lin, Q., and Willis, W.D. (2002). The effects of sympathectomy on capsaicin-evoked fos expression of spinal dorsal horn GABAergic neurons. *Brain Res 958*, 322–329.

22 Ascending Pathways for Itch

Steve Davidson, Hannah Moser, and Glenn Giesler

CONTENTS

A large number of studies published over the last 75 years indicate that, in addition to blocking pain, ventral lateral cordotomies (surgical disruption of the ventral lateral funiculus) also consistently block the sensation of itch. Such studies are taken as evidence supporting an important role for the spinothalamic tract (STT) in pruriception since its axons ascend within the VLF. Banzet (1927) first reported that cordotomy abolished pruritus. He noted that severe itch caused by irritation of the vulva was completely relieved following a ventral lateral cordotomy and recommended cordotomy as a treatment for the problem. Hyndman and Wolkin (1943) carefully studied the responses of ten cordotomy patients to the pruritic compound cowhage. They found that cowhage did not produce itch in the areas rendered analgesic in any of the patients but did produce itch in areas in which pain sensation was normal. These authors concluded that "it can be said with certainty that the sensation of itch is mediated through the spinothalamic tract" (Hyndman and Wolkin 1943, p. 130). White et al. (1950) reported that a patient who had undergone a cordotomy did not experience itch in the analgesic area of the body even following contact with poison ivy. These authors also noted that in another patient severe itching caused by a intramedullary neoplasm of the spinal cord was abolished by a cordotomy and that their patients were not "annoyed by the itch" following the bite of a mosquito. Foerster (1936), Graf (1960), and Taren and Kahn (1966) also reported that cordotomies blocked the sensation of itch. Therefore, these clinical studies showed that, in addition to information related to pain and temperature sensation, axons in the ventral lateral funiculus (VLF) convey information that is necessary for production of the sensation of itch. It should be noted that electrophysiological studies of neurons in other spinal pathways that send ascending axons within the VLF, such as the spinoparabrachial, spinomesencephalic, spinoreticular, and spinohypothalamic tracts, have not yet been carried out. Axons in any, or all, of these pathways may also carry pruriceptive information to any number of areas of the brain and thereby contribute to sensory-discriminative, autonomic, affective, or modulatory systems related to itch.

The responses of STT neurons to pruritic stimuli have been examined in several electrophysiological studies. Wei and Tuckett (1991) examined the responses of individual axons ascending within the VLF of cats to the nonhistaminergic pruritic agent cowhage. The spinal cords of the cats were cut at C1 to ensure that all axons that responded to the stimuli were ascending within the VLF. It was not possible in this study to determine whether the examined units were axons of the STT of any of the other pathways within the VLF. The responses of 34 axons classified as wide dynamic range (WDR) (i.e., responding to innocuous and noxious stimuli) were compared following application of inactive (boiled) and active cowhage to their receptive fields. Almost all units were activated by the insertion of either inactive or active cowhage. A majority of the examined units exhibited significantly increased levels of firing for minutes following removal of the active, but not the inactive, cowhage spicules. These findings showed that mechanically sensitive neurons are capable of carrying pruriceptive information and therefore support the idea that neurons that respond to nociceptive stimuli might also be capable of carrying pruriceptive information. It appears that no attempt was made in this study to determine whether any axons in the VLF were insensitive to mechanical stimuli. This study also provided additional evidence that pruriceptive information ascends within the VLF; this is indirect support for the idea that the STT is involved in pruriception.

Andrew and Craig (2001) examined responses of mechanically insensitive spinothalamic tract neurons in the lumbar spinal cord of cats. A small number of such cells were found. Four responded to iontophoretic application of histamine. Two of these were also activated by application of mustard oil, a noxious stimulus. The authors concluded sensory information about itch was carried by STT axons that responded only to pruriceptive stimuli. Responses to injections of capsaicin were not examined nor were the responses of mechanically sensitive nociceptive neurons to application of histamine.

Simone et al. (2004) examined the effects of intracutaneous injection of histamine (20 μg in 10 μL vehicle) and capsaicin (100 μg in the same volume) on the responses of STT neurons that projected to the ventral posterior lateral nucleus in monkeys. None of the examined neurons responded specifically to injections of histamine. That is, all examined STT neurons were activated by mechanical and thermal stimuli applied to their receptive fields. Virtually all examined neurons responded preferentially (WDR) or specifically to noxious mechanical stimuli. Approximately half of the examined neurons were activated at least weakly by injections of histamine (each was statistically above baseline), and each of the same neurons responded more intensely to injections of capsaicin.

Davidson et al. (2007) reported that pruriceptive primate STT neurons were almost always activated either by injections of histamine or application of cowhage spicules within their receptive fields but not by both stimuli. Figure 22.1 illustrates an example of an STT neuron that was activated by injection of histamine but failed to respond to application of cowhage. The results of this study indicated that within the monkey STT, separate axons carry information related to histaminergic and

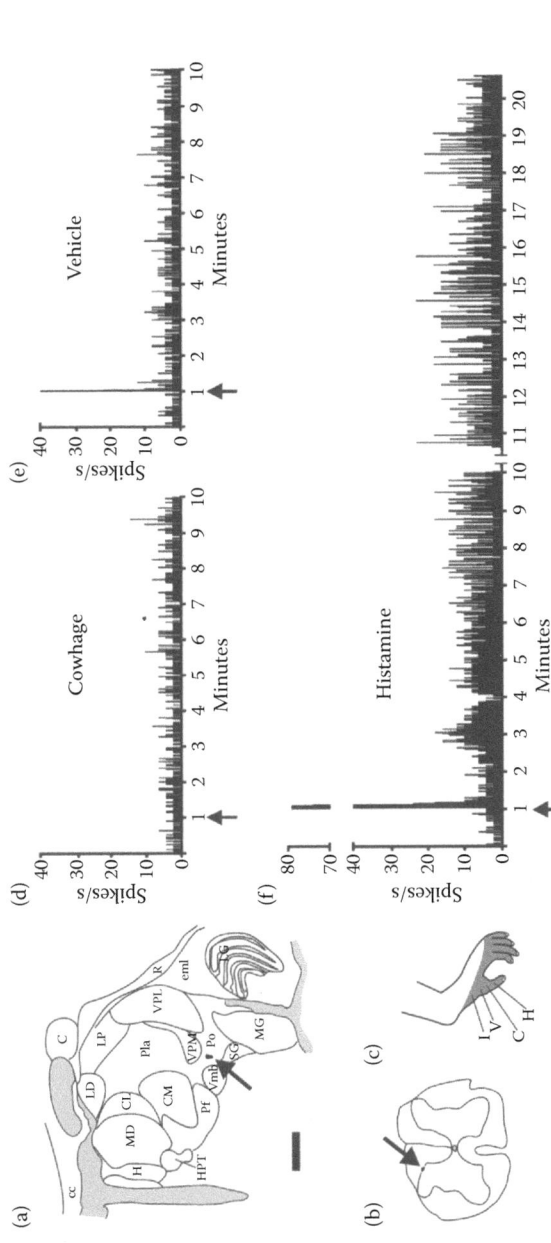

FIGURE 22.1 Primate spinothalamic tract neurons are activated by histamine or cowhage but almost never both, indicating the existence of separate histaminergic and nonhistaminergic pathways for itch within the central nervous system. (a) Axon termination area inside of the posterior nucleus of the thalamus for the neuron shown in (b) located in the marginal zone. (c) Receptive field of neuron in (a, b). (d) No change in activity after application of cowhage, or vehicle (e). (f) Histamine produced a robust and long-lasting response. (Adapted from Davidson, S. et al., *Journal of Neuroscience*, 27, 10007–10014, 2007.)

nonhistaminergic itch. In this study as well, all pruriceptive neurons were activated by noxious pinch.

Davidson et al. (2012) described responses of more than 100 primate STT neurons to innocuous and noxious mechanical stimuli, to heat, and to the pruritic stimuli histamine and cowhage. They found that roughly one third of lumbar STT neurons in monkeys were activated by either itch-producing stimulus and that roughly two thirds were not activated by either pruritogen. Approximately 13% of the examined STT neurons were activated by cowhage and 20% by intradermal injections of histamine. Each of these neurons also responded to noxious mechanical and often thermal stimuli. No evidence was found for the existence of primate STT neurons that responded exclusively to itch-producing stimuli. Axon termination zones for cowhage- and histamine-responsive STT neurons were tracked within the thalamus and were found to terminate within the ventrobasal complex and the posterior thalamic nuclei. STT neurons responsive to pruritogens discharged in a bursting pattern to pruritogens but also to capsaicin. Finally, it was found that responses to innocuous brushing were significantly increased in the presence of histamine, suggesting a potential mechanism for alloknesis.

22.1   PSYCHOPHYSICAL STUDIES OF SCRATCH

Itch and pain are experienced as separate sensations; yet they appear to share underlying neural substrates, such as discussed for STT neurons above. One clear difference between itch and pain is the effect of counterstimuli on each sensation. Scratching and other noxious stimuli serve to inhibit itch yet do not inhibit pain, as validated by a number of human psychophysical studies in which itchy stimuli are applied to human skin and subjects are asked to rate the intensity of itch during application of various counterstimuli. Counterstimuli that have been used in these psychophysical studies include scratching (Yosipovitch et al. 2007; Kosteletzky et al. 2009), pinpricks (Graham et al. 1951), pinching with forceps (Graham et al. 1951), electrical stimulation (Graham et al. 1951; Ward et al. 1996), noxious heat (Ward et al. 1996; Yosipovitch et al. 2007), noxious cold (Yosipovitch et al. 2007), and capsaicin (Brull et al. 1999). These counterstimuli cover a variety of stimulus modalities (mechanical, thermal, electrical, and chemical), and are all similar in being capable of activating nociceptors to cause pain. In general, innocuous stimuli have not been shown to reliably inhibit itch (Ward et al. 1996; Yosipovitch et al. 2007). Two main types of itchy stimuli have been used in human psychophysical studies of scratch and other counterstimuli: cowhage and histamine. Cowhage (*Mucuna pruriens*) is a tropical legume plant with pods covered in spicules containing mucunain, a cysteine protease that causes itch when the spicules are inserted into the skin (Shelley and Arthur 1955; Reddy et al. 2008). Itch can also be induced in human skin by histamine, which is generally applied in one of two ways: iontophoresis or intradermal injection. Iontophoresis of histamine is a method in which a small amount of electrical current is used to repel histamine ions out of solution and into the skin. Histamine and cowhage are known to activate separate populations of neurons in the periphery (Johanek et al. 2007) as well as in the spinal cord (Davidson et al. 2007), and while antihistamines, the most commonly prescribed treatment for itch, work to block histamine-induced itch, cowhage-induced itch is not affected by this class of drug (Shelley and Arthur 1955; Johanek et al. 2007). These qualities suggest that the simultaneous study and comparison of cowhage and histamine, and the effects of counterstimuli on each, may provide valuable insight into chronic itch conditions that are often resistant to inhibition of itch by counterstimuli and are not easily treated with antihistamine drugs. Few studies have compared the effects of scratching on cowhage versus histamine-induced itch. In one, it was found that while itch due to either stimulus is abolished after a period of scratching, cowhage-induced itch returns significantly more quickly after scratching subsides than does histamine-induced itch (Kosteletzky et al. 2009).

The time course of inhibition of itch by counterstimuli is difficult to determine due to the fact that the itch sensations induced by many stimuli can be transient or phasic in nature and are highly modulated by changes in attention or emotions, making consistent ratings of itch over long periods of time difficult to obtain. In humans, cowhage causes itch with a duration of 5–10 min (Johanek et al. 2007). Histamine applied iontophoretically (Magerl et al. 1990; Ward et al. 1996) or by injection into the skin (Simone et al. 1987; Johanek et al. 2007) can cause itch lasting 10 to 30 min. Histamine-induced itch sensation is reduced immediately upon application of a counterstimulus (Ward et al. 1996; Yosipovitch et al. 2007). After a period of scratching, itch intensity increases back toward prescratching levels; following a period of noxious heat or cold, however, itch intensity remains decreased for some time (Ward et al. 1996; Yosipovitch et al. 2007). Pinprick, a more intense noxious mechanical stimulus than scratching, also results in inhibition of itch that outlasts the duration of pain sensation caused by the counterstimulus (Graham et al. 1951). This is in accordance with many studies that have shown that noxious counterstimuli can be used to induce an "antipruritic state" in which normally itchy stimuli are incapable of producing itch when applied to an area of skin previously treated with a noxious stimulus. For example, noxious stimulation by application of transdermal current can reduce the intensity of itch caused by histamine up to 30 min after the counterstimulus (Ward et al. 1996). Other stimuli capable of producing an "antipruritic state" include insect bites, burns, and freezes (Bickford 1938). When a noxious stimulus is applied to the skin, a surrounding area of secondary hyperalgesia results in which pain thresholds are not altered but painful stimuli are perceived as causing more intense pain of longer duration. In areas of secondary hyperalgesia caused by electrical stimulation or pinching of the skin, cowhage is unable to produce itch (Graham et al. 1951); similarly, itch is greatly reduced when histamine is applied to an area of capsaicin-induced allodynia, in which normally innocuous stimuli induce pain (Brull et al. 1999). Therefore, it seems clear that conditions of increased pain result in reduced itch. As one might expect, reducing pain can lead to increased itch. Histamine induces itch of greater magnitude and longer duration when applied to an anesthetized area of skin in which nociception is greatly reduced (Atanassoff et al. 1999).

Several lines of evidence indicate that counterstimuli can reduce itch intensity even when applied to an area separate from the site of application of the itchy stimulus. Pinprick applied several centimeters from a site of cowhage application can inhibit cowhage-induced itch. The area in which pinprick abolishes itch grows with time following cowhage application, reaching the maximum area about 1 hour after cowhage application. Remarkably, it is reported that pinprick near the sternum can inhibit itching caused by application of cowhage to skin on the back, as long as the pinprick is administered within the same dermatome as the cowhage (Graham et al. 1951). These observations suggest that scratching and other counterstimuli may inhibit itch via a central, rather than peripheral, mechanism. Scratching itself has been suggested to be a reflexive reaction to the itch sensation, as spinalized animals are still able to exhibit scratch-like behaviors (Sherrington 1906). These experiments suggest that the circuitry underlying scratching is contained within the spinal cord.

22.2   INHIBITION ON INTERNEURONS IN THE SPINAL CORD

Interneurons within the rodent spinal cord have been shown to respond to a variety of itchy stimuli, including histamine as well as cowhage and the itch-producing protease it contains. Most spinal interneurons excited by itch-producing stimuli also show excitatory responses to noxious mechanical, thermal, and chemical stimuli when these stimuli are applied during a period when the cell is not responding to a pruritogen (Akiyama et al. 2009a, 2009b, 2011a). Several studies have addressed the effects of various noxious counterstimuli applied during responses to itch-producing stimuli in spinal interneurons in anesthetized animals. In accordance with human studies, innocuous stimulation has no effect on ongoing itch responses in rodent spinal neurons (Akiyama et al. 2011b). In rats, histamine-responsive spinal neurons were inhibited upon application of rubbing, scratching, or noxious heating of the receptive field (Carstens 1997). When applied during a response to histamine, each counterstimulus produced an additional increase in firing rate. When the counterstimulus was removed, however, the cell's firing rate decreased to levels lower than prior to the application of the counterstimulus. This reduction in firing rate in itch-responsive cells may be analogous to the reduction in itch intensity rating following application of a counterstimulus in humans; in contrast to the effects seen on interneurons in rats, itch ratings in humans decrease immediately upon application of a noxious counterstimulus (Ward et al. 1996; Yosipovitch et al. 2007). In addition to reducing activity during a response to an itchy stimulus, counterstimuli have been shown to contribute to an "antipruritic state" such as described in human psychophysical studies, in which application of a noxious stimulus leads to decreased responses to subsequent application of itchy stimuli. Responses to histamine in rat spinal interneurons are reduced when histamine is applied after application of noxious cold to the receptive field. Rewarming of the receptive field resulted in the histamine response returning toward baseline levels (Jinks and Carstens 1998). These rat studies have provided physiological data that closely match and help to explain important phenomena in human histaminergic itch. However, histamine does not induce itch-related behaviors in rats and may instead produce pain (Jinks and Carstens 2002; Klein et al. 2011).

Methods for scratching the receptive field while recording from spinal interneurons in anesthetized rodents have also provided valuable information regarding the effects of counterstimuli on physiological itch responses. In mice, spinal neuron activity induced by histamine or chloroquine, an antimalarial drug known to cause severe itch in some human patients, is reduced upon application of scratching (Akiyama et al. 2012). Activity induced by the noxious chemical stimulus capsaicin, however, is further enhanced by scratching. Therefore, scratch-evoked inhibition of chemical responses in spinal neurons appears to be dependent on the type of chemical stimulus and the sensation it evokes: responses to itch-producing stimuli are decreased while responses to pain-producing stimuli are increased, even within the same neuron (Davidson et al. 2009). In addition to being state dependent, the effects of scratching may also be site dependent. Scratching directed precisely to the site of pruritogen application had no effect on chloroquine-induced activity, but excited histamine-induced activity was followed by inhibition when the scratching stopped (Akiyama et al. 2012). Scratching at a site separated by several centimeters from the site of pruritogen application resulted in inhibition of spinal interneuron activity induced by either chloroquine or histamine. This ability of scratching to inhibit itch, even when delivered several centimeters away from the site of pruritogen application, is in accordance with the human psychophysical studies mentioned above in which application of a counterstimulus distant from application of the itchy stimulus has been shown to reduce itch intensity.

In addition to inhibiting responses to itch-producing chemical stimuli, scratching has also been shown to reduce itch-related activity in a mouse dry-skin model (Akiyama et al. 2011a, 2011b). In this model, drying agents are repetitively applied to a discrete area of skin, leading to a chronic itchy skin condition. Spinal interneurons with receptive fields corresponding to the area of dry skin exhibit increased firing rate levels, presumably because of an increase in itch signaling from the dry skin. Rather than being excited by noxious stimuli, these neurons are instead inhibited by scratching, pinching, and noxious heat (Figure 22.2). Unlike with histamine-induced activity in rats, inhibition of ongoing activity related to dry skin occurs immediately upon application of the counterstimulus. The ability of scratching to inhibit presumably itch-related signaling from dry skin is analogous to the ability of scratching and other counterstimuli to inhibit "spontaneous" itch, resulting from chronic dry skin conditions in humans (Yosipovitch et al. 2007). However, in patients suffering from chronic itch due to atopic dermatitis, noxious stimulation of an affected area of skin can actually induce more itch (Ikoma et al. 2004). Using the dry skin mouse model, it has been shown that noxious cold appears to have a mostly excitatory effect when applied to dry skin (Akiyama et al. 2011b). It is possible that chronic itchy skin conditions such as atopic dermatitis involve changes in neuronal signaling or circuitry that lead to excitation rather than inhibition of itch, such as seen with noxious cold in the mouse dry skin model.

From these studies in rodent itch models, it is clear that counterstimuli such as scratching can have a powerful inhibitory effect on neuronal signaling within the central nervous system. The two main types of inhibitory neurotransmitter within the spinal cord: glycine and GABA (γ-aminobutyric acid) have both been implicated in the ability of scratching to inhibit itch-related responses in mouse spinal interneurons (Akiyama et al. 2011b). Under control conditions, scratching inhibited spontaneous firing in the mouse dry skin model during the entire period that scratching is applied. When the glycine receptor antagonist strychnine was applied to the spinal cord, the ability of scratching to reduce spontaneous activity was abolished (Figure 22.3).

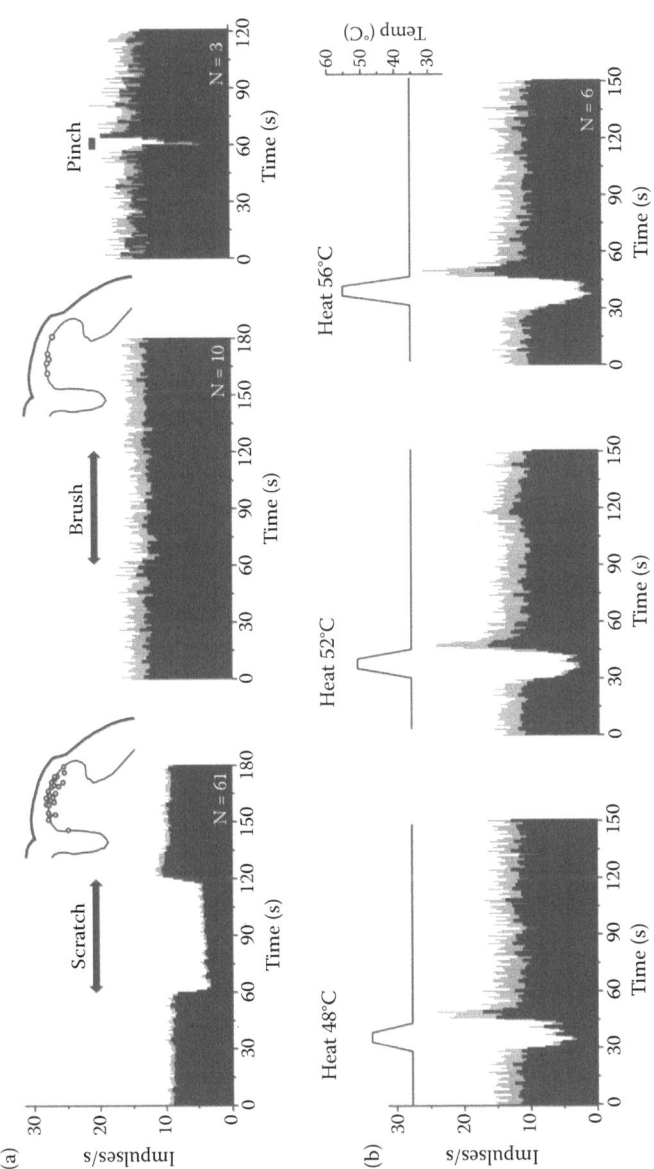

FIGURE 22.2 In a mouse model of dry skin, heightened pruritic activity in spinal interneurons is reduced upon application of noxious counterstimuli. (a) Scratching and pinching, but not innocuous brushing, of the receptive field decreases firing in cells recorded from the superficial dorsal horn (insets). (b) Noxious heat applied to the receptive field also decreases firing. (Adapted from Akiyama, T. et al., *PloS One*, 6, e22665, 2011.)

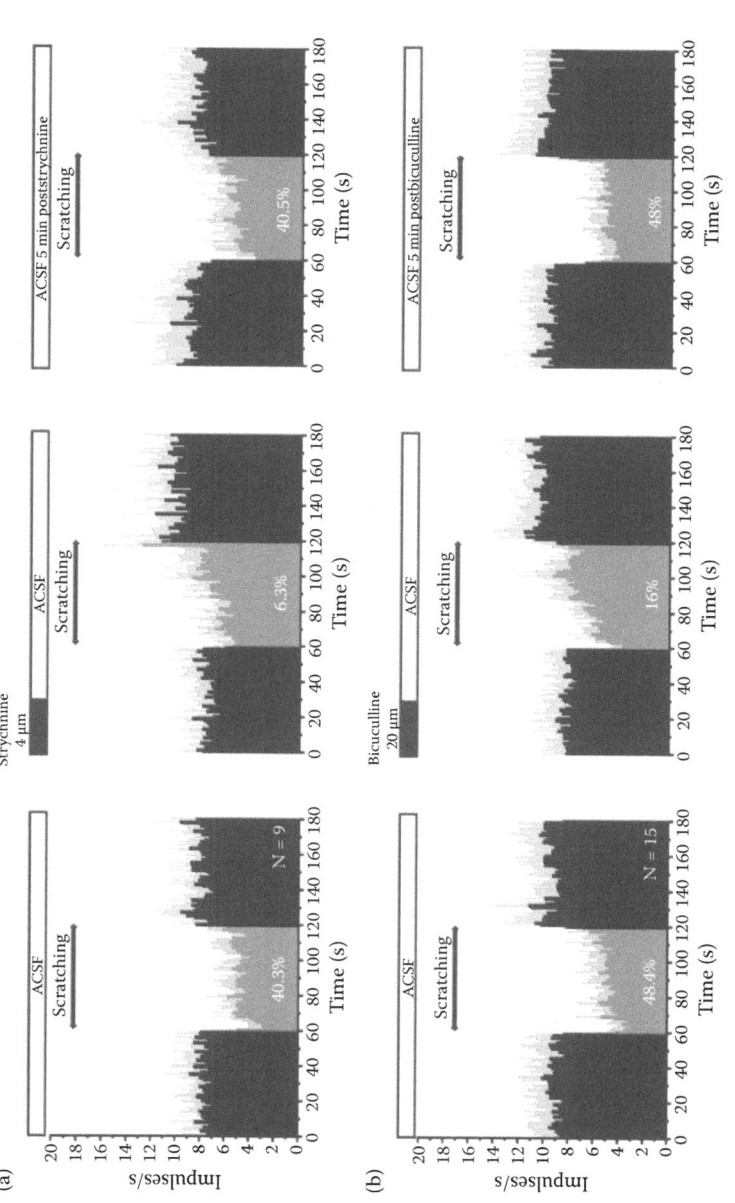

FIGURE 22.3 Scratch-induced inhibition of pruritic activity in mouse spinal interneurons depends on the inhibitory neurotransmitters glycine and GABA. (a) The ability of scratching to reduce activity related to itchy, dry skin is reduced in the presence of the glycine receptor antagonist strychnine. The effect is reversible upon washout of the antagonist. (b) The GABA receptor antagonist bicuculline also reduces the ability of scratching to inhibit itch responses. (Adapted from Akiyama, T. et al., *PloS One*, 6, e22665, 2011.)

Antagonists for the $GABA_A$ receptor (bicuculline) and $GABA_B$ receptor (saclofen) also reduced scratch-evoked inhibition of spontaneous activity. Therefore, inhibition of itch responses in spinal neurons by application of counterstimuli, such as scratching, is likely mediated by both glycine and GABA neurotransmitter systems.

22.3 PRURICEPTIVE STT NEURONS ARE INHIBITED BY SCRATCHING DURING ITCH

Several arguments point to a central rather than peripheral locus for the inhibition of itch by scratching and other counterstimuli. First, the primary afferent C-fibers that are activated by histamine in a manner that mirrors the itch sensation are mechanically insensitive (CMi) and are likely not activated by scratching (Schmelz et al. 1997, 2003). This suggests that scratching instead activates a group of mechanically sensitive fibers that engage a central circuit to inhibit itch. Second, counterstimuli applied centimeters away from the site of itching, so as to avoid directly affecting the transduction mechanisms in the itch-mediating nerve terminals, reduce the intensity of itching. For example, pinprick within the same dermatome obliterated cowhage-induced itch when delivered 24 cm from the cowhage application site (Graham et al. 1951). Likewise, scratching, heat, and electrical stimuli have each been shown to reduce itch when delivered several centimeters away from the site of pruritogen application (Nilsson et al. 1997; Yosipovitch et al. 2007; Kosteletzky et al. 2009). It should be noted that histamine-sensitive CMi fibers may extend into larger innervation territories than C-polymodal fibers, potentially allowing counterstimuli applied away from the histamine application site to impinge directly upon nerve endings (Schmelz et al. 1997; Schmidt et al. 2002). Finally, itching failed to develop when pruritogens were applied within an area of hyperalgesia, suggesting a spinally mediated block of itch (Bickford 1938; Graham et al. 1951; Brull et al. 1999). Interestingly, the neurogenic flare associated with histamine injection was not abolished within the sensitized skin, suggesting that histamine-responsive C-fibers continue to function despite the absence of itch sensation (Brull et al. 1999). Together these observations support the idea that itch signaling can be blocked by circuit mechanisms at the level of the spinal cord.

Primates respond to pruritogens similarly to humans as measured by site directed scratching (Johanek et al. 2008). Likewise, primate spinothalamic tract neurons respond to pruritogens with an activation profile similar to human psychophysical reports of itch (Simone et al. 2004; Davidson et al. 2007, 2012). Primate STT neurons are polymodal and can respond to a combination of mechanical, thermal, and chemical stimuli applied to their receptive fields. Synaptic inputs from multiple sources converge on STT neurons to generate action potentials that are transmitted from the spinal cord to the brain. Scratching on the cutaneous receptive fields of naïve lumbar STT neurons reliably evoked a discharge of action potentials (Davidson et al. 2009). In contrast, when STT neurons were engaged in an ongoing response to histamine, the same scratching stimulus produced inhibition of the ongoing response (Figure 22.4). In some cells, the inhibition of histamine-evoked activity began at the onset of the scratching and in others the inhibitory period began after a short delay.

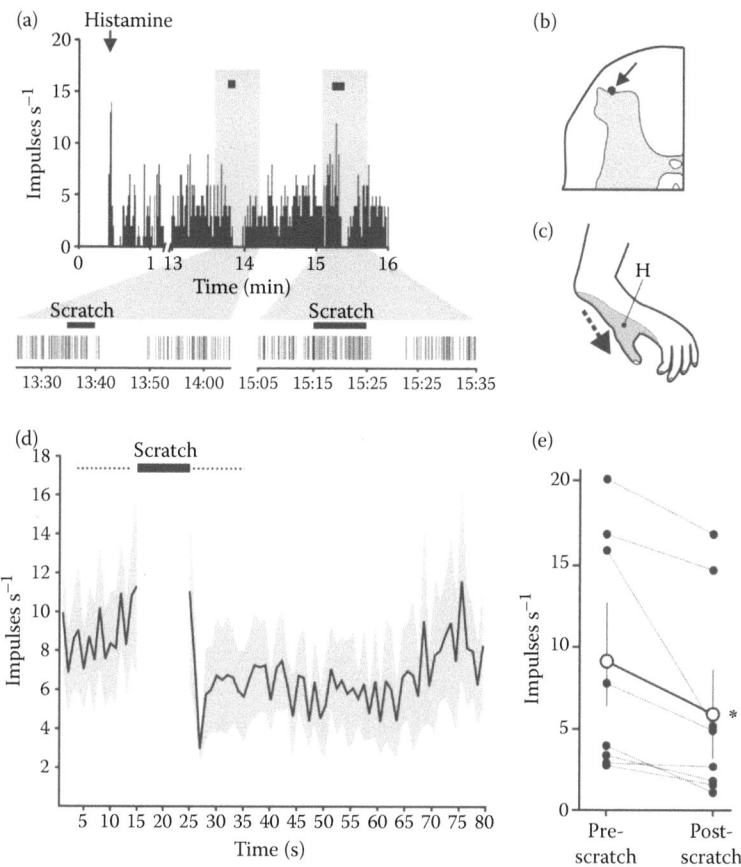

FIGURE 22.4 Primate spinothalamic tract neurons activated by histamine are inhibited by scratching. (a) Scratching at two separate time points inhibited the ongoing response to histamine. (b) This neuron was recorded in the marginal zone and had a receptive field (c) on the foot. Arrow shows the direction of scratching and histamine application site H is indicated. (d) Mean response of all spinothalamic tract neurons scratched during their response to histamine. There is a reduction in discharge lasting 30 s after scratching. (e) All cells in (d) comparing prescratch and postscratch discharge rates ($p < 0.05$, paired t-test). (Adapted from Davidson, S. et al., *Nature Neuroscience*, 12, 544–546, 2009.)

In all histamine-responsive STT neurons the activity after scratching was less than the ongoing histamine-evoked activity prior to scratching (Davidson et al. 2009). The reduced activity lasted about 30 s and then returned to prescratch levels. A transient reduction in histamine-evoked activity is consistent with observations made in psychophysical studies in which humans report a rebound of itch after scratching (Yosipovitch et al. 2007; Kosteletzky et al. 2009; Vierow et al. 2009). Spinothalamic tract neurons from anesthetized monkeys appear to reflect the histamine-induced itch sensation and subsequent relief of itch by scratching experienced by human subjects.

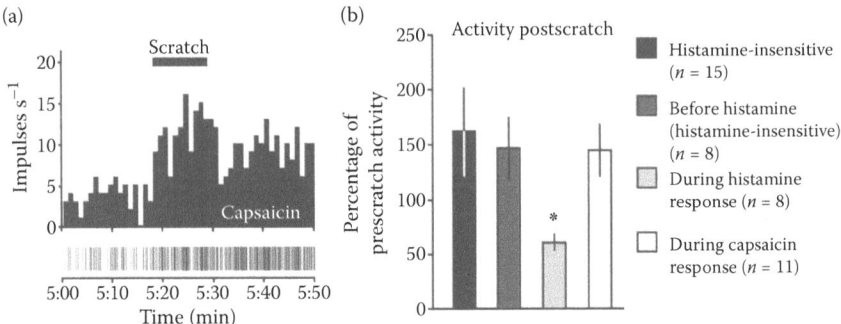

FIGURE 22.5 Primate spinothalamic tract neurons inhibited by scratching during a response to histamine are excited by scratching during a response to capsaicin. (a) Scratching the receptive field during a response to capsaicin in a histamine-responsive spinothalamic tract neuron leads to greater excitation. (b) Summary of activity change postscratch. Only cells with ongoing discharge to histamine were inhibited by scratching. These cells were not inhibited by scratching before the histamine response or during a response to capsaicin. Cells unresponsive to histamine showed no inhibition. (Adapted from Davidson, S. et al., *Nature Neuroscience*, 12, 544–546, 2009.)

As mentioned above, before histamine was delivered, scratching the receptive field always produced increased firing in primate STT neurons. This is interesting because it indicates that activation of STT neurons by histamine switched the way the neurons respond to mechanical stimulation. This switch is unique for histamine because the same cells showed no scratch-induced inhibition during a capsaicin response (Figure 22.5). In fact, scratching the receptive field of capsaicin-responsive STT neurons during the response to capsaicin significantly enhanced their discharge. These data suggest that pruritic agents can switch the state of spinal processing to allow scratching or other counterstimuli to engage an inhibitory circuit, while scratching during pain amplifies activity (Davidson et al. 2009).

The organization of the neural circuitry involved in producing relief from scratching includes a mechanism to block the itch signal within the spinal cord. In STT neurons, scratch-induced inhibition during the histamine response (but not capsaicin) suggests that ongoing itch generates state-dependent access for nociceptor activation to engage spinal inhibition. The reduced output of the spinothalamic tract during scratching an itch may have implications for activity in higher-order structures that are involved in the processing of itch. Counterstimulus-induced disruption of itch signaling in peripheral fibers cannot be ruled out for a role in inhibiting itch nor can inhibition of itch initiated at higher levels of the neuraxis.

22.4   SCRATCHING ON THE SKIN, ITCHING IN THE BRAIN

Neuron level physiological studies within the brain to identify the cells and circuits involved in the sensation of itch have not yet been performed. However, advances have been made in the identification of the anatomical structures of the brain involved in the sensation of itch (or the urge to scratch). Imaging techniques have been used to identify areas of increased and decreased cerebral activity after the application of pruritogens, and after counterstimuli. Blood-oxygen-level-dependent signals evoked by pruritic and noxious stimuli overlap extensively, but several distinctions between itch- and pain-induced areas of activity have been identified (Herde et al. 2007; Mochizuki et al. 2007).

To identify potential supraspinal regions involved in the suppression of itch, Mochizuki et al. (2003) scanned subjects' brains and coapplied a painfully cold counterstimulus that reduced the intensity of histamine-evoked itch. Cold pain during itch produced enhanced blood flow in the periaqueductal gray, an area that was not activated during itching or cooling alone. This suggested the possibility of a descending inhibitory pathway for itch with an anatomy similar to that described for pain (Figure 22.6a). More recently, the regions of the brain involved in suppression of itch by scratching have been examined. A number of brain areas were activated by nonautonomous (i.e., investigator mediated) scratching in the absence of itch (Yosipovitch et al. 2008). Direct comparison of cerebral activation during nonautonomous scratching, either in the presence or absence of itch, showed few differences between conditions (Vierow et al. 2009). A hypothesized decrease of the itch-evoked signal was not observed, possibly because the neural activity generated by scratching, which is a mild noxious stimulus, may have masked a decrease in the signal from the reduced itch. Interestingly, scratching during itch produced a unique and robust signal in the putamen, an area in which no clear activation was observed in the "no itch" or the "itch without scratching" conditions (Herde et al. 2007; Vierow et al. 2009). It was suggested that the putamen might be activated to coordinate motor behavior related to scratching (Vierow et al. 2009). Alternatively,

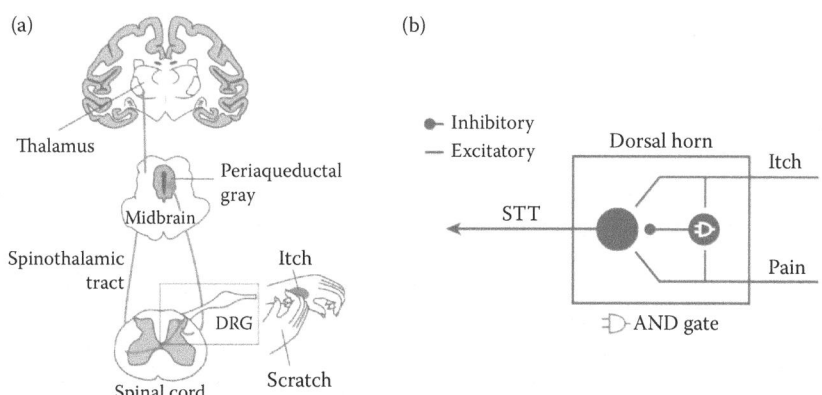

FIGURE 22.6 A schematic showing hypothesized circuitry regulating the inhibition of itch by scratching. (a) Descending modulatory system involving the periaqueductal gray may synapse onto dorsal horn inhibitory interneurons to inhibit itch. (b) Dorsal horn circuitry: the output from itch-signaling spinothalamic tract neurons could be inhibited by activation of an inhibitory interneuron. To satisfy the state-dependency of inhibition, this inhibitory interneuron could operate as an AND gate, discharging when summed inputs from itch-sensitive and nociceptive fibers occur together. (Adapted from Davidson, S. and Giesler, G. J., Jr., *Trends in Neuroscience*, 33, 550–8, 2010.)

putamen activity may reflect anticipation of the pleasure of scratching during itch (Bin Saif et al. 2012).

22.5   CIRCUITRY OF RELIEF

Melzack and Wall's (1965) influential gate control theory postulated a central neural circuit capable of modulating pain that can be adapted to also explain the inhibition of itch. Central to the theory was the proposal that functionally diverse inputs into the spinal cord activate not only ascending pathways but also a spinal inhibitory system that modulates the ascending output. Research into the nature of the inhibitory circuitry within the dorsal horn is beginning to provide the anatomical and functional evidence needed to understand the modulation of pruriceptive signals by counterstimuli. Spinal inhibition of itch by scratching is hypothesized to occur when C-fibers activated by scratching engage a central inhibitory interneuron which, in turn, inhibits the spinal neurons excited by the pruritogen activated C-fibers (Figure 22.6b). The existence of a fundamental central circuit permitting C-fiber mediated inhibition has been identified within the substantia gelatinosa in a series of elegant studies by Perl and colleagues (Lu and Perl 2003; Zheng et al. 2010). The activating stimuli of the peripheral fibers have not yet been determined, so that we do not yet know whether a pruritic agent engages this pathway. However, it was demonstrated that the postsynaptic signal from the slowest-conducting, high-threshold C-fibers was inhibited by activation of the more rapidly conducting, lower-threshold C-fibers. Because histamine-responsive C-fibers have been identified as some of the slowest-conducting and highest-threshold group (Schmelz et al. 1997), it seems reasonable to hypothesize that a spinal circuit to block itch could look similar.

The question of whether an inhibitory spinal neuron capable of suppressing itch exists was recently addressed in a study that identified the *bhlhb5* expressing neuron as playing just such a role (Ross et al. 2010). Loss of *bhlhb5* neurons in mice induced ongoing scratching behavior and heightened responses to applied itch-producing agents, suggesting that itch is under tonic inhibitory control in the spinal cord, and that disinhibition can lead to exacerbated itch and some forms of chronic pruritus (Ross et al. 2010). Further evidence for spinal inhibitory interneurons acting to suppress itch was established when it was determined that blocking GABA or glycine in the spinal cord reduced scratch-evoked suppression of ongoing activity in spinal neurons in a mouse model of itch (Akiyama et al. 2011b). Additionally, lesion of the rostral spinal cord also disrupted the ability for scratch to inhibit ongoing activity, suggesting a role for descending inhibition for itch (Akiyama et al. 2011b).

For spinal inhibitory interneurons to suppress itch, it is expected that they would be activated by nociceptive primary afferent fibers that respond to counterstimuli that suppress itch. To test the role of nociceptive input on itch, the presynaptic activity from nociceptive primary afferent fibers was reduced by genetic deletion of the vesicular glutamate transporter-2 (*vglut2*). This loss of glutamatergic input into the dorsal horn created a mouse model of ongoing pruritus, suggesting that a basal level of glutamatergic signaling establishes a tonic inhibition of itch (Lagerstrom et al. 2010; Liu et al. 2010). The loss of glutamate signaling from nociceptors also lead to reduced capsaicin-evoked c-fos expression in interneurons of the superficial dorsal horn expressing neuropeptide Y, a marker for spinal inhibitory interneurons (Liu et al. 2010). Together these studies provide evidence for the existence of a spinal inhibitory neuron that is located postsynaptic to nociceptive afferent fibers and that plays a critical role in the suppression of itch.

In addition to activation of an itch suppressing inhibitory interneuron by nociceptive fibers, there is evidence that such an inhibitory neuron may also be activated by pruriceptive fibers. Itch is enhanced if histamine is delivered to a locally anesthetized area, suggesting the anesthetic blocked histamine-sensitive fibers that decrease itch (Atanassoff et al. 1999). Additionally, scratching inhibited spinothalamic tract neurons only in the presence of ongoing itch, suggesting that activation of pruritic fibers is crucial in making an inhibitory neuron available to inhibit the itch signal (Davidson et al. 2009). These observations invite the hypothesis that inhibitory interneurons responsible for inhibiting itch may be postsynaptic to both itch and pain fibers and that simultaneous activation, such as through an "AND gate," is necessary to inhibit itch (Figure 22.6b).

REFERENCES

Akiyama, T., Carstens, M. I. and Carstens, E. 2009a. Excitation of mouse superficial dorsal horn neurons by histamine and/or PAR-2 agonist: Potential role in itch. *Journal of Neurophysiology* 102: 2176–83.

Akiyama, T., Merrill, A. W., Carstens, M. I. and Carstens, E. 2009b. Activation of superficial dorsal horn neurons in the mouse by a PAR-2 agonist and 5-HT: Potential role in itch. *Journal of Neuroscience* 29: 6691–9.

Akiyama, T., Carstens, M. I. and Carstens, E. 2011a. Enhanced responses of lumbar superficial dorsal horn neurons to intradermal PAR-2 agonist but not histamine in a mouse hindpaw dry skin itch model. *Journal of Neurophysiology* 105: 2811–7.

Akiyama, T., Carstens, M. I. and Carstens, E. 2011b. Transmitters and pathways mediating inhibition of spinal itch-signaling neurons by scratching and other counterstimuli. *PloS One* 6: e22665.

Akiyama, T., Tominaga, M., Carstens, M. I. and Carstens, E. E. 2012. Site-dependent and state-dependent inhibition of pruritogen-responsive spinal neurons by scratching. *European Journal of Neuroscience* 36: 2311–6.

Andrew, A. and Craig, A. D. 2001. Spinothalamic lamina I neurons selectively sensitive to histamine: A central neural pathway for itch. *Nature Neuroscience* 4: 72–7.

Atanassoff, P. G., Brull, S. J., Zhang, J., Greenquist, K., Silverman, D. G. and Lamotte, R. H. 1999. Enhancement of experimental pruritus and mechanically evoked dysesthesiae with local anesthesia. *Somatosensory and Motor Research* 16: 291–8.

Banzet, P. M. 1927. *La Cordotomie: Etude Anatomique, Technique, Clinique, et Physiologique.* Paris: Libraire Pouis Arnette.

Bickford, R. G. 1938. Experiments relating to the itch sensation, its peripheral mechanism and central pathways. *Clinical Science* 3: 377–86.

Bin Saif, G. A., Papoiu, A. D., Banari, L. et al. 2012. The pleasurability of scratching an itch: A psychophysical and topographical assessment. *British Journal of Dermatology* 166: 981–5.

Brull, S. J., Atanassoff, P. G., Silverman, D. G., Zhang, J. and Lamotte, R. H. 1999. Attenuation of experimental pruritus and mechanically evoked dysesthesiae in an area of cutaneous allodynia. *Somatosensory and Motor Research* 16: 299–303.

Carstens, E. 1997. Responses of rat spinal dorsal horn neurons to intracutaneous micro-injection of histamine, capsaicin, and other irritants. *Journal of Neurophysiology* 77: 2499–514.

Davidson, S., Zhang, X., Yoon, C. H., Khasabov, S. G., Simone, D. A. and Giesler, G. J., Jr. 2007. The itch-producing agents histamine and cowhage activate separate populations of primate spinothalamic tract neurons. *Journal of Neuroscience* 27: 10007–14.

Davidson, S., Zhang, X., Khasabov, S. G., Simone, D. A. and Giesler, G. J., Jr. 2009. Relief of itch by scratching: State-dependent inhibition of primate spinothalamic tract neurons. *Nature Neuroscience* 12: 544–6.

Davidson, S. and Giesler, G. J., Jr. 2010. The multiple pathways for itch and their interactions with pain. *Trends in Neuroscience* 33: 550–8.

Davidson, S., Zhang, X., Khasabov, S. G. et al. 2012. Pruriceptive spinothalamic tract neurons: Physiological properties and projection targets in the primate. *Journal of Neurophysiology* 108: 1711–23.

Foerster, O. 1936. Symptomatologic der erkankungen des ruckenmarks und seiner wurzeln. In *Handuch der Neurologied*, eds. O. Bumke and O. Foerster. Berlin: Springer-Verlag.

Graf, C. J. 1960. Consideration in loss of sensory level after bilateral cervical cordotomy. *Archives of Neurology* 3: 410–5.

Graham, D. T., Goodell, H. and Wolff, H. G. 1951. Neural mechanisms involved in itch, itchy skin, and tickle sensations. *Journal of Clinical Investigation* 30: 37–49.

Herde, L., Forster, C., Strupf, M. and Handwerker, H. O. 2007. Itch induced by a novel method leads to limbic deactivations a functional MRI study. *Journal of Neurophysiology* 98: 2347–56.

Hyndman, O. R. and Wolkin, J. 1943. Anterior cordotomy: Further observations on physiologic results and optimum manner of performance. *Archives of Neurology and Psychiatry* 50: 129–48.

Ikoma, A., Fartasch, M., Heyer, G., Miyachi, Y., Handwerker, H. and Schmelz, M. 2004. Painful stimuli evoke itch in patients with chronic pruritus: Central sensitization for itch. *Neurology* 62: 212–7.

Jinks, S. L. and Carstens, E. 1998. Skin cooling attenuates rat dorsal horn neuronal responses to intracutaneous histamine. *Neuroreport* 8: 4145–9.

Jinks, S. L. and Carstens, E. 2002. Responses of superficial dorsal horn neurons to intra-dermal serotonin and other irritants: Comparison with scratching behavior. *Journal of Neurophysiology* 87: 1280–9.

Johanek, L. M., Meyer, R. A., Hartke, T. et al. 2007. Psychophysical and physiological evidence for parallel afferent pathways mediating the sensation of itch. *Journal of Neuroscience* 27: 7490–7.

Johanek, L. M., Meyer, R. A., Friedman, R. M. et al. 2008. A role for polymodal C-fiber afferents in nonhistaminergic itch. *Journal of Neuroscience* 28: 7659–69.

Klein, A., Carstens, M. I. and Carstens, E. 2011. Facial injections of pruritogens or algogens elicit distinct behavior responses in rats and exite overlapping poopulations of primary sensory and trigeminal subnucleus cadudalis neurons. *Journal of Neurophysiology* 106: 1078–88.

Kosteletzky, F., Namer, B., Forster, C. and Handwerker, H. O. 2009. Impact of scratching on itch and sympathetic reflexes induced by cowhage (Mucuna pruriens) and histamine. *Acta Dermato-Venereologica* 89: 271–7.

Lagerstrom, M. C., Rogoz, K., Abrahamsen, B. et al. 2010. VGLUT2-dependent sensory neurons in the TRPV1 population regulate pain and itch. *Neuron* 68: 529–42.

Liu, Y., Abdel Samad, O., Zhang, L. et al. 2010. VGLUT2-dependent glutamate release from nociceptors is required to sense pain and suppress itch. *Neuron* 68: 543–56.

Lu, Y. and Perl, E. R. 2003. A specific inhibitory pathway between substantia gelatinosa neurons receiving direct C-fiber input. *Journal of Neuroscience* 23: 8752–8.

Magerl, W., Westerman, R. A., Möhner, B. and Handwerker, H. O. 1990. Properties of transdermal histamine iontophoresis: Differential effects of season, gender, and body region. *Journal of Investigative Dermatology* 94: 347–52.

Melzack, R. and Wall, P. D. 1965. Pain mechanisms: A new theory. *Science* 150: 971–9.

Mochizuki, H., Tashiro, M., Kano, M., Sakurada, Y., Itoh, M. and Yanai, K. 2003. Imaging of central itch modulation in the human brain using positron emission tomography. *Pain* 105: 339–46.

Mochizuki, H., Sadato, N., Saito, D. N. et al. 2007. Neural correlates of perceptual difference between itching and pain: A human fMRI study. *NeuroImage* 36: 706–17.

Nilsson, H. J., Levinsson, A. and Schouenborg, J. 1997. Cutaneous field stimulation (CFS): A new powerful method to combat itch. *Pain* 71: 49–55.

Reddy, V. B., Iuga, A. O., Shimada, S. G., LaMotte, R. H. and Lerner, E. A. 2008. Cowhage-evoked itch is mediated by a novel cysteine protease: A ligand of protease-activated receptors. *Journal of Neuroscience* 28: 4331–5.

Ross, S. E., Mardinly, A. R., McCord, A. E. et al. 2010. Loss of inhibitory interneurons in the dorsal spinal cord and elevated itch in Bhlhb5 mutant mice. *Neuron* 65: 886–98.

Schmelz, M., Schmidt, R., Bickel, A., Handwerker, H. O. and Torebjork, H. E. 1997. Specific C-receptors for itch in human skin. *Journal of Neuroscience* 17: 8003–8.

Schmelz, M., Schmidt, R., Weidner, C., Hilliges, M., Torebjork, H. E. and Handwerker, H. O. 2003. Chemical response pattern of different classes of C-nociceptors to pruritogens and algogens. *Journal of Neurophysiology* 89: 2441–8.

Schmidt, R., Schmelz, M., Weidner, C., Handwerker, H. O. and Torebjork, H. E. 2002. Innervation territories of mechano-insensitive C nociceptors in human skin. *Journal of Neurophysiology* 88: 1859–66.

Shelley, W. B. and Arthur, R. P. 1955. Studies on cowhage (Mucuna pruriens) and its pruritogenic proteinase, mucunain. *Archives of Dermatology* 72: 399–406.

Sherrington, C. S. 1906. Observations on the scratch-reflex in the spinal dog. *Journal of Physiology* 34: 1–50.

Simone, D. A., Ngeow, J. Y., Whitehouse, J., Becerra-Cabal, L., Putterman, G. J. and LaMotte, R. H. 1987. The magnitude and duration of itch produced by intracutaneous injections of histamine. *Somatosensory Research* 5: 81–92.

Simone, D. A., Zhang, X., Li, J. et al. 2004. Comparison of responses of primate spinothalamic tract neurons to pruritic and algogenic stimuli. *Journal of Neurophysiology* 91: 213–22.

Taren, J. A. and Kahn, E. A. 1966. Thoracic anterolateral cordotomy. In *Pain*, eds. R. S. Knighton and P. R. Dumke. Boston: Little Brown 299–310.

Vierow, V., Fukuoka, M., Ikoma, A., Dorfler, A., Handwerker, H. O. and Forster, C. 2009. Cerebral representation of the relief of itch by scratching. *Journal of Neurophysiology* 102: 3216–24.

Ward, L., Wright, E. and McMahon, S. B. 1996. A comparison of the effects of noxious and innocuous counterstimuli on experimentally induced itch and pain. *Pain* 64: 129–38.

Wei, J. Y. and Tuckett, R. P. 1991. Response of cat ventrolateral spinal axons to an itch-producing stimulus (cowhage). *Somatosensory and Motor Research* 8: 227–39.

White, J. C., Sweet, W. H., Hawkins, R. and Nilges, R. G. 1950. Anterolateral cordotomy: Results, complications and causes of failure. *Brain* 73: 346–67.

Yosipovitch, G., Duque, M. I., Fast, K., Dawn, A. G. and Coghill, R. C. 2007. Scratching and noxious heat stimuli inhibit itch in humans: A psychophysical study. *British Journal of Dermatology* 156: 629–34.

Yosipovitch, G., Ishiuji, Y., Patel, T. S. et al. 2008. The brain processing of scratching. *Journal of Investigative Dermatology* 128: 1806–11.

Zheng, J., Lu, Y. and Perl, E. R. 2010. Inhibitory neurones of the spinal substantia gelatinosa mediate interaction of signals from primary afferents. *Journal of Physiology* 588: 2065–75.

23 Brain Processing of Itch and Scratching

Hideki Mochizuki, Alexandru D.P. Papoiu, and Gil Yosipovitch

CONTENTS

23.1 OVERVIEW

In this chapter we discuss the findings of neuroimaging studies that investigated the processing of itch and scratching in the superior relays of central nervous system (CNS), when itch was induced experimentally or when pathological pruritus was studied as a preexisting condition in chronic diseases. Imaging the central processing of itch can lead to a better understanding of itch perception and its mechanisms, provide insight into potentially altered CNS processing of pruritus in disease states, and identify sources of central sensitization. Brain imaging can detect functional and anatomic changes in chronic itch and can ultimately indicate how itch can be inhibited for therapeutic purposes.

23.2 ITCH TRANSMISSION TO THE SUPERIOR CENTERS OF THE CENTRAL NERVOUS SYSTEM

23.2.1 From Spinal Cord to the Thalamus

Current models for neuronal itch transmission describe distinct pathways for histamine-mediated and nonhistaminergic forms of itch. To date, no consensus has been reached on whether itch is conveyed via a dedicated ("labeled") line or whether it is encoded by the differential participation of subpopulations of neurons expressing specific pruriceptors. It is possible that itch-specific information is primarily encoded at the spinal level by an interplay of inputs from peripheral afferents, spinal cord interneurons, inhibitory inputs from strictly nociceptive afferents, and top–bottom inhibitory inputs from higher CNS structures (such as the periaqueductal gray formation). It was also proposed that neurons in the thalamus can distinguish nociceptive versus pruriceptive information via a selection mechanism (Davidson and Giesler 2010). Two main neuronal pathways have been described for itch transmission: one mediated by histamine and the other by protease activated receptors PAR2(4) receptors that can be exogenously stimulated by spicules of cowhage (*Mucuna pruriens*). Upon skin contact cowhage releases a cysteine protease, mucunain, that acts as a ligand for PAR2 receptors. The histamine and cowhage itch modalities are conveyed via distinct peripheral C- and A-delta fibers that synapse in the dorsal horn of the spinal cord. The continuing spinothalamic pathways ascend via the lateral spinothalamic tract and (mostly) maintain their specialization, up to their thalamic (third neuron) stations (Davidson et al. 2009, 2012).

It is now clearly established that a major station for itch transmission is located in the thalamus. Histamine-responsive and cowhage-responsive spinothalamic neurons project in the ventral posterior lateral (VPL), ventral posterior inferior, and posterior nuclei, while cowhage-sensitive neurons additionally end in the suprageniculate and medial geniculate nuclei (Davidson et al. 2012). These studies using antidromic stimulation in nonhuman primates and neuroimaging studies in humans (Papoiu et al. 2012a) were in good agreement showing that cowhage itch evoked a more extensive activation of thalamic nuclei than histamine itch. Functional magnetic resonance imaging (MRI) data also suggested that cowhage itch activated more strongly a thalamic area consistent with the location of the mediodorsal nucleus (Leknes et al. 2007; Papoiu et al. 2012a), which is connected with the orbitofrontal area and the limbic system. Of note, a dysfunction of this circuit was suggested to occur in processing itch in atopic patients (Leknes et al. 2007).

23.2.2 From Thalamus to the Cortex

Because the exact projections of the third (thalamic) neurons conveying itch information from the thalamus to the cerebral cortex have not been identified, they can be inferred from brain imaging studies. Histamine has been the stimulus mostly used as the experimental pruritogen for brain imaging of itch (Hsieh et al. 1994; Darsow et al. 2000; Drzezga et al. 2001; Mochizuki et al. 2003, 2007, 2009; Herde et al. 2007; Valet et al. 2008; Leknes et al. 2007; Schneider et al. 2008; Ishiuji et al. 2009). Considering the lack of effectiveness of antihistamines in pathological forms of itch (e.g., atopic dermatitis and End-Stage-Renal-Disease [ESRD] pruritus) and the involvement of PAR2 in itch of atopic eczema, the PAR2/cowhage model has been proposed as a more suitable experimental model for chronic pruritus. Therefore, more recent neuroimaging studies employed both experimental models of itch induction, using histamine and cowhage for itch stimulation. When the patterns of brain activation evoked by these modalities were compared, common features, but also significant differences were observed (Papoiu et al. 2012a) — *vide infra* (part 3).

23.3 BRAIN AREAS ACTIVATED DURING ITCH PROCESSING

23.3.1 GENERAL OBSERVATIONS

The multidimensional character of itch experience is reflected in the complexity of itch processing in the brain. The experimental induction of itch leads to simultaneous and long-term activations in multiple brain areas. Itch processing activates somatosensory areas (S1, S2) and interoceptive areas such as insular cortex, and it is accompanied by an invariable emotional-affective response recruiting deep-seated areas of the limbic system, areas connected to craving, pleasure and addiction. Itch is a bothersome, intrusive, acute sensation, requesting an immediate action response; thus major arms of the cerebral itch response are involved in refocusing attention, planning the motor action and seeking itch relief. Numerous brain imaging studies—irrespective of the investigative modalities—agreed in their findings that these interconnected responses occur in the brain almost simultaneously.

23.3.2 PRINCIPAL AREAS IMPLICATED IN ITCH PROCESSING

The regions most frequently activated in the majority of brain imaging studies, using Positron Emission Tomography (PET) or functional Magnetic Resonance Imaging (MRI) techniques (Blood Oxygen Level Dependent [BOLD] or Arterial Spin Labeling [ASL]) were areas specialized in receiving somatosensory inputs: (S1), the associative/secondary somatosensory area S2, areas regulating emotional control and affective responses (pleasantness or unpleasantness), such as the cingulate cortex, in both its anterior and posterior divisions (ACC and PCC), and were linked with evaluative functions and decision-making, such as the dorsolateral prefrontal cortex (DLPFC). Premotor, motor, and supplementary motor areas as well as the cerebellum, which controls an effective/actual or planned response to initiate scratching, are also seen as inseparable from the sensory activations evoked by itch. The activation of premotor cortex and supplementary motor area (SMA) areas are seen in the majority of itch imaging studies, even when scratching itself is not allowed during the brain imaging experiments.

Areas related to memory retrieval, visuospatial processing and self-awareness such as the precuneus (medial parietal cortex) are prominently activated by the sensation of itch. The involvement of this area in itch processing is interesting, because it is not as heavily implicated in the processing of pain. The affective and emotional aspects of itch experience are significantly represented in the activation of ACC, amygdala, and in the subcallosal gray matter–nucleus accumbens (NAcc) (Leknes et al. 2007; Papoiu et al. 2012a, 2013). NAcc has been shown to be involved in itch processing experimentally induced-allergen itch in atopic patients and was found to be activated at rest (in baseline conditions) in ESRD patients on hemodialysis experiencing chronic pruritus.

The insular cortex has a paramount role in processing itch information. Insula is a cortical region linked to salience, self-awareness/interoception and addiction. Insula is considered a major hub for processing visceroceptive and interoceptive inputs, and it is also significantly involved in the processing of pain and especially in assessing stimulus intensity. Bilateral insula has been found to be activated in patients with ESRD pruritus at rest.

A discrete gray matter area whose role has been recently emphasized in itch processing is the claustrum. This area's functional specialization and connectivity seem very fitting for a region involved with itch sensing, because it has the capability to analyze, compare, and integrate sensory information from various inputs; it is connected to almost all areas of the cortex but especially with the somatosensory cortex, thalamus, and with the limbic structures: cingulate cortex, hippocampus and amygdala. The claustrum is closely linked with insula, anatomically and functionally. It has been shown to be activated more extensively by a complex, transient, fluctuating itch stimulus such as cowhage than by a rather constant stimulus like histamine. The activation of claustrum as well as of insula was largely correlated with the perceived itch intensity, while some discrete areas within the same structures activated irrespective of itch stimulus intensity. It is noteworthy that only very few brain areas insula, claustrum, S2, and the angular gyrus displayed this complex pattern of activations: encompassing regions that were capable to respond either in a manner directly correlated with itch intensity or responding invariably to itch, irrespective of stimulus intensity. Insula and claustrum, in particular, were activated continuously while itch intensity varied (slowly decreased) and were fully activated bilaterally when histamine and cowhage stimuli were administered at the same time. These features suggest a principal role in itch processing for these regions.

23.4 DIFFERENTIAL CEREBRAL PROCESSING OF HISTAMINE AND COWHAGE ITCHES

In healthy individuals, the cerebral representation of cowhage itch displayed a common core of activation with histamine itch, while showing certain features that were clearly unique (Figure 23.1). We contend that because the itches induced by histamine and cowhage shared important common features they consolidate a *central picture of activation* for itch processing in the brain. At the same time, in contrast to histamine, cowhage evoked a more extensive activation of the insular cortex, claustrum, globus pallidus, caudate body, putamen, and thalamic nuclei on the contralateral side of stimuli. It is conceivable that these differences may be related not only to an intrinsic specificity in cortical projection, but also to the fluctuating quality and associated nociceptive signaling such as stinging, burning elicited by cowhage. As in many cases of chronic itch these sensations are frequently reported (Yosipovitch et al. 2002).

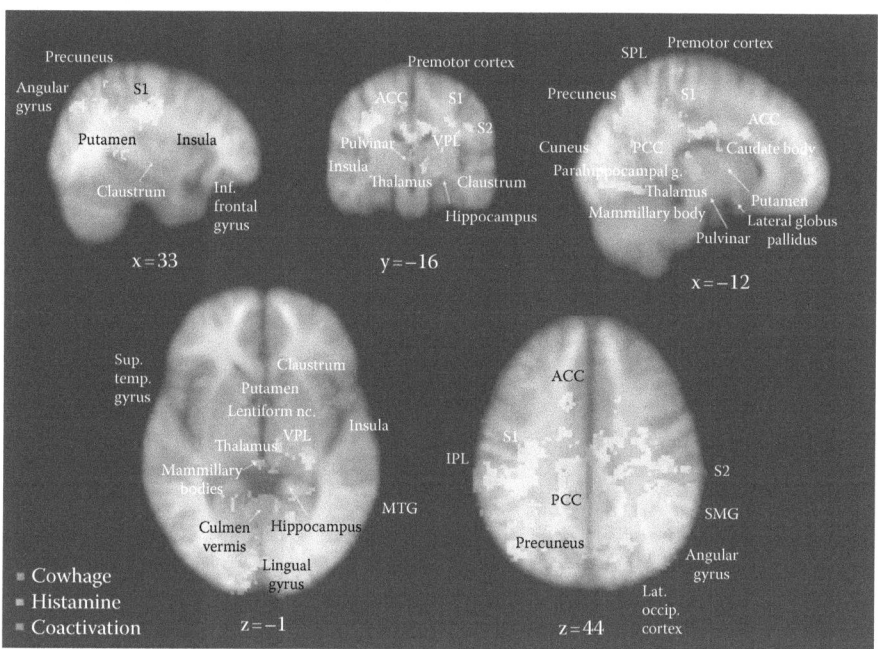

FIGURE 23.1 **(See color insert.)** Processing of cowhage and histamine itches in healthy individuals. The overlap of brain activations induced by histamine itch (in green) and by cowhage itch (in blue) illustrates the regions coactivated (in red) and distinct areas activated separately by the two itch pathways. Standard Talairach space coordinates. The color tones displayed correspond to Z score values as shown in the color bar. ACC, anterior cingulate cortex; PCC, posterior cingulate cortex; SPL, superior parietal lobule; M1, primary motor cortex; S1, primary somatosensory area; SMG, supramarginal gyrus; MTG, middle temporal gyrus; IPL, inferior parietal lobule; S2, secondary somatosensory area; VPL, ventral posterior lateral nucleus (of thalamus).

23.5 CHARACTERISTICS OF THE BRAIN PROCESSING OF PRURITUS IN CHRONIC ITCH CONDITIONS

The patterns of associations between activation of various brain areas and the perceived itch intensity appeared to vary with the underlying context of the disease and also to differ in their relationship with disease severity. For example, in atopic dermatitis patients, activation of certain areas such as ACC and dorsolateral prefrontal cortex (DLPFC) was directly correlated with disease severity (as measured by standardized clinical measures such as the EASI score), while histamine itch intensity correlated with activations in the ACC and insula. Overall, the pattern of associations between activations and perceived itch intensity was different than in healthy volunteers (Schneider et al. 2008; Ishiuji et al. 2009; Yosipovitch Group, unpublished).

The distinction between the patterns evoked by histamine and cowhage itches, clearly identified in healthy individuals appears to be blurred in chronic itch diseases

(ESRD and AD). Stimulus-specific patterns of activation appear differentially nuanced in a different pathological context. An exact cause for the blurring of this distinction is not apparent at the time of this writing.

An investigation of structural and functional perfusion differences between ESRD patients with chronic pruritus and healthy individuals found a significant thinning of gray matter in the thalamus, insula, ACC, precuneus, and caudate body in ESRD patients (areas involved in itch processing) as well as significant activations (persistent perfusion increases) at baseline in the insula, ACC, claustrum, amygdala, hippocampus and nucleus accumbens. Moreover, the processing of cowhage itch appeared altered in ESRD, while no significant differences could be demonstrated in processing histamine itch. In ESRD pruritus, multiple brain activations appeared to work either directly or inversely correlated with perceived itch intensity, suggesting a dual modulation of itch perception. These unique features could be facilitated by the reduced gray matter thickness in ESRD affecting critical areas involved in itch processing, thus revealing an unprecedented form of neocortical plasticity (and functional reorganization). In this condition, it appeared that the PAR-2 mediated itch pathway was already overstimulated, which could be linked to an overexpression of PAR-2 in the skin. This could lead to a tonic inhibition of the cortical processing of acute cowhage itch, when induced in the preexistent context of ESRD.

A working hypothesis for central top-to-bottom inhibition of itch is drawn from a parallel with the descending pathway for suppression of pain and proposes that the periaqueductal gray matter (PAG) modulates the activity of spinal interneurons. According to this model, descending inputs directed toward itch receptive neurons in the dorsal horn exert an inhibitory action, effectively silencing them (Carstens et al. 1997; Davidson and Giesler 2010). Another possibility is that the cortical projection of itch information into S1/S2 could be inhibited via cortico–cortical inhibitory loops, in a similar fashion with mechanisms that have been known to operate in chronic pain (reviewed in Henry et al. 2011). Our recent findings in ESRD patients on hemodialysis experiencing chronic pruritus suggest that a tonic inhibition may be exerted at the neocortical level to selectively limit the receptive fields for PAR-2-mediated itch processing in S1, precuneus and insula (Papoiu et al. 2013). The activation of these projection areas was significantly limited in contrast with healthy volunteers, although cowhage itch was perceived at a similar intensity. These findings were corroborated by a pattern of inverse correlations between activations in S1, S2, precuneus, and perceived itch intensity, supporting the notion that activation is modulated by negative feedback loops (at least in chronic pruritus of ESRD). These findings are of significant interest because they offer insight into mechanisms the brain may employ to process and modulate itch sensation. Contrary to a widely accepted paradigm in the neuroimaging literature, it is thus possible that a higher intensity itch does not necessarily translate into a higher or more extensive activation of the cerebral cortex. For example, cowhage-induced itch in ESRD patients is significantly more intense than histamine itch, but the cerebral activations evoked by cowhage are *less* extensive and as mentioned earlier, they are also significantly limited in comparison with healthy volunteers, in the regions of interest related to itch processing.

23.6 CONTAGIOUS ITCH

Contagious itch is an intriguing phenomenon that has been reported frequently in daily life and in medical circles and that has been confirmed in no less than four published studies in humans and one study in nonhuman primates (Niemeier et al. 2000; Papoiu et al. 2011; Holle et al. 2012; Lloyd et al. 2012; Feneran et al. 2012). The elucidation of central mechanisms underlying this intriguing phenomenon is of high interest because it could provide invaluable clues for the treatment of itch. It manifests as the endogenous induction of a surreptitious feeling of itch in an observer while looking at other people scratching or in people being presented visual cues merely suggestive of itch. It has been postulated that contagious itch is similar to other socially contagious behaviors (yawning), that it might be correlated with empathy or proneness to neurosis, and that it may work via "mirror neurons." Contagious itch was significantly easier to induce in atopic dermatitis sufferers (which could be already "sensitized" because they suffer from chronic itch) than in healthy volunteers. Interestingly, the itch induced by visual cues had a scattered, wide body distribution. The exact mechanism of itch "triggered by sight" or by mental suggestion is poorly understood. A recent brain imaging study set out to identify the neural brain networks involved in the generation of "contagious itch" sensation pointed toward Brodmann area 44 (BA44) and the premotor cortex (BA6) as areas activated during the process of "itch induction". These findings can be seen as a first step in attempting to elucidate this phenomenon (Holle et al. 2012). Just from a fundamental neuroscientific point of view, it is of significant interest to identify the brain centers that can support the generation of a somatic sensation in the absence of peripheral stimulation. In addition, it can provide insight on the specific central areas that can be targeted and used in a therapeutic approach to relieve itch. At this point in time, the existence of a single, specific itch processing center remains elusive, becoming increasingly clear that the complex neuronal processes involved in processing itch cannot be reduced to a single cortical or subcortical area.

23.7 CRAVING FOR ITCH RELIEF AND ITS CEREBRAL MECHANISMS

The sensation of itch and the immediate craving for itch relief manifested as the urge to scratch are inseparable. Reward circuits have been linked with the pleasurability of scratching and could play an important role in itch inhibition, as well in the formation of "vicious" itch scratch cycles, because it is not only the suppression of the sensory aspect of itching but also the quenching of bothersome-irritating emotional feelings that itch evokes, which is equally sought and ultimately rewarding. Novel findings suggest that the reward system of the midbrain, more specifically the ventral tegmentum (VTA), and substantia nigra, as well as nc. accumbens (NAcc), may play a role in the urge to scratch and the subsequent satisfaction (pleasure) derived from scratching, via their connections with the insula, ACC, and putamen. The implication of VTA and NAcc underscores the addictive nature of the itch–scratch cycle and also suggests a role for the dopaminergic system in itch relief (Papoiu et al. 2013). A possible mechanistic link to the alterations of sleep/wakefulness cycle—leading

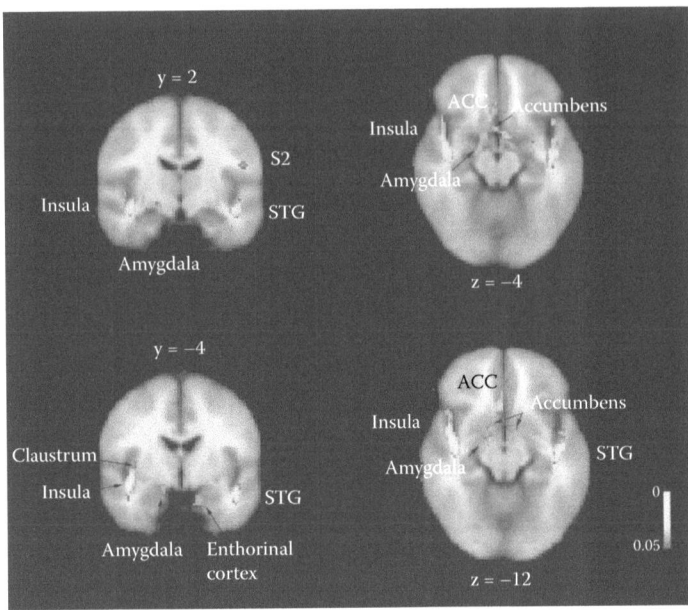

FIGURE 23.2 (**See color insert.**) Brain activations in patients with ESRD pruritus at baseline. Cerebral perfusion was higher at baseline in ESRD patients with chronic pruritus compared to healthy individuals in the insula, claustrum, ACC, amygdala, enthorinal cortex, subcallosal gray matter—nucleus accumbens. Arterial Spin Labeling fMRI; $p < 0.05$. STG, superior temporal gyrus; S2, secondary somatosensory area; ACC, anterior cingulate cortex.

to depression, mood changes frequently reported in chronic pruritus—could be explained by the connections of nucleus accumbens with thalamocortical circuits. An activation of NAcc was observed at rest in ESRD pruritus (Figure 23.2).

23.8 BRAIN PROCESSING OF SCRATCHING

In this section we briefly review neuroimaging studies that were focused on the brain mechanisms of scratching and discuss possible mechanisms of itch inhibition that were proposed to be exerted by scratching. We also address the relationship between pleasantness and scratching. It is interesting to observe that areas of the brain that are engaged by scratching appear to work excessively in chronic itch patients. Scratching was found to activate certain brain networks that can further lead to an amplification of this behavior. A working hypothesis is that the pleasant sensations evoked by scratching may induce and reinforce the unstoppable drive to continue scratching.

Several mechanisms have been proposed to explain the cathartic effect of scratching (Davidson and Giesler 2010). Most of them imply a regulatory control at the spinal level, leading to itch inhibition via the intercession of interneurons. These spinal relays are controlled in a "state-dependent" manner in such a way that scratching exerts an inhibition only when sensory itch input fibers are concomitantly stimulated

(Davidson et al. 2009). Much less is known about the supraspinal (cortical) areas capable of inhibiting itch. A reliable way to visualize in neuroimaging terms the successful relief of itch would translate into a deactivation of ACC and of the insular cortex. Our recent studies suggest that the action of self-scratching an itch is more rewarding compared to a "passive" form of scratching delivered exogenously by an investigator, as it induced a more robust deactivation of ACC, insula, and the ventral tegmental area (Papoiu et al. 2013, *PLoS One*, in press).

23.8.1 BRAIN REGIONS ASSOCIATED WITH THE DESIRE TO SCRATCH

Itch promptly evokes the desire to scratch. The responses elicited in the brain are strongly manifested in multiple regions that have a motor specialization. The activation of supplementary motor area (SMA) and premotor cortex (PM) during itch stimulation was frequently observed in most of the previous studies, irrespective of the particular neuroimaging techniques used (PET or fMRI) (Hsieh et al. 1994; Darsow et al. 2000; Drzezga et al. 2001; Mochizuki et al. 2003, 2007, 2009; Herde et al. 2007; Valet et al. 2008; Leknes et al. 2007; Schneider et al. 2008; Ishiuji et al. 2009; Papoiu et al. 2012a). The location of these regions in the brain is shown in Figure 23.3. In animal studies, it was reported that monkeys with lesions in the SMA and PM cortices showed a significant deterioration of task performances when the tasks required complex movement, such as reaching out their hands to target items, while avoiding obstacles and moving their hands and arms in a correct order (e.g., Brinkman 1984; Thaler et al. 1995; Moll and Kuypers 1979; Halsband and Passingham 1985). Thus, SMA and PM are considered to be associated with planning movement. In fact, neurons in SMA and PMC begin to activate before motor execution (Kurata and Wise 1988; Tanji and Shima 1994). Evidently, the cerebral processes of motor planning need to be performed before motor execution. Interestingly, functional neuroimaging studies in humans reported that SMA and PM were also activated when subjects just imagined moving their hands and arms, without execution (e.g., Lotze et al. 1999; Naito et al. 2002; Ehrsson et al. 2003; Meister et al. 2004; Ueno et al. 2010). Considering these previous findings, the activation of SMA and PM during itch stimulation may reflect the preparation for scratching. SMA and PM are anatomically and functionally connected to the striatum, which is also activated by itch (Herde et al. 2007; Leknes et al. 2007; Mochizuki et al. 2007; Papoiu et al. 2012a). The striatum may also play additional roles in itch perception and motivation. Motivation is an important factor in triggering actions and improving performance (Apicella et al. 1991; Schultz et al. 1992; Bowman et al. 1996; Breiter et al. 2001; Delgado et al. 2004; Harsay et al. 2011). Moreover, if the activation of the striatum during itch stimulation would reflect only motor planning, this region would be activated during an imagery task such as the imagination of finger movement. However, evidence showed that the striatum was not activated during motor imagery (e.g., Lotze et al. 1999; Naito et al. 2002; Ehrsson et al. 2003; Meister et al. 2004; Ueno et al. 2010). Therefore, the activation of the striatum during itch stimulation may reflect the motivation to scratch. The activity in the striatum in atopic dermatitis patients was significantly higher than in healthy subjects (Schneider et al. 2008; Ishiuji et al. 2009). In other words, the neural circuits associated with motivation

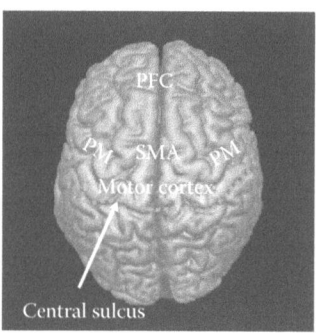

FIGURE 23.3 **(See color insert.)** Top view of MNI template of the human brain. PFC, prefrontal cortex; SMA, supplementary motor area; PM, premotor cortex; Motor cortex, primary motor cortex.

and motor planning of scratching were more strongly activated in atopic dermatitis patients when they perceived itch. Atopic patients are indeed troubled by severe skin damage and exacerbation of itch, since they scratch excessively and repeatedly. This phenomenon could be associated with the "excessive" activation of the striatum.

Previous PET and fMRI studies of itch have noted the activation of the primary motor cortex (M1) in the hemisphere contralateral to itch stimulation (Darsow et al. 2000; Drzezga et al. 2001; Ishiuji et al. 2009). A possible interpretation of this observation is that motor control was voluntarily exerted to refrain from moving the involved arm (Figure 23.3).

23.8.2 BRAIN IMAGING AND PROCESSING OF SCRATCHING REVEAL ITCH MODULATION MECHANISMS

A couple of fMRI studies have investigated the cerebral processing of the responses evoked by passive scratching, when investigators scratched the subjects' skin using small plastic brushes or copper plates (Yosipovitch et al. 2008; Vierow et al. 2009). The brain regions commonly observed in these studies were the prefrontal cortex (PFC), the anterior cingulate cortex (ACC), the insula, the secondary somatosensory cortex (S2), and the cerebellum (Figure 23.4). The significant activations of PFC and insula during scratching are interesting because these regions, in particular PFC, are less sensitive to vibrotactile stimuli (Gelnar et al. 1999; Coghill et al. 2000; Burton et al. 2004; Golaszewski et al. 2002; Seitz and Roland 1992; Hagen and Pardo 2002). The PFC was associated with the motivational aspects of action. Clinical studies in individuals suffering with addiction have also demonstrated the role of PFC in motivation, reporting that activity in PFC increased during craving (Brody et al. 2002; Olbrich et al. 2006; Franklin et al. 2007). Electrical stimulation of PFC using transcranial magnetic stimulation and transcranial direct current stimulation modulated craving reactions in these patients (Camprodon et al. 2007; Boggio et al. 2008; Amiaz et al. 2009), demonstrating that PFC plays a crucial role in controlling motivation-based actions. Thus, scratching may increase the motivation to scratch further by enhancing PFC activity.

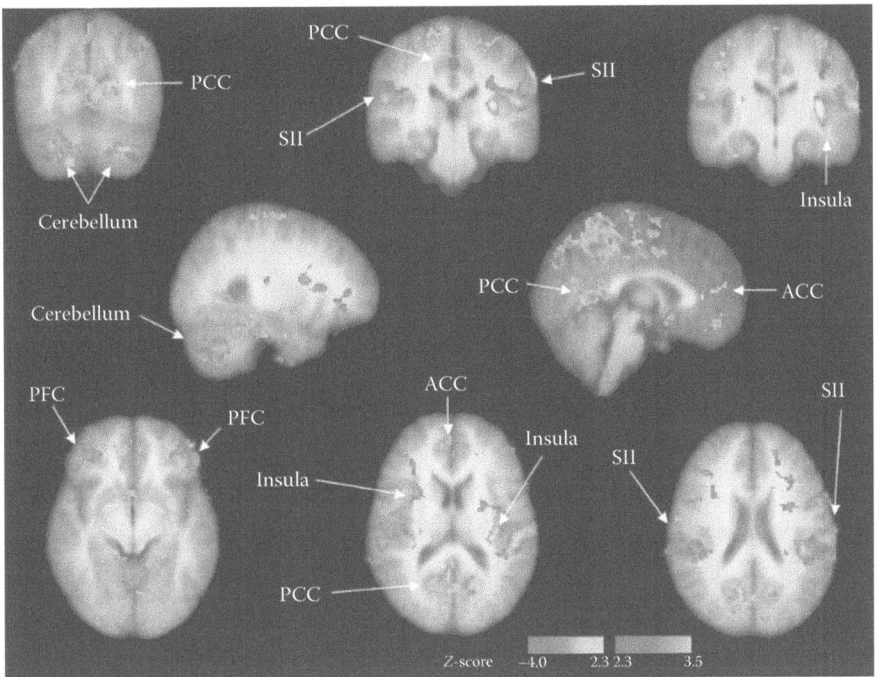

FIGURE 23.4 **(See color insert.)** Brain regions activated by scratching the skin. Red and blue regions were where neural activity was significantly increased and decreased during scratching, respectively. PCC, posterior cingulate cortex; SII, secondary somatosensory cortex; ACC, anterior cingulate cortex; PFC, prefrontal cortex.

The activity of the medial parietal cortex, including the posterior cingulate cortex and precuneus, was significantly decreased by scratching, while a significant activation of this region by itch was frequently observed (Herde et al. 2007; Mochizuki et al. 2007, 2009; Ishiuji et al. 2009; Papoiu et al. 2012a). In addition, the activity of this region was significantly correlated with itch intensity (Mochizuki et al. 2007). The medial parietal cortex is associated with memory and attention (Cavanna and Trimble 2006), and memory and attention can modulate perceptions.

Ruscheweyh and colleagues (2009) reported that pain evoked by physical stimuli such as heat, pinprick, or pressure was higher in subjects with a higher ability to imagine pain, an ability considered to be partly associated with memory. It was reported that itch and pain were reduced when subjects shifted their attention away (Leibovici et al. 2009; Schlereth et al. 2003). In addition, the medial parietal cortex is part of the brain network active during the resting state, known as the default mode network. It was reported that pain modulation depends on the strength of connectivity within the default mode network (Napadow et al. 2012). Activity in the medial parietal cortex significantly increased when pain hallucination was evoked in a coenesthesia patient (Bär et al. 2002). These studies suggest that the medial parietal cortex may have a role in modulating itch and pain. The reduced activity in the medial parietal cortex, which is induced by scratching, could diminish itch perception.

Scratching-induced reduction of activity was also observed in other brain regions such as the primary somatosensory cortex (S1), ACC, and motor areas such as SMA and PM (Figure 23.2) (Yosipovitch et al. 2008). S1 is considered to be associated with the sensory-discriminative aspects of itch (intensity, location, quality and duration), while ACC is linked with affective-motivational aspects, such as unpleasantness and the urge to scratch. Activity in these regions was significantly correlated with itch intensity (Darsow et al. 2000; Drzezga et al. 2001; Mochizuki et al. 2003, 2007; Leknes et al. 2007). However, not all fMRI studies of scratching observed significant activations in these regions. This discrepancy may have stemmed from differences in the pressure exerted during scratching. The scratching force applied in these studies was substantially different (0.29 and 2.65 N). However, even if the scratching was "less forceful," it was still effective to inhibit itch (Yosipovitch et al. 2007). Future studies examining active (self) scratching performed by healthy subjects and chronic itch patients themselves will enable us to better understand the brain mechanisms underlying the inhibition of itch.

Another mechanism of itch inhibition by scratching could be exerted by descending inhibitory pathways (Basbaum and Fields 1984; Millan et al. 2002). Descending pathways from the brain reach the dorsal horn of the spinal cord via the PAG, the raphe nuclei, and locus coeruleus. Neural signals conducted by the descending pathways inhibit nociceptive ascending signals at the spinal level. Electrical stimulation of PAG inhibited the neural responses in spinothalamic neurons evoked by application of pruritogens, such as histamine (Carstens et al. 1997). A human PET study reported that PAG activity increased while itch intensity was reduced, when pain stimuli were applied (Mochizuki et al. 2003). Thus, it is likely that the descending inhibitory control could inhibit itch. Previous fMRI studies of scratching did not show activation of PAG. It is conceivable that the intensity of activity (or the extent of the activated areas) engaged during pain or itch modulation could have been below the detection threshold. However, recent data from our group using a new ASL fMRI technique (3-D Grase Propeller) identified a significant deactivation of ventral tegmentum and of the raphe nucleus (interconnected with PAG) during itch inhibition by scratching (Papoiu et al. 2013, *PLoS One*, in press). Thus, further investigations using electrophysiological techniques and fMRI with a higher spatial and functional resolution at higher magnetic fields will be necessary to clarify in what way exactly the descending inhibitory control is associated with itch inhibition by scratching.

23.8.3 PLEASANTNESS AND SCRATCHING

Why are pleasant sensations evoked by scratching an itch? This is one of the great mysteries of itch. The pleasant sensations evoked by scratching are considered an exacerbating factor that could play an important role in the induction of the addictive itch–scratching cycle. A few neurophysiology or neuroimaging studies have addressed this phenomenon. It was speculated that the PFC and the orbitofrontal cortex are associated with the hedonic experience (Ikoma et al. 2006), because the medial PFC and the orbitofrontal cortex were previously linked with pleasure (Kringelback 2005; Ursu and Carter 2005; Kim et al. 2006). A couple of reports identified the striatum as one of the brain regions associated with the relief of itch associated with scratching (Vierow et al. 2009; Napadow et al. 2012). The striatum maintained a higher activity during scratching an itch, compared to scratching in the absence of itch. In addition, the activity in the striatum reached a maximum when itch was at a minimum, implying that the activation of the striatum was associated with scratching an itch, rather than with perceiving itch alone. This area was also found to play an important role in the pleasantness induced by pain relief (Leknes et al. 2011). Therefore, it is possible that the activation of the striatum during itch relief may have reflected hedonic aspects of scratching. A recently completed functional neuroimaging study addressed the question of why scratching is pleasurable and discovered that the activity of brain's reward circuits, especially in discrete formations of the midbrain including the ventral tegmental area, substantia nigra, and the raphe nucleus, was strongly correlated with itch relief (Papoiu et al. 2013).

23.9 INSIGHTS FOR A THERAPEUTIC STRATEGY TO RELIEVE ITCH DERIVED FROM BRAIN IMAGING STUDIES

Addressing the emotional and psychological suffering associated with itch are cornerstones for building a successful therapy for pruritus. The amplification of a vicious itch–scratch cycle is well known to aggravate the symptoms and the evolution of skin conditions marred by chronic pruritus, such as atopic eczema or psoriasis. Thus, an overactive limbic system (ACC-amygdala-nc. accumbens) may reflect a more intense, unbalanced craving for itch relief, accompanied by activations in the insular cortex (as seen in ESRD pruritus at baseline) that can lead only to the amplification of compulsive scratching behavior and more distress. It is not only the organic source of the problem that needs to be addressed according to its etiology, depending on the respective diagnosis, but also the subjective, profound impact that itch experience has upon the affected individuals "inner world." We surmise that neuroimaging studies provide insight not only on the physiological responses (or their dysfunction) but also serves as a window into the supramodal functions of the mind and psyche. Cognitive and emotional aspects of itch experience influence the higher level integration of physical stimuli. Brain areas involved in self-awareness (precuneus) and self-perception (insula) are getting immediately involved, confirming the observation that itch is a very intrusive and disturbing sensation, perturbing the well-being of the person. A successful treatment would need to target and interrupt the vicious itch–scratching cycle and offer a solution for the intense, "amplified" craving for relief. This raises the possibility that cognitive behavioral techniques could be helpful in limiting the emotional and affective impact of this bothersome symptom. Refocusing attention on tasks unrelated to itch could be one avenue worth exploring, because these approaches have been shown to be effective in diminishing the perception of pain (Zeidan et al. 2011). These "mindfulness" reframing techniques may prove to be even more effective for the subjective relief of itch, because itch (contrary to pain) can be easily generated via a central induction mechanism (the "contagious itch" phenomenology). Therefore, if there is a central "source" that is capable of producing itch sensation (in the absence of external pruritogenic stimuli), it must be a way to reverse the mechanism, turning it into a cure.

REFERENCES

Amiaz, R., Levy, D., Vainiger, D., Grunhaus, L. and Zangen, A. 2009. Repeated high-frequency transcranial magnetic stimulation over the dorsolateral prefrontal cortex reduces cigarette craving and consumption. *Addiction* 2104: 653–660.

Apicella, P., Ljungberg, T., Scarnati, E. and Schultz, W. 1991. Responses to reward in monkey dorsal and ventral striatum. *Exp. Brain Res.* 85: 491–500.

Basbaum, A. I. and Fields, H. L. 1984. Endogenous pain control systems: Brainstem spinal pathways and endorphin circuitry. *Annu Rev Neurosci.* 7: 309–338.

Bär, K. J., Gaser, C., Nenadic, I. and Sauer, H. 2002. Transient activation of a somatosensory area in painful hallucinations shown by fMRI. *Neuroreport* 13: 805–808.

Boggio, P. S., Sultani, N., Fecteau, S. et al. 2008. Prefrontal cortex modulation using transcranial DC stimulation reduces alcohol craving: A double-blind, sham-controlled study. *Drug Alcohol Depend.* 92: 55–60.

Bowman, E. M., Aigner, T. G. and Richmond, B. J. 1996. Neural signals in the monkey ventral striatum related to motivation for juice and cocaine rewards. *J. Neurophysiol.* 75: 1061–1073.

Breiter, H. C., Aharon, I., Kahneman, D., Dale, A. and Shizgal, P. 2001. Functional imaging of neural responses to expectancy and experience of monetary gains and losses. *Neuron* 30: 619–639.

Brinkman, C. 1984. Supplementary motor area of the monkey's cerebral cortex: Short- and long-term deficits after unilateral ablation and the effects of subsequent callosal section. *J. Neurosci.* 4: 918–929.

Brody, A. L., Mandelkern, M. A., London, E. D. et al. 2002. Brain metabolic changes during cigarette craving. *Arch. Gen. Psychiatry* 59: 1162–1172.

Burton, H., Sinclair, R. J. and McLaren, D. G. 2004. Cortical activity to vibrotactile stimulation: An fMRI study in blind and sighted individuals. *Hum. Brain Mapp.* 23: 210–228.

Camprodon, J. A., Martínez-Raga, J., Alonso-Alonso, M., Shih, M. C. and Pascual-Leone, A. 2007. One session of high frequency repetitive transcranial magnetic stimulation (rTMS) to the right prefrontal cortex transiently reduces cocaine craving. *Drug Alcohol Depend.* 86: 91–94.

Carstens, E. 1997. Responses of rat spinal dorsal horn neurons to intracutaneous microinjection of histamine, capsaicin, and other irritants. *J. Neurophysiol.* 77: 2499–2514.

Cavanna, A. E. and Trimble, M. R. 2006. The precuneus: A review of its functional anatomy and behavioural correlates. *Brain* 129: 564–583.

Coghill, R. C., Talbot, J. D., Evans, A. C. et al. 1994. Distributed processing of pain and vibration by the human brain. *J. Neurosci.* 14: 4095–4108.

Darsow, U., Drzezga, A., Frisch, M. et al. 2000. Processing of histamine-induced itch in the human cerebral cortex: A correlation analysis with dermal reactions. *J. Invest. Dermatol.* 115: 1029–1033.

Davidson, S. and Giesler, G. J. 2010. The multiple pathways for itch and their interactions with pain. *Trends Neurosci.* 33(12): 550–558.

Davidson, S., Zhang, X., Khasabov, S. G., Moser, H. R., Honda, C. N., Simone, D. A. and Giesler, G. J., Jr. 2012. Pruriceptive spinothalamic tract neurons: Physiological properties and projection targets in the primate. *J. Neurophysiol.* 108(6): 1711–1723.

Davidson, S., Zhang, X., Khasabov, S. G., Simone, D. A. and Giesler, G. J., Jr. 2009. Relief of itch by scratching: State-dependent inhibition of primate spinothalamic tract neurons. *Nat. Neurosci.* 12: 544–546.

Delgado, M. R., Stenger, V. A. and Fiez, J. A. 2004. Motivation-dependent responses in the human caudate nucleus. *Cereb. Cortex* 14: 1022–1030.

Drzezga, A., Darsow, U., Treede, R. D. et al. 2001. Central activation by histamine-induced itch: Analogies to pain processing: A correlational analysis of O-15 H_2O positron emission tomography studies. *Pain* 92: 295–305.

Ehrsson, H. H., Geyer, S. and Naito, E. 2003. Imagery of voluntary movement of fingers, toes, and tongue activates corresponding body-part-specific motor representations. *J. Neurophysiol.* 90: 3304–3316.

Feneran, A. N., O'Donnell, R., Press, A., Yosipovitch, G., Cline, M., Dugan, G., Papoiu, A. D., Nattkemper, L. A., Chan, Y. H. and Shively, C. A. 2012. Monkey see, monkey do: Contagious itch in nonhuman primates. *Acta Derm. Venereol.* doi:10.2340/00015555-1406.

Franklin, T. R., Wang, Z., Wang, J. et al. 2007. Limbic activation to cigarette smoking cues independent of nicotine withdrawal: A perfusion fMRI study. *Neuropsychopharmacology* 32: 2301–2309.

Gelnar, P. A., Krauss, B. R., Sheehe, P. R., Szeverenyi, N. M. and Apkarian, A. V. 1999. A comparative fMRI study of cortical representations for thermal painful, vibrotactile, and motor performance tasks. *Neuroimage* 10: 460–482.

Golaszewski, S. M., Siedentopf, C. M., Baldauf, E. et al. 2002. Functional magnetic resonance imaging of the human sensorimotor cortex using a novel vibrotactile stimulator. *Neuroimage* 17: 421–430.

Hagen, M. C. and Pardo, J. V. 2002. PET studies of somatosensory processing of light touch. *Behav. Brain Res.* 135: 133–140.

Halsband, U. and Passingham, R. E. 1985. Premotor cortex and the conditions for movement in monkeys (Macaca fascicularis). *Behav. Brain Res.* 18: 269–277.

Harsay, H. A., Cohen, M. X., Oosterhof, N. N., Forstmann, B. U., Mars, R. B. and Ridderinkhof, K. R. 2011. Functional connectivity of the striatum links motivation to action control in humans. *J. Neurosci.* 31: 10701–10711.

Henry, D. E., Chiodo, A. E. and Yang, W. 2011. Central nervous system reorganization in a variety of chronic pain states: A review. *Phys. Med. Rehab.* 3(12): 1116–1125.

Herde, L., Forster, C., Strupf, M. and Handwerker, H. O. 2007. Itch induced by a novel method leads to limbic deactivations a functional MRI study. *J. Neurophysiol.* 98: 2347–2356.

Holle, H., Warne, K., Seth, A. K., Critchley, H. D. and Ward, J. 2012. Neural basis of contagious itch and why some people are more prone to it. *Proc. Natl. Acad. Sci. USA* 109(48): 19816–19821.

Hsieh, J. C., Hägermark, O. and Stahle-Bäckdahl, M. 1994. Urge to scratch represented in the human cerebral cortex during itch. *J. Neurophysiol.* 72: 3004–3008.

Ikoma, A., Steinhoff, M., Ständer, S., Yosipovitch, G. and Schmelz, M. 2006. The neurobiology of itch. *Nat. Rev. Neurosci.* 7: 535–547.

Ishiuji, Y., Coghill, R. C., Patel, T. S., Oshiro, Y., Kraft, R. A. and Yosipovitch, G. 2009. Distinct patterns of brain activity evoked by histamine-induced itch reveal an association with itch intensity and disease severity in atopic dermatitis. *Br. J. Dermatol.* 161: 1072–1080.

Kim, H., Shimojo, S. and O'Doherty, J. P. 2006. Is avoiding an aversive outcome rewarding? Neural substrates of avoidance learning in the human brain. *PLoS Biol.* 4: e233.

Kringelbach, M. L. 2005. The human orbitofrontal cortex: Linking reward to hedonic experience. *Nat. Rev. Neurosci.* 6: 691–702.

Kurata, K. and Wise, S. P. 1988. Premotor cortex of rhesus monkeys: Set-related activity during two conditional motor tasks. *Exp. Brain Res.* 69: 327–343.

Leibovici, V., Magora, F., Cohen, S. and Ingber, A. 2009. Effects of virtual reality immersion and audiovisual distraction techniques for patients with pruritus. *Pain Res. Manag.* 14: 283–286.

Leknes, S. G., Bantick, S., Willis, C. M., Wilkinson, J. D., Wise, R. G. and Tracey, I. 2007. Itch and motivation to scratch: An investigation of the central and peripheral correlates of allergen- and histamine-induced itch in humans. *J. Neurophysiol.* 97: 415–422.

Leknes, S., Lee, M., Berna, C., Andersson, J. and Tracey, I. 2011. Relief as a reward: Hedonic and neural responses to safety from pain. *PLoS One* 6: e17870.

Lloyd, D. M., Hall, E., Hall, S. and McGlone, F. P. 2012. Can itch-related visual stimuli alone provoke a scratch response in healthy individuals? *Br. J. Dermatol.* doi:10.1111/bjd.12132.

Lotze, M., Montoya, P., Erb, M. et al. 1999. Activation of cortical and cerebellar motor areas during executed and imagined hand movements: An fMRI study. *J. Cogn. Neurosci.* 11: 491–501.

Meister, I. G., Krings, T., Foltys, H. et al. 2004. Playing piano in the mind—An fMRI study on music imagery and performance in pianists. *Brain Res. Cogn. Brain Res.* 19: 219–228.

Millan, M. J. 2002. Descending control of pain. *Prog Neurobiol.* 66(6): 355-474.

Mochizuki, H., Inui, K., Tanabe, H. C. et al. 2009. Time course of activity in itch-related brain regions: A combined MEG-fMRI study. *J. Neurophysiol.* 102: 2657–2666.

Mochizuki, H., Sadato, N., Saito, D. N. et al. 2007. Neural correlates of perceptual difference between itching and pain: A human fMRI study. *Neuroimage* 36: 706–717. Erratum in: 2008. *Neuroimage* 39: 911–912.

Mochizuki, H., Tashiro, M., Kano, M., Sakurada, Y., Itoh, M. and Yanai, K. 2003. Imaging of central itch modulation in the human brain using positron emission tomography. *Pain* 105: 339–346.

Moll, L. and Kuypers, H. G. 1979. Premotor cortical ablations in monkeys: Contralateral changes in visually guided reaching behavior. *Science* 198: 3117–3119.

Naito, E., Kochiyama, T., Kitada, R. et al. 2002. Internally simulated movement sensations during motor imagery activate cortical motor areas and the cerebellum. *J. Neurosci.* 22: 3683–3691.

Napadow, V., Kim, J., Clauw, D. J. and Harris, R. E. 2012. Decreased intrinsic brain connectivity is associated with reduced clinical pain in fibromyalgia. *Arthritis Rheum.* 64(7): 2398–2403.

Niemeier, V., Kupfer, J. and Gieler, U. 2000. Observations during an itch inducing lecture. *Dermatol. Psychosom.* 1(Suppl. 1): 15–18.

Olbrich, H. M., Valerius, G., Paris, C., Hagenbuch, F., Ebert, D. and Juengling, F. D. 2006. Brain activation during craving for alcohol measured by positron emission tomography. *Aust. N. Z. J. Psychiatry* 40: 171–178.

Papoiu, A. D. P., Coghill, R. C., Kraft, R. A., Wang, H. and Yosipovitch, G. 2012a. A tale of two itches. Common features and notable differences in brain activation evoked by cowhage and histamine induced itch. *Neuroimage* 59: 3611–3623.

Papoiu, A. D. P., Wang, H., Coghill, R. C., Chan, Y. H. and Yosipovitch, G. 2011. Contagious itch in humans: A study of visual "transmission" of itch in atopic dermatitis and healthy subjects. *Br. J. Dermatol.* 164(6): 1299–1303.

Papoiu, A. D., Nattkemper, L. A., Coghill, R. C., and Yosipovitch, G. 2012b. Visualizing the brain processing of the itch-scratch cycle by functional MRI. 75th Annual Meeting of the Society for Investigative Dermatology. Raleigh, N. C. *J Invest Dermatol.* 132: S94.

Papoiu, A. D., Nattkemper, L. A., Sanders, K. M., Kraft, R. A., Coghill, R. C. and Yosipovitch, G. 2013. Brain's reward circuits mediate itch relief. A functional MRI study of active scratching. PLoS One 2013, *in press* (provisionally accepted). *J. Neurosci.*

Ruscheweyh, R., Marziniak, M., Stumpenhorst, F., Reinholz, J. and Knecht, S. 2009. Pain sensitivity can be assessed by self-rating: Development and validation of the Pain Sensitivity Questionnaire. *Pain* 146: 65–74.

Schlereth, T., Baumgärtner, U., Magerl, W., Stoeter, P. and Treede, R. D. 2003. Left-hemisphere dominance in early nociceptive processing in the human parasylvian cortex. *NeuroImage* 20: 441–454.

Schneider, G., Ständer, S., Burgmer, M., Driesch, G., Heuft, G. and Weckesser, M. 2008. Significant differences in central imaging of histamine-induced itch between atopic dermatitis and healthy subjects. *Eur. J. Pain* 12: 834–841.

Schultz, W., Apicella, P., Scarnati, E. and Ljungberg, T. 1992. Neuronal activity in monkey ventral striatum related to the expectation of reward. *J. Neurosci.* 12: 4595–4610.

Seitz, R. J. and Roland, P. E. 1992. Vibratory stimulation increases and decreases the regional cerebral blood flow and oxidative metabolism: A positron emission tomography (PET) study. *Acta. Neurol. Scand.* 86: 60–67.

Tanji, J. and Shima, K. 1994. Role of the supplementary motor area cells in planning several movements ahead. *Nature* 371: 413–416.

Thaler, D., Chen, Y. C., Nixon, P. D., Stern, C. E. and Passingham, R. E. 1995. The functions of the medial premotor cortex. I. Simple learned movements. *Exp. Brain Res.* 102: 445–460.

Ueno, T., Inoue, M., Matsuoka, T., Abe, T., Maeda, H. and Morita, K. 2010. Comparison between a real sequential finger and imagery movements: An FMRI study revisited. *Brain Imaging Behav.* 4: 80–85.

Ursu, S. and Carter, C. S. 2005. Outcome representations, counterfactual comparisons and the human orbitofrontal cortex: Implications for neuroimaging studies of decision-making. *Cognitive Brain Research* 23: 51–60.

Valet, M., Pfab, F., Sprenger, T., Wöller, A., Zimmer, C., Behrendt, H., Ring, J., Darsow, U. and Tölle, T. R. 2008. Cerebral processing of histamine-induced itch using short-term alternating temperature modulation—An FMRI study. *J. Invest. Dermatol.* 128(2): 426–33.

Vierow, V., Fukuoka, M., Ikoma, A., Dörfler, A., Handwerker, H. O. and Forster, C. 2009. Cerebral representation of the relief of itch by scratching. *J. Neurophysiol.* 102: 3216–3224.

Yosipovitch, G., Duque, M. I., Fast, K., Dawn, A. G. and Coghill, R. C. 2007. Scratching and noxious heat stimuli inhibit itch in humans: A psychophysical study. *Br. J. Dermatol.* 156: 629–634.

Yosipovitch, G., Goon, A. T., Wee, J., Chan, Y. H., Zucker, I. and Goh, C. L. 2002. Itch characteristics in Chinese patients with atopic dermatitis using a new questionnaire for the assessment of pruritus. *Int. J. Dermatol.* 41: 212–216.

Yosipovitch, G., Ishiuji, Y., Patel, T. S. et al. 2008. The brain processing of scratching. *J. Invest. Dermatol.* 128: 1806–18011.

Zeidan, F., Martucci, K. T., Kraft, R. A., Gordon, N. S., McHaffie, J. G. and Coghill, R. C. 2011. Brain mechanisms supporting the modulation of pain by mindfulness meditation. *J. Neurosci.* 31: 5540–5548.

24 Central Nervous Processing of Itch and Pain

Clemens Forster and Hermann O. Handwerker

CONTENTS

24.1 BRAIN ACTIVATION STUDIES OF PAIN AND ITCH: SIMILARITIES AND DIFFERENCES

Since the ascent of functional imaging based on regional cerebral blood flow (rCBF), in particular of functional magnetic resonance imaging (fMRI), many studies have dealt with the processing of pain in the human central nervous system (for review, see Treede et al. 1999). Before the advent of modern cerebral imaging, it has been supposed that the experience of pain is multidimensional (Melzack and Casey 1968). Therefore, it was not surprising that no single central "pain center" has been found. From many functional imaging studies a general notion evolved that pain is processed in a "cerebral pain network." In a meta-analysis of 69 studies employing experimental pain stimuli, the following cerebral areas were encountered in descending order of frequency: insular cortex (both anterior and posterior), anterior and medial cingulate gyrus, somatosensory thalamus, somatosensory projection fields S2 (operculum) and S1 (gyrus postcentralis), and lateral and medial cortex fields (Apkarian et al. 2005). This "cortical pain network" can roughly be divided in a lateral portion (somatosensory projection areas) supposed to serve mainly the sensory evaluative dimension of pain and a medial portion consisting of limbic and prefrontal areas serving the emotional dimension (Treede et al. 1999).

This concept derived from functional imaging studies for the interpretation of functional imaging studies on experimental pain has been supported by observations of the consequences of cerebral lesions and by anatomical concepts. It is not undisputed, however. It has been shown that large parts of this network are also involved in the processing of other nonpainful stimuli, in particular, if they demand the full attention and/or emotional participation of the subjects. Iannetti and his group developed an alternative hypothesis identifying a "saliency network" instead of a pain network on the basis of detailed analysis of cortical evoked potentials (Legrain et al. 2011). These findings were supported by other groups. For example, our group has shown that in an fMRI study employing painful mechanical impacts and harmless touch stimuli both types of stimulation activated virtually the same cortical regions depending on the focusing of attention (Schoedel et al. 2008). These controversial interpretations of imaging studies cannot satisfyingly be resolved because the spatial resolution of available functional imaging methods is too low for analyzing the networks at a cellular level.

The neurobiology of itch was for a long time a neglected field. It remained even unclear if the signals of pruriceptors in the skin are transmitted to the CNS by a "labeled line" or by a subpopulation of nociceptors which project also into pain networks. Recently, it has been shown that two forms of itch exist: one depends on histamine while the other does not. Both forms are apparently transmitted to the thalamic projection sites by separate neuronal populations (Davidson et al. 2007). For a review of the development of concepts of itch processing, see (Handwerker 2013). The relationship of itch and pain seems to be complicated. On the one hand, it is well known from clinical observations and also proven by experimental studies (Brull et al. 1999) that pain suppresses itch. On the other hand, the clear separation of itch and pain mediators cannot be abided in the light of newer physiological and molecular findings. For example, histamine has been regarded as the prototypical itch mediator, though it has been known for a long time that subcutaneous applications of histamine induce pain rather than itch (Keele and Armstrong 1964). On the other side, capsaicin, the ligand of the TRPV1 membrane receptor, has been regarded as a prototypical pain mediator. However, it has been shown recently that histamine acts by binding to a G-protein coupled receptor (H1) that subsequently activates the TRPV1 receptor through a second messenger pathway (for review, see Ross 2011). These findings converge with the results of psychophysical experiments showing that tiny amounts of capsaicin brought into the skin by capsaicin-coated spicules induce a mixture of burning pain and itch (Sikand et al. 2009). Therefore, a "selectivity hypothesis" of itch is presently favored, assuming a "population coding" (Akiyama et al. 2009; Ma 2010). A smaller population of itch mediating primary afferents share receptor mechanisms with the larger population of nociceptors. By central synaptic processing, the "itch signals" are processed, if they account for the majority of the excited population, but itch is suppressed by inhibitory interneurons when the larger population of nociceptors induces pain (Handwerker 2013).

On this background it is of interest to explore with the help of functional imaging whether itch and pain are processed in identical or different cerebral networks. Pioneering was a positron emission tomography (PET) study in this context (Hsieh et al. 1994). Though no comparison with pain stimuli was made, this study emphasized the contribution of motor areas and of the cerebellum in the cerebral response to intracutaneous injections of histamine. This was interpreted as an expression of the urge to scratch. Later fMRI studies on the cerebral responses to histamine application did not compare itch with pain stimuli (for review, see Pfab et al. 2012).

One group compared electrically induced pain and itch (Mochizuki et al. 2007). Similar activation patterns were encountered, but activation of the precuneus was regarded as particularly important for itch processing. Electrical stimuli inducing highly synchronized patterns of cerebral activations are not comparable, however, with the typically asynchronous input from more "natural" itch and pain stimuli.

The analysis of the cerebral processing of natural (i.e., not electrically induced) itch sensations is difficult because typical itches, e.g., following histamine iontophoresis, histamine prick test, or insertion of cowhage spicules (to name only a few of the favored experimental methods), induce slowly developing itch which lasts for many minutes and has a tendency to waxing and waning. Studies using blood oxygenation level dependent (BOLD) fMRI demand, however, stimuli-inducing sensations with a clear on- and offset for statistical reasons. Several approaches have been described to handle this problem. One group used ASL-magnetic resonance imaging instead of the BOLD methodological approach (Papoiu et al. 2012). In another study a temporal segmentation approach was chosen with BOLD technology to overcome the problem of the sustained and slowly changing itch sensations (Kleyn et al. 2012). Another approach is by the use of interfering sensory stimuli that are known to suppress itch. For this approach it is important that the conditioning stimulus does not interfere with the itch stimulus at the level of nerve terminals. Itch suppression based on central nervous system (CNS) interactions has been demonstrated for cooling stimuli (Bromm et al. 1995; Mochizuki et al. 2003; Pfab et al. 2006; Valet et al. 2008). Our group has used scratch bouts to induce a temporary suppression of itch in imitation of the everyday behavior. It has been shown that the suppression of itch by scratch bouts is not due to an interaction at the nerve terminals (Yosipovich et al. 2007). Furthermore, we applied scratch not at the place of itch application, but nearby in our imaging experiments (Vierow et al. 2009). Experiments with interfering stimuli obviously raise interpretation problems: the conditioning stimulus also induces a sensation which suppresses itch. It is a question of the interpretation whether the observed BOLD effects are due to the itch suppression or to the genuine effect of the conditioning stimuli.

Therefore, we designed an experimental protocol in which itch was suppressed without interference of other stimuli by a mixture of histamine and codeine applied through intradermally positioned microdialysis fibers. The itch was terminated by lidocaine application through the same fiber and reappeared within a few seconds when the probe was flashed with saline. In a subsequent experiment performed in the same fMRI session, heat pain was provoked in the contralateral forearm with a Peltier thermode. During both experiments, activation clusters were found in brain areas that have been described previously to be frequently activated in response to painful stimuli. This includes prefrontal areas, supplementary motor areas (SMA), premotor cortex, anterior insular cortex, anterior midcingulate cortex, S1, S2, thalamus, basal ganglia, and cerebellum. In general, itch stimulation entailed more activation clusters, in particular, on the contralateral side of the brain. Only on itch, but not on heat pain, negative BOLD signals were found in the subgenual anterior cingulate cortex and the amygdala (Herde et al. 2007). The finding of negative limbic BOLD signals during itch has later been confirmed by another group (Kleyn et al. 2012). Amygdala deactivations may have been related to the urge for scratching but may also be due to the stressful character of itch stimulation (Herde et al. 2007).

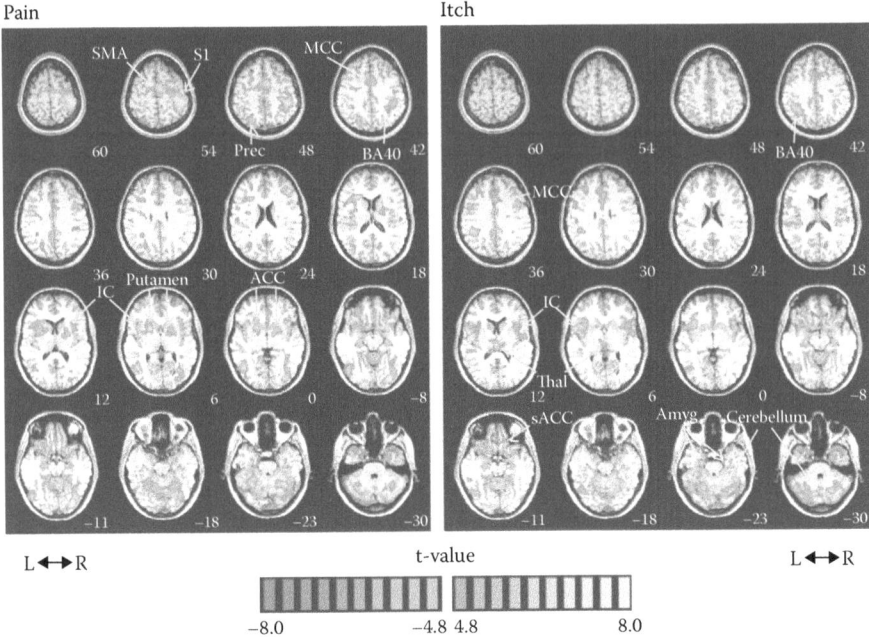

FIGURE 24.1   **(See color insert.)** Selected axial brain slices comparing the BOLD effects related to (left) pain and (right) itch. The white numbers mark the position of the brain slices within the Talairach space (z-values). The color scale depicts the strength of activation of a brain area as expressed by the t-value of the general linear model. Negative t-values (blue to green scale) mark brain areas in which the BOLD signal decreased during the stimulus, while the red to yellow scale depicts increased BOLD signals during the stimulus. SMA, supplementary motor area; MCC, medial cingulate cortex; Prec, precuneus; ACC, anterior cingulate cortex; IC, insular cortex; Thal, thalamus; sACC, subgenual anterior cingulate cortex; Amyg, amygdala.

Figure 24.1 shows brain slices comparing the BOLD patterns related to pain and itch stimulation in this experiment.

24.2   CONNECTIVITY STUDIES: A NEW OPTION FOR COMPARING CENTRAL ITCH AND PAIN PROCESSING

Comparing the pattern of activated brain areas during pain and itch reveals that there are many regions that are more or less similarly activated during both stimuli. From this, one has to assume that the information about the quality of the perceived stimulus is not coded alone in the activation of certain brain areas or a particular combination of activated brain regions. It is very likely that the information about the quality and the intensity of a stimulus or task is given by the degree to which the regions involved are coupled together in functionally connected circuits.

Biswal et al. (1995) were the first that showed that BOLD signals are correlated between regions of the motor network, both within and across hemispheres of the brain at rest. A finger-tapping task was used to map regions involved in a motor task and between these regions were then evaluated using BOLD data acquired without task performance. Later it has been shown that connectivity patterns differ as a function of cognitive task (Handwerker et al. 2012; Shirer et al. 2012) or can reflect the state of consciousness (Mhuircheartaigh et al. 2010). Recently, several fMRI studies have been published that enhanced our knowledge about the role of functional connectivity in pain perception (Taylor et al. 2009) and pain pathology (Jensen et al. 2012; Malinen et al. 2010). Jensen et al. (2012) even suggested using functional connectivity as a measure of pain dysregulation.

A variety of methods have been described to analyze the connectivity of brain regions based on PET or fMRI data (for review, see Rogers et al. 2007; Li et al. 2009). A common feature of all methods is the correlation of the rCBF changes between the regions of interest. In fMRI, rCBF is reflected by the BOLD signal and is taken as a measure of the local neural activity. A quite simple and straightforward

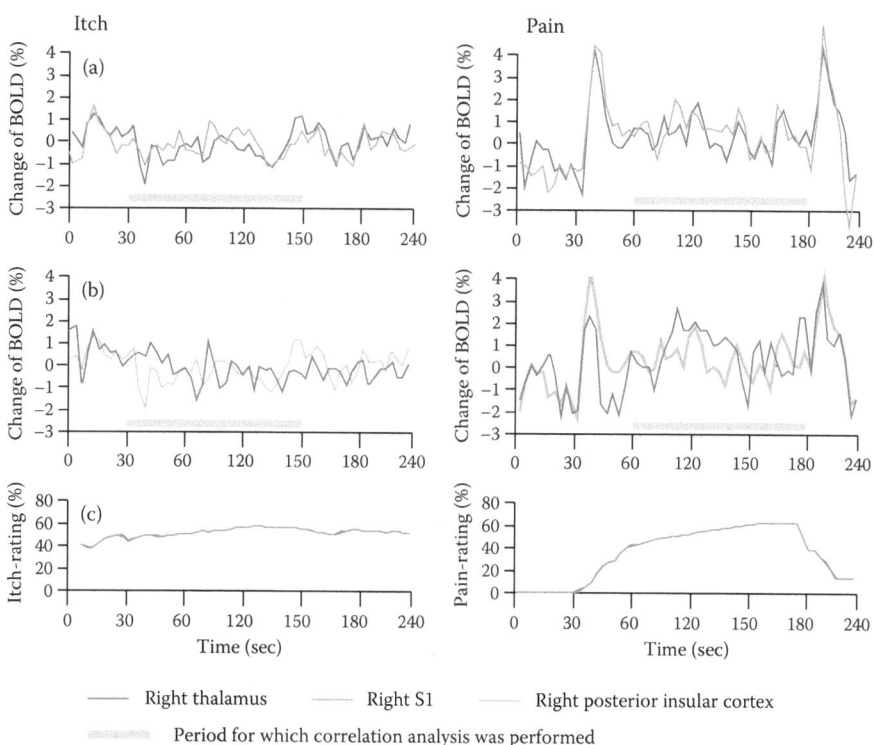

FIGURE 24.2 **(See color insert.)** Time courses of BOLD signals in three brain regions and rating. Plots A and B show the mean changes of the BOLD signals in three brain regions as paired time courses during itch and tonic pain. Itch and pain ratings are shown in panel C. Please note that the correlation analysis was done only for the periods marked by the gray bar. Therefore, the peaks in BOLD responses at the onset and at the end of a pain stimulus were not included. Most likely the peaks are induced by the salient changes of the pain stimulus that induced an arousal effect. Such salient changes usually do not occur during itch stimuli.

approach is the use of so-called "seed regions." The time course of the BOLD signal from this region is extracted and the correlation coefficients between this signal and that of all other regions of interest are separately calculated. To illustrate the principle, Figure 24.2 shows the sample of BOLD time courses in three different brain regions (averaged across all voxels within a region and across all subjects) during pain and itch. For this long lasting stimuli (2 min, see methods in the next section) have been selected to get more reliable correlation coefficients.

24.2.1 AN EXAMPLE OF A CONNECTIVITY ANALYSIS

Nine healthy volunteer, aged between 20 and 40 years, participated on an fMRI session. Two experiments were conducted during which either ongoing itch or three long lasting pain stimuli were induced.

Imaging was performed with a 3 Tesla Trio MRI scanner (Siemens, Erlangen, Germany). BOLD signals were collected using functional T2 weighted images obtained by the echo planar imaging (EPI) technique. The time resolution of these images was one block of 30 axial slices every 3 s. The subjects could watch a visual analog scale (VAS) outside the scanner via an overhead mirror. The rating scale was tuned by a turning knob with the right hand. Itch was induced by iontophoretic application of histamine. The technique was identical to that used by our group in several studies (Handwerker et al. 1987). Briefly, 1% histamine dihydrochloride was dissolved in a gel of 2.5% methylcellulose in distilled water. This jelly was filled into the 50 μL cavity of an acrylic applicator 6 mm in diameter. A current of 1 mA was delivered for 30 s from a silver-silver chloride electrode in this applicator to a large reference electrode applied to the skin. Usually, itching started 30 s after the beginning of the iontophoresis, on average, and lasted for about 10 min.

The second stimulus was strong tonic pain that was induced by squeezing the interdigital web between the second and third finger of the left hand (see Forster et al. 1988). Constant pressure was applied to the skin via two opposite probes with round edges of 6 mm in diameter. Three stimuli were applied, and each stimulus lasted 2 min, followed by a stimulus-free interval of 2 min. During the stimulus the force was kept constant and was set such that the subjects rated the pain between 60% and 80% on a VAS. This required a force to the probe that was between 10N and 14N.

The common preprocessing of the fMRI signal was performed (see Weigelt et al. 2010). BOLD time courses were extracted from brain regions of interest and were correlated to selected seed regions. To obtain comparable intervals, a time window of 2 min was used for all calculations during which the stimulus intensity (itch or pain, respectively) was at its maximum.

The results of three connectivity analyses are shown in Figure 24.3. It is obvious that the functional connectivity is more pronounced during the painful input as compared with the itch. This is true for nearly all brain regions between which correlations were performed, and may have its reason in a more synchronized activity during pain than during itch. The difference in synchronicity may already start in skin nerve endings where the total discharge rate during pain inducing noxious stimuli is often higher and the action potentials occur more regularly than during

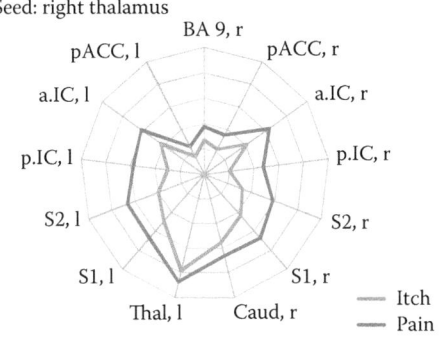

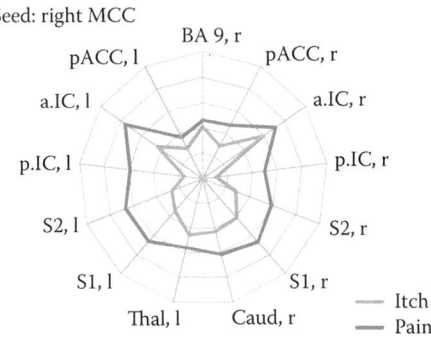

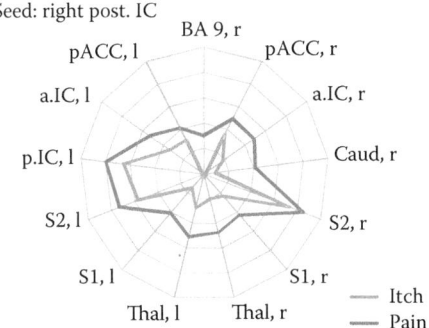

FIGURE 24.3 **(See color insert.)** Functional connectivity between selected brain regions during itch and pain. The connectivity is expressed as the correlation between the BOLD signals of the seed region and the regions depicted in the diagrams. The center of the diagram corresponds to a correlation coefficient of $r = 0$ between the seed region and the respective brain area, the outer border marks $r = 1$. It is obvious that pain generates higher functional connectivity than itch does. MCC, medial cingulate cortex; post. IC, posterior insular cortex; BA9, Brodman Area 9; pACC, anterior cingulate cortex, posterior part; a.IC, anterior insular cortex; p.IC, posterior insular cortex; S1, primary somatosensory cortex; S2, secondary somatosensory cortex; Caud, caudate nucleus; Thal, thalamus; l, left hemisphere; r, right hemisphere.

pruritic stimuli (Schmelz et al. 1997). Consequently, this effect is prominent in the lateral system (somatosensory cortex S1 and S2) that shows high synchronicity to the thalamus and is related to the sensory aspects in terms of quality, intensity, and timing of a stimulus (Apkarian et al. 2005). There is less synchronicity to the medial system and the frontal area during both stimulation conditions that may be caused by different time patterns of activations in these regions. A higher level of integration of the sensory experience dissociates the activation pattern in these regions from those cortical areas processing the primary input.

The posterior insular regions of both hemispheres have a high functional connectivity among each other (which is not very surprising) but also to both opercula (secondary somatosensory region, S2) with only a little difference between pain and itch. It has been assumed that the posterior insular cortex plays a unique role as the "interoceptive cortex" of the human brain (Craig 2002). This region receives strong input from the lamina 1 of the spinal cord to which various interoceptive inputs project in separate neuronal populations (Craig et al. 2000; Craig 2003). The interpretation of the posterior insular cortex as an area processing mainly the body-related aspects of various sensory inputs and not so much the "object-related" features makes this region unique. Activation of this region may indicate an interference of the body homeostasis, either by pain or by itch, leading to the desire to withdraw from the pain or to scratch, respectively (Leknes et al. 2007; Ruehle et al. 2006; Vierow et al. 2009). The strong functional connection between the posterior IC and the S2 regions may be related to the control of the effectiveness of planned actions against those nuisances.

While the differences in the functional connectivity of the posterior IC between the pain and itch conditions are comparatively small, these differences are considerably larger when the activation of the medial cingulate cortex (MCC) is used as the seed region. During pain, high correlation to anterior and posterior insular regions can be found while during itch especially the connectivity to the posterior IC nearly disappears. The MCC is engaged in motor planning (Vogt 2005) and hence may be involved in the generation of adequate motor responses. Because the subjects were not allowed to move or to scratch, active suppression of motor reflexes is required. In case of pain this is the withdrawal reflex of the left stimulated hand, during itch a scratch reflex executed by the contralateral right arm. This scratch reflex would require more planning than simple withdrawal. In particular, this suppression task could involve a network in which MCC and insular cortex are synchronized or, respectively, desynchronized during suppression.

Most studies on the central processing of itch did not find itch-related activation in the posterior insular cortex. It has been shown that histamine sensitive "itch fibers" in the cat project to lamina 1 of the spinal cord (Andrew and Craig 2001), is the area that provides the major input to the posterior insula via the thalamus (Craig et al. 2000). Only the study by Drzezga et al. (2001) have found activation in the left posterior insula that correlated with the unpleasantness of the itch. Interestingly, they did not find this area by the classical subtraction analysis (categorical comparisons between resting and activation conditions), but by correlation of the z-scores (as a measure of activity) with the stimulus strength and the perceived itch intensity. This indicates that this area does not show a strict on–off behavior corresponding to the stimulus but plays a role in the integration of sensations. Its high correlations with the operculum regions on both hemispheres and the contralateral posterior insula (see Figure 24.3) may indicate a network coding particular qualities of the sensation like intensity, unpleasantness, or realization of possible dangers. Napadow et al. (2010) showed that compared with healthy subjects chronic pain patients suffering from fibromyalgia had an increased connectivity of the posterior insula to areas that belong to the default-mode network (DMN). The DMN describes brain regions that usually decrease their activity during task performance that is interpreted such that they are important for the resting state of the brain (for review, see Fox and Raichle 2007). Ongoing input due to a chronic pain situation can consolidate this network that is expressed in an increased correlation of the activity between the brain regions belonging to the DMN, such as the posterior insula.

All interpretations of the connectivities found in this preliminary study are, of course, tentative and exploratory. They have to be confirmed in future experiments based on specific hypotheses. The demonstration of connectivity experiments in this short chapter on brain imaging shows, however, that the patterns of functional connectivity found between cortical areas involved in itch and pain processing are not so much qualitatively but quantitatively different. Some pairs of regions were found that changed their functional connectivities fundamentally between pain and itch. This is a further indication that these two types of sensations, itch and pain, are processed in similar cortical networks.

ACKNOWLEDGMENT

We thank Verena Vierow for her help with the analyses of the connectivity studies and the preparation of the figures.

REFERENCES

Akiyama T, Merrill AW, Carstens MI, Carstens E (2009) Activation of superficial dorsal horn neurons in the mouse by a PAR-2 agonist and 5-HT: Potential role in itch. *J Neurosci* 29:6691–6699.

Andrew D, Craig AD (2001) Spinothalamic lamina I neurons selectively sensitive to histamine: A central neural pathway for itch. *Nat Neurosci* 4:72–77.

Apkarian AV, Bushnell MC, Treede RD, Zubieta JK (2005) Human brain mechanisms of pain perception and regulation in health and disease. *Eur J Pain* 9:463–484.

Biswal B, Yetkin FZ, Haughton VM, Hyde JS (1995) Functional connectivity in the motor cortex of resting human brain using echo-planar MRI. *Magn Reson Med* 34:537–541.

Bromm B, Scharein E, Darsow U, Ring J (1995) Effects of menthol and cold on histamine-induced itch and skin reactions in man. *Neurosci Lett* 187:157–160.

Brull SJ, Atanassoff PG, Silverman DG, Zhang J, Lamotte RH (1999) Attenuation of experimental pruritus and mechanically evoked dysesthesiae in an area of cutaneous allodynia. *Somatosens Mot Res* 16:299–303.

Craig AD (2002) How do you feel? Interoception: The sense of the physiological condition of the body. *Nat Rev Neurosci* 3:655–666.

Craig AD (2003) A new view of pain as a homeostatic emotion. *Trends Neurosci* 26:303–307.

Craig AD, Chen K, Bandy D, Reiman EM (2000) Thermosensory activation of insular cortex. *Nat Neurosci* 3:184–190.

Davidson S, Zhang X, Yoon CH, Khasabov SG, Simone DA, Giesler GJ, Jr. (2007) The itch-producing agents histamine and cowhage activate separate populations of primate spinothalamic tract neurons. *J Neurosci* 27:10007–10014.

Drzezga A, Darsow U, Treede RD, Siebner H, Frisch M, Munz F, Weilke F, Ring J, Schwaiger M, Bartenstein P (2001) Central activation by histamine-induced itch: Analogies to pain processing: A correlational analysis of O-15 H$_2$O positron emission tomography studies. *Pain* 92:295–305.

Forster C, Anton F, Reeh PW, Weber E, Handwerker HO (1988) Measurement of the analgesic effects of aspirin with a new experimental algesimetric procedure. *Pain* 32:215–222.

Fox MD, Raichle ME (2007) Spontaneous fluctuations in brain activity observed with functional magnetic resonance imaging. *Nat Rev Neurosci* 8:700–711.

Handwerker DA, Roopchansingh V, Gonzalez-Castillo J, Bandettini PA (2012) Periodic changes in fMRI connectivity. *Neuroimage* 63:1712–1719.

Handwerker HO (2013) From pattern to specificity and to population coding. In: *Itch: Mechanisms and Treatment* (Akiyama T, Carstens E, eds.). Boca Raton, FL: CRC Press, pp. 1–8.

Handwerker HO, Magerl W, Klemm F, Lang E, Westeman RA (1987) Quantitative evaluation of itch sensation. In: *Fine Afferent Nerve Fibres and Pain* (Schmidt RF, Vahle-Hinz C, eds.). Weinheim: Wiley-VCH, pp. 463–473.

Herde L, Forster C, Strupf M, Handwerker HO (2007) Itch induced by a novel method leads to limbic deactivations a functional MRI study. *J Neurophysiol* 98:2347–2356.

Hsieh JC, Hagermark O, Stahle-Backdahl M, Ericson K, Eriksson L, Stone-Elander S, Ingvar M (1994) Urge to scratch represented in the human cerebral cortex during itch. *J Neurophysiol* 72:3004–3008.

Jensen KB, Loitoile R, Kosek E, Petzke F, Carville S, Fransson P, Marcus H, et al. (2012) Patients with fibromyalgia display less functional connectivity in the brain's pain inhibitory network. *Mol Pain* 8:32.

Keele CA, Armstrong D (1964) *Substances Producing Pain and Itch*. London: E. Arnold.

Kleyn CE, McKie S, Ross A, Elliott R, Griffiths CE (2012) A temporal analysis of the central neural processing of itch. *Br J Dermatol* 166:994–1001.

Legrain V, Iannetti GD, Plaghki L, Mouraux A (2011) The pain matrix reloaded: A salience detection system for the body. *Prog Neurobiol* 93:111–124.

Leknes SG, Bantick S, Willis CM, Wilkinson JD, Wise RG, Tracey I (2007) Itch and motivation to scratch: An investigation of the central and peripheral correlates of allergen- and histamine-induced itch in humans. *J Neurophysiol* 97:415–422.

Li K, Guo L, Nie J, Li G, Liu T (2009) Review of methods for functional brain connectivity detection using fMRI. *Comput Med Imaging Graph* 33:131–139.

Ma Q (2010) Labeled lines meet and talk: Population coding of somatic sensations. *J Clin Invest* 120:3773–3778.

Malinen S, Vartiainen N, Hlushchuk Y, Koskinen M, Ramkumar P, Forss N, Kalso E, Hari R (2010) Aberrant temporal and spatial brain activity during rest in patients with chronic pain. *Proc Natl Acad Sci USA* 107:6493–6497.

Melzack R, Casey KL (1968) Sensory, motivational, and central control determinants of pain. In: *The Skin Senses* (Kenshalo DR, ed.). Springfield, IL: Thomas. pp. 423–443.

Mhuircheartaigh RN, Rosenorn-Lanng D, Wise R, Jbabdi S, Rogers R, Tracey I (2010) Cortical and subcortical connectivity changes during decreasing levels of consciousness in humans: A functional magnetic resonance imaging study using propofol. *J Neurosci* 30:9095–9102.

Mochizuki H, Sadato N, Saito DN, Toyoda H, Tashiro M, Okamura N, Yanai K (2007) Neural correlates of perceptual difference between itching and pain: A human fMRI study. *Neuroimage* 36:706–717.

Mochizuki H, Tashiro M, Kano M, Sakurada Y, Itoh M, Yanai K (2003) Imaging of central itch modulation in the human brain using positron emission tomography. *Pain* 105:339–346.

Napadow V, LaCount L, Park K, As-Sanie S, Clauw DJ, Harris RE (2010) Intrinsic brain connectivity in fibromyalgia is associated with chronic pain intensity. *Arthritis Rheum* 62:2545–2555.

Papoiu AD, Coghill RC, Kraft RA, Wang H, Yosipovitch G (2012) A tale of two itches. Common features and notable differences in brain activation evoked by cowhage and histamine induced itch. *Neuroimage* 59:3611–3623.

Pfab F, Valet M, Napadow V, Tolle TR, Behrendt H, Ring J, Darsow U (2012) Itch and the brain. *Chem Immunol Allergy* 98:253–265.

Pfab F, Valet M, Sprenger T, Toelle TR, Athanasiadis GI, Behrendt H, Ring J, Darsow U (2006) Short-term alternating temperature enhances histamine-induced itch: A biphasic stimulus model. *J Invest Dermatol* 126:2673–2678.

Rogers BP, Morgan VL, Newton AT, Gore JC (2007) Assessing functional connectivity in the human brain by fMRI. *Magn Reson Imaging* 25:1347–1357.

Ross SE (2011) Pain and itch: Insights into the neural circuits of aversive somatosensation in health and disease. *Curr Opin Neurobiol* 21:880–887.

Ruehle BS, Handwerker HO, Ringler R, Forster C (2006) Brain activation during selective input from mechanoinsensitive and polymodal c-nociceptors. *J Neurosci* submitted.

Schmelz M, Schmidt R, Bickel A, Handwerker HO, Torebjork HE (1997) Specific C-receptors for itch in human skin. *J Neurosci* 17:8003–8008.

Schoedel AL, Zimmermann K, Handwerker HO, Forster C (2008) The influence of simultaneous ratings on cortical BOLD effects during painful and non-painful stimulation. *Pain* 135:131–141.

Shirer WR, Ryali S, Rykhlevskaia E, Menon V, Greicius MD (2012) Decoding subject-driven cognitive states with whole-brain connectivity patterns. *Cereb Cortex* 22:158–165.

Sikand P, Shimada SG, Green BG, Lamotte RH (2009) Similar itch and nociceptive sensations evoked by punctate cutaneous application of capsaicin, histamine and cowhage. *Pain* 144:66–75.

Taylor KS, Seminowicz DA, Davis KD (2009) Two systems of resting state connectivity between the insula and cingulate cortex. *Hum Brain Mapp* 30:2731–2745.

Treede RD, Kenshalo DR, Gracely RH, Jones AK (1999) The cortical representation of pain. *Pain* 79:105–111.

Valet M, Pfab F, Sprenger T, Woller A, Zimmer C, Behrendt H, Ring J, Darsow U, Tolle TR (2008) Cerebral processing of histamine-induced itch using short-term alternating temperature modulation—An FMRI study. *J Invest Dermatol* 128:426–433.

Vierow V, Fukuoka M, Ikoma A, Dorfler A, Handwerker HO, Forster C (2009) Cerebral representation of the relief of itch by scratching. *J Neurophysiol* 102:3216–3224.

Vogt BA (2005) Pain and emotion interactions in subregions of the cingulate gyrus. *Nat Rev Neurosci* 6:533–544.

Weigelt A, Terekhin P, Kemppainen P, Dorfler A, Forster C (2010) The representation of experimental tooth pain from upper and lower jaws in the human trigeminal pathway. *Pain* 149:529–538.

Yosipovitch G, Duque MI, Fast K, Dawn AG, Coghill RC (2007) Scratching and noxious heat stimuli inhibit itch in humans: A psychophysical study. *Br J Dermatol* 156:629–634.

25 Roles of Central Opioid Receptor Subtypes in Regulating Itch Sensation

Mei-Chuan Ko

CONTENTS

25.1 INTRODUCTION

Pruritus (itch sensation) was defined 350 years ago by a German physician, Samuel Hafenreffer, as an unpleasant sensation that elicits a desire or reflex to scratch. Itch/pruritus is the most common symptom in skin diseases. At any given time, one of three people in the United States has a skin disease (Johnson 2004; Thorpe et al. 2004). Skin disease is listed as one of the top 15 groups of medical conditions in which prevalence and healthcare spending increased the most between 1987 and 2000 in the United States (Thorpe et al. 2004). The estimated cost of skin diseases, including direct medical cost and the economic burden on quality of life to the American public in a single year 2004, was approximately US$96 billion (Bickers et al. 2006). Actually, the significance of skin diseases is a worldwide issue. For example, 70% of people living in developing countries suffer from skin diseases at some points in their lives, but of these 3 billion, people in 127 countries do not have access to basic skin medications (Ersser and Penzer 2000).

According to a recent review of epidemiological data, itch is also one of the key symptoms in patients suffering from a variety of systemic disorders including infectious, uremic, hepatic, and hematological diseases (Weisshaar and Dalgard 2009).

Although causes of itch depend on several factors such as disease, age, ethnicity, and characteristics of the regional healthcare systems, the symptom of itch is highly prevalent and represents a medical burden in the global community (Weisshaar and Dalgard 2009). Itch is not life threatening in most conditions. However, chronic or severe itch can cause a great deal of psychological distress, social isolation, and occupational difficulties, posing a significant financial burden in the form of physician visits, hospital care, prescription drugs, and over-the-counter products for treating the symptom (Yosipovitch et al. 2003; Bickers et al. 2006). Given that itch is a significant clinical problem afflicting a large worldwide population of humans, there is a strong need for more research on the cause, prevention, and treatment of itch, especially to advance the discovery of efficacious therapy for the treatment of itch originated from various nervous system disorders. The present review highlights the functions of central opioid receptor subtypes in regulating itch sensation.

25.2   OPIOID RECEPTOR SUBTYPES IN THE CENTRAL NERVOUS SYSTEM

Opioid receptors play important roles in a broad range of physiological functions, including pain modulation, reinforcing properties, release of neurotransmitters, and modulation of neuroendocrine responses (Bodnar 2012). They are seven transmembrane domain Gi/Go-coupled receptors and widely distributed in the central nervous system (Sanders et al. 2008; Dang and Christie 2012). Three classical opioid receptor subtypes, mu, kappa, and delta opioid receptors (i.e., MOP, KOP, and DOP), are well characterized and established by anatomical and pharmacological studies (Evans et al. 1992; Minami et al. 1994; Mansour et al. 1995; Simonin et al. 1995). In addition, the fourth member, nociceptin/orphanin FQ peptide (NOP) receptor, defined by the International Union of Pharmacology has also been studied extensively (Mollereau et al. 1994; Foord et al. 2005; Lambert 2008; Lin and Ko 2013). Following the development of agonists and antagonists selective for each opioid receptor subtype, it is clear that activation of each opioid receptor subtype produces distinct physiological profiles. Below is a brief overview of pharmacological evidence in terms of the functions of central opioid receptor subtypes in regulating itch sensation.

25.3   EVIDENCE OF CENTRAL MU OPIOID RECEPTORS INVOLVED IN ITCH

25.3.1   Endogenous Opioid Peptides

Itch/pruritus is a troublesome complication of chronic liver disease and often occurs in patients with cholestasis (Bergasa 2011; Bunchorntavakul and Reddy 2012). In the past 25 years, nonselective opioid receptor antagonists, including naloxone, naltrexone, and nalmefene, have been extensively used and evaluated for treating cholestasis-associated itch. In particular, several clinical studies have demonstrated that opioid receptor antagonists are able to ameliorate itch in patients with cholestasis (Thornton and Losowsky 1988; Bergasa et al. 1992; Bergasa 2011; Bunchorntavakul and Reddy 2012). These observations indicate that there is an increased opioidergic tone contributing to itch/pruritus in cholestatic patients. In other words, an increased release and circulation of endogenous opioid peptides is speculated, and opioid receptor antagonists can attenuate this increased endogenous opioid peptide-mediated itch. Interestingly, when plasma extracts from patients with pruritus of cholestasis were microadministered into the medullary dorsal horn of monkeys, it elicited facial scratching activity in monkeys. More importantly, this extract-induced scratching could be abolished or prevented by administration of an opioid receptor antagonist naloxone (Bergasa et al. 1993). These early findings strongly suggest that endogenous opioid peptides accumulate in the plasma of patients with cholestatic pruritus and that endogenous opioids contribute to the pruritus of cholestasis by activating the corresponding opioid receptors in primates.

Endogenous opioid peptides have not only been implicated in itch pruritus associated with patients with hepatic cholestasis but also in other clinical settings. For example, the opioid receptor antagonist naltrexone has been used to suppress itch in patients with chronic renal failure (Peer et al. 1996). Another example is chloroquine-induced itch. Chloroquine is considered as a widely used antirheumatic drug, and the clinical use of chloroquine evoked severe and generalized itch in approximately 70% of Black Africans during the treatment of malaria fever (Osifo 1984). Studies have found that naltrexone exerted antipruritic effects in patients with chloroquine-induced itch, indicating that endogenous opioids contribute to chloroquine-induced generalized itch in human malaria fever, as commonly seen in Africans (Ajayi et al. 2004). Although endogenous opioid peptides have been implied in itch derived from several clinical contexts, it is not clear how diverse opioid peptides differentially contribute to itch in primates.

25.3.2 Exogenous Opioid Agonists

Application of spinal opioids has become one of the most significant breakthroughs in pain management during the past three decades. For instance, intrathecal administration of morphine has been one of the most frequently used methods of pain management. However, the most common side effect of spinal morphine administration is itch, which sometimes is severe and lessens the value of spinal opioids for pain relief (Cousins and Mather 1984; Ganesh and Maxwell 2007). Morphine can release histamine from mast cells, but this is not its main mechanism underlying itch sensation. Several studies have shown that antihistamines are not effective in relieving morphine-induced itch (Dunteman et al. 1996). In addition, not all MOP agonists produce histamine release in humans. Intravenous administration of MOP agonists such as fentanyl produce itch sensation, but fentanyl-like analogs do not induce histamine release (Hermens et al. 1985; Stellato et al. 1992).

There are several animal studies that demonstrate the involvement of central MOP receptors in itch. First, microinjection of morphine into the medullary dorsal horn of monkeys produced dose-dependent increases in facial scratching that could be reversed by an opioid receptor antagonist naloxone, but not by an antihistamine chlorcyclizine (Thomas et al. 1993). Second, intrathecal administration of morphine at antinociceptive doses in monkeys produced long-lasting, profound body scratching responses, and antagonist studies demonstrated that the same spinal MOP receptors mediate both analgesic and itch/scratching responses in primates (Ko and Naughton 2000). Third, intracisternal, not intradermal, administration of a MOP agonist DAMGO produced facial scratching in mice (Kuraishi et al. 2000). Last, there is a large ratio of more than 1000-fold between the intrathecal and intravenous doses of MOP agonists in eliciting scratching behavior in monkeys (Ko et al. 2004). Taken together, these findings strongly indicate that central MOP receptors mediate itch sensation across species. Although it is known that the same MOP receptors mediate both morphine analgesia and itch and that MOP antagonists can be equally effective in attenuating itch and reversing analgesia, more studies are warranted to investigate the neurobiological mechanisms underlying sensory neurons expressing MOP receptors. Importantly, one recent study discovered that the MOP isoform MOP1D is required for morphine-induced itch scratching and that morphine-induced itch is separable from morphine-induced analgesia in mice (Liu et al. 2011). Such findings imply the importance of elucidating the functions of different MOP isoforms and indicate novel approaches by specifically disrupting G protein coupled receptor heterodimerization to potentially eliminate specific side effects.

25.4 SELECTIVITY AND EFFICACY OF MOP-MEDIATED ITCH

25.4.1 SELECTIVITY OF MOP-MEDIATED ITCH

Previously, researchers have established an experimental model of spinal morphine-induced itch scratching in monkeys (Ko and Naughton 2000). In particular, intrathecal morphine dose dependently produces antinociception simultaneously accompanied with hours of long-lasting, profound scratching responses. The spinal morphine-derived behavioral responses in monkeys parallel closely with analgesic and pruritic effects of spinal morphine in humans and provides a translational bridge for studying pain and itch (Bailey et al. 1993; Palmer et al. 1999). For example, nonhuman primates can be used to validate whether other opioid receptor subtype-selective analgesics possess an itch-eliciting property and what other types of neurotransmitters and receptors in the spinal cord mediate itch.

By using selective receptor agonists, pharmacologists are able to elucidate the role of each opioid receptor subtype in regulating pain and itch sensation in primates. It is known that KOP, DOP, and NOP agonists possess analgesic properties against diverse pain modalities following intrathecal and systemic administration (Butelman et al. 1993; Negus et al. 1998; Ko and Naughton 2009; Hu et al. 2010). However, these three types of agonists do not elicit scratching responses over a wide dose range (Ko et al. 2004, 2006, 2009; Molinari et al. 2013). These findings clearly indicate that only MOP agonists produce both analgesic and pruritic effects, but the other three opioid receptor subtype-selective agonists, i.e., KOP, DOP, and NOP agonists, only produce analgesic effects without eliciting itch. In other words, opioid-regulated pain and itch can be distinguished at the receptor level. The functional evidence from non-mu opioids such as KOP, DOP, and NOP agonists argue against an early view that opioids induce itch by the removal of pain inhibition (McMahon and Koltzenburg 1992; Ikoma et al. 2006). Perhaps, from the perspective of developing novel spinal analgesics, it is very exciting to discover that NOP agonists produce morphine-comparable analgesic effects without eliciting itch in primates following intrathecal administration, and such important findings may facilitate future advance of spinal analgesics (Ko et al. 2006; Hu et al. 2010; Molinari et al. 2013).

25.4.2 EFFICACY OF MOP RECEPTOR AGONISTS

Knowing that the scratching behavior represents an *in vivo* indicator of activating central MOP receptors in primates further facilitates the development of antipruritics targeting newly identified receptors. For example, diverse MOP agonists with different chemical structures produce different degrees of scratching responses. An MOP agonist DAMGO is known to have higher efficacy than morphine at the MOP receptors (Alt et al. 1998). Along with their differences in the efficacy, intrathecal DAMGO evoked a greater magnitude of scratching responses than intrathecal morphine (Ko et al. 2004). Therefore, both intrathecal DAMGO- and morphine-induced scratching models in monkeys can serve as itch models with different intensities to evaluate the antipruritic effectiveness of newly developed ligands (i.e., against moderate or profound itch intensity).

On the basis of efficacy differences and the occupancy of the same receptor population, low-efficacy MOP ligands are expected to antagonize scratching responses

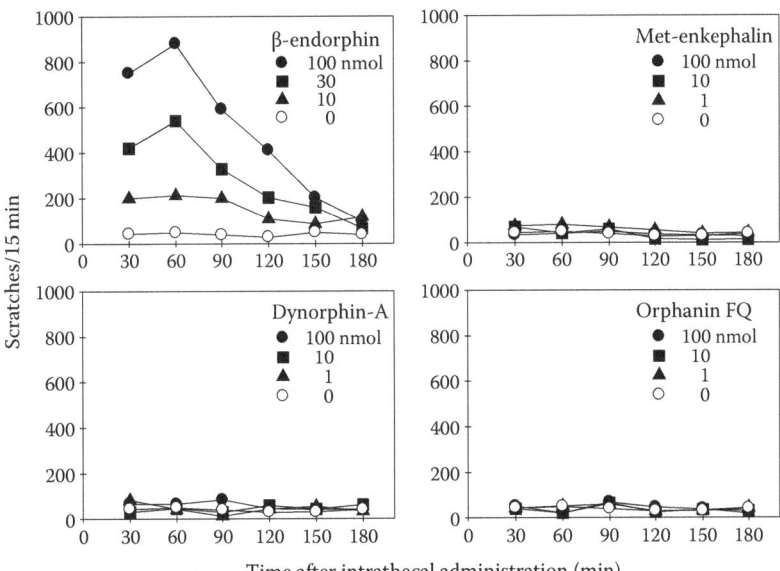

FIGURE 25.1 Effects of endogenous opioid peptides (i.e., MOP-preferring ligand (β-endorphin), KOP-preferring ligand (dynorphin-A), DOP-preferring (met-enkephalin), and NOP-preferring ligand (orphanin FQ)) following intrathecal administration on itch scratching responses in a single monkey. Symbols represent different dosing conditions for the same monkey.

induced by high-efficacy MOP ligands. This theoretical framework can be supported by both preclinical and clinical studies showing that partial MOP/KOP agonists such as nalbuphine and butorphanol, i.e., with mixed MOP antagonist/KOP agonist actions, are effective in attenuating spinal morphine-induced itch while maintaining spinal morphine analgesia (Cohen et al. 1992; Wang et al. 1998; Lee et al. 2007). Given that MOP receptors are abundant in the central nervous system, partial MOP agonists may have an advantage over MOP antagonists as supplemental agents to treat itch in patients receiving spinal opioid analgesics, or patients with speculated increased levels of endogenous opioid peptides.

On the other hand, by applying the procedure of intrathecal drug delivery in primates, we can better define itch-eliciting functions of diverse endogenous ligands that have different binding affinities on MOP receptors and different efficacies on activating MOP receptors. Figure 25.1 illustrates preliminary findings of comparing scratching eliciting effects of four endogenous opioid peptides with different preferring binding affinities in the same monkey. β-Endorphin seemed more efficient in eliciting scratching responses. It will be important to further compare scratching effects of β-endorphin with those of other endogenous MOP-preferring ligands and to identify the major endogenous ligands in mediating itch in primates.

25.5 POTENTIAL OF KOP RECEPTOR AGONISTS AS ANTIPRURITICS

The KOP receptor may be a prominent potential therapeutic target because several studies suggest that agonists acting at this receptor may be useful for treating refractory itch. One key relevant finding was that scratching was a prominent withdrawal sign in monkeys treated chronically with and withdrawn from a selective KOP agonist, U-50488H (Gmerek et al. 1987). Many withdrawal symptoms from opioids appear to be opposite to the acute effects of agonist administration (Martin et al. 1968; Heishman et al. 1989; Paronis and Woods 1997). Excessive scratching activity observed in monkeys during KOP agonist withdrawal suggested that acute administration of KOP agonists might have an antipruritic function. Animal studies seem to support this notion, as systemic administration of KOP agonists inhibited scratching behavior without interfering with locomotor activity in rodents (Togashi et al. 2002). Primate studies further demonstrated that nonsedative and nonantinociceptive doses of KOP agonists can attenuate intrathecal morphine-induced scratching without interfering with antinociception, supporting the clinical potential of KOP agonists as antipruritics in the presence of spinal opioid analgesics (Ko et al. 2003; Lee et al. 2007; Ko and Husbands 2009). More importantly, these animal studies have led to a successful clinical trial of a novel KOP agonist, nalfurafine, in hemodialysis patients suffering from uremic pruritus (Kumagai et al. 2009, 2012). A recent study in nonhuman primates has also demonstrated that antipruritic effects of nalfurafine were similar to other KOP agonists such as U-50488H, and they were mainly mediated by the activation of KOP in monkeys (Ko and Husbands 2009).

It is worth noting that opioid receptor antagonists produce parallel rightward shifts in the dose-response curves of central morphine-induced scratching (Ko and

Naughton 2000; Lee et al. 2003). These observations indicate that antipruritic effects of opioid receptor antagonists, naltrexone and nalmefene, are derived via the MOP through competitive and reversible MOP antagonism. In contrast, KOP agonists produce downward shifts in the dose-response curve of morphine-induced scratching and the antipruritic actions of KOP agonists can be reversed by a selective KOP antagonist (Ko et al. 2003; Lee et al. 2007; Ko and Husbands 2009). These observations indicate that KOP agonists do not produce MOP antagonism but inhibit MOP-mediated itch specifically through activating KOP receptors. Although the neurobiological mechanisms of the interaction between MOP and KOP in itch-selective neurons remain unclear, recent primate studies reveal exciting findings. For example, there are two separate populations of spinothalamic tract neurons responding to histamine versus nonhistaminergic pruritogenic agent in monkeys (Davidson et al. 2007). From the perspective of behavioral pharmacology, excessive scratching activity observed in monkeys during KOP agonist withdrawal may indicate that activation of KOP generally inhibits sensory neurons responding to different pruritogenic agents. Therefore, it is pivotal to pharmacologically investigate whether KOP agonists have a broader application as antipruritics in behaving primates. KOP agonists have different pharmacological properties, depending on whether they act centrally or peripherally (Ko et al. 1998; Butelman et al. 1998, 2001, 2004). Given the large body dermatomes affected by intrathecal administration, it is valuable to investigate to what extent intrathecally administered KOP agonists (effects currently unknown), as compared to systemically administered KOP agonists (Ko et al. 2003; Ko and Husbands 2009), can attenuate itch/scratching responses evoked centrally by different pruritogenic agents.

25.6 CONCLUSIONS

Collectively, functional studies of opioid receptor subtypes provide an interesting profile of each receptor's role in regulating pain and itch. Although agonists acting at MOP, KOP, DOP, or NOP receptors all produce analgesic effects, only MOP agonists elicit itch. In particular, excessive scratching activity is a selective behavioral end-point following activation of central MOP receptors. KOP, DOP, and NOP agonists do not evoke scratching responses, but KOP agonists have been shown to attenuate scratching elicited by different pruritogenic agents. By establishing centrally and peripherally evoked itch scratching models in animals, researchers can determine and compare the effectiveness of different pharmacological agents as antipruritics, in these different experimental itch models. Such studies will not only improve our understanding of the receptor mechanisms underlying itch/pruritus and antipruritics, but also offer functional evidence for partial MOP agonists, KOP agonists, and/or experimental compounds targeting novel receptors as a new generation of antipruritics in a broader context in humans.

ACKNOWLEDGMENT

Funding by the NIH-NIAMS grant (AR-059193) is gratefully acknowledged.

REFERENCES

Ajayi, A.A., Kolawole, B.A. and Udoh, S.J. (2004) Endogenous opioids, mu-opiate receptors and chloroquine-induced pruritus: A double-blind comparison of naltrexone and promethazine in patients with malaria fever who have an established history of generalized chloroquine-induced itch. *Int. J. Dermatol.* 43: 972–977.

Alt, A., Mansour, A., Akil, H. et al. (1998) Stimulation of guanosine-5′-O-(3-[35S]thio)triphosphate binding by endogenous opioids acting at a cloned mu receptor. *J. Pharmacol. Exp. Ther.* 286: 282–288.

Bailey, P.L., Rhondeau, S., Schafer, P.G. et al. (1993) Dose-response pharmacology of intrathecal morphine in human volunteers. *Anesthesiology* 79: 49–59.

Bergasa, N.V. (2011) The itch of liver disease. *Semin. Cutan. Med. Surg.* 30: 93–98.

Bergasa, N.V., Talbot, T.L., Alling, D.W. et al. (1992) A controlled trial of naloxone infusions for the pruritus of chronic cholestasis. *Gastroenterology* 102: 544–549.

Bergasa, N.V., Thomas, D.A., Vergalla, J. et al. (1993) Plasma from patients with the pruritus of cholestasis induces opioid receptor-mediated scratching in monkeys. *Life Sci.* 53: 1253–1277.

Bickers, D.R., Lim, H.W., Margolis, D. et al. (2006) The burden of skin diseases: 2004 A joint project of the American Academy of Dermatology Association and the Society for Investigative Dermatology. *J. Am. Acad. Dermatol.* 55: 490–500.

Bodnar, R.J. (2012) Endogenous opiates and behavior: 2011. *Peptides* 38: 463–522.

Bunchorntavakul, C. and Reddy, K.R. (2012) Pruritus in chronic cholestatic liver disease. *Clin. Liver Dis.* 16: 331–346.

Butelman, E.R., Ball, J.W. and Kreek, M.-J. (2004) Peripheral selectivity and apparent efficacy of dynorphins: Comparison to non-peptidic kappa-opioid agonists in rhesus monkeys. *Psychoneuroendocrinology* 29: 307–326.

Butelman, E.R., Ko, M.C., Sobczyk-kojiko, K. et al. (1998) Kappa-opioid receptor binding populations in rhesus monkey brain: Relationship to an assay of thermal antinociception. *J. Pharmacol. Exp. Ther.* 285: 595–601.

Butelman, E.R., Ko, M.C., Traynor, J.R. et al. (2001) GR89,696: A potent kappa-opioid agonist with subtype selectivity in rhesus monkeys. *J. Pharmacol. Exp. Ther.* 298: 1049–1059.

Butelman, E.R., Negus, S.S., Ai, Y. et al. (1993) Kappa opioid antagonist effects of systemically administered nor-binaltorphimine in a thermal antinociception assay in rhesus monkeys. *J. Pharmacol. Exp. Ther.* 267: 1269–1276.

Cohen, S.E., Ratner, E.F., Kreitzman, T.R. et al. (1992) Nalbuphine is better than naloxone for treatment of side effects after epidural morphine. *Anesth. Analg.* 75: 747–752.

Cousins, M.J. and Mather, L.E. (1984) Intrathecal and epidural administration of opioids. *Anesthesiology* 61: 276–310.

Dang, V.C. and Christie, M.J. (2012) Mechanisms of rapid opioid receptor desensitization, resensitization and tolerance in brain neurons. *Br. J. Pharmacol.* 165: 1704–1716.

Davidson, S., Zhang, X., Yoon, C.H. et al. (2007) The itch-producing agents histamine and cowhage activate separate populations of primate spinothalamic tract neurons. *J. Neurosci.* 27: 10007–10014.

Dunteman, E., Karanikolas, M. and Filos, K.S. (1996) Transnasal butorphanol for the treatment of opioid-induced pruritus unresponsive to antihistamines. *J. Pain Symptom. Manage.* 12: 255–260.

Ersser, S.J. and Penzer, R. (2000) Meeting patients' skin care needs: Harnessing nursing expertise at an international level. *Int. Nurs. Rev.* 47: 167–173.

Evans, C.J., Keith, D.E., Morrison, H. et al. (1992) Cloning of a delta opioid receptor by functional expression. *Science* 258: 1952–1955.

Foord, S.M., Bonner, T.I., Neubig, R.R. et al. (2005) International Union of Pharmacology: XLVI. G protein-coupled receptor list. *Pharmacol. Rev.* 57: 279–288.

Ganesh, A. and Maxwell, L.G. (2007) Pathophysiology and management of opioid-induced pruritus. *Drugs* 67: 2323–2333.

Gmerek, D.E., Dykstra, L.A. and Woods, J.H. (1987) Kappa opioids in rhesus monkeys. III. Dependence associated with chronic administration. *J. Pharmacol. Exp. Ther.* 242: 428–436.

Heishman, S.J., Stitzer, M.L., Bigelow, G.E. et al. (1989) Acute opioid physical dependence in postaddict humans: Naloxone dose effects after brief morphine exposure. *J. Pharmacol. Exp. Ther.* 248: 127–134.

Hermens, J.M., Ebertz, J.M., Hanifin, J.M. et al. (1985) Comparison of histamine release in human skin mast cells induced by morphine, fentanyl, and oxymorphone. *Anesthesiology* 62: 124–129.

Hu, E., Calo', G., Guerrini, R. et al. (2010) Long-lasting antinociceptive spinal effects in primates of the novel nociceptin/orphanin FQ receptor agonist UFP-112. *Pain* 148: 107–113.

Ikoma, A., Steinhoff, M., Stander, S. et al. (2006) The neurobiology of itch. *Nat. Rev. Neurosci.* 7: 535–547.

Johnson, M.L. (2004) Defining the burden of skin diseases in the United States—A historical perspective. *J. Invest. Dermatol. Symp. Proc.* 9: 108–110.

Ko, M.C., Butelman, E.R., Traynor, J.R. et al. (1998) Differentiation of kappa opioid agonist-induced antinociception by naltrexone apparent pA2 analysis in rhesus monkeys. *J. Pharmacol. Exp. Ther.* 285: 518–526.

Ko, M.C. and Husbands, S.M. (2009) Effects of atypical kappa-opioid receptor agonists on intrathecal morphine-induced itch and analgesia in primates. *J. Pharmacol. Exp. Ther.* 328: 193–200.

Ko, M.C. and Naughton, N.N. (2000) An experimental itch model in monkeys: Characterization of intrathecal morphine-induced scratching and antinociception. *Anesthesiology* 92: 795–805.

Ko, M.C. and Naughton, N.N. (2009) Antinociceptive effects of nociceptin/orphanin FQ administered intrathecally in monkeys. *J. Pain* 10: 509–516.

Ko, M.C., Song, M.S., Edwards, T. et al. (2004) The role of central mu opioid receptors in opioid-induced itch in primates. *J. Pharmacol. Exp. Ther.* 310: 169–176.

Ko, M.C., Wei, H., Woods, J.H. et al. (2006) Effects of intrathecally administered nociceptin/orphanin FQ in monkeys: Behavioral and mass spectrometric studies. *J. Pharmacol. Exp. Ther.* 318: 1257–1264.

Ko, M.C., Woods, J.H., Fantegrossi, W.E. et al. (2009) Behavioral effects of a synthetic agonist selective for nociceptin/orphanin FQ peptide receptors in monkeys. *Neuropsychopharmacology* 34: 2088–2096.

Ko, M.C.H., Lee, H., Song, M.S. et al. (2003) Activation of kappa-opioid receptors inhibits pruritus evoked by subcutaneous or intrathecal administration of morphine in monkeys. *J. Pharmacol. Exp. Ther.* 305: 173–179.

Kumagai, H., Ebara, T., Takemori, K. et al. (2009) Effect of a novel kappa-receptor agonist, nalfurafine hydrochloride, on severe itch in 337 haemodialysis patients: A phase III, randomized, double-blind, placebo-controlled study. *Nephrol. Dial. Transplant* 25: 1251–1257.

Kumagai, H., Ebata, T., Takamori, K. et al. (2012) Efficacy and safety of a novel k-agonist for managing intractable pruritus in dialysis patients. *Am. J. Nephrol.* 36: 175–183.

Kuraishi, Y., Yamaguchi, T. and Miyamoto, T. (2000) Itch-scratch responses induced by opioids through central mu opioid receptor in mice. *J. Biomed. Sci.* 7: 248–252.

Lambert, D.G. (2008) The nociceptin/orphanin FQ receptor: A target with broad therapeutic potential. *Nat. Rev. Drug Discov.* 7: 694–710.

Lee, H., Naughton, N.N., Woods, J.H. et al. (2003) Characterization of scratching responses in rats following centrally administered morphine or bombesin. *Behav. Pharmacol.* 14: 501–508.

Lee, H., Naughton, N.N., Woods, J.H. et al. (2007) Effects of butorphanol on morphine-induced itch and analgesia in primates. *Anesthesiology* 107: 478–485.

Lin, A.P. and Ko, M.C. (2013) The therapeutic potential of nociceptin/orphanin FQ receptor agonists as analgesics without abuse liability. *ACS Chem. Neurosci.* 4: 214–224.

Liu, X.Y., Liu, Z.C., Sun, Y.G. et al. (2011) Unidirectional cross-activation of GRPR by MOR1D uncouples itch and analgesia induced by opioids. *Cell* 147: 447–458.

Mansour, A., Fox, C.A., Burke, S. et al. (1995) Immunohistochemical localization of the cloned mu opioid receptor in the rat CNS. *J. Chem. Neruoanat.* 8: 283–305.

Martin, W.R., Jasinski, D.R., Sapira, J.D. et al. (1968) The respiratory effects of morphine during a cycle of dependence. *J. Pharmacol. Exp. Ther.* 62: 182–189.

McMahon, S.B. and Koltzenberg, M. (1992) Itch for an explanation. *Tr. Neurosci.* 15: 497–501.

Minami, M., Onogi, T., Toya, T. et al. (1994) Molecular cloning and in situ hybridization histochemistry for rat mu-opioid receptors. *Neurosci. Res.* 18: 315–322.

Molinari, S., Camarda, V., Rizzi, A. et al. (2013) [Dmt(1)]N/OFQ(1-13)-NH(2): A potent nociceptin/orphanin FQ and opioid receptor universal agonist. *Br. J. Pharmacol.* 168: 151–162.

Mollereau, C., Parmentier, M., Maiilleux, P. et al. (1994) ORL1, a novel member of the opioid receptor family: Cloning, functional expression and localization. *FASEB Lett.* 341: 33–38.

Negus, S.S., Gatch, M.B., Mello, N.K. et al. (1998) Behavioral effects of the delta-selective opioid agonist SNC80 and related compounds in rhesus monkeys. *J. Pharmacol. Exp. Ther.* 286: 362–375.

Osifo, N.G. (1984) Chloroquine-induced pruritus among patients with malaria. *Arch. Dermatol.* 120: 80–82.

Palmer, C.M., Emerson, S., Volgoropolous, D. et al. (1999) Dose-response relationship of intrathecal morphine for postcesarean analgesia. *Anesthesiology* 90: 437–444.

Paronis, C.A. and Woods, J.H. (1997) Ventilation in morphine-maintained rhesus monkeys. I: Effects of naltrexone and abstinence-associated withdrawal. *J. Pharmacol. Exp. Ther.* 282: 348–354.

Peer, G., Kivity, S., Agami, O. et al. (1996) Randomised crossover trial of naltrexone in uraemic pruritus. *Lancet* 348: 1552–1554.

Sanders, R.D., Brian, D. and Maze, M. (2008) G-protein-coupled receptors. *Handb. Exp. Pharmacol.* 182: 93–117.

Simonin, F., Gaveriaux-Ruff, C., Befort, K. et al. (1995) Kappa-opioid receptor in humans: cDNA and genomic cloning, chromosomal assignment, functional expression, pharmacology, and expression pattern in the central nervous system. *Proc. Natl. Acad. Sci. USA* 92: 7006–7010.

Stellato, C., Cirillo, R., de Paulis, A. et al. (1992) Human basophil/mast cell releasability. *Anesthesiology* 77: 932–940.

Thomas, D.A., Williams, G.M., Iwata, K. et al. (1993) The medullary dorsal horn. A site of action of morphine in producing facial scratching in monkeys. *Anesthesiology* 79: 548–554.

Thornton, J.R. and Losowsky, M.S. (1988) Opioid peptides and primary biliary cirrhosis. *BMJ* 297: 1501–1504.

Thorpe, K.E., Florence, C.S. and Joski, P. (2004) Which medical conditions account for the rise in health spending? *Health Aff.* 22:W4-437-445.

Togashi, Y., Umeuchi, H., Okano, K. et al. (2002) Antipruritic activity of the kappa-opioid receptor agonist, TRK-820. *Eur. J. Pharmacol.* 435: 259–264.

Wang, J.J., Ho, S.T. and Tzeng, J.I. (1998) Comparison of intravenous nalbuphine infusion versus naloxone in the prevention of epidural morphine-related side effects. *Reg. Anesth. Pain Med.* 23: 479–484.

Weisshaar, E. and Dalgard, F. (2009) Epidemiology of itch: Adding to the burden of skin morbidity. *Acta Derm. Venereol.* 89: 339–350.

Yosipovitch, G., Greaves, M.W. and Schmelz, M. (2003) Itch. *The Lancet* 361: 690–694.

26 Sensitization for Itch

Martin Schmelz

CONTENTS

Chronic localized itch is supposed to be based on spontaneous activity of primary afferent fibers that are excited by pruritic mediators in the skin. One might therefore conclude that continuous release of pruritic mediators underlies chronic itch. However, chemical responses in C-fibers are characterized by pronounced tachyphylaxis. Thus, naïve nociceptors can hardly sustain ongoing activity following prolonged chemical activation. This chapter is therefore focused on mechanisms changing the sensitivity of the neurons involved in itch processing, such that they can sustain chronic signaling of itch both in the periphery and in the spinal cord.

26.1 CHEMICALLY INDUCED ITCH

Classical inflammatory mediators such as bradykinin, serotonin, prostanoids, histamine, and low pH have been shown to activate and sensitize nociceptors with histamine being the best characterized pruritogen (Ikoma et al. 2006). There are only few mediators that can induce histamine-independent pruritus. Prostaglandins were found not only to enhance histamine-induced itch in the skin (Hägermark et al. 1977) but also to act directly as pruritogens in the conjunctiva (Woodward et al. 1995) and in human skin when applied via microdialysis fibers (Neisius et al. 2002). Acetylcholine has been identified as a pruritic in atopic dermatitis (AD) patients, whereas it induces pain in normal subjects (Heyer et al. 1997). This mechanism could probably explain the itch which many AD patients experience when they sweat. The potency of the well-established pruritics in normal skin can be defined as histamine > prostaglandin E2 > acetylcholine, serotonin (Schmelz et al. 2003b). In contrast, bradykinin and capsaicin application usually induce a pure pain sensation.

Cowhage spicules inserted into human skin produce itch in an intensity which is comparable to that following histamine application (LaMotte et al. 2009; Sikand et al. 2009). However, mechanoresponsive "polymodal" C-fiber afferents, the most common type of afferent C-fibers in human skin (Schmidt et al. 1995), can be activated by cowhage in the cat (Tuckett and Wei 1987) and, according to recent studies, also in nonhuman primates (Johanek et al. 2007, 2008) and human volunteers (Namer et al. 2008). These fibers are unresponsive to histamine iontophoresis (Schmelz et al. 2003b) and not involved in sustained axon reflex flare reactions (Schmelz et al. 2000). This is consistent with the observation that cowhage-induced itch is not accompanied by a widespread axon reflex flare (Johanek et al. 2007; Shelley and Arthur 1955, 1957). The active compound, the cysteine protease mucunain, has been identified lately and has shown to activate proteinase activated receptor 2 (PAR 2) and even more potently PAR 4 (Reddy et al. 2008). Most interestingly, when the algogen capsaicin was applied via inactivated cowhage spicules it also provoked itch in human volunteers (Sikand et al. 2009), indicating that itch can be induced by very localized activation of nociceptors.

In summary, several chemical mediators have been identified that can cause acute experimental itch in human volunteers. The key problem to translate this finding to chronic itch patients is the prolonged time course of the clinical symptom. Chemical responses of sensory nociceptor endings are generally characterized by a pronounced tachyphylaxis (Liang et al. 2001). Thus, prolonged application will typically cause initial pain or itch that fades in a couple of minutes even though the stimulation is ongoing (Ikoma et al. 2004).

26.2 PERIPHERAL SENSITIZATION

Sensitization of primary afferents has been suggested as an explanation for the mismatch between transient chemical responses of primary afferent fibers and chronic itch lasting for months and years (Schmelz 2010). Regulation of gene expression induced by trophic factors, such as nerve growth factor (NGF), has been shown to play a major role in persistently increased neuronal sensitivity. NGF is released in the periphery and specifically binds to trkA receptors located on nociceptive nerve endings. It is then conveyed via retrograde axonal transport to the dorsal root ganglion where gene expression of neuropeptides and receptor molecules, such as the vanilloid receptor (TRPV1), is increased. Trophic factors also initiate nerve fiber sprouting and thus change the morphology of sensory neurons. Sprouting of epidermal nerve fibers in combination with localized *pain* and hypersensitivity has been reported before (Bohm-Starke et al. 2001). Similar mechanisms are supposed to underlie chronic itch and chronic pain (Handwerker and Schmelz 2009; Yosipovitch et al. 2007).

Increased intradermal nerve fiber density has been found in patients with chronic pruritus (Urashima and Mihara 1998). In addition, increased epidermal levels of neurotrophin 4 (NT$_4$) have been found in patients with atopic dermatitis (Grewe et al. 2000; Yamaguchi et al. 2009), and massively increased serum levels of NGF and SP have been found to correlate with the severity of the disease in such patients (Toyoda et al. 2002). These similarities between localized painful and pruritic lesions might indicate that on a peripheral level similar mechanisms of nociceptor sprouting and sensitization exist.

Chronic itch has been associated with increased levels of epidermal NGF and enhanced neuronal signaling, for instance in AD patients (Tominaga et al. 2009; Yamaguchi et al. 2009). Experimentally, pruritus in AD can be elicited, independent of histamine (Ikoma et al. 2004). Spicules from *Mucuna pruriens* (cowhage) also produce sensations of itch when inserted into the epidermis of humans (Namer et al. 2008; Sikand et al. 2009). Cowhage activates mechanoresponsive polymodal C-nociceptors in humans and also Aδ fibers in monkeys (Ringkamp et al. 2011) and evokes itch sensations at an intensity comparable to those induced by histamine but without an accompanying axon reflex flare reaction.

NGF has been shown to sensitize cowhage- but not histamine-induced itch in human volunteers (Rukwied et al. 2013). The absence of sensitization to histamine at a time point showing the maximum increase in heat sensitivity is surprising. Apparently, sensitization of transient receptor potential (TRPV1) does not necessarily lead to increased histamine responses, even though histamine-induced itch is elevated inside the eczema of AD (Hosogi et al. 2006) and TRPV1 knockout mice showed reduced behavioral itch responses to histamine (Figure 26.1) (Imamachi et al. 2009).

Interestingly, sensitization to cowhage itch at the NGF-treated site correlated with the level of mechanical hyperalgesia. This correlation might suggest that sensitized mechanosensitive nociceptors underlie the increased cowhage-induced itch. Indeed, cowhage activates virtually all mechanosensitive polymodal units (Namer

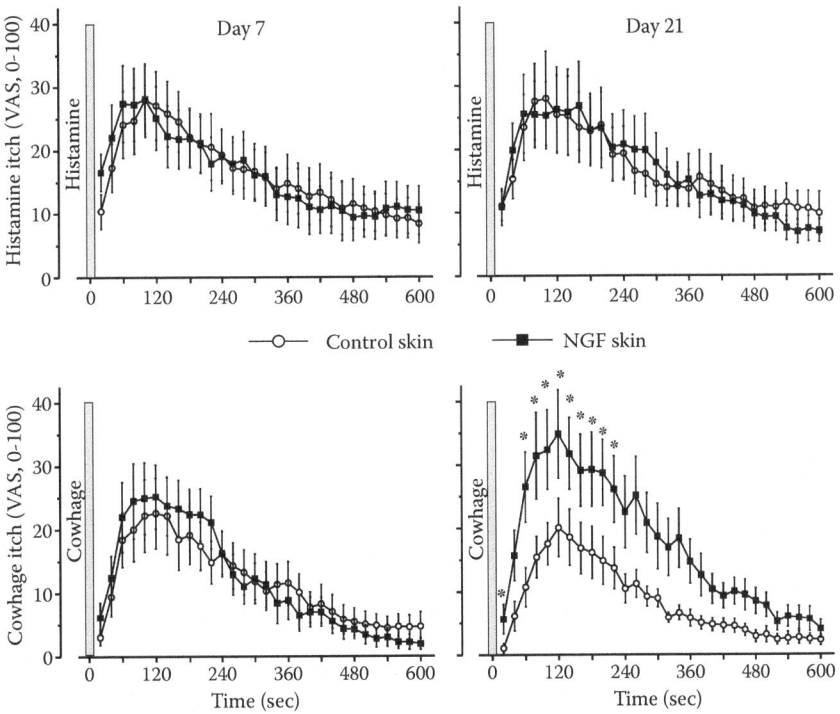

FIGURE 26.1 Nerve growth factor (NGF, 1 μg) was injected intracutaneously in human volunteers and sensitivity to histamine- (intophoresis) and cowhage-induced itch was tested at day 7 (left) and day 21 (right). Histamine itch ratings (visual analogue scale 0–100) were unchanged after NGF. In contrast, cowhage-induced itch was increased by NGF by about 50% at day 21. (Modified from Rukwied R et al., *J Invest Dermatol*, 2013.)

et al. 2008) and also some mechanosensitive Aδ fibers in the monkey (Ringkamp et al. 2011). Thus, mechanical stimulation at the NGF-sensitized site will cause an increased response of the local mechanosensitive nociceptors and thereby induce increased pain. However, following epidermal application of cowhage spicules, only very few receptive endings will be activated, whereas fibers in the immediate surrounding will remain silent. This mismatch pattern might be interpreted as itch in the central nervous system as discussed before (Schmelz 2009).

In addition to experimentally induced sensitization in volunteers (Rukwied 2010, 2013), elevated NGF levels in AD skin have been found (Tominaga et al. 2009; Yamaguchi et al. 2009), and thus NGF might sensitize primary afferents also in the patients and contribute to chronic pruritus. Sensitization of itch responses has already been investigated in atopic eczema patients comparing their responses to algogens and pruritics (Figure 26.2) (Hosogi et al. 2006).

The classic endogenous algogens serotonin and bradykinin were found to turn into potent pruritogens in lesional skin of patients with atopic dermatitis, with bradykinin being a particularly strong histamine-independent pruritogen. Of similar importance is the enhanced itch response to histamine application in the eczema (Hosogi et al. 2006; Ikoma et al. 2004). Enhanced itch responses in lesional skin of atopic dermatitis patients do not only reflect clinical importance of sensitization processes, but they also provide a simple experimental approach to quantify this sensitization in a clinical setting (Figure 26.3).

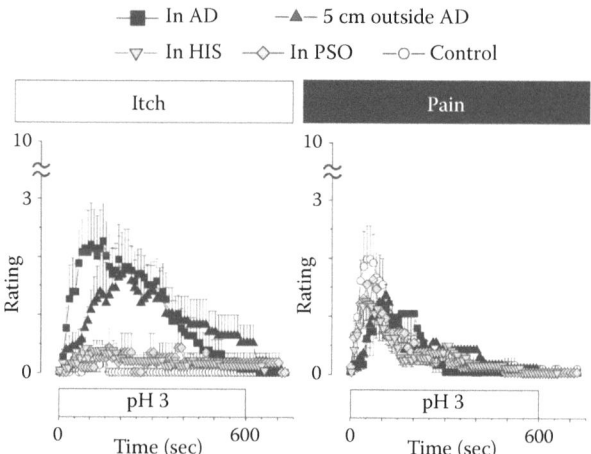

FIGURE 26.2 An acidic citrate buffer (pH 3) was perfused through intracutaneous microdialysis catheters for 10 min in healthy volunteers (control), patients with psoriasis and patients with atopic dermatitis (AD). Control subjects and patients with psoriasis (PSO) felt the acidic stimulus as transient pain (right panel; pain ratings on a scale from 0 to 10). In contrast, patients with atopic dermatitis felt the same stimulus as transient itch (left panel) both when applied inside the eczema (filled squares) and 5 cm outside the eczema (filled triangles). (Modified from Ikoma A et al., *Neurology* 62, 212–217, 2004.)

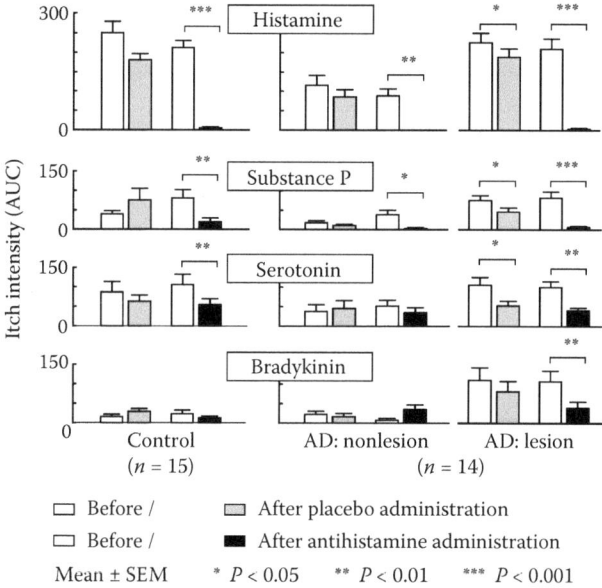

FIGURE 26.3 The areas under the curve (AUC) of itch ratings for 10 min after iontophoresis of histamine, substance P, serotonin, and bradykinin before (white columns) and after the administration of placebo (grey columns) or antihistamine (black columns). Repetition of histamine, substance P, and serotonin iontophoresis after a 3-h interval under the placebo condition reduced itch intensities significantly. Histamine- and substance P-induced itch was suppressed significantly by antihistamine in all groups, while serotonin- and bradykinin-induced itch was not in all groups. (Modified from Hosogi M et al., *Pain*, 126, 16–23, 2006.)

26.3 CENTRAL SENSITIZATION

There is a remarkable similarity between the phenomena associated with central sensitization to pain and itch. Activity in chemo-nociceptors does not only lead to acute pain but, in addition, can sensitize second-order neurons in the dorsal horn, thereby leading to touch-evoked pain (allodynia) and punctate hyperalgesia (Koltzenburg 2000). In itch processing, similar phenomena have been described: touch- or brush-evoked pruritus around an itching site has been termed "itchy skin" (Bickford 1938; Simone et al. 1991). Like touch-evoked pain, it requires ongoing activity in primary afferents and is most probably elicited by low threshold mechanoreceptors (A-β fibers) (Heyer et al. 1995; Simone et al. 1991). Also, more intense prick-induced itch sensations in the surroundings, called "hyperknesis," have been reported following histamine iontophoresis in healthy volunteers (Atanassoff et al. 1999).

The existence of central sensitization for itch can greatly improve our understanding of clinical itch. Under the condition of central sensitization leading to punctate hyperknesis, normally painful stimuli are perceived as itching. This phenomenon has already been described in patients suffering from atopic dermatitis who perceive normally painful electrical stimuli as itching when applied inside their lesional skin

(Nilsson and Schouenborg 1999). Furthermore, acetylcholine provokes itch instead of pain in patients with atopic dermatitis (Vogelsang et al. 1995), indicating that pain-induced inhibition of itch might be compromised in these patients. Similarly, recent data suggest that also diffuse noxious inhibitory control mechanisms are altered in chronic itch patients as painful electrical stimulation enhanced histamine-induced itch in psoriasis patients rather than decreasing it as in healthy controls (van Laarhoven et al. 2010).

The exact mechanisms and roles of central sensitization for itch in specific, clinical conditions still have to be explored, whereas a major role of central sensitization in patients with chronic pain is generally accepted. It should be noted that, in addition to the parallels between experimentally induced secondary sensitization phenomena, there is also emerging evidence for corresponding phenomena in patients with chronic pain and chronic itch. In patients with neuropathic pain, it has recently been reported that histamine iontophoresis resulted in burning pain instead of pure itch. In healthy volunteers, pure itch would be induced by this procedure (Baron et al. 2001; Birklein et al. 1997). This phenomenon is of special interest as it demonstrates spinal hypersensitivity to C-fiber input. Conversely, usually as painful perceived electrical, chemical, mechanical and thermal stimulation is perceived as itching when applied in or close to lesional skin of atopic dermatitis patients (Ikoma et al. 2003).

Long lasting activation of pruriceptors by histamine has been shown to experimentally induce central sensitization for itch in healthy volunteers (Ikoma et al. 2003): following the application of histamine via dermal microdialysis fibers, low pH stimulation of the skin close to the histamine site was perceived as itch instead of pain. Ongoing activity of pruriceptors, which might underlie the development of central sensitization for itch, has already been confirmed microneurographically in a patient with chronic pruritus (Schmelz et al. 2003a). Thus, there is emerging evidence for a role of central sensitization regarding itch in chronic pruritus. As there are many mediators and mechanisms, which are potentially algogenic in inflamed skin, many of them could provoke itch in a sensitized patient. Thus, a therapeutic approach, which targets only a single pruritic mediator, does not appear to be promising for patients with chronic itching diseases, e.g., atopic dermatitis. In contrast, the main therapeutic implication of this phenomenon is that a combination of centrally acting drugs counteracting the sensitization, and topically acting drugs counteracting the inflammation, should be more promising in ameliorating pruritus in these cases.

While there is obviously an antagonistic interaction between pain and itch under normal conditions, the patterns of spinal sensitization phenomena are surprisingly similar. It remains to be established whether this similarity will also include the underlying mechanism that would implicate similar therapeutic approaches such as gabapentin (Scheinfeld 2003) or clonidine (Elkersh et al. 2003) for the treatment of neuropathic itch.

26.4 PERSPECTIVES

Sensitization of nociceptive neurons both in the periphery and in the central nervous system has been identified as a key mechanism in chronic pain patients. Counteracting peripheral sensitization of nociceptors by capturing NGF has recently proven to be effective in reducing pain in osteoarthritis and chronic back pain (Katz et al. 2011; Lane et al. 2010), and clinically effective analgesics like gabapentin are known to reduce central sensitization (Gottrup et al. 2004). Thus, therapeutic approaches that counteract sensitization mechanisms in humans have been shown to be effective analgesics in chronic pain patients. On the basis of the similarity between pain and itch processing, it is tempting to suggest that a similar therapeutic success will be possible when sensitization mechanisms in chronic itch patients are targeted.

REFERENCES

Atanassoff PG, Brull SJ, Zhang J, Greenquist K, Silverman DG, LaMotte RH. 1999. Enhancement of experimental pruritus and mechanically evoked dysesthesiae with local anesthesia. *Somatosens Mot Res* 16:291–298.

Baron R, Schwarz K, Kleinert A, Schattschneider J, Wasner G. 2001. Histamine-induced itch converts into pain in neuropathic hyperalgesia. *Neuroreport* 12:3475–3478.

Bickford RGL. 1938. Experiments relating to itch sensation, its peripheral mechanism and central pathways. *Clin Sci* 3:377–386.

Birklein F, Claus D, Riedl B, Neundorfer B, Handwerker HO. 1997. Effects of cutaneous histamine application in patients with sympathetic reflex dystrophy. *Muscle Nerve* 20:1389–1395.

Bohm-Starke N, Hilliges M, Brodda-Jansen G, Rylander E, Torebjörk HE. 2001. Psychophysical evidence of nociceptor sensitization in vulvar vestibulitis syndrome. *Pain* 94:177–183.

Elkersh MA, Simopoulos TT, Malik AB, Cho EH, Bajwa ZH. 2003. Epidural clonidine relieves intractable neuropathic itch associated with herpes zoster-related pain. *Reg Anesth Pain Med* 28:344–346.

Gottrup H, Juhl G, Kristensen AD, Lai R, Chizh BA, Brown J, Bach FW, Jensen TS. 2004. Chronic oral gabapentin reduces elements of central sensitization in human experimental hyperalgesia. *Anesthesiology* 101:1400–1408.

Grewe M, Vogelsang K, Ruzicka T, Stege H, Krutmann J. 2000. Neurotrophin-4 production by human epidermal keratinocytes: Increased expression in atopic dermatitis. *J Invest Dermatol* 114:1108–1112.

Hägermark O, Strandberg K, Hamberg M. 1977. Potentiation of itch and flare responses in human skin by prostaglandins E2 and H2 and a prostaglandin endoperoxide analog. *J Invest Dermatol* 69:527–530.

Handwerker HO, Schmelz M. 2009. Pain: Itch without pain—A labeled line for itch sensation? *Nat Rev Neurol* 5:640–641.

Heyer G, Ulmer FJ, Schmitz J, Handwerker HO. 1995. Histamine-induced itch and alloknesis (itchy skin) in atopic eczema patients and controls. *Acta Derm Venereol* 75:348–352.

Heyer G, Vogelgang M, Hornstein OP. 1997. Acetylcholine is an inducer of itching in patients with atopic eczema. *J Dermatol* 24:621–625.

Hosogi M, Schmelz M, Miyachi Y, Ikoma A. 2006. Bradykinin is a potent pruritogen in atopic dermatitis: A switch from pain to itch. *Pain* 126:16–23.

Ikoma A, Fartasch M, Heyer G, Miyachi Y, Handwerker H, Schmelz M. 2004. Painful stimuli evoke itch in patients with chronic pruritus: Central sensitization for itch. *Neurology* 62:212–217.

Ikoma A, Rukwied R, Stander S, Steinhoff M, Miyachi Y, Schmelz M. 2003. Neuronal sensitization for histamine-induced itch in lesional skin of patients with atopic dermatitis. *Arch Dermatol* 139:1455–1458.

Ikoma A, Steinhoff M, Stander S, Yosipovitch G, Schmelz M. 2006. The neurobiology of itch. *Nat Rev Neurosci* 7:535–547.

Imamachi N, Park GH, Lee H, Anderson DJ, Simon MI, Basbaum AI, Han SK. 2009. TRPV1-expressing primary afferents generate behavioral responses to pruritogens via multiple mechanisms. *Proc Natl Acad Sci USA* 106:11330–11335.

Johanek LM, Meyer RA, Friedman RM, Greenquist KW, Shim B, Borzan J, Hartke T, LaMotte RH, Ringkamp M. 2008. A role for polymodal C-fiber afferents in nonhistaminergic itch. *J Neurosci* 28:7659–7669.

Johanek LM, Meyer RA, Hartke T, Hobelmann JG, Maine DN, LaMotte RH, Ringkamp M. 2007. Psychophysical and physiological evidence for parallel afferent pathways mediating the sensation of itch. *J Neurosci* 27:7490–7497.

Katz N, Borenstein DG, Birbara C, Bramson C, Nemeth MA, Smith MD, Brown MT. 2011. Efficacy and safety of tanezumab in the treatment of chronic low back pain. *Pain* 152:2248–2258.

Koltzenburg M. 2000. Neural mechanisms of cutaneous nociceptive pain. *Clin J Pain* 16:S131–S138.

LaMotte RH, Shimada SG, Green BG, Zelterman D. 2009. Pruritic and nociceptive sensations and dysesthesias from a spicule of cowhage. *J Neurophysiol* 101:1430–1443.

Lane NE, Schnitzer TJ, Birbara CA, Mokhtarani M, Shelton DL, Smith MD, Brown MT. 2010. Tanezumab for the treatment of pain from osteoarthritis of the knee. *N Engl J Med* 363:1521–1531.

Liang YF, Haake B, Reeh PW. 2001. Sustained sensitization and recruitment of rat cutaneous nociceptors by bradykinin and a novel theory of its excitatory action. *J Physiol* 532:229–239.

Namer B, Carr R, Johanek LM, Schmelz M, Handwerker HO, Ringkamp M. 2008. Separate peripheral pathways for pruritus in man. *J Neurophysiol* 100:2062–2069.

Neisius U, Olsson R, Rukwied R, Lischetzki G, Schmelz M. 2002. Prostaglandin E2 induces vasodilation and pruritus, but no protein extravasation in atopic dermatitis and controls. *J Am Acad Dermatol* 47:28–32.

Nilsson HJ, Schouenborg J. 1999. Differential inhibitory effect on human nociceptive skin senses induced by local stimulation of thin cutaneous fibers. *Pain* 80:103–112.

Reddy VB, Iuga AO, Shimada SG, LaMotte RH, Lerner EA. 2008. Cowhage-evoked itch is mediated by a novel cysteine protease: A ligand of protease-activated receptors. *J Neurosci* 28:4331–4335.

Ringkamp M, Schepers RJ, Shimada SG, Johanek LM, Hartke TV, Borzan J, Shim B, LaMotte RH, Meyer RA. 2011. A role for nociceptive, myelinated nerve fibers in itch sensation. *J Neurosci* 31:14841–14849.

Rukwied R, Main M, Weinkauf B, Schmelz M. 2013. NGF sensitizes nociceptors for cowhage- but not histamine-induced itch in human skin. *J Invest Dermatol* 133:268–270.

Rukwied R, Mayer A, Kluschina O, Obreja O, Schley M, Schmelz M. 2010. NGF induces non-inflammatory localized and lasting mechanical and thermal hypersensitivity in human skin. *Pain* 148:407–413.

Scheinfeld N. 2003. The role of gabapentin in treating diseases with cutaneous manifestations and pain. *Int J Dermatol* 42:491–495.

Schmelz M. 2009. How pain becomes itch. *Pain* 144:14–15.

Schmelz M. 2010. Itch and pain. *Neurosci Biobehav Rev* 34:171–176.

Schmelz M, Hilliges M, Schmidt R, Orstavik K, Vahlquist C, Weidner C, Handwerker HO, Torebjörk HE. 2003a. Active "itch fibers" in chronic pruritus. *Neurology* 61:564–566.

Schmelz M, Michael K, Weidner C, Schmidt R, Torebjörk HE, Handwerker HO. 2000. Which nerve fibers mediate the axon reflex flare in human skin? *Neuroreport* 11:645–648.

Schmelz M, Schmidt R, Weidner C, Hilliges M, Torebjörk HE, Handwerker HO. 2003b. Chemical response pattern of different classes of C-nociceptors to pruritogens and algogens. *J Neurophysiol* 89:2441–2448.

Schmidt R, Schmelz M, Forster C, Ringkamp M, Torebjörk HE, Handwerker HO. 1995. Novel classes of responsive and unresponsive C nociceptors in human skin. *J Neurosci* 15:333–341.

Shelley WB, Arthur RP. 1955. Mucunain, the active pruritogenic proteinase of cowhage. *Science* 122:469–470.

Shelley WB, Arthur RP. 1957. The neurohistology and neurophysiology of the itch sensation in man. *AMA Arch Dermatol* 76:296–323.

Sikand P, Shimada SG, Green BG, LaMotte RH. 2009. Similar itch and nociceptive sensations evoked by punctate cutaneous application of capsaicin, histamine and cowhage. *Pain* 144:66–75.

Simone DA, Alreja M, LaMotte RH. 1991. Psychophysical studies of the itch sensation and itchy skin ("alloknesis") produced by intracutaneous injection of histamine. *Somatosens Mot Res* 8:271–279.

Tominaga M, Tengara S, Kamo A, Ogawa H, Takamori K. 2009. Psoralen-ultraviolet A therapy alters epidermal Sema3A and NGF levels and modulates epidermal innervation in atopic dermatitis. *J Dermatol Sci* 55:40–46.

Toyoda M, Nakamura M, Makino T, Hino T, Kagoura M, Morohashi M. 2002. Nerve growth factor and substance P are useful plasma markers of disease activity in atopic dermatitis. *Br J Dermatol* 147:71–79.

Tuckett RP, Wei JY. 1987. Response to an itch-producing substance in cat. II. Cutaneous receptor population with unmyelinated axons. *Brain Research* 413:95–103.

Urashima R, Mihara M. 1998. Cutaneous nerves in atopic dermatitis—A histological, immuno-histochemical and electron microscopic study. *Virchows Arch Int J Pathol* 432:363–370.

Van Laarhoven AI, Kraaimaat FW, Wilder-Smith OH, Van De Kerkhof PC, Evers AW. 2010. Heterotopic pruritic conditioning and itch—Analogous to DNIC in pain? *Pain* 149:332–337.

Vogelsang M, Heyer G, Hornstein OP. 1995. Acetylcholine induces different cutaneous sensations in atopic and non-atopic subjects. *Acta Derm Venereol* 75:434–436.

Woodward DF, Nieves AL, Hawley SB, Joseph R, Merlino GF, Spada CS. 1995. The pruritogenic and inflammatory effects of prostanoids in the conjunctiva. *J Ocul Pharmacol Ther* 11:339–347.

Yamaguchi J, Aihara M, Kobayashi Y, Kambara T, Ikezawa Z. 2009. Quantitative analysis of nerve growth factor (NGF) in the atopic dermatitis and psoriasis horny layer and effect of treatment on NGF in atopic dermatitis. *J Dermatol Sci* 53:48–54.

Yosipovitch G, Carstens E, McGlone F. 2007. Chronic itch and chronic pain: Analogous mechanisms. *Pain* 131:4–7.

Index

Page numbers followed by f and t indicate figures and tables, respectively.

A

ACC (anterior cingulate cortex), in itch modulation, 400, 401f, 402
Acetylcholine, 4
  in AD patients, 431, 436
Acupuncture, 55, 56
  in CKD-aP, 55–56
AD (atopic dermatitis), 10–11, 19–27, 25–26, 41, 149, 156, 157
  acetylcholine in, 431, 436
  antihistamines for, 27, 165–167
  anti-inflammatory therapies for, 26–27
  atopic itch, 21–24
  chronic itch, 432–433
  eosinophils and nerves in, 24
  H₁R antihistamines for, 157, 159, 166
  IL-31 in, 25–26, 239, 240, 241–243, 242f, 244, 245, 248
  itch in, 25–26
  itch sensitization in. *See* Itch sensitization
  KOR in, 311–312
  MOR in, 308–311
  OSM in, 245
  PAR-2 in, 194, 199, 200, 201, 202, 204
  sphingomyelin deacylase in, 276
  target-specific therapy, 27
  therapy of, 26
A-fibers, in itch sensation, 137–138, 138f, 139f
AITC (allyl isothiocyanate), 285
Alcaftadine, 156
Alemtuzumab, in CTCL, 123
Allergic conjunctivitis, 156, 158, 162, 163t
  antihistamines in, 164–165
  H₁R antihistamines for, 165
Allergic disorders, 21
Allergic rhinitis, 155, 156, 157
  antihistamines in, 162–164, 163t
  H₁R antihistamines for, 158, 158t, 159, 162, 163–164, 163t
  pathogenesis, 162
  symptoms, 162
Allodynia (touch-evoked pain), 435
  mechanisms, 344

Alloknesis
  in chronic itch patients, 330
  in innocuous mechanical stimulation, 343
  mouse model of, 344
Allyl isothiocyanate (AITC), 285
ALT (anterolateral tract) neurons, 345, 347–348
Alvimopan, 309
Amelioration of itching, 51
American Dermatological Association, 149
Amitriptyline, 109
Amyotrophic lateral sclerosis, 91
Anesthesia dolorosa, 94, 109
Anesthetics, 79
  local, 110
Angiogenesis, 36
Anion adsorption, 76
Anterior cingulate cortex (ACC), in itch modulation, 400, 401f, 402
Anterolateral tract (ALT) neurons, 345, 347–348
Anticonvulsants, 55, 123
Antidepressants, 42, 77, 123, 166
Antihistamines, 26, 37, 42, 63, 144–147.
       See also specific entries
  for AD itch, 27
  in clinical practice, 156–170, 158t
    allergic conjunctivitis, 164–165
    allergic rhinitis, 162–164, 163t
    atopic dermatitis, 165–167
    cholestatic itch, 168
    contact dermatitis, 167
    CTCL, 168
    general safety considerations, 169–170
    insect bites, 165
    psoriasis, 167
    uremic pruritus, 168–169
    urticaria, 159–162, 161t
  corticosteroids in, 159
  diphenhydramine/Benadryl®, 145, 160, 170
  histaminergic and nonhistamineric itch, topical treatment, 131, 132f
  H₁ receptor (H₁R). *See* H₁ receptor (H₁R) antihistamines
  H₂ receptor (H₂R). *See* H₂ receptor (H₂R) antihistamines
  H₃ receptor (H₃R). *See* H₃ receptor (H₃R) antihistamines

441